© 2013, The Gangasas Press, LLC

No part of this publication may be reproduced, stored in a retrieval system, or transmitted in any form or by any means, electronic, mechanical, photocopying, recording or otherwise, without the prior permission of the Publishers. Permissions may be sought directly from the author.

First published in 2013
REV:Jan2013

ISBN: 978-0-9887452-0-9

Notice: The material in this text is a compendium of clinical and research information developed in the discipline of applied kinesiology (AK). There are references to the standard scientific literature that helps put AK procedures into perspective. The authors have taken care to make certain the information is compatible with information available in the field at the time of publication. The physician using this material must take into account all of the patient's findings including: clinical history, physical examination, laboratory tests, and other procedures when applicable to arrive at a final conclusion regarding the patient's condition and best approach to treatment. The author disclaims all responsibility for any liability, loss, injury, or damage incurred as a consequence, directly or indirectly, of the use and application of any of the contents of this volume.

TheGangasasPress.com

Reviews

Reader reviews on this series of AK textbooks:

"A must read for anyone interested in finding out about AK, or anyone interested in exploring different forms of physical medicine."

"Ultimate authority on AK."

"This book truly stands alone on the topic of Applied Kinesiology. Even if you have just a little understanding of anatomy, physiology, and chemistry you will find this to be clear enough and well-written enough to mushroom your knowledge. If you want to learn how to become adept at "assessing" health conditions and sub-clinical deficiencies and to incorporate AK into your work -- whether massage, chiropractic, osteopathy, kinesiology, PT, nutritional therapist, dentistry, standard medicine, or even just working with your family -- then THIS is the book you want. Packed with info that is straight to the point and simultaneously rich in details, it has been my guide into understanding how to assess and facilitate the healing and repair of the human body. It has served me extremely well and, in my opinion, it is the only book on the subject you will need. It does not disappoint."

"A thorough understanding of the material in this text will enable a physician, whatever his or her specialty, to more effectively handle problem patients and get to the basic underlying cause of a condition. Many health complaints are better understood with the information available in this text."

Acknowledgments

My sincere thanks go to the many chiropractors, osteopaths, manual medicine experts, physiotherapists, bodyworkers, nutritionists, acupuncturists and others whose work we have drawn on for this series of books. In particular I wish to pay tribute to the generous and warm collaboration of Dr. Anthony Rosner, who has done so much to bring not only chiropractic but specifically applied kinesiology methods into the era of evidence-based medicine. His review of these manuscripts have simply been invaluable.

From the chiropractic profession Drs. Schmitt, Illi, Janse, the Palmers father and son, the Masarskys husband and wife, Leaf, Walker, Maffetone and the entire membership of the International College of Applied Kinesiology -- as well as the many groups of professional and lay clinicians who derive their basic methodologies from AK -- who support and encourage applied kinesiology research and the development of clinical approaches to human health and illness...all deserve special thanks.

The osteopaths Sutherland, Magoun, Korr, Kuchera and Chaitow deserve outstanding credit as well.

Several giants of manual medicine, particularly Drs. Janet Travell and David G. Simons, have given us so many insights and so much impetus to the evolution of our concepts of myofascial dysfunction and particularly muscular inhibition resulting from muscle pain.

> David Gavin is the artist responsible for the illustrations in this new series of applied kinesiology texts. His work clarifies the material and makes for more pleasant reading. His illustrations for the lower extremity in this volume offer greater clarity to the AK evaluation of lower and upper body dysfunctions not previously published in any text. We recognize the importance of his talented work in helping us develop improved educational material in applied kinesiology. Some of this extremely accurate artwork was originally designed with Dr. Walther's guidance. The copyright for these original images was given to the ICAKUSA by Dr. Walther at the time of his passing, and are gratefully acknowledged by the present author.

Thanks also to those who gave other forms of assistance when needed, each in a distinctive and decisive form: Dr. Shaun Craig, our associate in practice whose enthusiasm for AK inspires; Dr. Don McDowall for his reassurance and problem-solving abilities; Jurgen Reinzuch and Dr. Mike Fuhrman for their lasting and wise counsel as friends.

I dedicate this book to the memory of Dr. David S. Walther and to our staff at the Chiropractic Health Center, PC, and particularly my secretary Sharon Schwab who endured my yoked-lifestyle in writing this book with extraordinary tolerance and support.

Lastly, a special debt of gratitude for my mother Pansy who has been and continues to be crucial to my well-being.

2013   SCC

Preface

The prototype of this textbook was developed by David S. Walther in 1976 in response to a perceived need for a single source that would provide entry-level knowledge in applied kinesiology (AK) manual muscle testing and assessment and treatment procedures; muscle physiology, muscle-joint interrelationships, and the coordinated muscular assessment tools for AK students. Preparation of this volume has also been spurred by the broad acceptance of Dr. Walther's *Applied Kinesiology, Synopsis: 1st and 2nd Editions*. The *Synopsis* has sold so well partly because practitioners who have learned to use it have provided relief to their patients spanning many of the professions that use AK including chiropractic, osteopathy, naturopathy, homeopathy, acupuncture, massage therapy and bodywork, nutritionists, allergists, energy psychologists, physiotherapists and many others. The *Synopsis* has been printed in English, translated into Japanese, Korean, Russian and Italian, and is scheduled to appear in Chinese and French.

The subject of this book is an enormous practical challenge and at the same time one that probes our common intellectual understanding. Population surveys reveal that large numbers of people are in pain of musculoskeletal origin, with nonspecific back pain of muscular origin showing life-time prevalence in the majority of people (60% to 85%). The incidence increases with age up to the level where there are few senior citizens who are not candidates for the diagnosis.

The human body may not have changed much since the beginning of the modern era of AK, however the technologies now employed to study it and the information gained about neuromuscular functioning has grown markedly. As in the previous editions of this series of AK textbooks we will attempt once more to strike the delicate balance between informing the reader of new research and hypotheses that are part of the ever-expanding developments in AK and complementary and alternative health care, while maintaining this textbook's role as a tool for professional education.

By carefully referencing our sources, we hope to encourage readers to seek additional resources as they pursue their professional education and development. The reader should note that we are using references from the very beginning of the study of manual muscle testing and treatment (going back as far as the late 19th century up through 2012), thereby presenting a scientific history of the evaluation and treatment of muscle dysfunction. When SCC first met David in his library, he was astonished to realize that he was visiting the largest chiropractic library since he had worked for the research department at the Palmer Chiropractic College in Iowa. Walther's library had approximately 7,000 journal papers in the database (2006), and 3,000 textbooks. This material was purchased by SCC for help in completing subsequent textbooks on AK, adding a historical element to our presentation. The extensive references in the book are used for two purposes: to support the statements made and to give credit to authors from whom we have derived information and ideas. The exception to this is George J. Goodheart, Jr., who developed most of the basis of applied kinesiology. We hope with this text to continue the tradition of offering an informative, readable and concise reference that will lay an appropriate foundation in your quest to understand human function through AK techniques.

We wished to insure that this addition includes important aspects of the large volume of research published in recent years while updating the clinically effective material from the decades past. The structured abstracts from the *Collected Papers of the International College of Applied Kinesiology* (published in the United States, Japan, Australia, the UK and Europe) have just been added to the world-wide database called MANTIS. What this means is that for the first time the research literature for AK, including outcomes data (published yearly since 1974), is now indexed and available to the scientific and lay-community worldwide. In addition, it is widely consulted in national and state politics as well as the insurance industry. The MANTIS database includes over 350,000 entries from 1,100 journals and is updated weekly.

If AK is to become an integral part of mainstream health care then conceptual bridges have to be built. Because integration depends upon effective lines of communication and the ability to complete learning loops, a good enough common language is an essential foundation.

Finding a scientific explanation for the principles behind some of the diagnostic and treatment approaches is one challenge. Although these theoretical objections may be barriers to integration within primary care, there should be no absolute objection to the pragmatic use of some AK approaches. After all, aspirin was used for decades before prostaglandins were discovered and it was only with the advent of the endorphins and pain-gate control theory that a scientifically coherent explanation for the acupuncture treatment of pain became possible.

Applied kinesiology, when practiced by a physician who is adequately trained and with a mild degree of prudence, is virtually risk-free, and it possesses the potential for great help. The outcomes described in these textbooks demonstrate these facts.

Since the last textbook in this series, Walther's *Synopsis*, 2nd Edition in 2000, we have not only updated content but also have expanded explanations to improve their clarity as well as supplementing the text with nearly 400 new figures and summary tables. Prepared by editors and coauthors of international renown in applied kinesiology, we have had to make difficult decisions about limiting the inclusion of new information. Details that might be useful to experienced clinicians have not been included unless they enhanced the basic understanding of the content without overwhelming the reader. Readers who wish to pursue such topics in greater depth are encouraged to continue their reading using the reference lists at the end of the chapters.

This first volume begins with lower body dysfunctions because there will be a major emphasis on structural balance which has a positive or negative effect on the spinal column and the function of the body as a whole. This text does is not intended to cover all aspects of lower body dysfunction and orthopedics. It does cover standard descriptions of orthopedic conditions and their diagnosis. There are some conditions which are not applicable to conservative treatment, and others that may be confused with conditions that are considered allopathic in nature which are favorably treated by conservative measures. Chapter one, on manual muscle testing of lower body muscles goes into great detail on each muscle covering anatomy, kinesiology, general characteristics, myofascial trigger point patterns of dysfunction, and proper muscle testing. This offers a comprehensive coverage of manual muscle testing for functional neurological diagnosis and is an expanded and updated version of the classic chapter Walther first wrote on this subject in 1981.

Through the years, the content was updated and broadened in response to ever-expanding research discoveries by Dr. George J. Goodheart, Jr. and the members of the ICAK and other clinicians around the world who use the manual muscle test and AK procedures, as well as the growth in AK scientific research and literature. What evolved was a multi-disciplinary text that not only encompasses the basic theory required to understand normal and abnormal function, but it also provides the foundation for understanding current trends in musculoskeletal evaluation and treatment.

The history of the present book reaches back to the year 2000, when the authors met for the first time in the clinic of the senior author (DSW) in Pueblo, Colorado. The author (SCC) came to Pueblo from Ireland in order to study AK methods with DSW. Contact between the authors was maintained and intensified during the ensuing decade. This allowed long and vivid discussions regarding all aspects of AK health care.

In writing this text, we have used the available international body of knowledge. However, the text is principally developed and guided by the outcomes of applied kinesiology research and its translation into clinical practice. Research is an ongoing process and we look forward to forthcoming developments in this field over the next thirty years.

We wanted to write a textbook that both preserves and supports Goodheart's teachings. This book is a tool for everyday practitioners; it is not meant to address all chronic pain syndromes or even all muscle imbalance syndromes. Instead it provides practical, relevant, and evidence-based information arranged into a systematic approach for lower body dysfunctions that can be implemented immediately and used along with other clinical techniques.

The authors have attempted to address musculoskeletal dysfunctions of the body from a particular perspective. This point of view relates to the diverse external influences to which the patient may be responding, broadly defined as biomechanical, biochemical and psychosocial. This is the famous "triad of health" model that has been part of the vitalistic perspective of healing for millennia, but which Goodheart brought into clinical examination with the manual muscle test.

Many of the assessment and treatment methods derive from the personal experiences of the authors, although the bulk emerges from the wonderfully rich AK research and inter-professional literature that makes up the AK professional movement, which has been trolled and studied in order to validate the information presented. In many instances direct quotes have been used since these could not be improved upon as they encapsulate perfectly what needed to be said. The authors thank most profoundly the many experts and clinicians cited, without whom much of this text would have represented personal opinions alone.

It has been my great pleasure to take up the mantle of AK researcher and gather within these pages the essence of how to achieve and maintain a health-enhancing state for one of the most common human ailments, lower body pain and dysfunction. The goal is to provide the clinician with a better way to evaluate and then remedy homeostatic balance within the human foundation.

Scott Cuthbert   Pueblo, CO 2013

About the Authors and Contributors

Scott Cuthbert BA DC has spent the past decade publishing AK outcome studies and literature reviews. His professional focus aims to advance education in all the allied professions to include applied kinesiology for acute, chronic, and functional health impairments and pain syndromes. A native of California, Dr. Scott first attended St. John's College, where he received his BA in the liberal arts. Dr. Scott then graduated from the Palmer College of Chiropractic in Davenport, Iowa. He migrated to Ireland and practiced chiropractic on Finn MacCool's island. He now resides in Pueblo, Colorado where he took over Dr. Walther's clinic after his passing, and continues to focus his writing about AK on the cliffs above the Arkansas River.

David S. Walther DC DIBAK is an internationally known and respected chiropractic practitioner and teacher of applied kinesiology methods. He is author of 3 classic textbooks on AK, the primary developer of the 100 hour basic course syllabus adopted by The International College of Applied Kinesiology (ICAK) as well as instruction workbooks for the revised 100 hour basic course syllabus adopted by the ICAK (1992). David Walther created over 10,000 slides for teaching applied kinesiology that are used by certified teachers of the ICAK, and has published nearly 100 patient-education pamphlets demonstrating the scope of practice of AK.

Anthony Rosner Ph.D., LL.D.[Hon.], LLC is a champion of interdisciplinary research in the health sciences, serving as Research Director of the International College of Applied Kinesiology, previously having been Director of Research and Education at the Foundation for Chiropractic Education and Research, Director of Research Initiatives at Parker College, Department Administrator in Chemistry at Brandeis University, and Technical Director at a teaching hospital of Harvard University and the Mayo Clinic. He was designated as Humanitarian of the Year in 2000 by the American Chiropractic Association and holds an honorary degree from the National University of Health Sciences. He obtained his Ph.D. from Harvard in Medical Sciences/Biochemistry in 1972.

Mark Force DC DIBAK is a diplomate and certified teacher for the International College of Applied Kinesiology (ICAK) and a fellow of the International Association of Medical Acupuncture (IAMA). He has published research, been technical editor for manuals on laboratory diagnosis and clinical practice, written a book on functional selfcare, contributed to the development of nutritional and herbal formulas, and taught applied kinesiology, clinical nutrition, craniosacral therapy, and functional medicine to chiropractic, osteopathic, naturopathic, and allopathic physicians. Dr. Force graduated from Western States Chiropractic College in 1984 and practices in Scottsdale, Arizona.

Table of Contents

Reviews iii
Preface iv
Acknowledgements vi
About the Authors and Contributors vii
Table of Contents viii

Chapter 1: Manual Muscle Testing for Lower Extremity Dysfunctions 1
David S. Walther
Contributions from Scott Cuthbert

- **Abductor Hallucis** 6
- **Extensor Digitorum Longus and Brevis** 7
- **Extensor Hallucis Longus and Brevis** 9
- **Flexor Digitorum Brevis** 11
- **Flexor Hallucis Longus and Brevis** 12-13
- **Gastrocnemius and Soleus** 14-16
- **Intrinsic Plantar Foot Muscles** 18
- **Lumbricales and Interossei** 19
- **Peroneus Longus and Brevis** 20
- **Peroneus Tertius** 22
- **Popliteus** 23
- **Tibialis Anterior** 25
- **Tibialis Posterior** 27

Chapter 2: Peripheral nerve entrapments (Introduction) 33
Scott Cuthbert
Contributions from David S. Walther

- **Entrapment Neurophysiology** 35
  - Case Report: AK management of Parsonage-Turner Syndrome
- **Cranial nerve entrapment** 35
- **Common sites of entrapment** 35
- **Cross Stimulation and Light Pressure** 43
- **Double Crush** 43
- **Ischemia** 44
- **Examination** 47
  - History 47
  - Differential Diagnosis 48
  - Medications with neuropathy as a potential side-effect 49
  - Palpation and Inspection 49
  - Cutaneous 49
  - Muscle 50
  - Nerve 51
  - General Physical 51
  - Sensibility Studies 51
  - Circulatory Evaluation 53
  - Plethysmography 54
  - Doppler 55
  - Doppler use in thoracic outlet 55
  - X-ray 55

Chapter 3: Peripheral nerve entrapments lower extremity 63
Scott Cuthbert
Contributions from David S. Walther

- **Lateral Femoral Cutaneous Nerve** 63
- **Obturator nerve** 66
- **Femoral nerve** 67
- **Saphenous nerve** 69
- **Piriformis Syndrome** 69
- **Clinical algorithm for AK treatment of piriformis syndrome** 76
  - **Peroneal Nerves** 76
  - Common Peroneal 76
  - Superficial Peroneal nerve 79
  - Deep Peroneal 80
- **Tibial nerve** 80
- **Tarsal tunnel syndrome** 80
  - Anatomy 81
  - Symptoms 81
  - Etiology 84
  - Examination 84
  - Treatment 85
- **Plantar Nerves** 86
  - Interdigital nerves 87
- **Plantar Interdigital Nerualgia (Morton's Neuroma)** 88
  - Anatomy 88
  - Symptomatic Pattern 88
  - Treatment 89

Chapter 4: Foot and Ankle 95
Scott Cuthbert
Contributions from David S. Walther

General examination and body language 98
  - Symptomatic 98
- **AK Foot and Ankle Examination Algorithm** 99
  - Appearance of the foot 102
  - Shoe wear 102
- **Meridian system influenced by foot and ankle dysfunction** 103
- **Non-weight bearing examination** 103
  - Palpation 106
  - Shock absorber test 106
- **Static weight-bearing examination** 107
  - Helbing's sign 107
  - Knee and leg position 107
  - Weight-bearing tests 108
- **Dynamic examination** 109
- **Foot and Ankle Anatomy and Physiology** .. 110
  - Divisions and Motions of the Feet 110
  - Classification of the Forefoot 110

- o Motions, Positions, and Fixed Structural Positions of the Foot 110
  - o Adduction 110
  - o Abduction 110
  - o Inversion 111
  - o Eversion 111
  - o Dorsiflextion 111
  - o Planter Flexion 111
  - o Pronation 111
  - o Supination 111
- **Foot Osteology** 111
  - o Hindfoot 112
  - o Calcaneus (os calcis) 112
  - o Talus (astragalus) 112
  - o Midfoot 113
  - o Navicular 113
  - o Cuboid 113
  - o Cuneiform Bones 113
  - o Forefoot 114
  - o Metatarsal Bones 114
  - o Phalanges 114
  - o Accessory Bones 116
- **Foot mechanics – The close examination** ... 116
- **Foot reflexes** 122
- **Ligaments of the foot** 123
- **Muscles of the foot and lower leg** 125
  - o Extrinsic Muscle System 125
  - o Intrinsic Muscle System 125
- **Arches of the Foot** 127
- **Bony architecture** 128
- **Ligaments and aponeurosis** 129
- **Muscle Role in Support of the Arch** 130
  - o Foot Weight Bearing 132
- **Ankle Anatomy and Function** 133
- **Ankle Stability in Active Function** 135
- **Ligament stretch reaction** 136
  - o Correction 138
- **Foot and Ankle Motion** 138
  - o Plantar Flexion and Dorsiflexion
- **Foot Reflexes and Reactions** 141
- **Positive support reaction** 141
  - o Examination of the Positive Support Reaction 143
- **Pronation** 144
  - o Gait 145
  - o Etiology 146
  - o Short Triceps Surae 146
  - o Tarsal Coalition 148
  - o Peroneal Spastic Flatfoot 149
  - o Internal tibial Torsion or Malleolar Torsion 149
  - o Ligamentous Laxity
  - o Hypermobility syndrome
  - o Prehallux or Accessory Navicular 149
  - o Talipes Calcaneovalgus 151
- **Postural Influence** 152
- **Examination** 152
  - o Static Stance 153
  - o Dynamic Evaluation 153
  - o Palpation for Pain 154
  - o Joint Motion and Challenge 155
  - o Muscle Evaluation 156
    - o Psoas Inihibition & Foot Pronation . 157
  - o X-Ray Evaluation 158
    - o Tarsal Coalition 158
    - o Tibiotalar Motion 159
- **Joint and Muscle Correction** 160
  - o Articular Adjusting 161
  - o Lateral or Medial Talus 162
  - o Rotated Cuboid 163
  - o Lateral Cuboid 163
  - o Inferior Cuboid 163
  - o Inferior Navicular 163
  - o Superior Mid-Tarsal bones 163
  - o Inferior Cuneiforms 164
- **Plantar Fasciitis and Heel Spurs** 164
  - o Etiology and Symptoms 164
  - o Treatment 167
- **Forefoot in extended pronation** 168
- **Transverse Arch in extended pronation** 168
- **Types of Metatarsalgia** 169
  - o Extended pronation in metatarsalgia
  - o Loss of Transverse Arch 170
  - o High arch
  - o Normal (medium) arch
  - o Footwear 171
  - o Cavus Foot 171
  - o Imbalance of Weight Distribution 171
  - o Sesamoiditis 172
  - o Metatarsalgia differential diagnosis 173
  - o Treatment 174
- **Freiberg's Infraction and Bone Fragmentation** 175
- **Stress Fractures** 176
- **Functional hallux limitus (FHL)** 177
  - o Foot stabilization 179
  - o Calcaneocuboid locking 179
  - o Truss and locking wedge effect 179
  - o Windlass effect 179
  - o Sequence of motion during gaiting 180
- **The effects of FHL** 180
  - o Symptoms 180
- **Examination** 181
- **Podiatry and FHL**
- **Applied kinesiology and FHL** 181
  - o Liver meridian 182
  - o Origin and insertion technique 182
  - o Repeated muscle activity patient induced 182
  - o Muscle stretch reaction 182
  - o Maximum muscle contraction 182
  - o Rib pump technique 182
  - o Deep peroneal nerve 182
  - o Support 182

- **Hallux rigidus** **183**
- **Hallux Valgus** **183**
  - Etiology 184
  - Congenital Bone Formation 184
  - Footwear 184
  - Muscle Role 185
  - Pronation 186
  - Examination and Treatment 186
  - First Metatarsal and 1st Cuneiform Adjustment 187
  - First Metatarsophalangeal Adjustment ... 188
  - Support 188
  - Surgery 189
- **Shoes** **189**
  - Proper Shoe Fit 192
  - Foot Support 194
  - Material for Padding and Taping 195
  - Medial Longitudinal Arch **195**
  - Versatility of AK Diagnosis 196
  - Metatarsal support **197**
  - Orthotics 198
- **Foot Rehabilitation** **200**
  - Triceps Surae Stretch 201
  - Foot rehabilitation with yoga 202
  - Thigh and Spinal Muscles 203
  - Planter Muscle Exercise 203
  - Obtaining Foot Flexibility 204

Chapter 5: Leg and Ankle 219
Scott Cuthbert
Contributions from David S. Walther

- **Ankle** **220**
- **Ankle Joint Strain** **220**
- **Tibial torsion** **220**
- **Ankle sprains** **221**
- **Types and severity of sprains** **222**
- **Examination** **223**
  - Orthopedic signs and tests for ankle involvements 224-225
- **X-Ray Examination** **226**
  - Nerve injury accompanying ankle sprain 227
  - Treatment 227
  - Rehabilitation
- **Prevention and optimal function** **230**
- **Exercise** **231**
- **Proprioceptive training** **232**
- **Tape and support** **233**
- **Peroneal tendon dislocation** **234**
- **Achilles Tendon** **235**
  - Achilles tendinitis 236
  - Diagnosis 237
  - Treatment 237
  - Achilles tendon rupture 238
  - Examination 239
  - Treatment 239
  - Achilles bursitis 240
  - Tibialis posterior tendinitis 241
- **Shin splints** **241**
  - Diagnosis 243
  - Treatment 243
- **Compartment syndrome** **244**
  - Chronic compartment syndrome 246
  - Acute compartment syndrome 246
  - Treatment 247
- **Restless leg syndrome (RLS)** **247**
  - Medical treatment for RLS 249
  - Side effects 249
  - AK treatment for RLS 249
  - Restless Legs Syndrome Patient List 251

Chapter 6: Applied kinesiology and systemic conditions of the lower body 261
Mark Force, Scott Cuthbert
Contributions from David S. Walther

- **Introduction** **261**
- **Testing biochemistry with AK** **262**
- **Viscerosomatic reflexes and Muscle-Organ-Gland relationships** **262**
- **Somatoautonomic nervous system, somatovisceral reflexes, and manual therapies** **263**
- **Myofascial release and somatovisceral effects** **264**
  - Percussion and myofascia 264
- **Therapeutic manipulation of visceral myofascial adhesions** **265**
- **Common areas of myofascial problems** **265**
- **Somatic fascia** **266**
- **Visceral myofascia** **267**
  - Visceral myofascial therapy by Barral 267
  - Visceral myofascial therapy by Walker ... 268
  - Visceral myofascial therapy by Chaitow (Specific Release Technique) 268
  - Visceral repositioning (Ptosis) 269
- **The pelvic region – chronic pelvic pain (CPP)** **270**
  - Inguinal hernia 270
- **Urinary tract dysfunction** **271**
  - Urinary incontinence 271
  - Urinary incontinence: AK case series report 271
- **Nocturia** **273**
- **Cystitis and interstitial cystitis** **273**
- **AK and digestive dysfunction** **274**
- **Stomach and small intestine** **274**
  - Hydrochloric acid 274
  - Hiatal hernia 274
  - Case Report: AK and severe digestive burning 275
  - Protein digestion 276
  - Protein and arthritis 276

- o Acid-Alkaline balance 278
- **AK and the large intestine** **279**
  - o Large intestine and rectal cancer 279
  - o Ileocecal valve syndrome 280
  - o Alimentary canal flora 281
  - o Inflammatory bowel disease (ulcerative colitis and Crohn's disease) 282
  - o Adrenal involvement in large intestine dysfunctions 282
  - o Irritable bowel syndrome (mucous colitis) 283
  - o Constipation 284
  - o Dehydration 284
  - o Diarrhea 285
  - o Infantile colic 285
  - o Obesity 285
  - o Post-antibiotic effects and candidiasis 285
  - o Fecal matter matters 286
- **Female reproductive system** **287**
  - o Infertility and amenorrhea 287
  - o Dysmenorrhea 288
  - o Pain, prostaglandins, and dysmenorrhea .. 288
  - o Premenstrual syndromes 289
  - o Estrogen imbalances 289
  - o Endometriosis 290
  - o Polycystic ovary disease 290
  - o Vaginitis/Vaginosis 290
  - o Benign uterine fibroid/leiomyoma 290
- **Male reproductive system** **290**
  - o Prostate disease 290
- **Aching pain, lactic acidosis, and mitochondrial dysfunction** **291**
  - o Urea-Guanidine cycle and deep aching pain 292
- o Inactivity aggravated pain and calcium metabolism 293
- o Burning pain, NSAIDs, and the prostaglandin system 293
- o Trans fatty acids 293
- o Itching, edematous pain and bradykinin 294
- **Systemic inflammatory pain** **294**
  - o Oxidation 294
  - o Methylation 294
  - o Homocysteine 295
  - o Leukotrienes 294
  - o Glycation 295
- **Focal occult infection** **295**
- **Joint pain** **296**
  - o Osteoarthritis 296
  - o Rheumatoid arthritis 296
- **Circulatory system** **296**
  - o Varicose veins 297
- **Edema** **297**
  - o Edema due to nutrient deficiencies 297
  - o Biochemical individuality and subclinical nutrient deficiency 297
  - o Impaired microcirculation 298
  - o Raynaud's syndrome 298
  - o Blood hyperviscosity syndromes 298
  - o Rouleaux formation 299
  - o Blood viscosity and Rouleaux interactions 299
- **Applied kinesiology's future in stress-related illness** **299**

> "Applied Kinesiology is not a departure from chiropractic but a penetration of chiropractic."
>
> "Anytime you are aware of part of your body there is something wrong with it."
>
> "There is no substitute for skill. Knowledge plus wisdom plus experience produces skill."
> ~ George J. Goodheart, Jr.

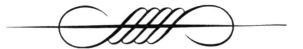

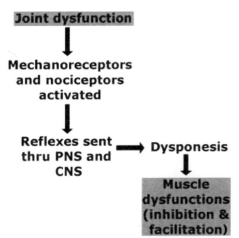

CHAPTER ONE

Manual Muscle Testing for Lower Body Dysfunctions

To an experienced applied kinesiologist, manual muscle testing can be of great value in the differential diagnosis of peripheral nerve entrapment. The first and obvious factor is failure of a muscle to perform normally when its efferent supply is disturbed. This is the primary consideration made in most texts dealing with muscle testing, (**Kendall et al., 2005; Daniels & Worthingham, 2002; Walther, 2000, 1981; Janda, 1983**) peripheral nerve entrapment, (**Russell, 2006; Staal et al., 1999; Kopell, 1980**) and general examination of the spine and extremities. (**Liebenson, 2007; Hammer, 2007; Hoppenfeld, 1976**) Their concern is how well the muscle can produce power. In applied kinesiology muscle testing, one is concerned primarily with how effectively the nervous system can control the muscle, as well as its ability to produce power. The applied kinesiologist observes for the muscle's locking capability on the initiation of the test. (**Walther, 2000, 1981**)

Schmitt (**Schmitt, 1986**) has described two types of muscle testing in applied kinesiology. The differentiation is whether the doctor initiates the muscle test by the patient resisting his pressure, or whether the patient applies pressure against the doctor's resistance prior to the doctor applying test pressure. These have been termed gamma I and gamma II muscle tests, respectively, relating to how the neuromuscular spindle cell operates in a test. Conable et al. (**Conable et al., 2005**) evaluated whether examiners trained in applied kinesiology could routinely and accurately assess whether patient-initiated or doctor-initiated MMT had been performed. Using surface EMG, Conable et al. found that the majority of testers in their study did near-simultaneous testing regardless of whether they tried to perform doctor-initiated vs. patient-initiated MMT. The examiner first tested the middle deltoid muscle of the subject in the normal fashion three times and identified the MMT style as "examiner-started" or "patient-started." The examiner was then asked to perform the other method of MMT. If the examiner said he did not know or did not differentiate which form of testing was initially done, he then performed one series each of examiner- and patient-started MMT. The results of this study showed that nine (approximately 43%) of testers identified their "normally done" muscle test as examiner-started, 4 (19%) as patient-started, and 8 (38%) as simultaneous or undifferentiated. In 64.5% of the MMT described as examiner-started, sEMG showed that the examiner's contraction started before the patient's. In tests identified as patient-started, 54% were indeed patient started. Undifferentiated tests were 45% patient-started, 45% examiner-started, and 10% exactly simultaneous. Near simultaneous contractions were observed in 55% of all tracings evaluated and 70% of undifferentiated tests. The examiner vs. subject initiated tests alone (as described by the examiners), as measured by sEMG, did not clearly differentiate between Schmitt's theorized forms of manual muscle testing.

The doctor–initiated or gamma I muscle test is the one most often used in applied kinesiology. The applied kinesiology methods of muscle testing appear to be much more geared to dynamics rather than simply evaluating whether the muscle is capable of producing power. In applied kinesiology, muscles are not graded as to exact level of strength; rather, consideration is made of whether the muscle is being effectively controlled by the nervous system. The applied kinesiology method of testing is thus very effective in determining subtle peripheral nerve entrapment.

Most other muscle testing methods have a grading system based on motion against gravity and the ability to hold against resistance. A grading system must be used when muscle testing is done to evaluate the nervous system for permanent impairment; it varies among authorities. Listed here is the system in *Guides to the Evaluation of Permanent Impairment, 5th edition* (**American Medical Association, 2001**) by the American Medical Association. In fact, most states have legislatively mandated its use.

(American Medical Association, 2001) Grading Scheme

1. Complete range of motion against gravity and full resistance, 0%.
2. Complete range of motion against gravity and some resistance, or reduced fine movements and motor control, 5–20%.
3. Complete range of motion against gravity and only without resistance, 25–50%.
4. Complete range of motion with gravity eliminated, 55–75%.
5. Slight contractility, but no joint motion, 80–90%.
6. No contractility, 100%.

In addition to the "weak" and "strong" muscles of applied kinesiology and the strength grading scheme of manual muscle testing for impairment ratings is the effort to quantitate strength with dynamometers. There is considerable difference between these methods of muscle testing. (**Kendall et al., 2005; Walther, 2000; Harms-Ringdahl, 1993**) First, applied kinesiology's doctor–initiated manual muscle testing overpowers the maximum isometric contraction and takes the muscle into an eccentric contraction. This appears to evaluate the nervous system's ability to adapt the muscle to combat the changing force of the examiner's test; thus the muscle must be controlled by the nervous system to react properly to information being received from the neuromuscular spindle cell. When force is applied slowly, as in impairment rating or against a fixed strain gauge, there is more time for the nervous system to adapt. This appears to be more a function of whether the alpha motor neuron impulse can be transmitted to the muscle. Simply producing force also appears to be the main factor in testing against isokinetic dynamometers. (**Blaich & Mendenhall, 1984; Blaich, 1981**) Testing a muscle's ability to produce pure strength, as against a strain gauge, often does not reveal the same findings that manual muscle testing does.

Dynamic MMT Assessment

Simons et al (**Simons et al., 1998**) suggest that weakness needs to be evaluated both statically and dynamically, confirming the methods developed by Goodheart. In static testing, a single muscle is being evaluated as the patient attempts a voluntary contraction and the process is under cortical control. In dynamic testing, which involves muscular effort relative to a normal functional movement and where a degree of coordinated muscular effort is required, there is a greater degree of "vulnerability to reflex inhibition", (**Simons et al., 1998**) for example, involving trigger points. Dynamic activity is under less direct cortical control and often involves coordinated patterns of integrated neural and muscular function which are semi-automatic, largely under cerebellar control.

Because of manual muscle testing's versatility it is possible to test patients in many different positions and states (including walking, sitting, supine, prone and moving any number of parts of the nervous systems while the body is being tested). In the AK approach during the manual muscle test, the focus is on the mutual interactions between the patient, their muscular system, and their environment. A large portion of this approach deals with the capabilities of the muscle system to produce movements that are influenced ("challenged") by the patient's sensory or environmental information that coexists with their pain and dysfunction.

One of the most important real world factors that need to be kept in mind when assessing function is that muscle testing and movement assessments should reproduce those actually performed in daily life, particularly those movements that produce pain for the patient. It is appropriate to evaluate a single direction of motion (for example, abduction of the arm) in order to gain information about specific muscles. In daily life, however, abduction of the arm is a movement seldom performed on its own; this movement is usually accompanied by flexion or extension and some degree of internal or external rotation, depending on the reason for the movement. This highlights the fact that most body movements are compound, and a great many have a spiral nature (to bring a cup to the mouth requires adduction, flexion and internal rotation of the arm). These observations reinforce the need when performing manual muscle tests, to take account of movement patterns which approximate real life activities, most of which are multi-directional.

Specificity of the AK MMT

Manual muscle testing, when expertly done, is better able to maximally isolate the muscle, with the examiner able to observe recruitment of synergistic muscles. An extensive checklist for effective MMT against which all publications which purport to describe AK has been presented by Schmitt and Cuthbert. **(Schmitt & Cuthbert, 2008)** A general evaluation of hand strength with a dynamometer has even less ability to reveal the same information that manual muscle testing does if the entrapment is of a single nerve.

For example, a general evaluation of neck strength with a dynamometer may not reveal the same information that manual muscle testing does if, for instance, the injury is to a single nerve, an individual muscle, or even a portion of a muscle. Several human muscles, including the upper trapezius and anterior scalenes, have broad fan-like attachments dividing the muscle into serial segments, and each section of a dysfunctional muscle may be assessed for strength using the manual muscle test with precise patient positioning (and a dynamometer cannot do this).

Even in a severe neck pain syndrome, only a portion of the fibers may be involved, allowing some muscles to function very well. These may be the muscles primarily tested by instruments evaluating the gross muscle strength in group muscle tests; yet significant changes in individual muscles may be present when manually tested. **(Cuthbert et al., 2011)**

If the involvement is at the thoracic outlet, the chances of the hand–held dynamometer revealing greater strength after correction improve in comparison with an involvement at the hand or wrist, such as a carpal tunnel syndrome. This is because, in some types of thoracic outlet syndrome, the entire neurovascular bundle may be compressed, involving many muscles that may weaken the general hand grip. Only one nerve will be involved in entrapment with more distal syndromes, weakening only some of the muscles affecting the general hand grip.

Even in a severe thoracic outlet syndrome, only a portion of the fibers may be involved, allowing some muscles to function very well. These may be the muscles primarily tested by the hand-held dynamometer. As will be discussed later, challenge, therapy localization, and orthopedic tests are used in conjunction with manual muscle testing to evaluate peripheral nerve entrapment. These may show no change in muscle strength in the extremity when tested with a hand–held dynamometer; yet significant changes in individual muscles may be present when manually tested. Likewise, effective correction at the thoracic outlet may also show no change with a hand–held dynamometer for the same reason.

Familiarity with the course of the muscle's nerve supply and possible locations of entrapment enables the physician to evaluate various muscles supplied by the nerve above and below the possible area of lesion. For example, the flexor pollicis longus muscle receives median nerve supply proximal to the carpal tunnel. The opponens pollicis muscle receives median nerve supply distal to the carpal tunnel. Manual muscle testing revealing the flexor pollicis longus to be strong but the opponens pollicis muscle to be weak gives strong indication that there is a nerve entrapment distal to the innervation of the flexor pollicis longus and proximal to that of the opponens pollicis. Other applied kinesiology procedures, such as challenge and therapy localization, add to this information. If a challenge to the carpal tunnel or the radius and ulna improves the strength of the opponens pollicis muscle, there is added indication of a carpal tunnel syndrome. If therapy localization to structures at the carpal tunnel also strengthens the opponens pollicis muscle, there is even further indication the involvement is a result of entrapment in the carpal tunnel. This information is almost pathognomonic of the condition; added to clinical findings such as paresthesia and history, it provides a firm diagnosis. (Carpal Tunnel Syndrome will be covered in the next volume of this series.)

When a nerve entrapment is more proximal, such as in some types of thoracic outlet syndrome, many muscles in the extremity may be weak because the entire neurovascular bundle may be involved. Still, challenge and therapy localization can help locate the area of involvement. When the correct approach is found for this more central entrapment, many muscles in the extremity will strengthen.

As noted throughout the MMT literature, a muscle must function from a stable base to test strong. For instance, stability of the clavicle and/or scapula is essential in shoulder muscle function, and if during examination of the shoulder there is weakness during shoulder MMT, re-evaluate the test by stabilizing the clavicle or scapula. For example, if the pectoralis (clavicular division) tests weak, stabilize the clavicle and re-test the muscle. If it now tests strong, test for subluxations of the sternoclavicular and acromioclavicular articulations. When lack of clavicular or scapular stability is causing a shoulder muscle to test weak, determining the reason for the instability goes a long way toward correcting shoulder dysfunction. This is an example of the AK challenge and therapy localization methods. When the examiner stabilizes the scapula with the AK sensorimotor challenge system of testing, and the dysfunctional muscles of the shoulder strengthen, an important diagnostic clue has just been offered to the examiner. These tools along with the functional evaluation of muscles provide valuable additional information for the physician in the evaluation of the total extremity joint complex.

Ideally, correction is directed toward primary factors. In the example above, if the pectoralis (clavicular division) muscle tests weak and challenge to the clavicle or some other bone causes it to test strong, the subluxation is primary. Although the pectoralis (clavicular division) could probably be strengthened by stimulating the neurolymphatic or neurovascular or acupuncture reflexes or with other types of rehabilitative exercises, it is likely it would immediately lose its correction with gait or some other structural stress to the clavicle. If so, the primary cause of the pectoralis (clavicular division) weakness appears to be the result of improper stimulation to receptors by the subluxated joint. It might be due to the muscle vainly trying to stabilize the clavicle or some other bone in subluxation, and weakening because of its inability to do so.

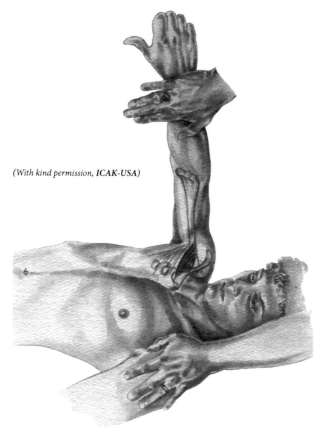

(With kind permission, ICAK-USA)

MMT pectoralis major(clavicular division) muscle

There will not always be muscle weakness with peripheral nerve entrapment. The neuropathy may be of a purely sensory nerve with pain the primary complaint. Or sometimes muscle weakness is not observed because synergistic muscles are recruited and are primary in the test. This may be due to poor muscle testing ability on the examiner's part, or because the muscle cannot be "isolated" well. If the nerve involved is mixed and sensory fibers are involved, there will nearly always be an involvement of the motor fibers also, probably because the nerves supplying muscles are more susceptible to compression than cutaneous nerves. **(Russell, 2006; Aguayo, 1975; Haldeman & Meyer 1970; Sunderland 1945; Denny—Brown & Brenner, 1944; Gasser & Erlanger, 1929)**

Vertebral fixations and subluxations or radiculopathy are not considered in this discussion of peripheral entrapment. Nerve encroachment at the radix, such as from degenerative joint disease or the intervertebral disc, causes the same type of predictable involvement of the motor and sensory systems. Vertebral subluxations and fixations do not have that same predictability. Some may argue that it is impossible to have degenerative joint disease, disc involvement, and other local problems without a chiropractic subluxation being present. This argument is well-taken and is probably correct. Long-term and well-established clinical evidence indicates a muscle that tests weak does not necessarily relate to a specific spinal level. Associating the level of spinal innervation to a muscle that tests weak is not an adequate method for determining subluxation levels. As an example, lower extremity muscle weaknesses found on manual muscle testing are often strengthened with upper cervical subluxation correction. This lack of specificity probably relates to the subluxation's influence on the central nervous system, as well as its local segmental nerve level. General disorganization within the neuronal pools may develop.

Manual muscle testing to evaluate peripheral nerve involvement relates more specifically to the nerve involved. In most instances, challenge, therapy localization, and clinical findings described in the chapters on the foot, leg and ankle, and peripheral nerve entrapments in the lower extremity will locate the area of entrapment. There are occasions where an apparent motor nerve entrapment cannot be demonstrated by weakness of the muscles it supplies, yet it seems obvious that there is peripheral nerve entrapment at a specific location. Further evaluation may reveal there is a neurologic correlation. It appears that sometimes the afferent supply of a muscle is cross-stimulated by the entrapment, sending a signal of agonist contraction to the muscle's antagonist when, in reality, it is not contracting. In this case the antagonist would be inhibited and test weak on manual muscle testing. The same methods of testing with challenge and therapy localization will identify the antagonist muscle dysfunction. Correction of the entrapment will improve antagonist and agonist muscle function.

With the complexity of symptoms on display in the typical patient, including pain and dysfunctional tissues, joints, fear about their pain and others, where would it be most appropriate to initiate treatment? The manual muscle test offers a great diagnostic advantage because it identifies the dysfunctional muscle tissues and the process of therapy localization and challenge allows for the identification of the precise anatomical area (articular, muscular, reflex, craniosacral, nutritional, and/or psychological impairment, whether local or remote) that will improve that finding.

Occasionally when efforts are made to clear a peripheral nerve entrapment, the muscles supplied by the nerve do not immediately change from weak to strong on manual muscle testing. It appears there may be actual physiologic change in the muscle from nerve entrapment, as well as a change in the motor control. Physical trauma to muscles and nerves can damage the mechanoreceptors and axons resulting in localized proprioceptive losses. **(Kolosova et al., 2004; Myers & Lephart, 2002; Sharma, 1999)** These proprioceptive deficits may be very small in the muscles but their clinical significance very large. A study done by Christiansen and Meyer **(Christiansen & Meyer, 1987)** observed a change in muscle enzymes after implanting silastic pellets transverse to the sciatic nerve in mice. Further study is needed to understand the physiologic role of peripheral nerve entrapment. The treatment of the tissue in which the receptors are embedded is also important for proprioceptive recovery. **(Leinonen et al., 2003)** In the treatment of musculoskeletal disorders, what we are confronted with is not "pathology" in the classic Vercovian sense of that term, where each disease has a verifiable tissue injury or biochemical disorder, but rather a disturbance of the normal rhythms of the musculo-skeletal system – a dynamic dysfunctional state.

Therapy localization can identify an area of possible peripheral nerve entrapment by causing an indicator muscle to weaken when the patient touches the suspected area. Therapy localization can also sometimes cause the muscle that tests weak due to peripheral nerve entrapment to test strong when the patient touches the area of entrapment. In fact, this was Goodheart's original observation of therapy localization. As will be discussed under carpal tunnel syndrome in the subsequent text of this series, when a patient squeezes the distal radius and ulnar bones together in certain types of carpal tunnel syndrome, the opponens pollicis muscle will become strong when it previously tested weak. To demonstrate the effect of the radius and ulnar bones separating, the patient is often asked to squeeze the bones together while the examiner demonstrates the improved strength of the opponens pollicis muscle. On one occasion Goodheart was explaining to the patient that when she held the bones together it released the nerve so the muscle became strong. Surprisingly, the patient said, "I didn't understand you wanted me to squeeze the bones together; I was simply touching the area" — yet the muscle became strong. This unusual experience was the first observation of therapy localization.

Challenge is usually mechanically moving a bone or other structure to a position that temporarily improves the nerve entrapment or irritation. For example, in the tarsal tunnel syndrome the calcaneus is usually subluxated posteriorly. An anterior challenge of the calcaneus will often strengthen the intrinsic muscles innervated by the posterior tibial nerve past the point of entrapment. **(See Foot chapter)** This gives positive indication not only of the entrapment but also of the direction of subluxation correction. As with other extraspinal subluxations, the bone is adjusted in a direction that causes the weak muscle to strengthen.

There are numerous orthopedic provocative tests that are used in evaluating peripheral nerve entrapment in the lower body. Most of these can be used in conjunction with manual muscle testing. An example is Adson's test for scalenus anticus syndrome. In this condition, the ulnar nerve is usually involved. A previously strong triceps muscle may test weak in this position, indicating entrapment by the scalene muscle(s), 1st rib, cervical rib, or anomalous fibrous bands. Conversely, when the triceps muscle tests weak, positions that relax the scalene muscles or otherwise change the position of the thoracic outlet may cause the triceps to become strong, providing evidence of entrapment at the thoracic outlet.

Cross–stimulation of the nerve from entrapment or stimulation to the joint receptors often accompanies peripheral nerve entrapment. The improper afferent signals could go to any muscle associated with the agonist or joint. This neurologic model appears to explain many clinical observations from the evaluation and correction of peripheral nerve entrapment. A common example is neck flexor weakness associated with a tarsal tunnel syndrome. Proprioceptive information from the articulations and intrinsic muscles of the foot, mediated through the central nervous system, provides control for facilitation and inhibition of the neck flexor muscles when walking. This model proposes that the afferent supply from the ligaments, tendons, muscles, fascia, and skin of the foot can be disturbed by entrapment in the tarsal tunnel, creating information to be sent to the central nervous system that is not in keeping with the current actions of the foot. For this reason, the neck flexors — or any other associated muscle in the gait system for that matter — may be inappropriately inhibited; they immediately regain normal function when the tarsal tunnel entrapment is corrected.

Although peripheral nerve entrapment can have widespread influence in the muscular system, the most common — and easiest — use of manual muscle testing to evaluate peripheral nerve entrapment is to test the muscles innervated by the nerve distal to the area of suspected entrapment; this will generally provide the diagnosis. When the structure is corrected and the involvement released, there will often be many additional benefits as remote muscles improve in their function. On the unusual occasion when the suspected peripheral entrapment cannot be evaluated in this manner, the physician must look to synergistic, fixator, and antagonistic muscles that are controlled by the afferent supply of the suspected nerve.

Examination and treatment of all extremity joint disorders must insure that agonist and synergist muscles are contracting at full strength; that there is appropriate timing of the contracting muscles, and that the antagonist muscles are releasing at the appropriate time. The only method of diagnostic evaluation that can measure each of these components of muscle function and their interactions in the clinical setting is the MMT. **(Schmitt & Cuthbert, 2008; Cuthbert & Goodheart, 2007)**

With today's knowledge we see that extremity dysfunction can influence the total body more than just the spine and its nervous system relationship, as described early in the chiropractic profession's history by Janse, **(Janse, 1976)** Goodheart and Walther in applied kinesiology, and many others in the osteopathic, naturopathic, and manual medicine fields, **(Page et al., 2010; Chaitow et al., 2008; Lewit, 1999)** all of whom have written extensively about the closed kinematic chain and structural integration of the body.

A major reason that the MMT for all of the peripheral muscles should be added to the standard diagnostic methods used by the professions and taught in the chiropractic, osteopathic, and manual medicine colleges is that patients with extremity disorders demonstrate joint instability, ligament strain, and muscle inhibition. **(Leaf, 2010; Maffetone, 1999; Logan, 1995, 1994)** Each of these causative factors in extremity joint dysfunction—of obvious interest to the health professions—can be specifically diagnosed and treated with MMT and manipulative treatment. Applied kinesiology assessment adds to lower extremity examination because it detects and specifically treats these muscle impairments that either cause or perpetuate extremity dysfunctions.

From *Applied Kinesiology Volume 1* (1981)
with updates

Abductor Hallucis

Attachments: From the medial process of the calcaneus, flexor retinaculum, plantar aponeurosis, and intermuscular septum with the medial tendon of the flexor hallucis brevis into the medial side of the base of the proximal phalanx of the great toe.

Action: Abducts the great toe (from median line of the foot).

Testing position and stabilization: The patient abducts the toe. This is best accomplished by asking the patient to spread his toes like fan. It is often found that a patient cannot abduct the toe into the testing position. The examiner stabilizes the foot at the heel and lateral aspect, and does not grasp the foot over the abductor hallucis.

Test: Pressure is directed to the medial aspect of the great toe in a direction of adduction. During the test, the examiner continues to observe visually -- and possibly with palpation -- for activity of the muscle belly. Since the muscle is often congenitally inserted so as to be incapable of true abduction, this portion of the test helps delineate whether hallux valgus is due to muscle weakness, or lack of the muscle's ability to abduct the great toe.

Body Language of Weakness:

Testing position: Patient is unable to abduct the great toe into the testing position. This inability has to be evaluated carefully by the examiner because many individuals cannot isolate the muscle activity. Palpation for the presence of the muscle and its ability to contract is often necessary.

Postural imbalances: Hallux valgus position, and evidence in the general foot of a tarsal tunnel syndrome.

Special Notes: The evaluation of the abductor hallucis is difficult until the examiner has evaluated this muscle in many individuals, both normal and abnormal. The examiner tends not to test this muscle unless there is body language of its weakness or absence. It is necessary to test normal individuals to learn palpation of the muscle belly, and how to motivate individuals to abduct the great toe into the testing position.

Kerr and Basmajian (**Basmajian, 1978**) studied 22 adult feet by dissection, with emphasis on the insertion of the abductor hallucis to determine its ability in abduction. They concluded that the muscle is so attached as to be capable of true abduction in only 20%. The greatest indication was that the muscle would flex the great toe. When the applied kinesiology examiner is evaluating the abductor hallucis in conditions of hallux valgus, it is necessary to palpate the belly of the muscle to determine if contraction is taking place and to observe the motion of the great toe. Obviously, efforts to balance the abductor and adductor hallucis and flexor hallucis brevis will not improve hallux valgus if the abductor hallucis is inserted in such a way that cannot abduct the great toe. (**Brenner, 1999**)

Abductor hallucis crosses the passageway of the plantar vessels and nerves which serve the sole of the foot and it may entrap these nerves against the medial tarsal bones. (**Travell & Simons, 1992**) Trigger points in the abductor hallucis referred pain to the medial aspect of the heel and foot and the taut bands associated with trigger points in this muscle may be responsible for tarsal tunnel syndrome itself. (**Travell & Simons, 1992**)

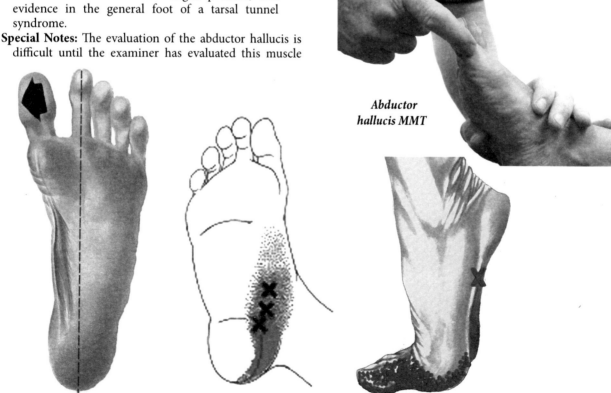

Abductor hallucis MMT

Abductor hallucis action   *Abductor hallucis MTrPs*   *Abductor hallucis MTrP*

Extensor Digitorum Longus and Brevis

Extensor Digitorum Longus

Attachments: From the lateral tibial condyle, proximal three-fourths of anterior fibula body, interosseous membrane, deep fascia, and intermuscular septa to the lateral four toes, middle and distal phalanges dorsal surface.

Action: Extends toes and with continued action assists in foot and ankle dorsiflexion and eversion; is a strong dorsiflexor and pronator of the foot.

Nerve supply: Peroneal, L4, **5, S1**.

Extensor Digitorum Brevis

Attachments: From the anterior portion of the superolateral surface of the calcaneus, lateral talocalcaneal ligament, and cruciate crural ligament to the first tendon to the dorsal surface of the base of the great toe proximal phalanx. This muscular slip and tendon are often called the extensor hallucis brevis. The other three tendons insert to the lateral sides of the extensor digitorum longus tendons.

Action: Extends phalanges of the medial four toes.

Nerve supply: Deep peroneal, L4, **5**, S1.

Testing Position: The patient extends his toes to the maximum amount.

Patient Fixation Requirements: The ankle should be stabilized by the patient's and examiner's efforts.

Stabilization: The examiner stabilizes the foot at the calcaneus, with his thumb over the dorsal surface of the foot to prevent the patient from dorsiflexing the ankle.

Test of longus and brevis: The examiner stabilizes the foot in slight plantar flexion, and the patient's toes are extended. The test force is to bring the toes into flexion. Usually the middle three toes are tested; the hallux is tested separately. The most common purpose for examination of these muscles is to determine the relative strength of the toe extensors to the flexors, to evaluate the extension of the metatarsophalangeal articulations in the "hammer toe" position. In the anterior tarsal tunnel syndrome, the most common motor weakness is in the extensor digitorum brevis muscle; the foot should be placed into full plantar flexion in some cases to find this inhibition pattern.

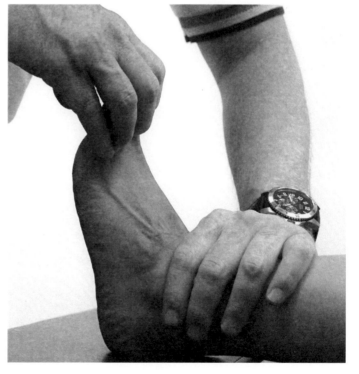

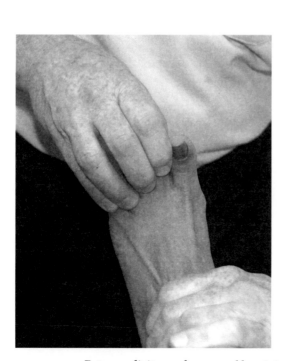

Extensor digitorum longus and brevis tested together

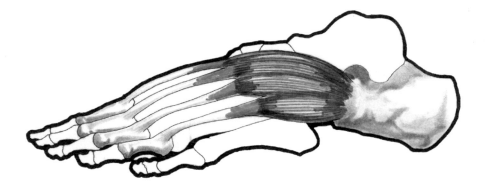

Frequently in this syndrome, an anterior talus subluxation will be pressing the deep peroneal nerve into the extensor retinaculum of the foot. Additionally, inhibition in this muscle is an important clinical parameter for L5-S1 radiculopathy, with associated sensory and motor impairments as well as atrophy in this muscle. (**Sinanovic & Custovic, 2010**)

Trigger points in the extensor digitorum longus (most commonly in the upper part of the muscle approximately a hand's width below the head of the fibula) refer pain into the dorsum of the foot or ankle and into the three middle toes. MTrPs in this muscle may also compress the deep peroneal nerve against the head of the fibula with the development of weakness of the toes and a foot drop.

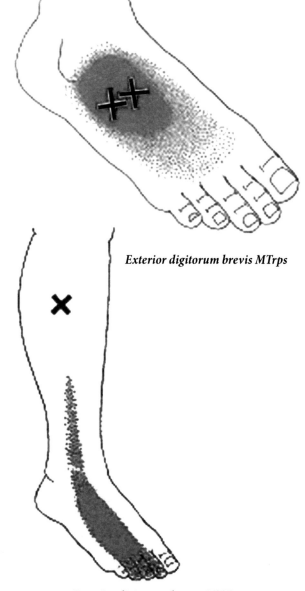

Exterior digitorum brevis MTrps

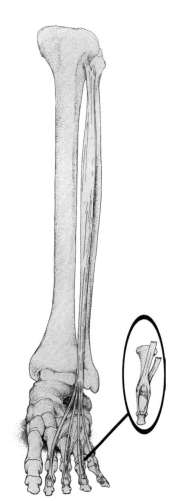

Attachment to the toes

Exterior digitorum longus MTrP

Extensor Hallucis Longus and Brevis

Extensor Hallucis Longus

Attachments: From the middle half of anterior fibular surface and adjacent interosseous membrane to the base of distal phalanx of great toe.
Action: Extends the great toe; continued action assists in dorsiflexion and inversion of the foot and ankle.
Nerve supply: Deep peroneal, L4, 5, S1.

Extensor Hallucis Brevis

The extensor hallucis brevis is the medial slip of the extensor digitorum brevis.
Attachments: From the anterior portion of the superomedial surface of the calcaneus, lateral talocalcaneal ligament, and cruciate crural ligament to the dorsal surface of the great toe's proximal phalanx.
Action: Extends great toe's proximal phalanx.
Nerve supply: Deep peroneal, L4, **5, S1.**
Testing Position: Patient extends the great toe, with the ankle in a neutral position.
Stabilization: Stabilize the calcaneus with the thumb over the dorsal surface of the foot.
Test for longus and brevis: With the patient sitting or standing, his great toe is placed in extension and the examiner stabilizes the foot and grasps the great toe to flex it while the patient resists.
Body Language of Weakness: Inability to hold the testing position. Significant weakness of this muscle is not nearly as common as other foot muscles. Weakness in the presence of lower lumbar spinal conditions and sciatic neuralgia indicates that evaluation of the fourth lumbar disc is needed.

Special Notes: An important condition affecting the foot and great toe was first described by Dananberg, **(1986)** called functional hallux limitus (FHL). It is the inability of the hallux to fully extended at the metatarsophalangeal articulation at the proper stage of the gait stance phase. The widespread adaptive and dysfunctional patterns which results from FHL demonstrate how foot and ankle imbalances can affect the rest of the body. Dananberg notes that "FHL... because of its asymptomatic nature

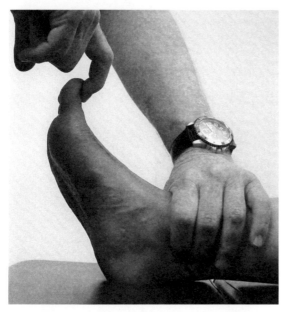

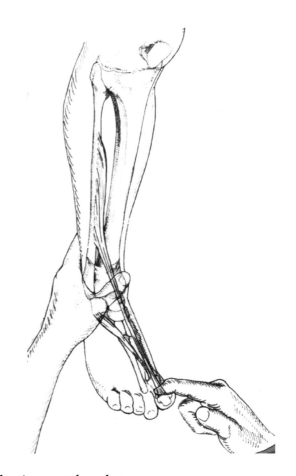

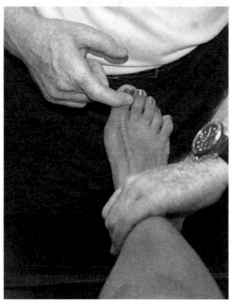

Exterior hallucis longus and brevis are tested together

and remote location, has hidden itself as an etiological source of postural degeneration."

The extensor hallucis longus and brevis muscle will usually test weak in the presence of FHL. This weakness causes secondary contraction of the flexor hallucis longus and brevis muscles. The contracted muscle is secondary to the weak antagonist muscle, a familiar finding in applied kinesiology and consistent with contemporary thinking regarding the pain-adaptation model of muscle dysfunction. **(Lund et al., 1991)** Applied kinesiology treatment for FHL is directed to returning the weak extensor hallucis muscles to normal. **(See Foot and Ankle Chapter) (Walther, 2000)**

Trigger points in the extensor hallucis longus refers pain across the dorsum of the foot and strongly into the first metatarsal and great toe areas. Travell and Simons **(1992)** note that an L4-L5 radiculopathy may also lead to MTrPs in the long extensor muscles of the toes.

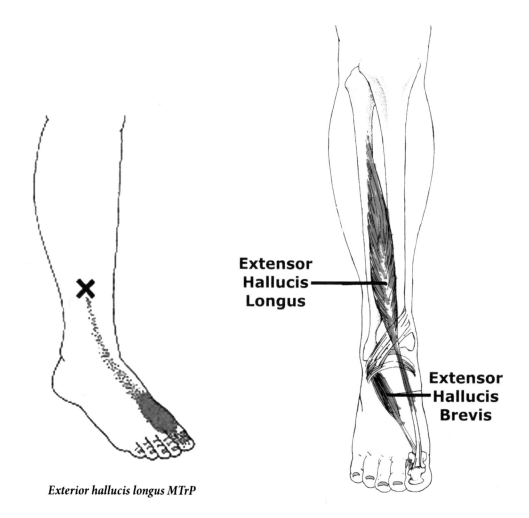

Exterior hallucis longus MTrP

Flexor Digitorum Brevis

Attachments: From the medial process of the calcaneus, central part of the plantar aponeurosis, and the intermuscular septa between it and the adjacent muscles to attach to the four tendons of the middle phalanges of the 2nd-5th toes. The entire muscle belly is firmly united with the plantar aponeurosis.

Action: Initial action flexes the middle phalanges on the proximal; continued action flexes the proximal phalanges on the metatarsals.

Testing Position: Patient flexes the 2nd-5th toes, with emphasis toward the proximal middle phalanx articulation.

Stabilization: Examiner stabilizes across the top of the metatarsals with the foot and ankle in a neutral position.

Synergist: Flexor digitorum longus.

Test: Examiner contacts all four toes on the plantar surface, and directs pressure toward extension.

Body Language of Weakness:
  During Test: Effort to flex toes into testing position pulls distal phalanx into flexion, but does not efficiently flex metatarsophalangeal articulations.
Movement aberrations: As the patient walks or stands, the distal phalanx will point down toward the floor and anterior sway will cause the distal phalanx to dig into the floor. The middle phalanx will rise.
Postural imbalances: The toes will be in a "hammer toe" position in which the metatarsophalangeal articulation will be in extension while the interphalangeal articulations are in flexion. There will be weight bearing evidence on the tips of the toes, and failure of the entire toe to extend into a neutral position.

Nerve supply: Medial plantar, L4, **5**, S1

Special Notes: The tendon of the 5th digit is congenitally absent in 23% of cases. **(Gray's Anatomy, 2004)** This muscle frequently requires treatment to the neuromuscular spindle cell, Golgi tendon organ, origin/insertion technique, or percussion. Attempts to exercise this muscle are clinically ineffective if there is a tarsal tunnel syndrome present. Atrophy of the muscle indicates the probability of a tarsal tunnel syndrome.

The way in which the tendons of the flexor digitorum brevis divide and attach to the phalanges is, according to Gray's Anatomy, **(2004)** identical to that of the tendons of the flexor digitorum superficialis in the hand.

Trigger points in the flexor digitorum brevis refer to the plantar surface of the foot, mostly to the area of the heads of the four lesser metatarsals. Travell and Simons **(Travell & Simons, 1992)** associate these trigger points with those found in the flexor digitorum longus muscle.

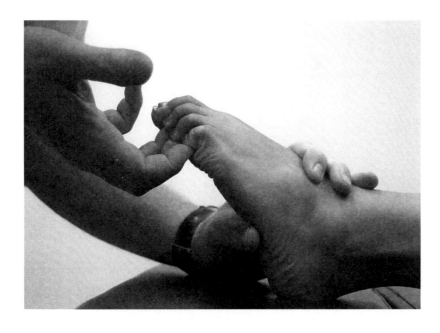

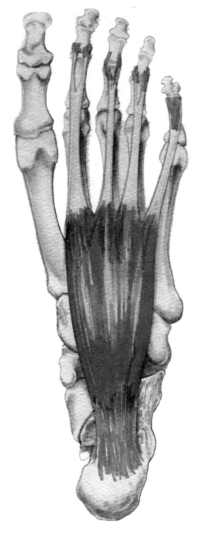

Flexor Hallucis Longus

Attachments: From lower two-thirds of posterior fibula, interosseous membrane and adjacent intermuscular septa and fascia to the plantar surface of distal phalanx of great toe.

Action: Flexes great toe; continued action aids in plantar flexing the foot; helps give medial ankle stabilization.

Testing position and stabilization: With the patient supine, the examiner stabilizes the metatarsophalangeal articulation in slight extension and holds the foot halfway between dorsal and plantar flexion. The patient flexes the distal phalanx of the great toe.

Synergists: The muscles that flex the distal phalanges of the toes and fingers are the only ones which can be 100% isolated for muscle testing. The flexor hallucis brevis attaches to the proximal phalanx and is the reason stabilization and slight extension are necessary between the proximal phalanx and the first metatarsal.

Test: From this testing position of flexion between the proximal and distal phalanx, the examiner directs pressure against the distal phalanx of the great toe in the direction of extension.

Nerve supply: Tibial, L5, S1, 2.

Neurolymphatic:

Anterior: Inferior to the symphysis pubis at the height of the obturator bilaterally (same as peroneus longus and brevis).

Posterior: Bilaterally between PSIS and L5 spinous.

Neurovascular: Bilateral frontal bone eminences.

Nutrition: Raw bone concentrate correlating with tarsal tunnel syndrome or other subluxations of the foot.

Meridian association: Circulation sex.

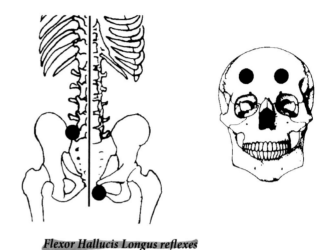

Flexor Hallucis Longus reflexes

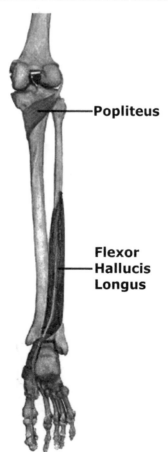

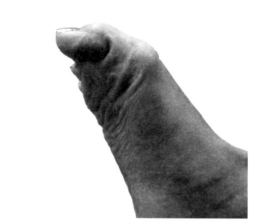

Flexor Hallucis Longus MMT

Flexor Hallucis Brevis

Attachments: Medial portion of the plantar surface of the cuboid bone, adjacent portion of the lateral cuneiform bone, and from prolongation of the tendon of the tibialis posterior to the medial and lateral sides of proximal phalanx of the great toe.

Action: Flexes metatarsophalangeal articulation of great toe.

Testing Position and Stabilization: Bringing the patient into the test position for the flexor hallucis brevis is best done in three steps. First, plantar flexes the 2nd, 3rd, and 4th digits to keep them out of the test. Second, fully dorsiflex the 1st and 2nd phalangeal articulations of the great toe to help remove the action of the flexor hallucis longus. Third, keeping the interphalangeal articulation in complete dorsiflexion, plantar flex the metatarsophalangeal articulation into testing position. While holding this position, the examiner stabilizes the foot, maintaining a neutral position between dorsiflexion and plantar flexion of the ankle.

Test: The examiner stabilizes the interphalangeal articulation of the great toe and places the metatarsophalangeal articulation in flexion for the starting test position. While maintaining hyperextension of the interphalangeal articulation, the examiner directs pressure against the plantar surface of the proximal phalanx toward extension.

Body Language of Weakness:

Movement aberrations: When the patient attempts to flex the great toe and there is weakness of the flexor hallucis brevis but strength of the flexor hallucis longus, the proximal phalanx may hyperextend in the distal phalanx will flex. The toe appears to remain straight while walking.

Postural imbalances: When the foot is relaxed, the toe will probably be in a hammer-toe position because of the failure of the flexor hallucis brevis – which inserts on the proximal phalanx -- to hold that phalanx toward flexion.

Nerve supply: Tibial, L4, **5, S1**, 2.

Neurolymphatic:

Anterior: Inferior to the symphysis pubis at the height of the obturator (same as peroneus longus and brevis).

Posterior: Between PSIS and L5 spinous.

Neurovascular: Bilateral frontal bone eminences.

Nutrition: Raw bone concentrate correlating to tarsal tunnel syndrome or other subluxations of the foot.

Meridian association: Circulation sex.

Special Notes: The flexor hallucis longus and brevis are most frequently tested in relation to the tarsal tunnel syndrome. Beardall (**Beardall, 1985**) made an improvement in the methods of testing these two muscles. When testing the flexor hallucis longus, stabilization of the proximal phalanx should be made by the doctor as he tests the plantar flexion ability on the distal phalanx. It was once thought that, in the tarsal tunnel syndrome, both the flexor hallucis longus and brevis were weak; however, with the improved testing methods it became obvious that the flexor hallucis usually remains strong while the brevis shows weakness. The difference in strength of the flexor hallucis longus and brevis in the tarsal tunnel syndrome is realistic, because the longus is in the upper leg and is not innervated by the nerves going through the tarsal tunnel.

Weakness in this muscle can be associated with the weakness of the tibialis posterior due to an inferior displacement of the navicular bone. Ramsak and Gerz (**Ramsak & Gerz, 2002**) recommend shifting the body weight onto the forefoot when testing these muscles in the standing position in order to uncover these problems. This method of testing will increase the stress on the tarsal tunnel during the MMT.

Trigger points in the flexor hallucis longus referred pain and tenderness into the plantar surface of the first metatarsal and great toe. Overactivity of the toe flexor muscles contributes to the development of hammer toes, claw toes and other deforming foot conditions as they attempt to stabilize the foot during weight bearing. (**Travell & Simons, 1992**)

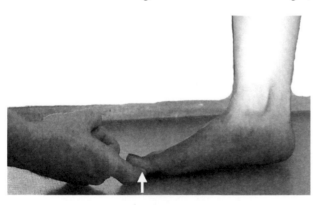

Flexor Hallucis Brevis MMT, standing

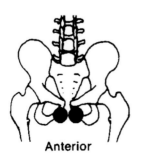
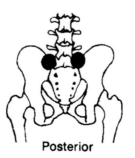

Flexor Hallicus Brevis reflexes

Gastrocnemius

Attachments: Two heads, one from the lateral and the medial condyles, the surface of the femur and capsule of the knee to merge distally with the soleus muscle to form the Achilles tendon which attaches to the posterior surface of the calcaneus.

Action: Plantar flexes foot.

Change of action: Dorsiflexion of the foot increases knee flexion capability. Since the gastrocnemius originates above the knee and the soleus below the knee, the differentiating factor in testing the two muscles is the knee position during the MMT.

Synergists: Soleus, plantaris, tibialis posterior, peroneus longus and brevis, flexor hallucis longus, flexor digitorum longus.

Test: The medial and lateral heads of the gastrocnemius can be tested as described by Beardall. **(Beardall, 1985)** The test must be correlated with hamstring strength because they are significantly synergistic in the test. For both medial and lateral heads of the gastrocnemius, the supine patient flexes the knee to approximately 110° and maximally plantar flexes the foot. For the medial head, the leg is internally rotated; for the lateral test, it is externally rotated. The examiner stabilizes the knee while extending it by pulling on the calcaneus contact.

Alternate Testing Method: All plantar flexors can be tested with the patient standing. The patient stabilizes herself with her hand on a table or wall, but does not use it to aid the test. The patient raises directly up on her toes with one foot, while the examiner observes for capability of elevating to the toes without bending the knee or leaning forward, indicating the recruitment of synergistic muscles due to inhibition of the plantar flexors.

Nerve supply: Tibial, L4, 5, **S1, 2**.

Neurolymphatic:

Anterior: 2" above umbilicus and 1" from midline.

Posterior: Between T11, 12 bilaterally near laminae.

Neurovascular: Lambda.

Nutrition: Adrenal concentrate or nucleoprotein extract.

Meridian association: Circulation sex.

Organ association: adrenal.

Special Notes: The gastrocnemius muscle must be correlated with the soleus muscle for evaluation of its strength. The maximum shortening of the gastrocnemius and soleus varies considerably. The soleus is capable of shortening 44 mm., the gastrocnemius only 39 mm. This is important in the action of these two mucles. Because the gastrocnemius originates proximal to the knee, it is lengthened or shortened with extension or flexion of the knee. When the knee is flexed, the gastrocnemius is shortened, equal to or exceeding its length of contraction. **(Kapandji, 2010)**

The origin of the gastrocnemius above the knee is very important in the muscle's role in the gait mechanism. As the knee is extended by quadriceps action, the gastrocnemius is lengthened to its most advantageous functional

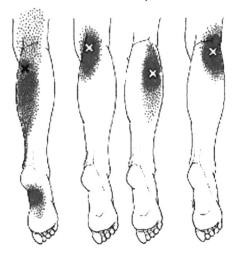

Gastrocnemius MTrp referral

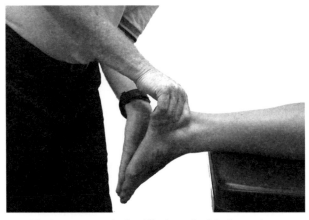

Testing pressure is dorsiflexion of ankle & traction on the calcaneus.

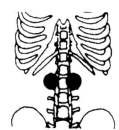
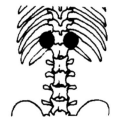

Neurolymphatic

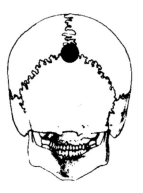
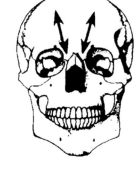

Neurovascular   *Stress receptor*

Gastrocnemius reflexes

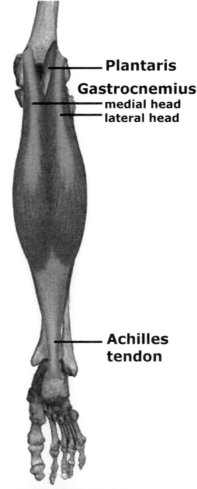

position. Thus the action of the quadriceps transfers more power to the ankle by the gastrocnemius. **(Kapandji, 2010)**

The postural position of hyperextension of the knee in the presence of a weak gastrocnemius compensates for failure of its normal action in maintaining the knee in a slight degree of flexion. The medial head of the gastrocnemius is more active than the lateral in maintaining this position. **(Houtz & Walsh, 1959)**

The gastrocnemius and soleus combination is not often tested in applied kinesiology in relation to its glandular correlation with the adrenal glands because of the difficulty in testing these very powerful muscles with the limited leverage available. The gastrocnemius supplies most of its information about glandular correlation from the associated tenderness that usually develops in the muscle when there is adrenal involvement. Frequently, insalivation of a useful adrenal support for the patient will reduce this specific palpation tenderness **(Goodheart, 1998)**

The gastrocnemius is often involved in athletic injuries on a reactive muscle basis. The muscle dysfunction may be primary, causing a weakness in another area, or it may be secondary and develop weakness from another muscle previously contracting. The popliteus and gastrocnemius are often involved with the gastrocnemius on a reactive muscle basis. Usually when these muscles are involved, there is a knee disturbance manifested when running and with consequent cutting action.

Persistent wearing of high or slippery heels may cause a shortness of the gastrocnemius, soleus, and Achilles tendon. The Achilles tendon plays an important part in improving the energy cost and efficiency of locomotion by storing energy elastically and releasing it later in the gait cycle. Goniometer measurement should show 15 degrees of dorsiflexion. When there is limited range of motion, differential diagnosis between shortness of the gastrocnemius and soleus can be obtained by comparing foot dorsiflexion. If the gastrocnemius is short and the foot is dorsiflexed, there will be a restriction of knee extension. If the knee is extended, there will be a restriction of dorsiflexion.

In normal walking, gastrocnemius restrains the tibia from rotating on the talus as the weight is shifted from the heel to the ball of the foot during the stance phase. **(Travell & Simons, 1992)**

Trigger points in the gastrocnemius refer pain to the posterior-inferior thigh, posterior leg, posterior knee and to the arch of the foot. **(Travell & Simons, 1992)** Some of these trigger points may be associated with nocturnal calf cramps (restless legs), although vitamin (especially riboflavin, niacin, and B6) and mineral imbalances (especially potassium) can be responsible for this disturbing condition. **(Cuthbert & Rosner, 2010; Travell & Simons, 1992)**

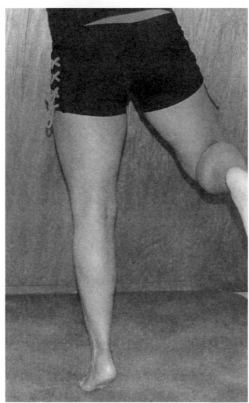

Standing test for all plantar extensor muscles

Soleus

Attachments: From the posterior surface of the head and upper one-third of the shaft of the fibula; middle one-third of the medial border of the tibia; tendinous arch between tibia and fibula into calcaneus with gastrocnemius by way of the Achilles tendon which attaches to the posterior surface of the calcaneus.

Action: Plantar flexes foot.

Change of action: When the individual is standing, the calcaneus becomes the fixed origin of the muscle. The muscle's action is important in stabilizing the tibia on the calcaneus in the standing position, and limiting forward sway.

Test: The prone patient flexes the knee to 90° and plantar flexes the foot. The examiner directs traction on the calcaneus and pressure on the forefoot in a direction of dorsiflexion. The knee flexion helps take the gastrocnemius out of the test. Because of the great strength of the soleus and its limited leverage, this muscle is difficult to evaluate.

Stabilization: Added differential diagnostic ability between the gastrocnemius and soleus MMT can be obtained by having an assistant stabilize the leg in the 90 degree flexed position. The patient then adds some attempt to extend the knee by quadriceps action, which causes reciprocal inhibition of the gastrocnemius muscle to better isolate the soleus during the MMT.

Synergists: Gastrocnemius, plantaris, tibialis posterior, peroneus longus and brevis, flexor hallucis longus, and flexor digitorum longus.

Body Language of Weakness:

Testing position: As the patient plantar flexes the foot, the gastrocnemius-soleus action on the Achilles tendon gives straight plantar flexion if the entire group of muscles is functioning normally. If the foot moves into inversion with plantar flexion, it indicates either recruitment of the synergist tibialis posterior and toe flexors, or weakness of the peroneus longus and brevis. If the foot moves into plantar flexion with eversion, it indicates recruitment of the synergist peroneus longus and brevis, or weakness of the medial muscles, the tibialis posterior and toe flexors.

During test: An effort to include inversion in the test indicates recruitment of one or all of the tibialis posterior, flexor hallucis longus, or flexor digitorum longus muscles. Attempts to evert the foot, along with plantar flexion, indicate recruitment of the peroneus longus and/or brevis.

Movement aberrations: Difficulty in rising onto the toes or walking on them.

Nerve supply: Tibial, L4, 5, **S1, 2**.

Neurolymphatic:

Anterior: 2" above umbilicus and 1" from midline.

Posterior: Between T11, 12 bilaterally near laminae.

Neurovascular: Lambda.

Nutrition: Adrenal concentrate or nucleoprotein extract.

Meridian association: Circulation sex.

Gland association: Adrenal.

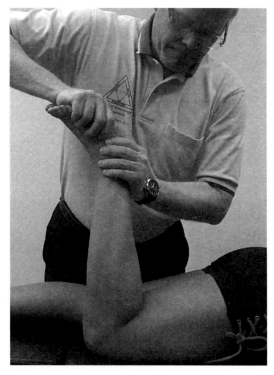

Soleus muscle origin insertion on the leg with the knee bent, stretching affects soleus most

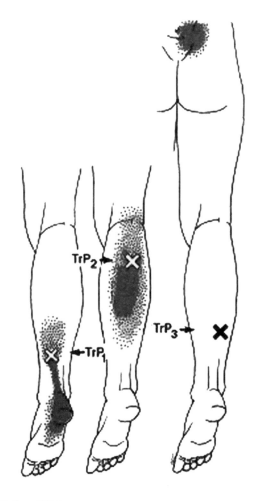

Soleus MTrPs

MMT for Lower Body Dysfunctions

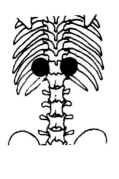
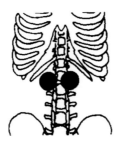
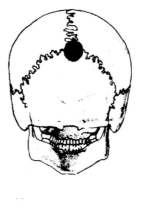
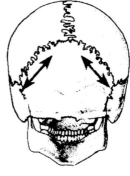

Neurolymphatic *Neurovascular* *Stress receptor*
Soleus reflexes

Special Notes: The soleus is not usually tested in applied kinesiology unless there is the possibility of gross muscular weakness, or an evaluation is being made for reactive muscle patterns.

The gastrocnemius and soleus have typically been known as the triceps surae because the two heads of the gastrocnemius and the soleus insert into a common tendon. It is the strongest plantarflexor of the foot. Campbell et al. **(Campbell, 1973)** have demonstrated electromyographically that these muscles, in reality, act as a quadriceps surae since the medial and lateral aspects of the soleus are capable of acting independently. O'Connell, **(O'Connell, 1958)** using surface and needle electrodes, also found that the medial and lateral aspects of the soleus act independently. This medial and lateral action of the gastrocnemius and soleus correlates with a medial or lateral calcaneus subluxation, which is found in applied kinesiology by using the sensorimotor challenge mechanism. Usually the subluxation correlates with a tarsal tunnel syndrome, where the calcaneus is also posterior. The calcaneus will usually be lateral, and there will be weakness on the medial head of the gastrocnemius and the medial aspect of the soleus. This gives poor medial support to the calcaneus, allowing it to deviate laterally. As with other structural distortions, it is necessary to return the muscles to normal balance to obtain maximum correction. Since the gastrocnemius and soleus are very difficult to evaluate by standard MMT, it is best to evaluate the medial and lateral aspects by therapy localization directly to the muscle. The most common disturbance is dysfunction of the neuromuscular spindle cell or Golgi tendon organ, uncovered with "pincer palpation" followed by MMT. **(Leaf, 2010; Cuthbert, 2002)** There may also be a muscle stretch response on the tight side of the muscles, indicating the presence of myofascial trigger points.

When the triceps surae is short, a surgical procedure to lengthen the Achilles tendon is sometimes done. Ryerson **(Ryerson, 1948)** points out, "The tendo-Achilles is not short. The muscle bellies are short." He recommends lengthening the muscle. In applied kinesiology there are many techniques to do this, including fascial release and stretch and spray techniques, and treatment to the muscle proprioceptors. A review of the literature also indicates three studies showing that treatment of MTrPs in the lower limb can improve reduced ankle ROM. **(Grieve et al., 2011; Wu et al., 2006; Grieve, 2006).**

The gastrocnemius has a higher concentration of fast muscle fibers than the soleus. The gastrocnemius is primarily a muscle for fast, quick action, whereas the soleus is a postural muscle for prolonged use. The fatigue characteristics of the gastrocnemius and soleus were studied by Ochs et al. **(Ochs et al., 1977)** with electromyography. The soleus muscle has a mean of about 70% slow muscle fibers, which are the oxidation type, and 30% fast glycolytic fibers. The gastrocnemius is equally divided, with 50% of each. The action potential of the gastrocnemius decreased more rapidly than the soleus, indicating a faster fatigue of the gastrocnemius.

The gastrocnemius and soleus are in their neutral

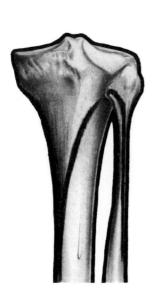

Soleus MMT

positions when there is about 30 degrees of plantar flexion; consequently, when in an upright position, the muscles are slightly stretched, taking the slack out of the tendon and keeping the muscles in a position of functional readiness. **(Houtz & Walsh, 1959)** In the standing position, the gastrocnemius and soleus' attachments are reversed, making the foot the base and the origin for muscle action. In the relaxed standing position, the stretch reflex is constantly monitoring the anterior and posterior sway of the body, activating the soleus and anterior tibial for balance. During normal standing, the line of gravity is in front of the ankle joint; the soleus muscle maintains a tonic contraction to prevent the body from falling forward. **(O'Connell, 1958)**

The glandular association of the gastrocnemius and soleus with the adrenal provides body language to indicate when adrenal involvement may be present. The patient may complain of aching in the calves of his legs, especially after being on his feet for a period of time. There will be tenderness in the soleus-gastrocnemius complex when squeezed by the examiner. Of course, this tenderness must be differentiated from vascular conditions and direct metabolic problems with the muscle, such as calcium deficiency causing hypertonicity.

Travell and Simons **(Travell & Simons, 1992)** suggest that the soleus muscle provides the major pumping action for the return of blood from the lower limb to the heart. When the soleus muscle is functioning properly its strong contractions compress the venous sinuses so that its venous blood is forced upward. They call this pumping action of the muscle "the body's second heart" and note that this mechanism depends on competent valves in the popliteal veins.

"Valves in the veins to prevent reflux of the blood are most numerous in the veins of the lower limbs where the vessels must return blood against high hydrostatic pressure. The popliteal vein usually contains four valves. Deeper veins that are subject to the pumping action of muscle contraction are more richly provided with valves."

Additional causes of impairment of the blood supply to these muscles include wearing socks with tight elastic tops; pressure exerted on the calf from sitting with it pressed against the edge of a chair or leg rest, and arterial obstruction such as peripheral arterial disease. **(Baldry, 2001)** The gastrocnemius muscle is the one in which ischemia-induced MTrP activity most often takes place, followed by the gluteus medius, soleus, and tibialis anterior. **(Baldry, 2001)** Trigger points in the soleus muscle refer to the heel, Achilles tendon and ipsilateral sacroiliac joint. Travell and Simons also describe a rare trigger point that is known to refer to the ipsilateral face and jaw, possibly altering occlusion of the teeth.

If injecting a local anesthetic into the MTrP of the gastrocnemius is the method of treatment chosen, it should be explained to the patient that there is a high incidence of post-injection soreness that often lasts for as long as a week **(Travell & Simons, 1992)** In fact, the soreness can be so severe that they warn the procedure should never be carried out in both legs at the same time. Additionally, using this method of treatment in the popliteal region, caution must be exercised to avoid damaging the popliteal artery. These potential complications are avoided by using the techniques employed in AK.

Intrinsic Plantar Foot Muscles

The intrinsic muscles of the foot are usually evaluated in applied kinesiology when there are subluxations or pain present. The muscles appear to stabilize the articulations in the foot and return them to a balanced position when functioning normally, similar to the way the intrinsic muscles of the spinal column are responsible for the articulations in that area. The usual treatment for the intrinsic muscles is origin/insertion technique, neuromuscular spindle cell, Golgi tendon organ, and percussion. **(Fulford, 1996; Cuthbert, 2002)**

There is no listing for neurolymphatic, neurovascular, organ-gland, meridian, or nutritional association for these muscles. Beardall, from his intensive clinical and manual muscle testing experience, has offered these correlations. The intrinsic plantar foot muscles appear to respond clinically to these specific AK manipulative approaches; however, the association is better established for the larger muscles whose association has been established over the past 46 years in applied kinesiology by the thousands of practitioners in the International College of Applied Kinesiology around the world.

In the normal foot, the intrinsic muscles do not appear to have an arch supporting role. **(Basmajian & Stecko, 1971)** "The muscles are spared when the ligaments suffice." **(Basmajian, 1961)** Although the specific intrinsic muscles of the arch were not studied by Gray **(Gray, 1969)** in his comparison of the normal and the flat foot, it seems likely that they are active in the flat foot because of the lack of ligamentous support. In the presence of a chronic tarsal tunnel syndrome, there is clinically observed atrophy of the plantar intrinsic muscles. It is valuable to exercise these muscles after the tarsal tunnel syndrome has been corrected. In many cases, exercise improves function of the foot so that the patient may discontinue wearing orthopedic supports for the arches. **(See Foot and Ankle chapter)**

Travell and Simons **(Travell & Simons, 1992)** note that trigger points in the plantar intrinsic muscles are exacerbated by the wearing tight, poorly fitting or designed shoes, ankle and foot injuries, structural imbalances of the foot, subluxation or loss of structural integrity of the joints of the foot, walking on sandy or sloped surfaces, and conditions where the feet can get chilled as well as systemic conditions (gout, rheumatoid arthritis, diabetes, among others).

It should be remembered that manual treatment to the plantar surface of the foot must be applied through the plantar fascia. The integrity of this fascia is important to the arch system of the foot. The plantar aponeurosis is tensionally loaded and in this way helps maintain the plantar arch. When injured by subluxation or soft tissue dysfunction, this issue may develop inflammation which is commonly termed plantar fasciitis. **(Cailliet, 1997)**

Myers **(Myers, 2001)** offers a simple test that can demonstrate to the clinician and the patient the interrelatedness of the plantar muscles of the foot to the rest of the body. Ask the patient to do a forward bend while keeping the knees straight. Measure the resting distance the patient can reach down the legs and the contour of the back. Then ask the patient to roll a golf or tennis ball deeply into the plantar surface of the foot on one side only, working throughout the foot deeply and slowly. Do this for several minutes, making sure the entire foot from the toes to the ball of the foot to the front edge of the heel are treated. Then ask the patient to do the forward bend again and note the bilateral differences in hand position and back contour and ease of movement. In most people this will produce an obvious change in function and show them how working on the intrinsic muscles and fascia of the foot can affect the functioning, comfort, and flexibility of the whole person.

It is obvious to practitioners experienced in applied kinesiology that the methods of challenge and therapy localization to the foot can make this kind of dramatic, measurable improvements in muscle strength, postural function and ease of use with greater rapidity and specificity.

Lumbricales and Dorsal and Plantar Interossei

Lumbricales

Attachments: Between the tendons of the flexor digitorum longus, except for the first, which arises from the medial side of the first tendon of the flexor digitorum longus to attach on the medial side of the proximal phalanx and into the expansions of the tendons of the extensor digitorum longus of the 2nd-5th toes.

Action: Flex the proximal phalanges on the metatarsals, and extend the two distal phalanges of the four small toes.

Synergists: Flexor digitorum brevis and longus for the metatarsophalangeal articulation, and extensor digitorum longus and brevis for the extensor function; dorsal interossei and plantar interossei.

Testing Position: There are two tests for the lumbricales. Described here is the flexing action of the metatarsophalangeal articulation. The extension of the middle and distal phalangeal articulations is tested with the extensor digitorum longus and brevis.

Testing position for flexion of the metatarsophalangeal articulation is difficult to obtain in most individuals. The optimum testing position is flexion of the metatarsophalangeal articulation, with neutral position of the middle and distal interphalangeal articulations.

Stabilization: The examiner stabilizes across the dorsal surface of the metatarsals.

Test: Examiner contacts plantar surface of the proximal phalanges of the 2nd-5th toes, and directs pressure toward extension of the metatarsophalangeal articulations.

Body Language of Weakness:

Testing Position: The patient has difficulty in actively achieving the testing position. When a patient cannot move into the testing position, the examiner can passively place the toes into the position to determine if the patient can hold it. Care must be taken, because many individuals cannot isolate muscle activity well enough to obtain or maintain the testing position, even though the muscle is not weak; only control is lacking.

Movement aberrations: Similar to those described for the flexor digitorum brevis.

Postural imbalances:

Nerve supply:

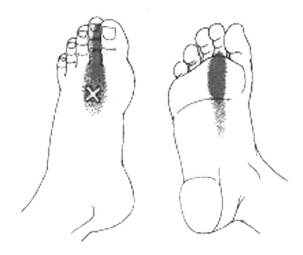

Dorsal Interossei MTrPs

1st **Lumbricalis:** medial plantar nerve, L4, 5
2nd, 3rd, 4th **lumbricales:** Lateral plantar nerve, S1, 2

Dorsal Interossei

Attachments: There are 4 dorsal interossei which arise by double pennate fibers from the bases and sides of the bodies of adjacent metatarsal bones and attach to the base of the proximal phalanx and aponeurosis of the tendons of the extensor digitorum longus. The 1st dorsal interosseous (arising from the 1st and 2nd metatarsals) inserts into the 2nd toe. The 2nd-4th dorsal interossei insert into the lateral sides of the 2nd, 3rd, and 4th toes.

Action: Assist in flexing the proximal phalanx and extending the middle and distal phalanges. Abduct the toes from the longitudinal axis of the 2nd toe.

Nerve supply: Lateral plantar

Lumbricales

Plantar Interossei

Attachments: There are 3 plantar interossei arising from the base and medial plantar surface of the 3rd, 4th, and 5th metatarsal bones that attach to the medial side of the base of the 1st phalanx of the same toe, and into the tendons of the extensor digitorum longus.
Action: Flex the proximal and extend the distal phalanges, and adduct toes toward the axis of the 2nd toe.
Nerve supply: Lateral plantar

Special Notes:
The lumbricales and interossei muscles are not often tested in applied kinesiology. Knowledge of the action and location of the muscles is informed for application of direct treatment to them in the case of subluxations and soft tissue dysfunction, as for instance myofascial trigger points, especially if they are recurrent.

The interossei act to stabilize the foot in uneven or rough terrain and stabilize the toes during gait.

The location of the interossei muscles indicates they are important in the gait mechanism. The acupuncture points which are treated with gait dysfunction in AK are in the location of these muscles. It is clinically observed that when there is a gait dysfunction, there is an exquisite tenderness on digital pressure in the area indicated for gait treatment. It seems likely that the muscle in the location of the apparently dysfunctioning acupuncture point is involved in some manner with the problem. Failure to correct any soft tissue involvement or subluxations seems to cause the active acupuncture point to return after it has apparently been effectively treated.

Travell and Simons (1992) note that the lumbricales trigger point referral patterns are likely to be similar to the interossei; however the patterns for this muscle have not been confirmed. Travell and Simons add that trigger points in the first dorsal interosseous muscle may produce tingling in the great toe as well as disturbances of sensation that can extend to the dorsum of the foot and the lower shin.

Peroneus Longus and Brevis

The peroneus longus and brevis muscles will be considered together because they are tested simultaneously. The primary requirement for differentiating these two muscles is the origin and insertion for tendon treatment, or attention to the golgi tendon organs or neuromuscular spindle cell for the individual muscle.

Peroneus Brevis

Attachments: From the lower two-thirds of fibula on lateral side and adjacent intermuscular septa to the lateral side of proximal end of 5th metatarsal.
Action: Plantar flexes foot and everts it; gives lateral stability to the ankle.

Peroneus Longus

Attachments: From the lateral condyle of tibia, head and upper two-thirds of lateral surface of fibula, intermuscular septa and adjacent fascia to the proximal end of the 1st metatarsal and medial cuneiform on their lateral portions.
Action: Plantar flexes foot and everts it; gives lateral stability to the ankle.

Peroneus Longus and Brevis

Change of Action: When the foot is stabilized, as in standing, the peroneus longus and brevis stabilize the leg on the foot. They are synergistic to the gastrocnemius and soleus muscles in extending the tibia and fibula at the ankle when in the standing position.
Testing Position and Stabilizaton: Supine patient fully plantar flexes the foot and then everts it as much as possible. The toes should be kept neutral or in flexion to limit action of the long muscles of toe flexion and extension. The testing position should be such that the muscle and tendon passing behind the lateral malleolus are in as straight a line as possible. The examiner stabilizes the leg above the ankle.
Test: The supine patient maximally plantar flexes the foot and everts it, with the toes kept in flexion or neutral position. Testing pressure is directed to the side of the foot in the direction of inversion. The test must start from the maximum eversion allowed when the foot is in complete plantar flexion. The range of motion in this test is limited. No dorsiflexion of the foot should be allowed, nor should there be any extension of the toes. The examiner should observe the tendon that courses behind the external malleolus as evidence of maximum isolation of the muscles.
Body language of weakness:
Testing position: It is not unusual to have such great weakness of these muscles that the patient cannot bring the foot into the testing position.
During test: The patient's effort to recruit other muscles into this test is dramatic, and sometimes extremely difficult to stop. The patient will attempt dorsiflexion of the foot and extension of the toes. Even if the examiner prevents the patient from dorsiflexing his foot and extending the toes by holding the position, the patient's effort will change the test significantly. If dorsiflexion of the foot or extension of the toes is allowed, a strong

Peroneus Longus & Brevis MMT

MMT for Lower Body Dysfunctions

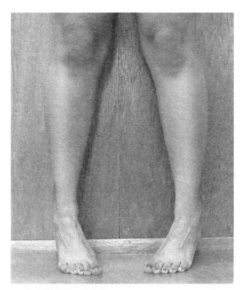

Peroneal Exercises

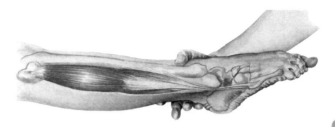

Peroneus longus & brevis muscle

peroneus longus and brevis may be indicated; in reality, the muscles may be exquisitely weak. In order to eliminate the dorsiflexion and toe extension, it may be necessary to restart the test several times. Once the examiner has felt the dramatic difference between testing a weak peroneus longus and brevis correctly, and the dramatic change the patient can make with dorsiflexion or toe extension, it becomes much easier to perform this test properly.

Postural balance: Balance of the ankle muscles is best observed when the patient is supine. Weakness of the peroneus longus and brevis causes the foot to invert. An imaginary line extending down the tibial ridge should extend over the second toe. Weakness of the peroneus longus and brevis causes this line to be lateral of the second toe or to miss the foot entirely. When evaluating the structural balance of the foot in the supine position, care must be taken that the foot is not resting on the table in such a manner that the table pressure deviates the foot from its relaxed position. It is best to have the foot hang over the edge of the table, or be supported by an ankle rest such as is present on most chiropractic tables.

Nerve supply: Peroneal, L4, **5, S1**.
Neurolymphatic:
Anterior: Inferior symphysis pubis, bilaterally.
Posterior: Bilaterally between posterior superior iliac spine and L5 spinous process.
Neurovascular: Bilateral frontal bone eminences.
Nutrition: Calcium, vitamin B complex. Avoid oxalic acid foods.
Meridian association: Bladder.
Organ association: Urinary bladder.

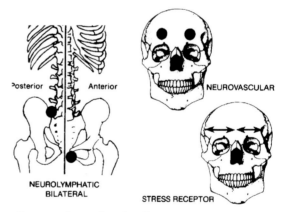

Peroneus longus brevis reflexes

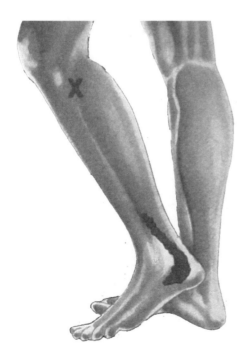

Peroneus Longus MTrP

Special Notes:
The peroneus longus and brevis are very important in maintaining normal foot and ankle function; this relates to the entire gait mechanism of the body. Weakness of these muscles is often due to foot subluxations. This can be easily determined using applied kinesiology by challenging the foot structures and then re-testing the peroneus longus and brevis for improved function. These muscles are important lateral ankle stabilizers. No other therapeutic method that we know of offers this kind of immediate evaluation tool to determine the measures necessary to improve ankle muscle function so rapidly and accurately. Myofascial trigger points (MTrPs) in the peroneus longus muscle may lead to development of taut bands in it and these will generate the muscle stretch reaction which in AK identifies the presence of MTrPs. These taut bands are able to compress the common peroneal nerve and/or its superficial and deep branches against the fibula. Compression of motor fibers in

these nerves may cause a foot drop to develop, as well as numbness and tingling on the dorsum of the foot between the first and second toes.

The peroneus longus and brevis may be injured from trauma, such as when the ankle is twisted in an inversion sprain. The trauma to the muscle may require origin/insertion technique or treatment to the proprioceptors of the muscles. After the ankle strain or sprain has recovered, evaluation of the muscles should be made for the possible need for treatment. Many recurrent twisted ankles are due to weakness of these muscles and their failure to fully recover. Richie (**Richie, 2001**) has reviewed the syndrome of functional ankle instability and finds that the majority of patients with this condition have a loss of neuromuscular control. The components of neuromuscular control include proprioception, muscle strength and reaction time and postural control during movement. Proprioceptive deficits lead to a delay in peroneal reaction time.

Some consider that the peroneus longus and brevis -- acting together with the peroneus tertius and tibialis posterior and anterior -- function as a sling mechanism to support the arch of the foot.

Peroneus Tertius

Attachments: From lower one-third of the anterior surface of the fibula and adjacent intermuscular septum to the dorsal surface of the base of the 5th metatarsal.
Action: Dorsiflexes and everts the foot.
Testing position and stabilization: The patient is best tested in the supine position. He brings the foot into dorsiflexion and eversion, with the toes kept in neutral position or toward flexion.
Synergists: The extensor digitorum longus assists in the peroneus tertius test. The peroneus tertius is a part of the extensor digitorum longus and might be described as its 5th tendon. The extensor digitorum longus is best kept out of the test by having the toes neutral or slightly flexed.
Test: The supine patient dorsiflexes and everts the foot with the toes kept in neutral position, or toward flexion. Examining pressure is directed against the dorsal lateral surface of the 5th metatarsal in the direction of plantar flexion and inversion. The examiner should evaluate the tendon of the peroneus tertius and the tendons of the extensor digitorum longus for best direction to maximize the effect of the peroneus tertius and minimize that of the toe extensors.
Body Language of Weakness:
Testing position: When the peroneus tertius is weak, it is difficult for the patient to move into the testing position without extension of the toes
During test: The toes attempt to extend from the neutral or slightly flexed position.
Postural imbalances: The peroneus tertius gives some lateral stabilization to the ankle. Lack of lateral stabilization is best seen when the patient is in the supine position. The foot inverts, causing an imaginary line drawn along the tibial ridge to intersect the foot lateral to the second toe. There may also be a lack of support to the lateral longitudinal arch of the foot. In the normal foot, there is no requirement of the peroneus tertius to support the lateral longitudinal arch. In a flat-footed individual, there is greater requirement of all the muscles which support the arch. (**Gray, 1969**)
Nerve supply: Peroneal, L4, 5, S1.
Neurolymphatic:
Anterior: Inferior ramus of pubic bones.
Posterior: Between L5 transverse and sacrum.
Neurovascular: Bilateral frontal bone eminences.
Nutrition: Calcium, B complex. Avoid oxalic acid foods, e.g., caffeine, cranberries, plums, and others.
Meridian ssociation: Bladder.
Organ association: Urinary bladder.

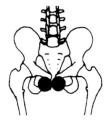

Peroneus Tertius MTrP

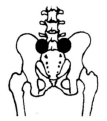  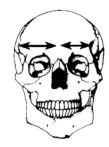

Peroneus Tertius (reflexes)

Special Notes:
The peroneus tertius is absent in between or 4.4% to 8.2% of individuals. **(Gray's Anatomy, 2004; Travell & Simons, 1992).** One cause of weakness in the peroneus tertius muscle is a subluxation of the proximal and distal tibiofibular joint, often following an inversion sprain. A compression support, e.g. an approximation challenge over the lateral and medial malleoli or taping over the lateral and medial malleoli, may immediately strengthen the peroneus tertius. Trigger points in the peroneus tertius refer pain into the anterolateral ankle and into the lateral malleolus and the heel. **(Travell & Simons, 1992)**

Ramsak & Gerz **(2002)** recommend the following exercise for the peronei. The starting position is with both feet in 45 degree internal rotation. Slowly rise up on toes; then lower the heels back to starting position. "If more than 40 repetitions are possible without a problem, the exercise should be done on one leg only, for athletes maybe with additional weights."

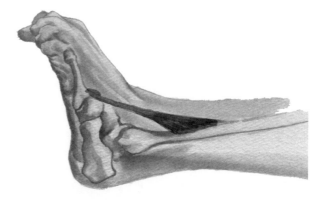
Peroneus Tertius Muscle

Peroneus Tertius MMT

Popliteus

Attachments: From the lateral condyle of femur, posterior horn of lateral meniscus and fibular head to the triangular area on posterior surface of tibia above soleal line.

Action: Rotates the tibia internally on the femur or the femur externally on the tibia, depending upon the one fixed; withdraws the meniscus during flexion, and provides rotatory stability to the femur on the tibia; brings the knee out of the "screw-home" position of full extension; helps with posterior stability of the knee.

Testing Position: The prone patient flexes the knee to 90° and medially rotates the tibia on the femur.

Patient Fixation Requirements: The foot is used to impart rotation to the tibia. The ankle must be fixed, and there must be no pathology in the ankle, foot, or knee to cause pain to the patient.

Stabilization: The examiner must make certain that the patient does not rotate the femur, or flex or extend the knee.

Test: With the patient's knee flexed to 90°, pressure is directed on the distal medial foot, with counter-pressure on the calcaneus to impart lateral rotation of the tibia on the femur. The actual testing motion is slight and can be evaluated only by observing the tibia rotating on the femur and watching for motion of the tibial tubercle. It is quite possible for the examiner to obtain foot rotation, appearing to be a weak popliteus; in fact, it may be a twisting of the tibia and fibula.

Body Language of Weakness:

Testing position: In the presence of extreme weakness, the patient will be unable to rotate the tibia medially on the femur.

During test: The patient may attempt to change the parameters of the test by knee flexion, extension, or femur rotation.

Postural imbalances: The patient may stand with a

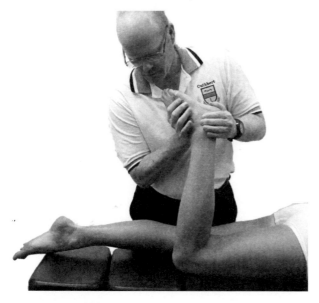
Popliteus MMT

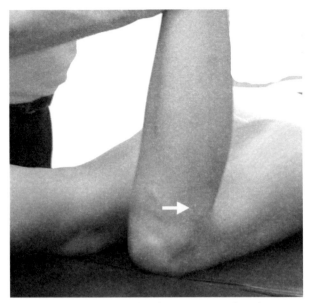

Movement of tibial tubercle in presence of weak popliteus

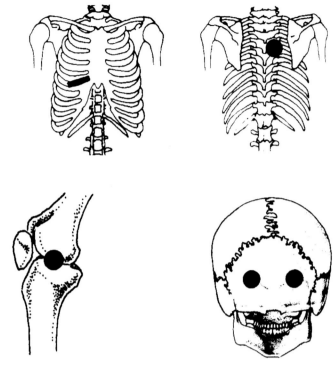

Popliteus reflexes

hyperextend knee as compensation for failure of the popliteus to offer good posterior knee stabilization. There will be lateral rotation of the tibia on the femur, giving an appearance of external rotation of the entire leg; actually, it is of the tibia and foot only. Weakness of the popliteus is best observed on a postural basis when the patient is in a relaxed, seated position with his leg hanging freely from the examination table. Weakness will cause the tibial tubercle to be lateral to the leg's midline.

Reactive Muscle Correlation: Gastrocnemius, hamstrings, and upper trapezius.
Nerve supply: Tibial, **L4, 5, S1**.
Neurolymphatic:
Anterior: 5th intercostal space from mid-mamillary line to sternum on the right.
Posterior: Between T5-6 laminae on right.
Neurovascular: Medial aspect of knee at meniscus.
Nutrition: Vitamin A, bile salts, beet leaf extracts.
Meridian Association: Gallbladder.
Organ Association: Gallbladder.
Special Notes:

Bilateral popliteus weakness indicates a probable midcervical functional fixation of the vertebrae. Correction of this fixation will immediately restore strength to the muscles.

The popliteus origin at the posterior horn of the lateral meniscus of the knee is a source of considerable involvement when this muscle is weak.

Basmajian and Lovejoy, using bipolar fine-wire electromyography, demonstrated with dynamic studies that the popliteus is more involved with rotation than with flexion. They also concluded that the popliteus is important in drawing the lateral meniscus posterolateral during flexion of the knee and medial rotation of the tibia. This helps prevent forward dislocation of the femur on the tibia during flexion of the knee, and is accomplished by continuous marked activity of the popliteus in the semi-crouched, knee-bent position.

The popliteus muscle is attached to the lateral meniscus as the semimembranosus muscle is to the medial meniscus. The posterior movement of the menisci produced by these muscles during knee flexion helps decrease the chance of meniscus entrapment and the resultant limitation of knee flexion which would result. Travell and Simons (**Travell & Simons, 1992**) note that the popliteus prevents forward displacement of the femur on the tibial plateau. The trigger point referral patterns that they describe, go primarily into the back of the knee.

Baker's cysts are enlargements of bursa in the posterior knee and are continuous with the synovial cavity. The Baker's cyst is a collection of synovial fluid that has escaped from the knee joint to form a 'cyst' in the popliteal space. It is often the result of knee injury or disease, such as a meniscal tear or rheumatoid arthritis. (**Travell & Simons, 1992**) If applied kinesiology's conservative measures fail to reduce the swelling, surgical removal may be required, especially when the cyst encroaches on the neurovascular tissues that course through the popliteal fossa.

Tibialis Anterior

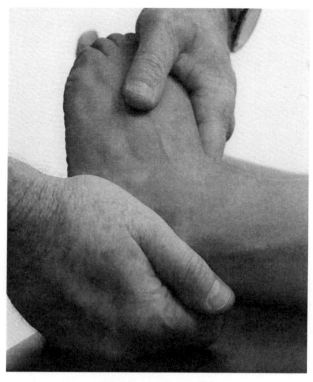
Tibialis anterior MMT

Attachments: Lateral condyle of tibia, proximal two-thirds of lateral surface of tibia, interosseous membrane, deep fascia and lateral intermuscular septum to attach to the medial and plantar surface of medial cuneiform, and base of 1st metatarsal.

Action: Dorsiflexes foot and inverts it.

Change of Action with standing: When in the standing position, the foot is fixed and becomes the origin for the muscle. Action causes forward body lean antagonistic to the plantar flexion of the soleus and gastrocnemius. Active in the balance mechanism of anterior and posterior sway. **(Gray & Basmajian, 1968; O'Connell, 1958)**

Stabilization: The examiner stabilizes the leg above the ankle.

Synergist: Extensor hallucis longus, extensor digitorum longus.

Antagonist: If the gastrocnemius is shortened, the test should be performed with the knee in the flexed position to allow greater ankle dorsiflexion.

Test: The supine patient inverts and dorsiflexes the foot, with the toes kept in flexion. The examiner applies pressure against the medial dorsal surface of the foot in the direction of plantar flexion and eversion. The examiner should see effective contraction of tibialis anterior as indicated by the tendon elevation during the test.

Alternate Testing Methods: The test can be performed in the sitting position, or in the prone position when the knee is flexed. This may be helpful when testing for reactive muscles. Ramsak & Gerz **(Ramsak & Gerz, 2002)** suggest the muscle will be inhibited more frequently when tested standing or sitting because of lumbar disc problems.

Body Language of Weakness:

During Test: Patient attempts to extend the toes to recruit synergistic action of the extensor digitorum longus and extensor hallucis longus when the tibialis anterior is weak. The ideal test minimizes the contraction of these muscles, as indicated by their tendons not rising or the toes extending. Pain in the muscle is often associated with shin splints.

Movement aberrations: Tendency toward foot drop.

MTrPs in the muscle causes it to become weakened, with impairment of dorsiflexion and, as a consequence, the development of a foot drop. Foot drop must be significant before shoe wear will give an indication of it. A peak of electromyographic activity occurs at toe-off of the stance phase of gait. This is apparently to dorsiflex the ankle, permitting the toes to clear the floor. **(Basmajian, 1978)**

Postural imbalances: The tibialis anterior is a medial ankle stabilizer. In the supine and relaxed position, there will be deviation of the foot laterally. It can be evaluated with an imaginary line drawn down the anterior tibial ridge, which should extend into the second toe. When evaluating the foot for balance in the supine position, care must be taken that the foot is not being deviated from its relaxed position by pressure on the calcaneus by the table. It is best to have the patient in a position where the foot is hanging freely over the end of the table.

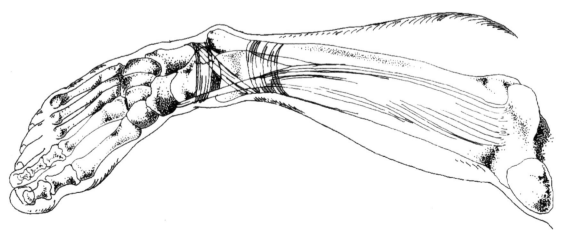
Tibialis anterior action

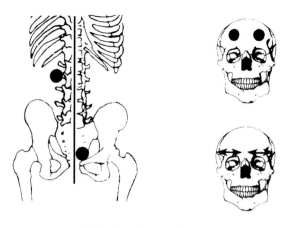

Tibialis Anterior reflexes

Nerve supply: Peroneal, **L4, 5**, S1.
Neurolymphatic:
Anterior: 3/4" above symphysis pubis bilaterally.
Posterior: L2 transverse process.
Neurovascular: Bilateral frontal bone eminences.
Nutrition: Vitamin A.
Meridian association: Bladder.
Organ association: Urinary bladder.
Special Notes:

This muscle is often associated with shin splints. When present, the treatment is frequently origin/insertion, proprioceptive, percussion, or fascial release technique. When shin splints are present, there will usually be pain during the muscle test; this should be dramatically relieved after treatment.

The response of the tibialis anterior muscle to a "proprioceptive technique" used in applied kinesiology was investigated by Perot et al (**Perot et al., 1991**) during manual muscle testing using a graphical registration of both mechanical and electromyographic parameters. Experiments were conducted blind on ten subjects. Each subject was tested with MMT and EMG ten times, five times as a reference, and five times after proprioceptive techniques to inhibit the muscle. Results indicated that after treatment an inhibition was easily registered. The reliability of the proposed procedure is mostly dependent upon satisfactory subject-examiner coordination that is also necessary in standard manual muscle testing.

This study demonstrated that there was a significant difference in electrical activity in the muscle, and that this corresponded with the difference found between "strong" versus "weak" muscle testing outcomes by the examiners. It further demonstrated that these outcomes were not attributable to increased or decreased testing force from the doctor during the tests. In addition, the study showed that manual treatment methods used in AK to reduce the level of muscle tone in the spindle cells of the muscle are in fact capable of creating a reduction in tone of the muscle, as had been observed clinically. (**Perot, 1991**)

Another study by Costa and de Araujo showed that by needling the accupuncture point ST36, functional changes (decreased strength) in the tibialis anterior muscle was produced as measured by EMG. (**Costa & de Araujo, 2008**) According to AK, the tibialis anterior muscle corresponds to the Bladder meridian. This sedation point stimulation of the Bladder meridian, and its weakening effect upon the tibialis anterior muscle, confirms one of the approaches AK has used for decades in evaluating the meridian system.

In the relaxed standing position, the only activity of the tibialis anterior is in controlling sway. (**Gray & Basmajian, 1968; O'Connell, 1958**) In a normal foot, there is no electrical activity of this muscle to support the arch when in the standing position. Basmajian and Stecko (**Basmajian & Stecko, 1963**) studied the muscles of the normal foot and leg and their influence on arch maintenance using electromyography while weight-loading the foot. The subject was in a seated position, and the load was placed on the knee to transmit weight through the vertical leg and into the foot; thus there was no possible influence of postural sway. Activity of the tibialis anterior was demonstrated with a weight of 400 pounds placed on the knee, but not with all subjects. Gray (**Gray, 1969**) studied both normal and flat feet, and found that in the normal standing position there was marked activity on EMG in 23 out of 27 subjects with flat feet -- a significant contrast to 6 subjects tested with normal feet which revealed only one subject with slight activity; 5 showed none.

In free movement of dorsiflexion, electromyography shows that the tibialis anterior begins the motion, followed by contraction of the extensor digitorum longus. Near the end of the motion, the extensor hallucis longus participates in the action. (**Suzuki, 1956**) This seems to indicate that the primary synergist to the tibialis anterior muscle is the extensor digitorum longus. The extensor hallucis longus probably is more active in the foot inversion position in which the tibialis anterior is tested than it is in straight

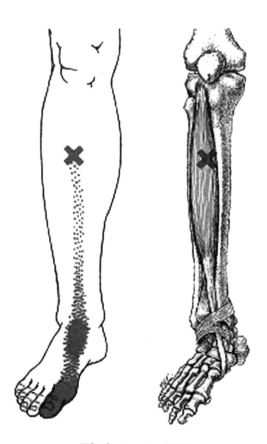

Tibialis Anterior MTrP

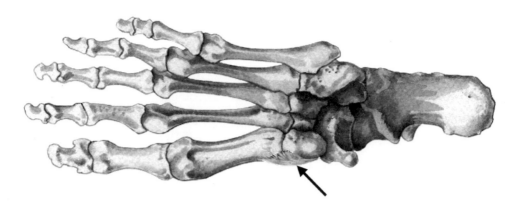
Tibialis Anterior attachment

dorsiflexion. The examiner can limit these muscles in the test by being certain there is no toe extension during the procedure.

Decreased ankle dorsiflexion is frequently caused by excess tone in the gastrocnemius and soleus muscles, and this may inhibit the tibialis anterior muscle on MMT. **(Maffetone, 2003)**

Ramsak and Gerz **(Ramsak & Gerz, 2002)** observe that pathologic or functional disturbances of the bladder may not cause the tibialis anterior to show weakness on MMT, and suggest instead that a previously strong indicator muscle be tested with therapy localization to the acupuncture point Conception Vessel 3 instead (alarm point for the bladder).

Trigger points in the tibialis anterior muscle referred pain and tenderness of the mid-shin region to the distal end of the great toe, and strongest to the ankle and toe. These trigger points may be activated by ankle injuries of many kinds, or by walking on sloping surfaces or rough terrain. **(Travell & Simons, 1992)**

Tibialis Posterior

Attachments: From the lateral part of posterior surface of tibia, medial two-thirds of fibula, interosseous membrane, intermuscular septa, and deep fascia to attach to the tuberosity of navicular bone, plantar surface of all cuneiforms, plantar surface of base of 2nd, 3rd, and 4th metatarsal bones, cuboid bone, and sustentaculum tali.
Action: Inverts and plantar flexes foot; medial ankle stabilizer.
Test: The supine patient maximally plantar flexes the foot and then inverts it, keeping the toes in a flexed position. The examiner places his hand on the medial side and over the foot. Pressure is directed against the medial side of the foot in the direction of eversion. The

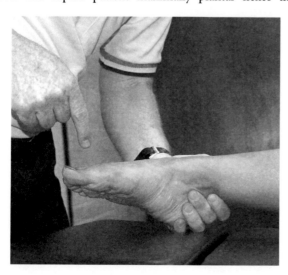
Foot is first brought into full plantar flexion and inversion, with toes in flexion. Observe for attempts at toe extensor dorsiflexion

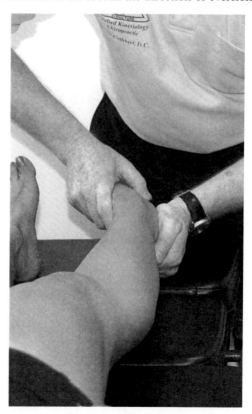
Tibialis Anterior MMT

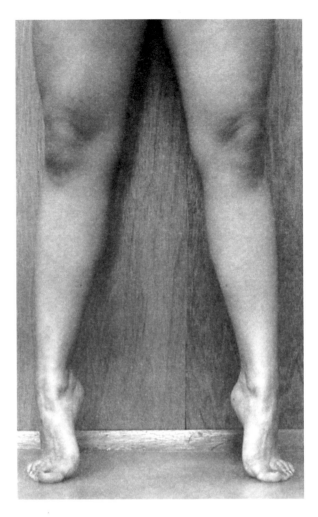

Tibialis Posterior exercises

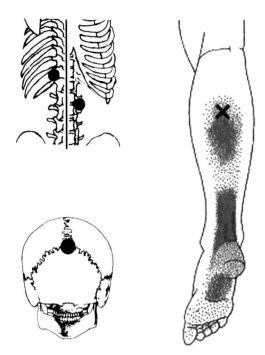

Tibialis Posterior reflexes   *Tibialis Posterior MTrPs*

examiner should observe for the rising tendon of the tibialis posterior when the muscle contracts. Care should be taken that the patient does not dorsiflex the foot to change the parameters of the test.

Nerve supply: Tibial nerve (**L4-S1**).
Neurolymphatic:
Anterior: 2" above the umbilicus and 1" from the midline bilaterally.
Posterior: Between T11, 12 bilaterally by laminae.
Neurovascular: Lambda.
Nutrition: Adrenal concentrate or nucleoprotein extract.
Meridian association: Circulation sex.
Organ/gland association: Adrenal; possibly urinary bladder.
Special Notes:

Tibialis posterior is the deepest muscle of the posterior compartment. **(See compartment syndromes, Chapter 5)** When the body is supported on one leg the supinator action of the tibialis posterior, exerted from below, helps to maintain balance by resisting any tendency to sway laterally.

The tibialis posterior is often found weak during accurate MMT. It contributes much to foot dysfunction and is often involved with foot pronation syndromes. Proper function of this muscle is necessary for sports requiring athletes to rise on their toes. Maffetone (**Maffetone, 2003**) observes that tibialis posterior inhibition may be followed by secondary tightness of the gastrocnemius and soleus and may be responsible for posterior shin splints.

As with other muscles associated with the foot, the primary weakness of the tibialis posterior is often due to a foot subluxation. When the muscles of the foot are weak, it is of value to challenge the foot in its many different aspects in an attempt to find a subluxation affecting the muscle. If a positive challenge is found which strengthens the muscle, a subluxation is probably the primary cause of weakness and should be corrected first. Neurolymphatic reflex, neurovascular reflex, acupuncture meridian involvements, and other treatment approaches will probably strengthen the muscle; however, if a foot subluxation is primarily responsible for the weakness, the other corrections will not hold. As soon as the patient walks, the muscle weakness will usually return until the subluxation in the foot is corrected.

This muscle will often test weak the presence of a caudal subluxation of the navicular bone, as well as in pronation of the foot. Examination of the tibialis posterior muscle should

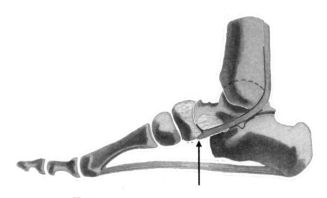

Tibialis Posterior action on the navicular

always include palpation of the muscle belly, where painful sites will frequently be found. (**Ramsak & Gerz, 2002**)

It is frequently found that a more long-term approach to therapy is required if the muscle weakness is associated with adrenal dysfunction.

Activity of the tibialis posterior is important in stabilization of the ankle mortise. Kapandji points out that the distal tibiofibular articulation is a syndesmosis without articular cartilage, and the two bones are not in contact with each other. The change of relationship of the medial and lateral malleolus is necessary to accommodate the trochlear surface of the talus, which is wider at the anterior than the posterior by approximately 5 mm. In plantar flexion, the intermalleolar space must narrow to maintain articular stability on the narrower portion of the trochlear surface of the talus. The movement of the distal tibiofibular articulation is efficiently accomplished in plantar flexion by the pennate arrangement of the muscle fibers. As the posterior tibialis contracts, it pulls through its pennate arrangement on the tibia and fibula, approximating them, which reduces the intermalleolar space and keeps the ankle mortise tight and secure. In dorsiflexion, the tibialis posterior is relaxed, allowing the intermalleolar space to widen to accommodate the wider portion of the trochlear surface of the tibia.

Individuals with active trigger points in the tibialis posterior muscle will complain of pain in the foot when running or walking. The pain is particularly felt in the sole of the foot and Achilles tendon, as well as the midcalf and heel. Differential diagnosis of the causative factor in cases of foot and ankle pain, using referred pain as your only guide, is therefore troublesome.

Tibialis posterior trigger points are difficult to treat with manual techniques, or injections, due to be overlying muscles and adjacent neurovascular structures. The authors have found spray and stretched techniques, ice stripping, described by Travell & Simons (**1999**), and particularly percussion techniques described by Fulford (**Cuthbert, 2002; Fulford, 1996**) to be an effective treatment method. With these approaches, in addition to the correction of associated articular, muscular, and myofascial conditions, in addition to the appropriate home-treatment program, successful reduction in pain and dysfunction from these trigger points are likely.

References

- Aguayo AJ. Neuropathy due to compression and entrapment. In: Peripheral Neuropathy, Vol I. Eds. Dyck PJ, Thomas PK, Lambert EH. W.B. Saunders Co.:Philadelphia; 1975.
- American Medical Association: Guides to the Evaluation of Permanent Impairment, 5th Ed. 2001:510.
- Baldry PE. Myofascial pain and fibromyalgia symptoms: A clinical guide to diagnosis and management. Churchill Livingstone: Edinburgh; 2001.
- Basmajian JV, Lovejoy JF. Functions of the popliteus muscle in man. J Bone Joint Surg. 1971:53A(3).
- Basmajian JV, Stecko G. The role of muscles in arch support of the foot — an electromyographic study. J Bone Joint Surg. 1963;45:1184-90.
- Basmajian JV, Stecko. The role of muscles in arch support of the foot - an electromyographic study. J. Bone Joint Surgery. 1971;45A(6).
- Basmajian JV. Muscles Alive, 4th ed. Williams & Wilkins Co: Baltimore, MD;1978.
- Basmajian JV. Weight-bearing by ligaments and muscles. Can J Surg. 1961;4:166-70.
- Beardall AG. Clinical Kinesiology, Vol V: Muscles of the Lower Extremities — Calf and Foot. Privately published: Lake Oswego, OR; 1985.
- Blaich RM,. Mendenhall EI. Manual muscle testing and Cybex machine muscle testing, a search for a correlation. In: Selected Papers of the International College of Applied Kinesiology (1984). Ed. Schmitt WH, Jr. ICAK USA: Park City, UT; 1984.
- Blaich RM,. Mendenhall EI. Manual muscle testing and Cybex machine muscle testing, a search for a correlation. In: Selected Papers of the International College of Applied Kinesiology (1984). Ed. Schmitt WH, Jr. ICAK USA: Park City, UT; 1984.
- Blaich RM,. Mendenhall EI. Manual muscle testing and Cybex machine muscle testing, a search for a correlation. In: Selected Papers of the International College of Applied Kinesiology (1984). Ed. Schmitt WH, Jr. ICAK USA: Park City, UT; 1984.
- Brenner E. Insertion of the abductor hallucis muscle in feet with and without hallux valgus. Anat Rec. 1999;254(3):429-34.
- Cailliet R. Foot and Ankle Pain, 3rd Ed. F.A. Davis Co: Philadelphia; 1997.
- Campbell KM, Biggs NL, Blanton PL, Lehr RP. Electromyographic investigation of the relative activity among four components of the triceps surae. Am J Phys Med. 1973;52(1):30-41.
- Chaitow L, et al. Naturopathic Physical Medicine: Theory and Practice for Manual Therapists and Naturopaths. Churchill Livingstone: Edinburgh; 2008.
- Christiansen JA, Meyer JJ. Altered metabolic enzyme activities in fast and slow twitch muscles due to induced sciatic neuropathy in the rat. J Manip Physiol Ther. 1987;10(5):227-231.
- Conable K, Corneal J, Hambrick T, Marquina N, Zhang J. Investigation of Methods and Styles of Manual Muscle Testing by AK Practitioners. J Chiropr Med. 2005;4(1):1-10.
- Costa LA, de Araujo JE. The immediate effects of local and adjacent acupuncture on the tibialis anterior muscle: a human study. Chin Med. 2008;3(1):17.
- Cuthbert S, Rosner A, McDowall D. Association of manual muscle tests and mechanical neck pain: Results from a prospective pilot study. J Bodyw Mov Ther. 2011;15(2):192-200.
- Cuthbert S, Rosner A. Applied Kinesiology Methods for Sciatic and Restless Leg Syndrome. ICS Review. 2010;2(1):6-9.
- Cuthbert S. Applied Kinesiology and the Myofascia. Int J AK and Kinesio Med, 2002;13-14.
- Cuthbert SC, Goodheart GJ Jr. On the reliability and validity of manual muscle testing: a literature review. Chiropr Osteopat. 2007;15:4.
- Dananberg HJ. Functional hallux limitus and its relationship to gait efficiency. J Amer Podiatric Med Assoc. 1986;76(11):648-52.
- Daniels L, Worthingham K: Muscle Testing – Techniques of Manual Examination, 7th Edition. Philadelphia, PA: W.B. Saunders Co.; 2002.
- Denny—Brown D, Brenner C. Paralysis of nerve induced by direct pressure and by tourniquet. Arch Neurol Psychol; 1944;51(1).
- Fulford R. Dr. Fulford's Touch of Life. Pocket Books: New York; 1996.
- Gasser HS, Erlanger J. The role of fiber size in the establishment of a nerve block by pressure or cocaine. Am J Physiol. 1929;88.
- Goodheart GJ, Jr. Applied Kinesiology Research Manuals. Detroit, MI: Privately published yearly; 1998-1964.
- Gray EG, Basmajian JV. Electromyography and cinematography of leg and foot ("normal" and flat) during walking. Anat Rec. 1968;161(1):1-15.
- Gray ER. The role of leg muscles in variations of the arches in normal and flat feet. Physical Therapy. 1969;49.
- Gray's Anatomy: The Anatomical Basis of Clinical Practice. Churchill Livingstone: Edinburgh; 2004.
- Grieve R, Clark J, Pearson E, Bullock S, Boyer C, Jarrett A. The immediate effect of soleus trigger point pressure release on restricted ankle joint dorsiflexion: A pilot randomised controlled trial. J Bodywork and Movement Therapies 2011;15(1):42-49.
- Grieve R. Proximal hamstring rupture, restoration of function without surgical intervention: a case study on myofascial trigger point pressure release. J Bodywork and Movement Therapies. 2006;10:99–104.
- Haldeman S, Meyer BJ. The effect of constriction on the conduction of the action potential in the sciatic nerve. S Afr Med J. 1970;44(31):903-906.
- Hammer W. Functional Soft-Tissue Examination and Treatment by Manual Methods, 3rd Ed. Jones and Bartlett Publishers, Inc.; 2007.
- Harms-Ringdahl K, Ed.. Muscle Strength. Edinburgh: Churchill Livingstone; 1993.
- Hoppenfeld S. Orthopaedic Neurology — A Diagnostic Guide to Neurologic Levels. J.B. Lippincott Co: Philadelphia, PA; 1977.
- Hoppenfeld S. Physical Examination of the Spine and Extremities. Appleton–Century–Crofts: New York; 1976.
- Houtz SJ, Walsh F. Electromyographic analysis of function of muscles acting on the ankle during weight—bearing with special reference to triceps surae. J Bone Joint Surg. 1959;41A:1469-81.
- Janda V: Muscle Function Testing. Butterworths: London; 1983.
- Kapandji IA. The Physiology of the Joints, Vol. 2 – The Lower Limb, 6th Ed. Churchill Livingstone: New York; 2010.
- Kendall FP, McCreary EK, Provance PG, Rodgers MM, Romani WA. Muscles: Testing and Function, with Posture and Pain. Williams & Wilkins: Baltimore; 2005.

- Kolosova LI, Nozdrachev AD, Moiseeva AB, Ryabchikova OV, Turchaninova LN. Recovery of mechanoreception at the initial stage of regeneration of injured sciatic nerve in rats in conditions of central axotomy of sensory neurons. Neurosci Behav Physiol. 2004;34(8):817-20.
- Kopell HP. Lower extremity lesions. In: Management of Peripheral Nerve Problems. Eds. Omer GE, Spinner M. W.B. Saunders Co: Philadelphia; 1980.
- Leaf D. Applied Kinesiology Flowchart Manual, 4th Ed. Privately Published: David W. Leaf: Plymouth, MA; 2010.
- Leaf D. Applied Kinesiology Flowchart Manual, 4th Ed. Privately Published: David W. Leaf: Plymouth, MA; 2010.
- Leinonen V, Kankaanpää M, Luukkonen M, Kansanen M, Hänninen O, Airaksinen O, Taimela S. Lumbar paraspinal muscle function, perception of lumbar position, and postural control in disc herniation-related back pain. Spine. 2003;28(8):842-8.
- Lewit K: Manipulative Therapy in Rehabilitation of the Locomotor System, 3rd ed. London: Butterworths; 1999.
- Liebenson C. Ed: Rehabilitation of the Spine: A Practitioner's Manual, 2nd ed. Lippincott, Williams & Wilkins: Philadelphia; 2007.
- Logan AL. The Foot and Ankle: Clinical Applications. Gaithersburg, MD: Aspen Publishers, Inc; 1995.
- Logan AL. The Knee: Clinical Applications. Gaithersburg, MD: Aspen Publishers, Inc; 1994.
- Lund JP, Donga R, Widmer CG, Stohler CS. The pain-adaptation model: a discussion of the relationship between chronic musculoskeletal pain and motor activity. Can J Physiol Pharmacol. 1991;69(5):683-94.
- Maffetone P. Complementary Sports Medicine: Balancing traditional and nontraditional treatments. Human Kinetics: Champaign, IL; 1999.
- Maffetone P. Fix Your Feet. The Lyons Press: Guilford, CT; 2003.
- Myers JB, Lephart SM. Sensorimotor deficits contributing to glenohumeral instability. Clin Orthop. 2002;400:98-104.
- Myers TW. Anatomy Trains: Myofascial Meridians for Manual and Movement Therapists. Churchill Livingstone: Edinburgh; 2001.
- O'Connell AL. Electromyographic study of certain leg muscles during movements of the free foot and during standing. Am J Phys Med. 1958;37(6):289-301.
- Ochs RM, Smith JL, Edgerton VR. Fatigue characteristics of human gastrocnemius and soleus muscles. Electromyogr Clin Neurophysiol. 1977;17(3-4):297-306.
- Page P, Frank CC, Lardner R. Assessment and treatment of muscle imbalance: The Janda Approach. Human Kinetics: Champaign, IL; 2010.
- Perot C, Meldener R, Goubel F. Objective measurement of proprioceptive technique consequences on muscular maximal voluntary contraction during manual muscle testing. Agressologie. 1991;32(10 Spec No):471-4.
- Ramsak I, Gerz W. AK Muscle Tests At A Glance. AKSE: Oberhaching; 2002.
- Richie DH Jr. Functional instability of the ankle and the role of neuromuscular control: a comprehensive review. J Foot Ankle Surg. 2001;40(4):240-51.
- Russell SM. Examination of Peripheral Nerve Injuries: An Anatomical Approach. Thieme Medical Publishers, Inc.: New York; 2006.
- Ryerson EW. Discussion of article by Harris RI, Beath T. Hyermobile flatfoot with short tendo Achillis. J Bone Joint Surg. 1948;30A(1).
- Schmitt WH, Cuthbert SC. Common errors and clinical guidelines for manual muscle testing: The "arm test" and other inaccurate procedures. Chiropr Osteopat. 2008;16:16.
- Schmitt WH, Cuthbert SC. Common errors and clinical guidelines for manual muscle testing: The "arm test" and other inaccurate procedures. Chiropr Osteopat. 2008;16:16.
- Schmitt WH, Jr. Muscle testing as functional neurology differentiating functional upper motorneuron and functional lower motorneuron problems. 1986 Selected Papers of the International College of Applied Kinesiology. ICAK: Shawnee Mission, KS; 1986.
- Schmitt WH. But what if there's no water in the hose? Collected Papers International College of Applied Kinesiology. 1986;Winter:125-144.
- Sharma L. Proprioceptive impairment in knee osteoarthritis. Rheum Dis Clin N Am. 1999; 25(2):299-314.
- Simons D, Travell J, Simons L. Myofascial pain and dysfunction: the trigger point manual: Upper half of body. Vol. 1, 2nd Ed. Williams & Wilkins: Baltimore; 1998.
- Sinanovic O, Custovic N. Musculus extensor digitorum brevis is clinical and electrophysiological marker for L5/S1 radicular lesions. Med Arh. 2010;64(4):223-4.
- Staal A, van Gijn J, Spaans, F. Mononeuropathies: Examination, Diagnosis and Treatment. WB Saunders: London; 1999.
- Sunderland S. Traumatic injuries of peripheral nerves. I — Simple compression injuries of the radial nerve. Brain. 1945;68.
- Suzuki R. Function of the leg and foot muscles from the viewpoint of the electromyogram. Journal of the Japanese Orthopedic Surgery Society. 1956;30.
- Travell JG, Simons DG. Myofascial Pain and Dysfunction: The Trigger Point Manual. Baltimore, MD: Williams & Wilkins;1992.
- Walther DS. Applied Kinesiology — Synopsis, 2nd Ed. International College of Applied Kinesiology: Shawnee Mission, KS; 2000
- Walther DS. Applied Kinesiology, Volume I — Basic Procedures and Muscle Testing (Pueblo, CO: Systems DC, 1981.
- Wu S, Hong C, You J, Chen C, Wang L, Su F. Therapeutic effect on the change of gait performance in chronic calf myofascial pain syndrome: a time series case study. Journal of Musculoskeletal Pain. 2006;13(3):33–43.

"In the last decades some (orthopaedic) surgeons have developed a deplorable tendency to advocate operation for pain in the absence of any neurological abnormality whatsoever, on the nebulous assumption that the symptoms should be attributed to compression of a nerve that happens to be in the vicinity of the painful area. As a general rule these are 'pseudo-entrapment syndromes', the actual cause being overused muscles, bands, ligaments, joints, or a somatisation syndrome on the basis of a depressive illness or personality disorder. Of course several of these factors may coexist and interact."
(Staal et al., 1999)

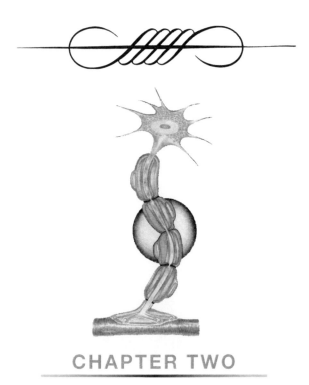

CHAPTER TWO
Peripheral Nerve Entrapment

Introduction

Peripheral nerve entrapment — sometimes called pressure neuropathy **(Akuthota & Herring, 2009; Pham & Gupta, 2009; Guyton & Hall, 2005)** — relates to some form of pressure or irritation on a peripheral nerve that changes its ability to function normally. It is generally considered to cause injury and an inflammatory response in the nerve. Peripheral nerve entrapment ranges from causing severe pain and muscle atrophy to mild, intermittent paresthesia and/or muscle weakness. As observed in applied kinesiology, the muscle dysfunction may not even be noticed by the patient, but it may be adversely influencing body posture, locomotion, or in some other way contributing to dysfunction.

Peripheral nerve entrapment was introduced into applied kinesiology in Goodheart's discussions of the carpal tunnel **(Goodheart, 1967)** and tarsal tunnel syndromes. **(Goodheart, 1971)** Walther's early review of peripheral nerve entrapment in 1982 broadened the subject in applied kinesiology, **(Walther, 1982)** and has served as the initial outline of this chapter. Cranial nerve entrapment syndromes, treated with applied kinesiology methods, have also been discussed in the recent peer-reviewed literature. **(Cuthbert & Rosner, 2010; Cuthbert & Barras, 2009; Blum & Cuthbert, 2006; Cuthbert & Blum, 2005)** Since muscular weakness found in routine applied kinesiology examination may be due to peripheral nerve entrapment, it is particularly important that the physician be aware of and able to differentially diagnose different types of the condition. Failure to return muscle weakness to normal function may result from undetected peripheral nerve entrapment. Because applied kinesiologists routinely test muscles to evaluate function, it is not uncommon to come across the more subtle types of peripheral nerve entrapment. Subtle entrapment may cause major symptoms to the patient that interfere with normal function and create remote problems.

Applied kinesiology is particularly effective in finding remote factors that contribute to peripheral nerve entrapment. The diagnosis of a piriformis syndrome indicates an entrapment of the sciatic nerve as it passes under or through the piriformis muscle. Direct treatment may be needed to the piriformis muscle, but a more in-depth study that takes the total body into consideration may find remote skeletal factors causing the piriformis tension. Another factor that might cause piriformis imbalance is disturbance in the organ and/or gland associated in applied kinesiology with the muscle. Treatment may need to be directed toward an overactive meridian, or any other of the five factors of the IVF.

It is important to recognize that like Breig with the central nervous system, many researchers have put forward evidence that for the peripheral nervous system, the clinical consequences of nerve injuries are greatly underestimated. Many experts now argue that peripheral nerve problems are far more common than clinicians believe. (**Mense & Simons, 2001; Butler, 2000; Staal et al., 1999; Devor & Seltzer, 1999; Loeser, 1985; Sunderland, 1978**) It is remarkable that only 50 years have passed since Phalen's description of carpal tunnel syndrome made this the easily recognizable clinical entity that it is today. (**Phalen, 1972**)

Cyriax was able to develop without a laboratory analysis, but by paying careful attention to his patients, the notion of "dural pain". Cyriax stated in 1948 that back pain comes from pressure and irritation to the dura mater. (**Cyriax, 1948**) He also observed that dura mater does not obey the rules of segmental reference, i.e. does not follow the familiar myotomal nor dermatomal pattern. In retrospect, and judging by the number of recent citations in the literature, Cyriax and Breig were far ahead of their time with their concepts on the biomechanics of the central nervous system and with their insistence that we are only beginning to realize the neurophysiological consequences of adverse tension in the central nervous system – and the peripheral nervous system is similarly prone to substantial adverse tensions that impair human function. Sunderland's work on the internal structure of the peripheral nerve and the role of ischemia in entrapment neuropathies also stands out. His classic text, *Nerves and Nerve Injuries* is as relevant for manual physicians as it is for surgeons. (**Sunderland, 1978**)

Much of this and the following chapter is directed at the peripheral nervous system (PNS), probably mirroring the available research work and our present understanding of the central nervous system (CNS). More is known about the PNS. It is more accessible, has far better regenerative powers and is more amenable to movement than the more protected CNS. However, despite intensive research attention to the PNS, due respect must be paid to the CNS as a contributing factor to symptoms, signs and treatment responses for PNS disorders. It is a sobering thought that, for every axon in the PNS, there are one-thousand in the CNS.

Butler has elegantly extended the early work of Breig to encompass the peripheral nervous system. (**Butler, 2000, 1991**) Butler and others have exhaustively documented how soft tissues which surround neural structures (muscle, tendon, disc, ligament, fascia, skin, or direct osseous pressure like an arthritic spur, for example), and which move independently of the nervous system, have a critical influence on neural function. These tissues have been labeled the "mechanical interface" of the nervous system and have been evaluated specifically in applied kinesiology from the very beginning. The concept, examination and treatment of "dural tension" have been presented by Goodheart and Walther for several decades. (**Walther, 2000; Goodheart, 1983**) Butler and Maitland have shown how any pathophysiology in the mechanical interface may produce tension on neural structures, with unpredictable results (motor, sensory, and autonomic). (**Maitland, 2001**)

Travell and Simons (**1999**) note that a degree of nerve compression that causes identifiable neuropathic and electromyographic changes may be associated with an increase in the number of active MTrPs. Travell and Simons (**1992**) have described how peripheral nerve entrapments can occur due to myofascial trigger points (MTrPs) in the following muscles:

Abductor hallucis
Adductor magnus
Extensor digitorum longus
Gastrocnemius
Gluteus maximus
Iliopsoas
Obturator externus and internus
Paraspinal muscles
Peroneus longus
Piriformis
Plantaris
Quadratus plantae
Soleus

Although the cause of peripheral nerve entrapment can usually be determined in an applied kinesiology examination, the precise diagnosis may never be exactly made, even though the entrapment is effectively treated. Visual observation of the entrapment at surgery is sometimes necessary for correct diagnosis. This is particularly true in conditions such as the thoracic outlet syndrome, in which various levels and causes of entrapment may be present.

Entrapment neuropathy may be subclinical, i.e., not causing symptoms of which the patient is aware. (**Akuthota & Herring, 2009; Russell, 2006**) Neary et al. (**Neary 1975**) studied the median and ulnar nerves obtained by routine autopsies from patients without known disease of the peripheral nervous system. There was enlargement in the ulnar groove at the elbow compared with the nerve proximal and distal to the site. At the carpal tunnel there was no visible swelling, but the median nerve was indented at the flexor retinaculum. Localized change of the nerve was found under the flexor retinaculum in some of the cases. Although frank clinical entrapment was not listed in the case histories of the subjects who died from unrelated factors, one wonders what an applied kinesiology examination prior to death would have revealed. Although manual muscle testing results are often confirmed by nerve conduction studies, it appears that some entrapment not documented by nerve conduction studies is observable by manual muscle testing. This is supported by the change in function after manipulative efforts to bony, ligamentous, or muscle structures, or in some other way treating the patient to release entrapment. Clinically the idea of subclinical entrapment arises when a patient complains of an old symptom 'reoccurring' by a new injury.

Today it is rare that an advanced applied kinesiologist cannot find the reason a muscle tests weak. Proper function is usually easily restored. Occasionally all efforts to locate the cause of dysfunction fail until one observes peripheral nerve entrapment affecting the muscle. This, too, usually responds, but sometimes conservative treatment to the area of entrapment fails to bring back normal muscle function. Failure may indicate a nerve entrapment needing surgical release.

The cause of muscle weakness can sometimes be enigmatic. Spinner and Spencer (**Spinner & Spencer, 1974**) state, "It is not satisfactory to consider muscle paralysis with or without sensory loss as idiopathic, spontaneous or viral, when it can be explained on the basis of a nerve entrapment." Staal et al. (**Staal et al., 1999**) expand on this and say "In the last decades some (orthopaedic) surgeons have developed a deplorable tendency to advocate operation for pain in the absence of any neurological abnormality whatsoever, on the nebulous assumption that the symptoms should be attributed to compression of a nerve that happens to be in the vicinity of the painful area. As a general rule these are 'pseudo-entrapment syndromes', the actual cause being overused muscles, bands, ligaments, joints, or a somatisation syndrome on the basis of a depressive illness or personality disorder. Of course several of these factors may coexist and interact." Applied kinesiology, with its many methods for evaluating muscle, joint, and other soft-tissue dysfunctions, can usually find the cause even prior to the development of many of the classic neurologic signs and symptoms. Using this method, conditions can be corrected before they advance to the point of paralysis, atrophy, or major sensory loss.

Cranial nerve entrapments have also been found in applied kinesiology to effect the function of cranial nerves. The cranial nerves carry dural sleeves with them for some distance; therefore any abnormal meningeal tension may be transmitted to a nerve and affect its function. Tension anywhere along the contiguous meninges can therefore be transmitted to the cranial nerves. This is because the peripheral and the central nervous systems are a continuous tissue tract, a fundamental insight of the work of Breig, Sutherland, Goodheart, DeJarnette, Chaitow and others. (**Chaitow, 2005; Goodheart, 1998-1964; Sutherland, 1998; DeJarnette, 1979; Breig, 1978**)

The neuropathies that may result from cranial bone dysfunction can be motor and/or sensory, and their severity depends on the amount of compression and neural irritation as well as the amount of ischemic radiculopathy. Breig has shown that problems come about primarily because of the entrapment neuropathy's effects on the vasculature of the nerve root. (**Breig et al., 1966**) "Pain is the cry of the nerve deprived of its blood supply", Sir Henry Head wrote. The effects of ischemia on cranial and peripheral nerve tissue have been well studied, and increasing interest in the pathophysiology of nerve compression has indicated that any rise in intrafascicular pressure – as a result of edema, compression, or torsion of the nerve root, for example – can also be damaging to neural tissue and function. (**Pham & Gupta, 2009; Patten, 1995; Moller, 1991; Dyck & Thomas, 1993; Lundborg, 1988**)

Throughout the cranium there are a number of sites where cranial nerves may be impinged upon by soft tissue at bony ridges or foraminal openings. (**Walther, 1983**) These sites may reflect mechanical or physiological changes in neural function, leading to a mechanical subset of cranial neuropathies that have been or can be successfully treated clinically by cranial practitioners. (**Blum & Cuthbert, 2006**)

Impingement of a nerve usually occurs where the nerve traverses a confining space such as the osteofibrous carpal tunnel, through a muscle such as the supinator, or between a muscle and a stable structure such as bone, as in the pectoralis minor, ulnar nerve, and piriformis syndromes. There are also a number of other, more individualized, types of impingement. (**Russell, 2006; Staal et al., 1999**)

Common entrapment syndromes are:
- Median and radial nerves at the elbow, forearm, and wrist (**Saratsiotis & Myriokefalitakis, 2010; Leaf, 2010; Hogg, 1995; Walther, 1982**)
- Ulnar nerve at the elbow and wrist (**Robertson & Saratsiotis, 2005; Alis, 2004**)
- Interdigital nerves in the hands (**Heidrich, 1995**)

Case Report: AK management of Parsonage-Turner Syndrome

A 30-year-old male with Parsonage-Turner Syndrome – acute brachial neuritis – presented for AK treatment involving right arm contracture, atrophy of the right arm and thenar eminence, and weakness producing a general paralysis of the forearm and index finger. (**Charles, 2011**). Electromyographic investigation led his orthopedist to recommend surgical release of the nerve entrapment in the pronator teres muscle, with no improvement; in fact the arm went from being immobilized at 45% flexion, and able to make only small rocking movements, to being paralyzed at 90% elbow flexion, unable to move his arm in any direction. The patient responded favorably in 8 treatment sessions guided by AK examination for the spinal manipulation, soft tissue treatment to myofascial trigger points, exercises and prescribed stretches, whereas previous surgical and pharmacological interventions had failed. Improved range of motion was experienced after the first treatment session, and by the eighth treatment, the patient was able to fully straighten the arm. Three years later, the patient was able to mountain climb with his arm fully functional and pain-free.

- Suprascapular nerve at the coracoid process **(Kharrazian, 2001)**
- Iliohypogastric nerve in the inguinal canal **(Leaf, 2010; Goodheart, 1998)**
- Lateral femoral cutaneous nerve in the pelvis **(Leaf, 2010; Skaggs et al., 2006)**
- Saphenous nerve at Hunter's canal in the leg **(Leaf, 2010; Cuthbert, 2003; Goodheart, 1998)**
- Peroneal nerves in the soleus muscle and ankle **(Walther, 2000; Travell & Simons, 1992; Heidrich, 1993)**
- Tibial nerve in the tarsal tunnel **(Walther, 2000)**
- Intertarsal nerves in the foot **(Travell & Simons, 1992)**
- Intercostals nerves at the ribs and from the intercostals muscles **(Bronston & Larson, 1990)**
- Pudendal nerve in Alcock's canal of the pelvis **(Cuthbert & Rosner, 2012; Browning, 1995)**
- Double or Triple Crush syndromes **(covered later)**

Entrapment may begin with specific trauma, but often it is very subtle; the patient may not be aware of when the problem actually began. Sometimes an acute episode will be blamed on a hard day's work, with the expectation that rest will "cure" the problem. The symptoms go away but the basic fault continues to be present, with exacerbation of the condition occurring with the next phase of heavy activity.

The actual extent by which surgery is avoided by correcting these conditions in their early developing state is unknown. However, much of the suffering from chronic pain is preventable if the acute pain is controlled promptly and effectively. Clinical examples of the importance of this principle are increasing rapidly. Specifically with regard to myofascial trigger points, Hong and Simons demonstrated that the length of treatment required for patients who had developed a pectoralis myofascial trigger point syndrome as a result of whiplash injury was directly related to the length of time between the accident and the beginning of trigger point treatment. **(Hong & Simons, 1993)** With longer initial delay, more treatments were required and the likelihood of complete symptom relief was decreased. Certainly, on a clinical basis, many applied kinesiologists obtain symptomatic relief in peripheral nerve entrapment conditions, as discussed in the next chapter and the subsequent volumes of this series. Those patients who obtain relief rarely have exacerbations that end in need of surgical release. Unfortunately, there is no method for determining what future problems are prevented by obtaining correction in the earlier states of the condition.

Entrapment neuropathies can have an exogenous or an endogenous etiology. An example of an exogenous etiology is trauma to the wrist, which causes subluxations of the carpal bones and perhaps separation of the radius and ulna. This stretches the flexor retinaculum, narrowing the carpal tunnel and producing entrapment of the median nerve. Muscle strain from overuse can also cause entrapment. Repeated forceful pronation of the forearm may cause hypertrophy and hypertonicity of the pronator teres muscle, in turn causing irritation on the median nerve as it travels through the pronator teres muscle. **(Discussed in subsequent volume)**

An endogenous involvement may be a systemic health problem that makes a person more susceptible to nerve entrapment. One should specifically look for peripheral nerve entrapment when patients have a condition with widespread effect on the small vessels, such as diabetes mellitus, periarteritis nodosa, certain types of amyloidosis, and severe arteriosclerosis with peripheral vascular insufficiency. **(Akuthota & Herring, 2009; Kopell 1980; Asbury, 1970)** Diseases like diabetes can raise the intrafascicular pressure throughout the body. **(Veves et al., 2002; Myers & Powell, 1981)**

In addition to disease processes that affect the small blood vessels, control of the blood vascular beds by the autonomic nervous system must be considered. There may be an imbalance of the autonomic nervous system affecting the peripheral circulation, causing the person to be more susceptible to peripheral nerve entrapment. Lombardini et al, **(Lombardini et al., 2009)** at the University of Perugia, conducted a study that assessed the effects of OMT (osteopathic manipulative treatment) for patients with peripheral artery disease (PAD) - and with symptoms of intermittent claudication.

> "Peripheral arterial disease (PAD) is a manifestation of systemic atherosclerosis associated with impaired endothelial function and intermittent claudication is the hallmark symptom. Hypothesizing that osteopathic manipulative treatment (OMT) may represent a non-pharmacological therapeutic option in PAD, we examined endothelial function and lifestyle modifications in 15 intermittent claudication patients receiving osteopathic treatment (OMT group) and 15 intermittent claudication patients matched for age, sex and medical treatment (control group). *Compared to the control group, the OMT group had a significant increase in brachial flow-mediated vasodilation, ankle/brachial pressure index, treadmill testing and physical health component of life quality.*"

This study suggests that practitioners of manual approaches may be able to offer safe, effective, integrated health care even for patients with peripheral arterial disease, over and above the attention they offer for musculoskeletal pain and dysfunction. Johnson et al. **(Johnson et al., 1984)** discuss the influence of nervous system dysfunctions upon the vascular and cardiovascular systems specifically. Watanabe has demonstrated the clinical and neurophysiological responses of the cardiovascular system to chiropractic treatment. **(Watanabe & Polus, 2007)**

Other clinical examples would be in cases of resolution of the vascular component of cervicogenic headaches, **(Cramer & Darby, 1995)** or the positive influence upon a dysfunctional lumbar spinal segment associated with premenstrual syndromes or dysmenorrhea. – conditions that are functional in nature (pathophysiological) rather than pathological. **(Walsh & Polus, 1999; Browning, 1995)**

The "double crush" condition is discussed later, in which an individual has two problems, such as a circulatory deficiency and entrapment of the nerve. Either condition may be minimal and incapable of producing symptoms, but when found together, symptoms occur. On the other hand, circulatory problems may be secondary to entrapment. Certain types of entrapment may produce Raynaud's phenomenon. This secondary circulatory deficiency is not contributing to the peripheral nerve entrapment; rather, it is caused by it.

Premenstrual and pregnancy fluid retention are systemic problems that can contribute to entrapment. **(Wetz et al., 2006)** The excess fluid appears to cause additional pressure in confined areas through which the nerve must pass. Studies have documented carpal tunnel syndrome **(Smith et al., 2008)** and tarsal tunnel syndrome due to pregnancy by electrodiagnostic methods. The entrapment was sufficient to cause some complaint of leg pain in 69.5% of the women. **(Helm et al., 1971)** Of those studied following pregnancy, 92.4% reverted to normal values at postpartum follow-up. Peripheral nerve entrapment is often associated with an ileocecal valve syndrome in applied kinesiology. This condition causes toxicity to which the body has a natural reaction of retaining fluid to dilute the toxicity. **(Walther, 2000; Goodheart, 1998-1964)**

Applied kinesiology methods for evaluating peripheral nerve entrapment have revealed the importance of studying the entire postural balance, muscle organization, and movement patterns. The basic underlying cause of an entrapment may be remote from the actual area of involvement, e.g., a pelvic fault can cause a compensating shift in the shoulder girdle that results in a thoracic outlet entrapment. Using electromyography Mooney et al. **(Mooney et al., 2001)** found that in cases of sacroiliac joint dysfunction, hypertonicity of the gluteus maximus was found ipsilaterally and in the latissimus dorsi contralaterally. The integration between the shoulder girdle and sacroiliac joint was obvious. Giggey & Tepe **(Giggey & Tepe, 2009)** report that treatment of sacroiliac dysfunction using the sacro-occipital technique's padded-wedges (as used in applied kinesiology as well) showed significant improvements in strength of the cervical extensor muscles pre- and post-treatment; once again indicating the functional connection between the pelvis and the neck and shoulder muscles.

Failure to correct peripheral nerve entrapment by conservative methods often results from diagnosing only the area of local entrapment and failing to observe remote factors causing the problem. In this case, treating the local area of entrapment is just treating the symptoms.

Some peripheral nerve entrapment may be the result of old fractures, surgical adhesions, or congenital anomalies. Moncayo & Moncayo has recently demonstrated the usefulness of the AK treatment approach for scar-tissue. **(Moncayo & Moncayo, 2009)** Kobesova **(Kobesova, 2007)** suggests that scars may develop adhesive properties that disturb tissue tension, alter proprioceptive input, and create functional changes similar to active myofascial trigger points. This may create faulty sensory input that can result in disturbed motor output leading to adaptive postural patterns, impaired muscular strength, disturbed neurovascular activity, and pain syndromes. The diagnosis of the muscular consequences of scar-tissue has been described in the reports of Moncayo, Goodheart, Garten and Gerz. **(Moncayo, 2007; Goodheart, 1998; Garten, 2004; Gerz, 2001)**

Repeated trauma to a nerve can produce a tumor-like mass composed of dense, fibrous tissue adjacent to the nerve. This may cause symptoms of nerve entrapment. If the nerve is superficial, the mass can be palpated and sometimes mistaken for a neuroma. An example of this is continued irritation to the thumb from the trauma of a bowling ball on the bowler's thumb. **(Swanson et al., 2009; Marmor, 1966)**

The term "occupational nerve lesions" refers to pressure neuropathies caused by repeated trauma, which is often microtrauma. The Panel on Musculoskeletal Disorders in the Workplace **(National Research Council, Institute of Medicine, 2001)** states that nerve lesions caused by a single accident should not be called occupational nerve lesions. These injuries may not be noticed by the worker until there is significant disability because of the insidious nature of the developing condition. The correlation between the work and the illness is easily overlooked. Efforts are being made to study occupational causes of peripheral nerve entrapment, **(Ian'shina et al., 2009; Punnett, Keyserling 1987, Punnett, 1985)** but considerable work remains to be done. An example of this research was a study where electromyography examinations were performed on 102 patients with occupational hand disorders. 87% of the patients showed signs of tunnel syndromes: 44% of those had one tunnel syndrome, 36% two syndromes, and 7% multiple syndromes. The study concluded that "peripheral nerve disorders in residual defects with occupational hands diseases are persistent due to compression within physiologic tunnels." **(Ian'shina et al., 2009)**

The term "peripheral nerve entrapment" includes the radix of the nerve at the intervertebral foramen. Other than for differential diagnosis, radiculitis is not considered here. It is very important to correlate the material presented in this section on peripheral nerve entrapment with that in previous applied kinesiology textbooks on the spine and pelvis. **(ICAK Bookstores, 2012)** The "pseudoradicular syndrome" is peripheral nerve entrapment in the lower extremity that appears to be a lumbar radicular syndrome, but the actual causes are muscular, ligamentous, psychosomatic, and/or joint dysfunction. **(Barral & Croibier, 2007; Staal et al., 1999; Simons et al., 1998)**

Saal et al. **(Saal, 1988)** studied thirty-six cases that had peripheral nerve entrapment of the lower extremity instead of lumbar radiculopathy, for which they were referred. Forty-four percent of the cases had a positive nerve root tension sign, such as straight leg raise or femoral stretch test. Twenty-four percent had spinal range of motion abnormalities. History of lower extremity injury helped direct attention to the possible peripheral nerve lesion. Most helpful was local nerve tenderness at the lesion site. Electrodiagnostic studies and peripheral nerve blocks were used to make a final decision regarding the peripheral entrapment.

A similar situation is often present in the upper extremity. **(Discussed in the subsequent volume of this series)** Imaging studies may show spondylosis in the cervical spine, and one quickly attributes the patient's arm or hand problem to radiculitis. Many asymptomatic individuals from mid-life on have spondylosis. The symptoms may be due to double crush, or a thorough investigation may reveal a localized peripheral nerve entrapment without the spondylosis contributing to the pain.

When an area of entrapment is found, the assessment should consider all other possible factors. The patient may have more than one condition. In a study of 140 patients diagnosed with carpal tunnel syndrome and surgically decompressed, it was found that some of the symptoms were relieved in 20% of the patients, but continued pain was present. Upon close examination it was determined that a second condition was present. DeQuervain's stenosing tendinitis, ulnar compression neuritis, acute tenosynovitis, and epicondylitis were diagnosed. **(Crymble, 1968)** This is a cautionary tale about not plunging into surgery too soon without knowing the complete picture of a patient's inter-related pathophysiologic process.

Another example illustrating a secondary area of involvement is dysfunction in the foot that causes an imbalance in the piriformis. This in turn may create a sciatic neuralgia and also cause a sacral subluxation because of the piriformis' role in sacral stabilization. **(Lee, 2004; Retzlaff et al., 1974; Goodheart, 1972)** One cannot assume that the problem is local until remote components have been evaluated. If the piriformis muscle is not being influenced remotely, it may have been traumatized by an exogenous source, **(Steiner et al., 1987; TePoorten, 1969; Edwards, 1962; Robinson, 1947)** creating muscle imbalance and making it the primary cause of the peripheral nerve entrapment. Another advantage of the applied kinesiology approach to muscle system evaluation is that the piriformis muscles can be assessed in the prone, supine, and side lying as well as the standing positions; this position challenges the piriformis muscles' function with the positive support mechanisms in the feet and assesses the muscle's functionality in the positions in which the patient actually lives. The applied kinesiology approach of manual muscle testing patients while they are moving, in positions of daily living, or in positions of physical stress greatly adds to the evaluation of muscle function.

The amount of nerve compression and its relationship to symptoms has a wide range. On the one hand, as noted previously, entrapment can be observed at necropsy when there was no history of symptoms during life. **(Pham & Gupta, 2009; Guyton & Hall, 2005; Staal et al., 1999; Patten, 1995; Neary, Ochoa, Gilliatt, 1975)** On the other hand, Lusskin **(Lusskin, 1982)** states, "…the insidious and mild nature of the compression phenomenon is out of proportion to the annoying and often disabling symptoms." In severe entrapment, such as caused by space-occupying lesions or bony or fibrous-band congenital anomalies, the compression may become severe, requiring surgical intervention. This type of involvement is out of the scope of this text, other than for purposes of differential diagnosis. The thorough diagnostic evaluation taught in applied kinesiology attempts to eliminate these conditions from conservative treatment. When in doubt, a conservative approach is justified if there is no evidence of conditions such as space-occupying lesions and severe vascular or nerve compression. In general, the condition should improve within a maximum of three to four months. It is possible that continued delay in the presence of nerve entrapment can cause additional nerve damage, **(Butler, 2000; Lundborg, 1975; Sunderland, 1945; Denny—Brown & Brenner, 1944)** with lessened results from surgical decompression. This is not a universal rule. In a study of 313 patients who received carpal tunnel surgery, the duration of symptoms had little bearing on the degree of recovery. When the symptoms had been present for less than one year, there was 84% recovery; when present more than five years, 79% recovered. **(Cseuz et al., 1966)** The consideration of surgical decompression must be on an individual basis when a therapeutic trial fails to produce results. **(Kline & Hudson, 2007)**

Peripheral nerve entrapment must be differentially diagnosed from a spinal involvement such as a subluxation, intervertebral disc, or foraminal encroachment from degenerative joint disease. This requires diagnosing the level of nerve involvement. In order to make a differential diagnosis, numerous areas where the nerve can be irritated must be evaluated. Various levels can cause a similar set of symptoms that must be objectively evaluated.

Disease of the nerve — such as a neuroma — must also be differentiated. It becomes obvious that all these differentiations are necessary, because failing to find the basic underlying cause of the problem and directing attention to the wrong pathological or functional process or to the wrong area of the nerve leaves the patient essentially untreated.

Entrapment Neurophysiology

Nerves are typically classified as motor, sensory, or mixed. A nerve may be classified as purely motor or sensory, which is erroneous since there are no "pure" nerves. Every muscle is supplied by both motor and sensory fibers. In this way the CNS is aware of the state of tension in muscle fibers and tendons. The automatic and coordinated interaction of agonist and antagonist muscles during movement requires innervation from the sensory nerve fibers. Because every nerve needs a certain portion of sensory fibers, there are no pure motor nerves. Motor nerves contain afferent supply from the neuromuscular spindle cell, and sensory nerves contain autonomic fibers. Within a peripheral nerve trunk there may be three main groups of fibers. (1) Motor (efferent) fibers supply the skeletal muscles for voluntary control of their activity. Their cell bodies are located in the gray matter of the spinal cord and brain stem. (2) Sensory (afferent) fibers are stimulated by various types of receptors, such as in skin, muscles, and special sensory organs. These impulses are interpreted in the central nervous system as sensations. (3) Autonomic fibers (efferent) control the smooth muscle of organs, glandular activities, and certain trophic functions of the body. **(Guyton & Hall, 2005)**

Peripheral nerves are made up of fibers from dorsal and ventral roots. On each side of each spinal segment these join together to form spinal nerves, which after only a short distance across the intervertebral foramina again divide into dorsal and ventral rami; the dorsal rami innervate the paraspinal muscles and skin, and the ventral rami of the cervical and lumbosacral spinal nerves combine to form the

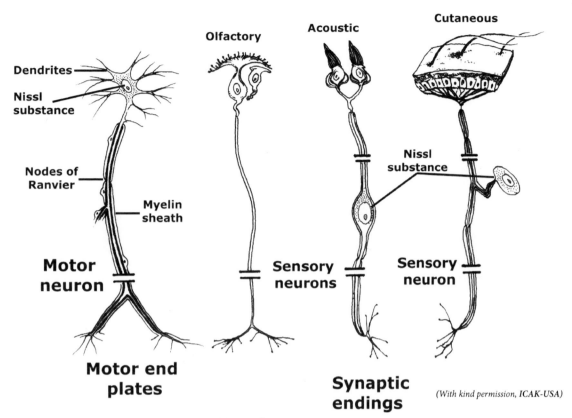

Types of Neuron

(With kind permission, ICAK-USA)

brachial and lumbosacral plexuses. The individual nerves to the limbs originate from these plexuses. The ventral rami of the thoracic origin form the intercostal nerves, except that from the first thoracic nerve, which joins the brachial plexus.

Many receptors and their axons have a lower tensile strength compared to the tissues in which they are embedded. Physical trauma to tissues and nerve trunks can damage the mechanoreceptors and their axons resulting in localized proprioceptive losses. (**Kolosova, et al., 2004**)

Reduced physical activity may result in a kind of sensory disuse, affecting the entire sensory apparatus from the receptors to their representation in the brain. In the peripheral nervous system, immobilization can lead to muscle spindle atrophy and changes in its sensitivity and firing rate. (**Halasi et al., 2005**) More centrally, it has been shown that tactile impoverishment and sensorimotor restriction of an animal's paw causes deterioration in the cortical sensory map representing that area. (**Coq & Xerri, 1999**)

Each nerve fiber has a cell body, a greatly elongated fiber called the axon, and — finally — the initiating and terminal aspects, which are the dendrites and teledendria, respectively. The dendrite or dendritic zone is the receptor membrane of a neuron. The cell body is usually located here, although it may lie within the axon (e.g., auditory neurons) or be attached to the side of the axon (e.g., bipolar neuron). The cell body containing a nucleus is the controlling metabolic center of the nerve cell. Severe injury of a nerve fiber causes degeneration of the distal segment no longer in contact with the cell body. The nerve fiber may be myelinated or unmyelinated. In myelinated nerves, the myelin sheath envelops the axon except at the endings and at periodic constrictions known as the nodes of Ranvier. The axon is maintained by materials formed in the cell body. Some of these materials are enzymes, polypeptides, polysaccharides, free amino acids, neurosecretory granules, mitochondria, and tubulin sub-units. (**Pham & Gupta, 2009; Butler, 2000; Hurst, Weissberg & Carroll, 1985**) The movement of these materials along the axon is called axoplasmic transport. Efficient transport of the material to the nerve fiber is essential for its normal structure and function, including normal growth, regeneration after injury, normal conduction, and normal transmission at the neuromuscular junctions and sensory receptors.

Nerve conduction is not explicitly an electrical current but instead continues along myelinated fibers in the form of consecutive depolarizations of one node of Ranvier to the next. If the nerve or axons are injured, transmission along the nerve is disturbed; if the chemical environment around the nerve is lacking sufficient electrolytes and other substances necessary for repolarization, nerve transmission and function will also be disturbed. Schmitt (**Schmitt, 1986**) has elegantly shown why this part of functional neurological assessment in applied kinesiology is so necessary to the complete diagnosis and treatment of nerve injury: the detection of the chemical milieu in the patient may be critical to the transmission of nerve impulses throughout the organism.

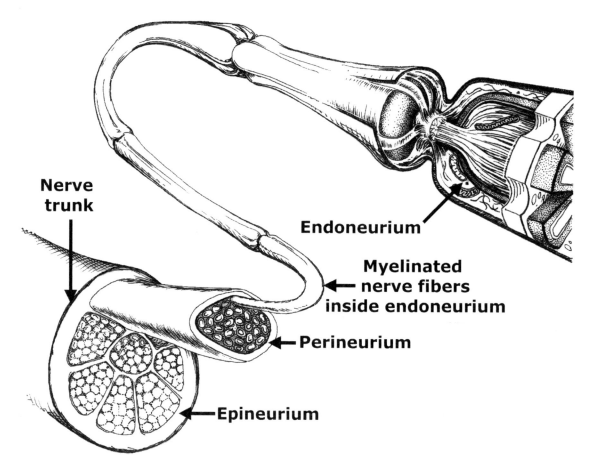

Nerves are subject to direct pressure, stretch, angulation, friction, torsion, ischemia, edema, and malnutrition. **(Maitland, 2001; Butler, 2000; Dawson, Hallett, Millender, 1983; Korr, 1976)** The strength and stability of peripheral nerves is incredible. It must be remembered that axons can be over 3 feet long, have differing sources of blood supply, bend and move constantly, rub on various tissues hard and soft and yet the axon is just one cell! Nerves can also be injured by cutting, too much squeezing and pulling, by irritating chemicals around the nerve, and by sustained reduction in blood supply. **(Pham & Gupta, 2009; Butler, 2000)** All around the body, nerves slide as the body moves. Injury or disease which alters posture or movement may lead to pain with movement. Real-time ultrasound imaging of fascial planes sliding and gliding upon each other during normal movement (flexion and extension) in asymptomatic individuals has shown how this smooth gliding motion is impaired or stuck in individuals with nerve pain. One such film was made by Dr Guimberteau, a plastic surgeon who works on hand reconstruction, taken with micro-cameras inserted during surgery, and was presented at the 7th World Low Back and Pelvic Pain Congress in Los Angeles. **(2010)**

Films like these showing the free-moving nature of myofascial motion in living human tissue are revelatory of how the body actually works. These myofascial planes can get stuck (suffer myofascial gelosis) as a result of trauma, overuse, inflammation, and aging among others, and this corresponds with measurable densification, pain and limitation of movement. **(Myers, 2001; Mense & Simons, 2001)**

It appears that at least a part of the work that clinicians do achieves its results via restoring gliding potentials to these superficial layers of densely innervated (mechanoreceptors, nociceptors, etc.) myofascial tissues.

Any of these factors can cause peripheral nerve entrapment or cause the nerve to be more susceptible to entrapment. The peripheral nerve, unlike a blood vessel, does not easily stretch. It must be able to move freely in relation to its neighboring structures to avoid irritation. **(Butler, 2000; Kopell and Thompson, 1976)** Another potential mechanism that may lead to long-term sensory and motor losses is a physical change in injured tissues in which the receptor is embedded. The proprioceptive receptors and nerves may be fully intact, but their ability to detect movement may be disrupted by changes in the surrounding tissues, in the form of adhesions or shortening. **(Kolosova, et al., 2004; Brumagne et al., 2000)**

From its point of leaving the spinal cord to its destination, the nerve is vulnerable to entrapment or to factors that make it more susceptible to entrapment. Korr **(Korr, 1976)** summarizes these factors as "…compression by narrowing of the foramen; adhesions between roots and sleeves, causing angulation, shearing and constriction; shearing forces acting upon nerves passing through fascia; compression (for example, of posterior rami of spinal nerves) by sustained contraction of the paravertebral muscles through which the nerves pass; constriction at duroarachnoid junctions of root pouches; compression within foramina secondary to venus congestion (compression of spinal and radicular veins). Hypoxia, pH shifts, and other chemical changes in the environments of the nerves due to ischemia (compression of spinal arteries, sustained contraction of muscles through which nerves

Nerve Trauma

Normal -- Normal nerve transmission may be disturbed from high pressure exerted for a short time, from moderate or low pressure exerted for intermittent or long periods. The subtlety of pressure that can cause entrapment neuropathy is illustrated by fluid accumulation causing entrapment at the carpal or tarsal tunnel, in pregnancy, premenstrual syndrome, or the ileocecal valve syndrome. The irritation creating symptoms is often long-standing; it causes an inflammatory response in the nerve and maintains its improper function. Short-term, subtle irritation on a nerve does not appear capable of changing the characteristics of its function on a constant or permanent basis.
It is necessary to diagnose the entrapment early so that conservative treatment will be effective. Chronicity increases the chance that surgical decompression may be necessary.

Neurapraxia is the lesser of the types of nerve involvement. It is a segmental block of axonal conduction due to a focal region of demyelinization of the nerve. The nerve has continuity, but conduction cannot be carried out over the demyelinated area. With only slight myelin damage there may be conduction, but it is slowed, requiring greater time to activate the widened nodal region. This is probably the type of nerve involvement present in the more subtle types of entrapment often observed in routine applied kinesiology examination, when the patient is not aware of symptoms.

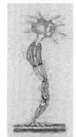

Axonotmesis is the loss of continuity of nerve axons, but with continuity of the connective tissue sheath. This leads to wallerian degeneration of the distal part of the nerve. Following Wallerian degeneration, the proximal part of the nerve attempts to regrow. "This growth occurs at the rate of approximately 1 mm per day, 1 cm per week, or 1 inch per month." Obvious symptoms and signs of peripheral nerve entrapment are present in this condition. When the entrapment is released, repair begins and the patient can expect to return to normal or nearly normal. The main factor is time for regrowth and maintenance of an entrapment-free condition.

Neurotmesis is the separation of the axon and is the most advanced involvement. The separation of the nerve from its nutritive sources causes Wallerian degeneration, the fatty degeneration of the nerve fiber. Regeneration may still take place and will result in regaining continuity or in forming a neuroma illustrated below.

Neurotomesis regaining continuity. Because of the loss of continuity of the nerve's supporting elements there is only a chance that the nerve will connect with the proximal portion. If it does the slow process of regeneration can take place. The amount of regeneration to normal is variable.

Neurotomesis failing to regain continuity is illustrated. The continued nerve regeneration causes a neuroma to develop.

(With kind permission, ICAK-USA)

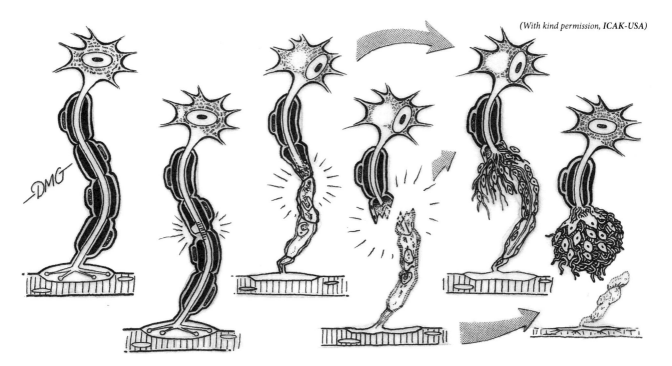

(With kind permission, ICAK-USA)

Sequence of peripheral nerve injury

pass, et cetera) are also important factors in the alteration of axonal excitation and conduction."

There may be damage from high pressure exerted for a short time, or from moderate or low pressure exerted for long periods or intermittently. The subtlety of pressure that can cause entrapment neuropathy is illustrated by fluid accumulation causing entrapment at the carpal (**Akuthota & Herring, 2009; Wetz et al., 2006; Spaans, 1970**) or tarsal (**Smith, 2008; Helm, Nepomuceno, & Crane, 1971**) tunnel in pregnancy, premenstrual syndrome, or the ileocecal valve syndrome. The irritation creating symptoms is often long-standing; it causes an inflammatory response in the nerve and maintains its improper function. Short-term, subtle irritation on a nerve does not appear capable of changing the characteristics of its function on a constant or permanent basis. It is necessary to diagnose the entrapment early so that conservative treatment will be effective. Chronicity increases the chance that surgical decompression may be necessary, and that more permanent long term damage may result.

The three types of nerve injury are neurapraxia, axonotmesis, and neurotmesis. Neurapraxia is the lesser involvement. It is a segmental block of axonal conduction due to a focal region of demyelinization of the nerve. The nerve has continuity, but conduction cannot be carried out over the demyelinated area. With only slight myelin damage there may be conduction, but it is slowed, requiring greater time to activate the widened nodal region. This is probably the type of nerve involvement present in the more subtle types of entrapment often observed in routine applied kinesiology examination, when the patient is not aware of symptoms. Treatment of this type of nerve conduction injury can speed up the recovery time and facilitate a more complete recovery.

Axonotmesis is the loss of continuity of nerve axons, but with continuity of the connective tissue sheath (the neurolemma or sheath of Schwann remains intact). This leads to Wallerian degeneration of the distal part of the nerve. Following Wallerian degeneration, the proximal part of the nerve attempts to regrow. This growth occurs at the rate of approximately 1 mm per day, 1 cm per week, or 1 inch per month. Obvious symptoms and signs of peripheral nerve entrapment are present in this condition. When the entrapment is released, repair begins and the patient can expect to return to normal or near-normal. The main factor is time for regrowth and maintenance of an entrapment-free condition. (**Chien et al., 2003; Staal et al., 1999; Dawson, 1983**)

The most advanced involvement is neurotmesis, in which the connective tissue and the axons are destroyed. Following the Wallerian degeneration of the axons, there is less chance for regrowth in the appropriate direction due to the loss of continuity of the nerve's supporting elements. This failure of direction may cause the nerve regrowth to ball up into a neuroma. This injury goes beyond the possibilities of applied kinesiology therapy; a surgical procedure is usually required for repair here.

In peripheral nerve entrapment that is applicable to applied kinesiology's conservative treatment, we are not concerned with the catastrophic situations in which whole nerves or roots are crushed, or even in which conduction has been blocked in all or most of the axons. This complete — or almost complete — interruption of the axoplasmic continuity would cause a near-total or total loss of neural function, with Wallerian degeneration distal to the disruption. In severe injuries the axoplasm will actually leak out of the damaged or torn nerve. Research has shown that even minor injury to a peripheral nerve as well as to the environment surrounding the nerve will have important consequences for the rate of axoplasmic flow and the quality of the axoplasm. (**Butler, 2000; Dahlin et al., 1986; Rydevik et al., 1980**)

In the more moderate situation in which conservative

treatment is effective, there is a conduction block in some of the fibers in a nerve, causing a corresponding loss of sensory and motor function. This loss may be transient or fluctuating; in some cases, the sensory or motor deficits would not even be perceived by the patient. Referring to this type of entrapment neuropathy, Korr **(1976)** states, "However, since some types of fibers are more susceptible to deformation block than others, garbled sensory input and incomplete and uncoordinated efferent output may be the clinically more significant consequences." This correlates well with some of the unusual functional findings observed in applied kinesiology. Sometimes the dysfunction does not seem to follow the expected neurologic pathways. Release of an entrapment may increase range of motion in remote muscles as they improve their strength or as they relax. This may take place as muscles supplied by the pathway strengthen. Intermingling of impulses between neurons appears to be a common occurrence in peripheral nerve entrapment.

Cross Stimulation and Light Pressure

There is a great variation in impairment due to nerve compression. **(Barral & Croibier, 2007; Russell, 2006; Denny—Brown and Brenner, 1944)** Light forces can cause disturbance in peripheral nerves, even though the nerves can absorb large forces with their associated structures. Nerve roots, in comparison, are mechanically frail due to lack of perineurium. **(Barral & Croibier, 2007; Luttges, Stodieck, & Beel, 1986)** The epineurium proves to be a particularly reactive tissue that is not difficult to injure. Slight trauma such as mild compression or intermittent irritation to a peripheral nerve can cause non-degenerative, inflammatory changes and epineural edema that result in decreased nerve conduction velocity and progressive facilitation during early refractory periods. **(Butler, 2000, 1991; Rydevik et al., 1984; Triano & Luttges, 1980)** The nerve may be even more susceptible to entrapment at the radix.

In simple compression trauma to a nerve, the motor fibers are more vulnerable than the sensory ones. **(Russell, 2006; Haldeman & Meyer, 1970; Sunderland, 1945; Gasser & Erlanger, 1929)** If significant sensory defects appear, there will be delayed restoration of full motor function. In compression injuries, such as sleep paralysis of the radial nerve, an individual is more vulnerable if general health is reduced, or other factors — such as undue fatigue, drugs, or alcohol — are involved.

Pressure applied to a nerve can cause an artificial synapse between nerve fibers. This is referred to as "cross stimulation." **(Russell, 2006)** Gardner **(Gardner, 1967)** relates to Hering's experiment in which "Cross stimulation was produced by a pressure applied so gently that it did not impair conduction of the original impulse. The investigators then relieved the pressure (i.e., decompressed the nerve) and irrigated the nerve with a saline solution, which caused interruption of the artificial synapse...."

In a similar study by Granit et al. **(Granit, Leksell, & Skoglund, 1944)** pressure was applied to a nerve, causing an artificial synapse between motor and sensory fibers of the nerve. In addition to the cross stimulation between motor and sensory fibers, the original impulse was maintained. They then removed the pressure and irrigated the nerve with a saline solution; after a brief time the nerve returned to normal function.

In normal situations, nerve impulses begin with stimulation at the central or cellular ends of the nerve and travel in only one direction; that is, efferent impulses begin at the central nervous system and travel toward the periphery, and sensory impulses begin at the periphery and travel toward the central nervous system. In peripheral nerve entrapment, there may be impulses generated at the deformation site that travel in both directions. Impulses that do not begin at the end of the nerve fiber are additional ones superimposed on the usual nerve transmission. As normal impulses pass through the area of nerve deformation, there may be an amplification and prolongation of the impulse, changing the effect centrally or peripherally. Cross-talk between fibers may take place, sending an originally propagated impulse to an area different than is normally intended. "This lateral, side-to-side (ephaptic) transmission is usually from large fibers to small fibers." **(Korr, 1976)**

An example of cross stimulation resulting from nerve compression is meralgia paresthetica. **(Travell & Simons, 1992)** Stimulating a small area of skin may result in widespread burning pain in the distribution of the lateral femoral cutaneous nerve. This is referred to as "cross talk." "Back talk" is a result of cross stimulation that travels in the opposite direction, that is, efferent supply due to cross stimulation affecting an afferent fiber. **(Russell, 2006; Butler, 1991; Gardner, 1967)**

Double Crush

Some patients have multiple nerve compression syndromes in the same extremity, indicating that there may be a patient population that is at high risk for the development of compressive neuropathies. **(Butler, 2000, 1991; Staal et al., 1999; Ritts, Wood, & Linscheid, 1987)** This may be explained by "double crush," which consists of two entrapments or one entrapment with some other disturbance that interferes with normal nerve transmission, such as diabetes, malnutrition, or ischemia. Metabolic and endocrine disorders have been implicated in neuropathic syndromes, including hypo- and hyperthyroidism, acromegaly, and others. **(Staal et al., 1999)** The factors may be insignificant individually, but together there are significant symptoms at the distal area.

The recognition of double crush syndromes encourages clinicians to undertake a more global investigation of peripheral nerve injuries and pain syndromes, and requires the clinician to employ methods that can evaluate many tissues and their mutual influences.

In a series of 115 patients with an electrophysiologically-proven entrapment neuropathy, Upton and McComas **(Upton, & McComas, 1973)** found evidence of a cervical root lesion in addition to the one in the extremity. 70% of these patients with either carpal tunnel syndrome or lesions of the ulnar nerve at the elbow showed clear electrophysiological and clinical evidence of neural lesions

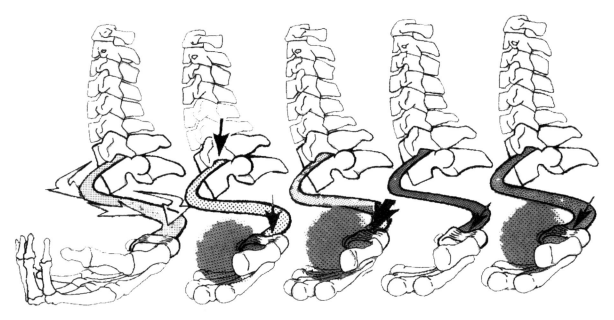

(With kind permission, ICAK-USA)

Double crush

A. Normal nerve function at the nerve root and carpal tunnel syndrome. B. Light entrapment at both the nerve root level and carpal tunnel causes symptoms. C. Heavy entrapment at the carpal tunnel causes symptoms. D. Light entrapment at the carpal tunnel causes no symptoms. E. Vitamin B6 deficiency, ischemia, or some other factor that causes reduced nerve function and light entrapment at the carpal tunnel produces symptoms. Heavy entrapment at any one area can produce symptoms, whereas light entrapment at two locations causes symptoms, but the light entrapment at any one area would be symptom-free.

in the neck. Upton and McComas developed the double crush hypothesis, indicating that the proximal lesion in the cervical spine reduced the axoplasmic transport. They proposed that the functional integrity of the axon and its target structures is impaired by the proximal disturbance, making the distal peripheral nerve more vulnerable to entrapment. Amadio (**Amadio, 1987**) comments on this work, "In the double crush there are two separate compressive foci on the same nerve, each of which individually might cause only minor symptoms, but which in combination cause a significant neuropathy at the distal focus. The most classic example of this is the double crush which occurs with a combined cervical spondylosis and carpal tunnel syndrome." In a retrospective study of 1,000 cases of carpal tunnel syndrome, Hurst et al. (**Hurst, Weissberg & Carroll, 1985**) found that the double crush effect was statistically significant in patients with carpal tunnel syndrome and either diabetes or cervical spondylosis. It is unknown whether the diabetic correlation is due to circulatory disturbance or reduced health of the nerve cell body in producing the necessary growth factors to be distributed to the nerve fiber by axoplasmic transport.

The reduction of axoplasmic transport is observed by the significant degree of narrowing of axons distal to the area of entrapment. Proximal to the entrapment, there is considerable swelling due to the damming of axoplasm. (**Korr, 1976**) Lundborg (**Lundborg, 1988**) proposed a reversed double crush syndrome where nerve injuries at the wrist could predispose proximal nerve tissues to injury. Dahlin (**Dahlin, 1987**) extended the double crush hypothesis by showing that compression as seen in carpal tunnel syndromes will block the retrograde axonal flow and induce changes in the cell bodies in the dorsal root ganglia. It must be recognized that multiple nerve compressions, including nerve roots, will significantly influence the CNS, causing facilitation and upregulation, and will produce antidromic CNS originated reflexes. Antidromic impulses travel from proximal to distal along a C fiber (i.e. the wrong way).

Since the introduction of the double crush concept it is no longer sensible to examine patients only at the site of pain or only the nerve that is capable of referring to the anatomical site of pain. The double crush concept also helps us understand patients who have a poor result after surgical release for neuropathy. (**Idler, 1996**)

Ischemia

Although this section is primarily on peripheral nerve entrapment, circulatory dysfunction plays a primary role. Everyone has had the common experience of having a leg go to sleep after crossing one's knees. Within a relatively short period, heaviness and a strong pins–and–needles sensation develop in the limb. The condition is rapidly reversible with movement, and there is no recognizable pathology. "Ischemia may play some role in chronic entrapment and may well be responsible for some of the frequently seen axonal loss and the connective-tissue changes." (**Staal et al., 1999; Dawson, Hallett, & Millender, 1983**) The symptoms of a chronic entrapment can be worsened by ischemia, which may be the factor that causes nocturnal symptoms in carpal tunnel syndrome (**Staal et al., 1999; Lundborg, 1975; Fullerton, 1963; Gilliatt & Wilson, 1954**) because during sleep the arterial blood pressure falls, pulse rate decreases, and the metabolic rate lowers. (**Guyton & Hall, 2005**)

This is seen in arterial embolism and accompanying small vessel embolism, as well as in other systemic or localized conditions that block the arterial flow. (**Veves et al., 2002; Lusskin, 1982**) A major consideration in the past has been evaluation of the thoracic outlet by provocative tests that decrease the radial pulse. As will be discussed later, these tests are not as important today in evaluating peripheral nerve entrapment; however, one should consider and do the appropriate tests for circulatory dysfunction in all cases appearing to be peripheral nerve entrapment. Symptoms of vascular occlusion should never be overlooked. (**Butler, 1991; Rob & Standeven, 1958**)

Ischemia can cause a true neuropathy independently as nerve fibers are dependent upon an uninterrupted supply of blood for normal function. (**Butler, 1991; Sunderland, 1978**) Large nerve fibers suffer earlier than smaller fibers from compression and ischemia. (**Ochoa, 1980**) The cells of the central nervous system are extremely sensitive to oxygen deprivation; however their branches in the form of peripheral nerves are less so. However if the ischemia lasts too long, the nerves are in danger of real damage. This can manifest as permanent motor or sensory dysfunction. (**Lundborg, 1988**) Anoxia also has a destructive impact on the endothelium of the endoneural blood vessels.

Commonly occurring with peripheral nerve entrapment is circulatory deficiency at the capillary level. If ischemia causes the nerve to be more susceptible to peripheral nerve entrapment, the capillary deficiency can be due to a local pressure effect that is not only on the nerve itself but also on its intrinsic blood supply. (**Barral & Croibier, 2007; Staal et al., 1999; Lusskin, 1982**) Peripheral neuropathy at a more proximal location can cause capillary constriction due to the post-ganglionic sympathetic fibers being stimulated as they pass intermingled with the myelinated nerve fibers. Adson (**Adson, 1951**) considers this the cause of vasospasm and hyperhidrosis that sometimes predominate over other neurologic symptoms in thoracic outlet syndrome.

Both mechanical compression and ischemia are responsible to varying degrees in peripheral nerve entrapment syndromes. It is critically important that the applied kinesiologist remember a fundamental rule of neurophysiology: alpha fibers are more susceptible to mechanical compression; beta fibers are relatively more resistant, but they are more susceptible to anoxia from ischemia. For this reason, when one develops paresthesia from crossing his knees, *motor function is not greatly impaired; however, when there is peripheral nerve compression, muscle function is impaired to a greater degree than the paresthesia that develops.*

For practical purposes, peripheral nerve entrapment is compression or other mechanical disturbance to the nerve combined with impairment of blood circulation. The peripheral nerve is normally highly vascularized, with an abundance of collateral sources of flow. Experimentally it is difficult to adequately shut off the flow to cause ischemic changes in a peripheral nerve by ligation of nutrient vessels. It takes very high pressure applied to a nerve alone to diminish conduction. When the pressure is applied to cause anoxia by diminishing local blood flow, much less pressure is needed. (**Akuthota & Herring, 2009; Butler, 1991; Asbury, 1970**) When there is prolonged ischemia, recovery from entrapment is slower. (**Staal et al., 1999; Kopell, 1980**) Ischemia without pressure can produce Wallerian degeneration and paranodal demyelination, the former being more extensive. (**Hess, 1979**)

Compression ischemia, in general, first causes mild dysesthetic sensation, followed by loss of touch, then pain sensation and finally — much later — loss of motor power. Asbury (**Asbury, 1970**) observes that this pattern of symptoms does not correlate with electrophysiologic findings, which indicate that heavily myelinated fibers providing motor supply stop conducting before the lightly myelinated and unmyelinated fibers. This indicates that motor loss would be first. *Clinical observation in applied kinesiology indicates that motor weakness, as observed by the manual muscle test, is almost always present with sensory or ischemic disturbance with entrapment of mixed nerves.* One must recognize that the manual muscle test does not evaluate muscle function in the same manner as the muscle producing power against a dynamometer; (**Nicholas et al., 1976**) however the concurrent validity of the manual muscle test has been found to be excellent in 11 controlled clinical trials when compared to muscle strength testing instruments. (**Cuthbert & Goodheart, 2007**)

The effect of ischemia in peripheral nerve entrapment is demonstrated by tests designed to block the arterial flow to the area of entrapment. These tests will be described later, but in general they cause increased symptoms more rapidly in those with entrapment neuropathy than in normal subjects. (**Butler, 1991; Fullerton, 1963; Gilliatt & Wilson, 1954**)

Further evidence that ischemia contributes to peripheral nerve entrapment is seen where measurements of the amount of pressure in a confined area in some cases are not significantly different from normal subjects, indicating that some other factor is contributing to it, such as ischemia. (**Zhao et al., 2011; Gelberman et al., 1981**) Ischemia, therefore, seems to be another factor in the double crush syndrome; the single entrapment may not be adequate to cause symptoms, but when the health of the nerve is impaired for some reason, an additional mechanical factor can cause neuropathy.

Although ischemia contributes to peripheral nerve entrapment, the mechanical factors are the critical ones in chronic entrapment lesions. (**Staal et al., 1999; Butler, 1991; Dawson, Hallett, Millender, 1983; Kopell, 1980**) In the paresthesia that develops from crossing one's knees or from the application of a suprasystolic cuff to a limb, the lack of nerve function is almost immediately reversible upon release of the cuff or uncrossing the legs; there is no recognizable microscopic or ultrascopic pathology. This failure of nerve conduction is strictly due to ischemia. (**Kopell, 1980**)

The sympathetic nervous system should be evaluated when examining for peripheral nerve entrapment. Vasoconstriction of blood vessels may be caused by irritation of the sympathetic trunk. The perineurium and epineurium are sympathetically innervated. (**Lundborg, 1988**) Selander et al (**Selander et al., 1985**) measured blood flow in the sciatic nerves of rabbits and found that by stimulation of the lumbar sympathetic trunk, intraneural blood flow could be reduced to 10% of control values. It was also found that an injection of noradrenalin into the aorta reduced blood flow by 40%.

It should be noted that muscle spindle intrafusal fibers receive innervation from branches of the sympathetic nervous system, an idea which has found experimental verification specifically regarding TMD. Sympathetic stimulation has been shown to reduce type Ia and II motor output. **(Passatore et al., 1985)** This research indicates that sympathetic action on muscle spindles – including the fight-or-flight reaction involving Selye's General Adaptation Syndrome and adrenal stress disorders – may be one of the mechanisms by which physical and emotional stress inhibit the muscular system. **(Wilson, 2002)** Patients with chronic pain syndromes often assume protective postures that may tense the sympathetic chain.

Interruption of the sympathetic supply changes sweat gland secretion, depending on the level and type of nerve involvement. If there is severance of the nerve root proximal to the sympathetic ganglia, an absence of sweating after central physiologic stimulation will result. An interruption distal to or within the ganglia will cause no sweat response to stimuli or to cholinergic substances. A peripheral irritation will cause no change or excessive sweating. **(Korting & Denk, 1976)** When the autonomic nervous system is irritated, a true causalgia may develop. In the early stage, the extremity is red, warm, dry, and swollen, later becoming cool, clammy, hot with pallor, and/or cyanotic. Raynaud's phenomenon may occur. **(Mondelli et al., 2009; Sternschein et al., 1975)**

Examination

History

Differential diagnosis of peripheral nerve entrapment is sometimes simple and clear-cut; on the other hand, it may be difficult, with many potential traps for the unwary physician. One must be constantly alert for the double crush syndrome. This may include two areas of entrapment, or one area of entrapment plus a general neuropathy from nutritional deficiency or drug side effects, or other factors. Finding only part of the problem leads to disappointing treatment results. It is an old maxim of neurology that the cause of a disorder is hidden in the patient's history and that the site of the dysfunction is detected by examination. Proper diagnosis begins with an in-depth patient history and report of current status.

The characteristic symptoms, signs, and dysfunctions of peripheral nerve entrapment provide the first clues that this may be the cause — or partial cause — of the patient's problem. The physician's clear mental picture of various areas in the body that are susceptible to peripheral nerve entrapment, and the distribution of the nerve involved, should indicate the possible areas of entrapment. Most peripheral nerve entrapment is located in more common places, such as at the carpal tunnel, tarsal tunnel, and piriformis. The less common entrapments often relate with sports or occupational activities. Often one can determine the site of the entrapment by asking about habits or posture that is relevant to the nerve or anatomical region in question. It is important to locate the area of entrapment and determine the cause. Often analysis of the activity will reveal changes that can help eliminate future problems. (**Leaf, 2010; Russell, 2006; Staal et al., 1999; Goodman, 1983**)

Etiology of peripheral nerve entrapment often relates to sports, occupational activity, and habit patterns; thus, history and consultation should delve deeply into the way the patient lives. Sports that can cause entrapment neuropathy are hiking with backpacks, gymnastics, baseball, volleyball, horseback riding, and skiing, among others. (**Toth, 2008; Hirasawa & Sakaida, 1983**)

Activities performed during one's occupation are common causes of peripheral nerve entrapment, often called occupational or craft palsies. (**National Research Council, 2001**) Throughout the discussion of various types of peripheral nerve entrapment, comments will be made about particular occupations or sports that may cause entrapment. The list of these contributing factors is exhaustive, (**National Research Council, 2001; Murphy, 1955**) and not all can be considered. The important factor is to examine the patient in the manner in which he lives and works. Consider localized factors, such as how an individual holds the tools with which he works, (**National Research Council, 2001; Charash, 1982**] the use of vibrating tools, and whether there is jerking and jostling of the body or constrained postural work positions.

When neuropathy develops rapidly, the patient usually notices the change and seeks help. Symptoms of peripheral nerve entrapment may be evident early, or become prominent many years after the initial injury. (**Staal et al., 1999; Kopell, 1980**) Often craft, occupational, and some sports activities cause conditions to develop insidiously, with slow loss of sensation and muscle function. This is often ignored by the patient, and treatment is sought for some functional deficit such as tripping, turning an ankle, or catching a toe on small rises in the substrate. Poor weight distribution and structural deformity in the foot may cause corns, calluses, alterations, or painful areas. (**Langer, 2007; Maffetone, 2003; Greenawalt, 1988; Lusskin, 1982**) The patient may first notice a problem in the upper extremity because of dropping items, poor dexterity of the hand, and inability to get the arm into certain positions. (**Goodheart, 1971**) Often no specific traumatic episode will be described during consultation because many peripheral entrapment syndromes are chronic, having developed symptoms insidiously.

The symptomatic picture presented will depend greatly on the type of nerve involved. The three potential types are cutaneous sensory, motor, and mixed. Although the first two classifications are often referred to as "pure," there are no truly pure peripheral nerves. The cutaneous sensory nerves carry afferent impulses and also supply efferent fibers to various structures in the skin. The motor nerves return afferent impulses from the proprioceptors of the muscle, joints, and associated connective tissues. The motor nerve to a muscle returns sensation from the joint(s) upon which that muscle acts. A mixed nerve is a combination of cutaneous sensory and motor nerves in one trunk. Complicating the picture is Hilton's Law which states that the nerve supplying a joint also supplies the muscles which move the joint and the skin covering the articular insertion of those muscles. A good way to think of the symptomatic pattern of peripheral nerve entrapment is sensory, motor, and sympathetic. (**Lusskin, 1982**)

Patients with acute or chronic entrapment neuropathies will often complain of feeling that their joints or muscles are weak and they fatigue easily. This experience may sometimes persist long after the pain has been alleviated and repair seems to be fully resolved. There is a biological reason for this mechanism. Forceful muscle activation will raise the intramuscular as well as the intracapsular pressure and may result in further damage to the nerve and its surrounding tissues. (**Racinais et al., 2008**)

In cutaneous sensory nerve involvement, the paresthesia is usually precisely defined by the patient. It may be a sharp, well-described pain, numbness, or altered sensations such as pruritus.

Motor nerve involvement will affect muscle activity which, if chronic, may be demonstrated as atrophy. The associated pain is more general and less well-defined by the patient. It will usually involve the muscle, its associated joints, and connective tissue in agreement with Hilton's Law.

Most neuropathies of either a motor or cutaneous sensory nerve will have some characteristics of both types. The often-involved mixed nerve may have characteristics of only the motor or cutaneous sensory fibers involved. There may be chronic sensory disturbance with no muscle atrophy, or vice versa.

Differential Diagnosis

There are two concerns in differential diagnosis of peripheral nerve entrapment. Obviously one is to differentiate conditions that simulate peripheral nerve entrapment, such as other forms of peripheral neuropathy, reflex sympathetic dystrophy, and radiculopathy. Sometimes the differentiation is to find the entire problem. This is what has already been discussed as the double crush syndrome. Double crush was originally related to two levels of peripheral nerve entrapment. **(Upton & McComas, 1973)** The same basic principle of reduced axoplasmic transport relates to conditions that cause the nerve cell body to have a lowered level of health. This can be diabetes, **(Veves et al., 2002; Hurst, Weissberg, & Carroll, 1985)** malnutrition, or other systemic conditions. Sometimes a local peripheral nerve entrapment is diagnosed, but a proximal one or a systemic problem is missed. The area primarily diagnosed may be a low level of entrapment that would not cause symptoms if the more proximal or systemic problem were not present.

When numerous factors are involved in peripheral nerve entrapment, the multiple diagnosis may be rather difficult. When two or more factors contribute to cause a symptomatic peripheral nerve entrapment, the individual factors making up the combined problem may be rather insignificant individually. For example, when there are two levels of nerve entrapment, either may be insufficient to cause symptoms but combined they do. **(Butler, 1991)** Spinner and Spencer state, "EMG studies are not always sufficiently sensitive to localize a double lesion." **(Spinner & Spencer, 1974)** When simultaneous entrapment or a diffuse peripheral neuropathy is suspected, one can evaluate other peripheral nerves to help differentiate a localized process from a systemic one. In applied kinesiology, various methods such as challenge, therapy localization, and nutritional testing can help discover all components responsible for the patient's condition.

Peripheral nerve entrapment must be differentiated from other forms of peripheral neuropathy. Metabolic disorders, such as diabetes mellitus, alcoholism, uremia, malnutrition, or vitamin deficiency may be confused with peripheral nerve entrapment or be contributing to a double crush condition. **(Travell & Simons, 1983)** These metabolic problems usually cause a symmetrical polyneuropathy, but an asymmetrical one may be present. In symmetrical neuropathy, the distal areas are involved first, with tingling and numbness beginning in the feet and progressing up the legs. The tips of the fingers may become involved. If there is muscle weakness it follows the course of the paresthesias, beginning in the feet.

Peripheral entrapment neuropathy is usually asymmetrical and localized, confining itself to a particular nerve and its branches. The complicating factor in differential diagnosis is that one who has a polyneuropathy is more subject to entrapment than a patient without the generalized problem is. Diabetics are good examples of this, as they are more susceptible to classic entrapment such as the carpal tunnel syndrome. **(Akuthota & Herring, 2009)**

Two common causes of vascular neuropathy — diabetes and vasculitis — should be easily diagnosed. Most patients with vasculitis are systemically ill with malaise, fever, weight loss, or an active disease process such as arthritis or renal disease. This can usually be easily distinguished with blood and urine laboratory tests.

Peripheral nerve entrapment must be differentiated from radiculopathy. There are usually spinal symptoms and signs indicating the possible involvement. The usual testing procedures described in Walther's *Applied Kinesiology -- Synopsis* **(Walther, 2000)** on the vertebral column help make the differential diagnosis, but keep in mind the double crush syndrome.

Dawson et al. **(Dawson, Hallett, Millender, 1983)** describe the following paresthesias and muscle weaknesses with levels of the cervical spine: C5 — deltoid and infraspinatus; C6 — biceps and wrist extensors, with paresthesias in the thumb and index finger; C7 — triceps, long finger flexors, and finger extensors, with paresthesias in the forearm and dorsum of the hand; C8 — intrinsic muscles of the hand and wrist flexors, with paresthesias in the little finger. L4 radiculopathy correlates with quadriceps muscle weakness and a reduced knee reflex. This must be distinguished from femoral neuropathy. L5 radiculopathy may resemble peroneal nerve entrapment. The tibialis posterior will be strong in peroneal nerve entrapment since it is innervated by the tibial nerve. "Tarsal tunnel syndrome and S1 radiculopathy may both cause pain in the heel and sole of the foot, whereas weakness of the knee flexors and foot plantar flexors points to the presence of radiculopathy." **(Dawson, Hallett, Millender, 1983)**

Amyotrophic lateral sclerosis may show its first symptoms in a localized area, such as the hand, causing it to be confused with peripheral nerve entrapment. There may be weakness of the hand, but muscle fasciculations are usually widespread and occur in muscles not yet clinically weak. Although they may not be seen by the examiner, the patient can usually describe and locate them. "A feature of the illness that is particularly pathognomonic is increased tendon reflexes in a muscle group that is fasciculating and atrophic." **(Dawson, Hallett, Millender, 1983)**

Reflex sympathetic dystrophy (RSD) has a similar local pain, tingling numbness, and atrophy as peripheral nerve entrapment. Early vasomotor changes, in most cases, include hyperhidrosis, hyperthermia, and erythema. These changes depend on the degree of sympathetic outflow. Mense and Simons **(Mense & Simons, 2001)** have summarized the basic science and clinical aspects of RSD, and Muir and Vernon have reviewed the chiropractic approach to RSD. **(Muir & Vernon, 2000)**

Drug toxicity can cause peripheral neuropathies that result in paresthesias and hypoesthesias. **(www.drugs.com www.)** The drug companies have listed the possible side-effects of their products. You can check these side-effects by reading the drug companies' literature at www.drugs.com or www.rxlist.com. A wide variety of drugs have been implicated in peripheral neuropathies.

Finally, long-term exposure to statins may also substantially increase the risk of polyneuropathy. **(Gaist et al., 2002)** Gaist et al's report in the journal *Neurology* expressed concern for the increased susceptibility to neuropathy among diabetics placed on statin drugs. They estimated that diabetics had as much as a sixteen fold increase in risk of neuropathy when statin drugs are used but suggested that non-diabetics are also susceptible.

There is a widely-recognized "drug side-effect epidemic" at work among all the patients physicians see today. The *American Society of Clinical Pharmacology and Therapeutics* stated (**2000**) that the percentage of elderly patients receiving nine or more medications was 27%, compared to 17% in 1997. This figure continues to mushroom. Most drugs have simply not been tested for use with other drugs, so that the "polypharmacy" seen in many patients must be confronted by the holistic physician. A recent article in **Current Psychiatry** (**Werder, 2003**) states "most individuals who are prescribed five or more drugs are taking unique combinations, representing an uncontrolled experiment with effects that cannot be predicted in the literature." The problem is so prevalent that some physicians "seem indifferent to so-called 'minor' side effects such as headaches, abdominal pain, fatigue, ringing in the ears, joint pain, insomnia, constipation, and hundreds of others that make people miserable." (**Cohen, 2001**)

It should be obvious that with a formidable list of such possible causes, it is necessary to discover what prescribed and over-the-counter medications the patient may be taking in any apparent peripheral nerve entrapment problem.

Palpation and Inspection

Palpation and inspection primarily deal with three types of evaluation for a specific nerve suspected of involvement, including inspection of the dermatome, muscles supplied by the nerve and of the nerve along its course.

Cutaneous

Deane Juhan (**Juhan, 1987**) writes about the importance of the skin:

> "The skin is no more separated from the brain than the surface of a lake is separated from its depths; the two are different locations in a continuous medium. 'Peripheral' and 'central' are merely spatial distinctions, distinctions which do more harm than good if they lure us into forgetting that the brain is a single functional unit, from cortex to fingertips and toes."

Table 1
Medications with neuropathy as a side-effect

Medications	Antibiotics Blood Pressure	Statins	Anti-Anxiety Anti-Depressant
Cymbalta®, Duloxetine hydrochloride®, Lyrica®, Neurontin®, Pregabalin®, Allopurinol®, Aminodipinberglate®, Amiodarone®, Amiodipine®, Amiodarone HCL®, Amitriphyline®, Besylate®, Cordarone®, Flagyl®, Lipitor®, Lotril®, Metrogl® Metrinidosole®, Metrofuranton®, Metronidazole®, Norvaso®, Pechexiline®, Vitorin®, and Zyloprin®.	Cipro®, Flagyl®, Metronidazole® **B.P.** Amiodarone®, Atenolol®, Aceon®, Altace®, Cozaar®, Hydralazine®, Hydrochlorothiazide (HCT)®, Hydrodiuril®, Hyzaar®, Lisinopril®, Micardis®, Norvasc®, Perindopril®, Perhexiline®, Prazosin®, Prinivil®, Ramipril®, Zestril®.	Advicor®, Altocor®, Atorvastatin®, Baycol®, Caduet®, Cerivastatin®, Crestor®, Fluvastatin®, Lescol®, Lescol XL®, Lipex®, Lipitor®, Lipobay®, Lopid®, Lovastatin®, Mevacor®, Pravachol®, Pravastatin®, Pravigard Pac®, Rosuvastatin®, Simvastatin®, Vytorin®, and Zocor®.	Ambien (Zolpidem)®, BuSpar®, Klonopin (Clonazepam)®, Xanax®, Celexa (Citalopram)®, Cymbalta (Duloxetine)®, Effexor (Venlafaxine)®, Effexor XR®, Nortriptyline®, Zoloft®

Evaluation of the skin deals first with observation of its texture. There may be a palpable change in the area of nerve distribution, or the change may be only visually observed. Texture change gives some indication of how much involvement there is of the sympathetic fibers. **(Chaitow, 1987)** This may influence perspiration and circulation into the area, as the sweat glands are controlled by the sympathetic portion of the autonomic nervous system. If the nerve is severed proximal to the sympathetic ganglia, there will be an absence of sweating with central physiologic stimulation. Severance of a nerve on a peripheral basis causes lack of response to any stimulation, and there will be no response to cholinergic drugs. A peripheral irritation causes undiminished or excessive sweat reactions. **(Guyton & Hall, 2005)** The secretory pores of the sweat glands are controlled by acetylcholine, a neurotransmitter which also influences the contraction and relaxation of muscles.

The skin is also evaluated for varying temperatures by visualization and palpation. Accurate palpation also depends upon the experience of the examiner, his hydration, peripheral circulatory efficiency, sympathetic nervous system activity, and the ambient humidity and temperature of the room. The physician can best feel for cool or warm areas by running the dorsal aspect of his hand over the various dermatomes being evaluated. When evaluating the skin for excessive or diminished thermal areas, remember that imbalance of the meridian system can create hot (Re) and cold (Han) spots in specific locations. **(Walther, 2000; Stux & Pomeranz, 1987)** Barral has written a book on manual thermal diagnosis and it is recommended for further refinements of this approach. **(Barral, 1996)**

There is a wide range of electronic thermometers available that readily measure skin temperature. The temperature of a dermatome can be used to initially diagnose the cause of involvement, or to evaluate treatment effectiveness. Dermatomes with abnormal temperature can be identified by comparing the symptomatic and asymptomatic sides. When a dermatome of involvement is identified, tape the thermometer in place and allow the recording device to stabilize. If a two-thermometer recording device is available, tape the second thermometer to the normal bilateral counterpart in unilateral conditions. Make a comparison between the two sides and then put the patient through various provocative tests, which will be described with many of the conditions in the following chapter. Allow 2-3 minutes in the position to observe for temperature change.

In a similar manner, thermal recording can be used to evaluate the effectiveness of treatment. Place the thermometer either unilaterally or bilaterally as described above. After allowing the recording device to stabilize, apply the therapeutic effort, which may be structural, nutritional, or meridian system balancing, among others. Allow a few minutes to pass and observe temperature change. This is most effective as an evaluation method when both the normal and abnormal sides are being recorded. Ordinarily one expects no change in the normal side, and normalization of the involved side.

Plethora can be observed by digital pressure on the skin to blanch it, and then noting the length of time it takes for balanced color to return. The plethora may be systemic or localized in an extremity. If localized, areas of possible venous entrapment must be evaluated. Thorough evaluation of the circulation and sensibility of the dermatomes is part of the skin evaluation; this will be discussed later in the specific areas.

Methods for evaluating structural abnormalities in the skin have been developed in applied kinesiology. **(Leaf, 2010; Walther, 2000; Goodheart, 1986, 1983)** The cutaneous receptors are stimulated with joint motion. For example, when the knee flexes, the skin over the quadriceps and anterior knee stretches. When the knee extends, the skin over the hamstrings and popliteus muscle stretches. The muscle underlying the area of stretched skin is normally inhibited with the joint motion. The contribution of the cutaneous receptors to the neurologic organization of muscles in joint movement can be observed by manual muscle testing in applied kinesiology. For instance, the examiner can stretch the skin over the quadriceps muscle in alignment with the muscle fibers. This may be done by simply taking a pinch of skin between the thumb and forefinger of each hand and pulling the skin apart, taking care not to heavily pinch the skin with the fingers. Immediately after stretching the skin, manual muscle test the quadriceps muscle group. Under normal circumstances, the muscles will test weak for a variable length of time. Stretching the skin in this manner simulates knee flexion, as if the hamstrings had contracted. This would cause reciprocal inhibition of the quadriceps muscles, which is what is observed when the muscle tests weak.

Clinical evidence in applied kinesiology indicates that in some patients the cutaneous receptors may be inappropriately stimulated or react inappropriately to stimulation, sending information not in keeping with the joint motion. When a muscle tests weak as a result of improper signaling from the cutaneous receptors, it can immediately be made to test normal by skin stretching. For example, if the hamstring muscles test weak, stretch the skin over the quadriceps group in the direction in which it would normally be stretched if the hamstrings were contracting. The amount of stretch is the amount the skin will yield without damage or severe pain. If dysfunction of the cutaneous receptor is associated with the dysfunctioning hamstring muscles, they will test strong immediately after the skin stretch. Since no structure other than the skin is manipulated or stimulated, it appears that the improved muscle function results from stimulation of the cutaneous receptors. Making a muscle strong in this manner is only a diagnostic test for apparent cutaneous receptor dysfunction. The strengthening of the hamstrings will only last for twenty or thirty seconds, regardless of how vigorous or long-lasting the skin stretch is. Treatment of the cutaneous dysfunction (soft-tissue, instrumental, nutritional, cold-laser and other techniques in applied kinesiology) will correct this problem. **(Kharrazian, 2002; Ramsak, 1997; Dauphine, 1994; Burstein, 1990)**

Muscle.

The muscles innervated by the nerve being evaluated should be inspected and palpated for consistency, pain, and size. When a motor nerve is involved, the pain will typically be throughout the muscle and in the joints served by that muscle. The patient will usually complain of a dull, deep, general-type pain rather than the sharper, delineated pain of skin sensory nerve involvement. The muscle may have atrophy that is observed only by astute palpation, especially in the earlier stages of entrapment. In this case the muscle will feel like it has lost consistency, which is often called tone. Comparison with the bilateral counterpart helps determine this loss.

The circumference of the extremity should be measured bilaterally. Standard levels of measurement are established, such as 4" above the olecranon process with the arm extended, so that repeated measurements will be accurate. It is important to document muscle size on an initial examination, especially in personal injury and worker's compensation cases. Progressive atrophy from the time of injury to maximum medical improvement is an important documentation in permanent impairment rating. Muscle wasting can begin surprisingly quickly in patients with chronic low back and neck pain. The cervical and lumbar multifidus, as well as the psoas muscles have been observed to experience muscle wasting within 24 hours of the onset of pain. (**Danneels et al., 2009; Fernandez-de-las-Penas et al., 2008; Hides et al., 1994**)

There may be muscle dysfunction that is not observable by palpation and inspection, but it is observed by manual muscle testing as used in applied kinesiology. Specific challenge, therapy localization, and types of muscle tests will be explained with the various syndromes.

Nerve.

Palpation and inspection of the course of the nerve provide much useful information. In many areas, the nerve is rather superficial and can easily be palpated, which may reveal tumors or pain along the nerve.

Pain along the nerve may be proximal or distal to the point of entrapment. This tenderness is known as the Valleix phenomenon. (**Barral & Croibier, 2007; Kopell & Thompson, 1976**) Nerves have been described as small, thin, slightly twisted strings or cords. The site of involvement may be localized by observation of swelling at a potential area of entrapment. The swelling may be a result of trauma to the area, or it may be systemic from a toxic condition, hormone imbalance, or disease process.

Tinel's sign (**Evans, 2008; Wilkins & Brody, 1971**) refers to pain or tingling of a nerve when it is struck over an area of neuropathy. Tinel's sign was originally discovered almost simultaneously and independently by a German doctor, Paul Hoffmann, and a French doctor, J. Tinel, during World War I. Tinel's description indicates that pressure on a damaged nerve trunk often produces a tingling sensation projected to the periphery of the nerve and localized to a very exact cutaneous area. He differentiates this tingling sensation from that of pain produced by pressure on an injured nerve. Pain is a sign of nerve irritation; tingling is a sign of regeneration or, more precisely, it indicates the presence of young axons in the process of growing. When a nerve is relatively superficial, Tinel's sign should only be elicited by a gentle tap or pressure over the nerve. A sharp blow over the nerve of a normal individual will cause pain or tingling and should not be interpreted as a positive sign. (**Phalen, 1972**) Today Tinel's sign is usually elicited by tapping with a reflex hammer over the nerve site being evaluated, instead of the tapping finger described by Hoffmann and Tinel. (**Evans, 2008**)

General Physical Examination

A general physical examination is required to determine if there are any systemic health problems. Toxicity from conditions such as the ileocecal valve syndrome seems to concentrate swelling in areas of generalized weakness. If a patient has subclinical disturbance of structure — a tight osteofibrous tunnel or weak intervertebral disc — swelling may cause an additional enlargement of the structures that creates a symptomatic involvement; thus, both the systemic and localized problems must be treated. Other conditions such as hypothyroidism, pregnancy, rheumatoid arthritis, acromegaly because of proliferations of tissues narrowing the osteofibrous tunnel, (**Bastron, 1975**) hypoadrenia, (**Wilson, 2002**) and generalized structural imbalance can cause an individual to be more susceptible to peripheral nerve entrapment. In systemic problems such as nutritional deficiency, the distal peripheral nerves are more vulnerable to dysfunction. In these cases the entrapment is more likely to be found in distal areas such as the carpal tunnel, tarsal tunnel, and intermetacarpal tunnel.

Sensibility Studies

Sensory examination can help define cortical and thalamic lesions and cord involvement, as well as peripheral neuropathies. Here we will deal primarily with peripheral nerve entrapment.

The patient must be cooperative and alert, and the physician must be careful to avoid suggestions that influence the patient's response. If a thorough sensory evaluation is made, it can be very time-consuming and tiring to both the physician and the patient. It is best, then, to intersperse this activity with other examination procedures to break up the monotony of the examination.

The tools of examination include a wisp of cotton for evaluating soft touch, a pin with a sharp point and dull head mounted on an applicator or a Wartenberg pinwheel for pain, vials containing both hot and cold water for thermal testing, a tuning fork for deep sensation, and a compass with two dull points for two-point discrimination.

In addition to the general examination tools mentioned above, there are very sensitive, graded-hair test kits to evaluate sensory loss. In general, a relatively gross method of examination for sensory loss is adequate. In a study by Gilliatt and Wilson, (**Gilliatt & Wilson, 1953**) the very sensitive, graded-hair method was used to determine sensory loss in comparison with massive stimulus, such as the examiner's finger touching the patient's skin. It was found that there was little advantage in using the sensitive approach, such as a wisp of cotton or graded hairs. They concluded, "From the practical point of view there is every advantage to be gained from recording merely the subjective sensation in response to a massive stimulus, as this is easily determined in unintelligent patients, and is much less affected by horny skin or cold hands than more elaborate methods of sensory testing."

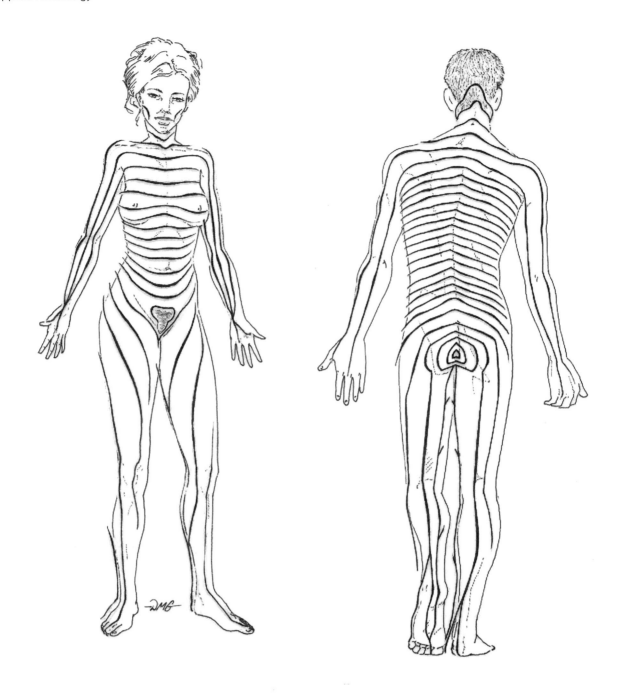

Dermatone distribution

It is necessary for the examiner to have a general knowledge of dermatome distribution and the congenital variations of the peripheral nervous system. A chart should be available for mapping the examination findings. When a test is to be done, the patient should be familiar with its procedure and how to respond to the stimulus to avoid any misunderstanding. To ensure that the patient does understand the test, go through a demonstration and have him or her respond to the stimulus.

When the testing procedures actually begin, it is best for the patient to close his eyes and relax, concentrating on the sensation testing. Various procedures should be alternated and timed at different rates. This not only helps break the monotony of the examination, it also keeps the patient from responding to an expected stimulus without the sensation being perceived. Although many types of responses are suggested for perceived sensation, a simple "yes" is satisfactory. It also adds another depth that may sometimes be very revealing when the neurotic, unsophisticated, or malingering patient answers "no" to a stimulus.

It is necessary to evaluate the body bilaterally when testing for peripheral nerve involvement. Usually there is partial rather than total loss of sensation. This can best be observed by comparing the two sides. Bilateral comparison is easily done by simultaneously stimulating both sides, using two Wartenberg pinwheels or wisps of cotton and asking the patient to make a comparison.

When areas of sensation loss are located, the borders must be defined. In typical peripheral nerve entrapment, the borders will be abrupt and will correspond in general to the appropriate dermatome. In peripheral neuropathy resulting from ischemia, the border is ill-defined; the hypoesthetic area will probably cross many dermatomes. The same is often true when a large number of nerves is involved, such as the neurovascular bundle in certain types of thoracic outlet syndrome. When there is hypoesthesia as a result of malnutrition, the distal aspects of the extremities are most involved, causing a general loss of sensation that is greatest distally and decreases proximally. This is the so-called "glove" or "stocking" type of sensory loss. It must be differentiated from sensory loss due to ischemia.

The important aspect of the sensibility of the hand is how it can do the many tasks required in performing necessary functions. It is also important to be able to evaluate increasing or decreasing loss of function. Moberg (**Moberg, 1958**) developed a test he calls the "picking-up test." The patient is asked to pick up numerous small, familiar objects from a table and put them in a container, first with one hand and then the other. After becoming familiar with the task, the subject is then asked to repeat the process blindfolded. During both visual and blindfolded performances he is timed, and his efficiency in picking up the objects is evaluated. Under normal conditions of stereognosis, he is able to identify the objects when doing the task blindfolded. Usually the objects are picked up with median nerve-innervated fingers. If there is median nerve loss, the subject will often divert the task to ulnar-innervated fingers combined with the thumb.

Loss of sensory peripheral nerve function is accompanied by loss of sweat gland function. Moberg (**Moberg, 1958**) also devised an objective test that depends on the sweating of the skin of the fingers — the ninhydrin fingerprint method — and with it he was able to objectively show the parts of the skin with intact sensation and those that were defective. He also observed that "…it was possible to see changes in an area of defective sensibility with the naked eye. Thus it could be seen that the pulp was atrophied, the ridges were more or less eradicated and the colour was changed." One can often see these changes when comparing the normal with the abnormal hand.

A simple test to objectively document sensory loss has been described by O'Riain. (**O'Riain, 1973**) The characteristic wrinkling of the skin of the fingers when immersed in warm water is absent in the presence of sensory denervation. The procedure is especially valuable in examining young children who may not have the ability to cooperate. The extremity is immersed in 40°C (104°F) water for thirty minutes. Normally the skin of all fingers wrinkles or "shrivels"; denervated fingers remain smooth.

Circulatory Evaluation

Disturbance of blood circulation may be a result of peripheral neuropathy or a cause of it; consequently, differential diagnosis of circulation must be done when considering peripheral nerve entrapment. A disturbance in circulation may be secondary to peripheral nerve entrapment as a result of involvement of the sympathetic fibers; this is neurogenic ischemia. On the other hand, nerve tissue is very susceptible to ischemia; consequently, entrapment may be more significant secondary to a diminished vascular supply.

It must be remembered that the diameter of the arteries and veins is controlled by the nerves and nervous system. The vasomotor effect of manipulative treatment is achieved either by the surrounding nerves of the blood vessels or by the sympathetic nervous system. After chiropractic treatment to the cervical spine and brachial plexus, we have consistently seen changes in the terminal branches of the brachial plexus which can be recorded on the sphygmomanometer. (**Mc Knight, DeBoer, 1988; Einaudi, 1959; Wright, 1956, 1955**)

There is a wide range of diminished circulation. Some conditions require the immediate attention of a vascular surgeon; others are effectively treated with applied kinesiology conservative methods. Obviously, if the ischemia is due to an occlusion of the blood vessel, circulation must be returned to normal before the nerve will function properly. The ischemia may be a result of thrombosis, embolism, neoplasms, orthopedic faults, or other factors that occlude the vessel. Also, consideration should be given to vascular diseases, such as arterio- and atherosclerosis, and systemic health problems such as diabetes.

When there is evidence of circulatory disturbance, a thorough work-up must be done. One should not jump to the conclusion that a cervical rib is causing circulatory deficiency. Many people have cervical ribs that are asymptomatic. A patient may have both a vascular condition and a cervical rib that is no more than coincidental. Symptoms of vascular occlusion should never be overlooked. "The end-result may be loss of fingers, ischaemic contracture of forearm and hand, or a major amputation, following occlusion of the main vascular channels by embolism or thrombosis.

Physical tests must be put into proper perspective relative to what they are capable of showing and what they fail to reveal. In general, there are no pathognomonic provocative tests for peripheral nerve entrapment or vascular deficiency.

Measurement of the extremities provides information regarding venous or lymph drainage. Comparison can be made after an individual has been recumbent for a period of time, such as before rising in the morning, with the end of the day. Continued comparison over days, weeks, or months indicates treatment effectiveness.

It takes very high pressure applied to a nerve alone to diminish conduction, but when pressure is applied to cause anoxia by diminishing blood flow, much less pressure is needed. Compression ischemia, in general, first causes dysesthetic sensations, followed by loss of touch, then pain sensation, and finally — much later — loss of motor power. With motor weakness appearing last, it is last to return in persistent deficits. (**Russell, 2006; Asbury, 1970**)

There are several variations of the "tourniquet test," in which a blood pressure cuff is inflated above systolic level to create ischemia or additional ischemia in the area being evaluated. (**Nitz & Dobner, 1989**) In patients who have peripheral nerve entrapment, the increase in paresthesia or pain will develop more rapidly than in those who do not. Fullerton (**Fullerton, 1963**) studied nerve conduction in individuals diagnosed as having carpal tunnel syndrome when artificial ischemia was produced by a supra-systolic blood pressure cuff applied to the upper arm. The pressure was maintained for 30 minutes at 200–220 mm Hg. In control subjects the action potential did not change after 20 minutes ischemia. After 25 minutes there was still more than 70% of the initial size in all subjects. At the end of 30 minutes, all subjects had at least 50% of initial value. Patients with carpal tunnel syndrome were more susceptible to the ischemia. Occasionally the action potential had almost disappeared after 10 minutes; in some it had fallen to less than 40% after 25 minutes. In this study there did not appear to be any difference in recovery between the control subjects and patients after release of the blood pressure cuff. Both ischemia and direct pressure appeared to contribute to the abnormality of nerve function. The recovery rate here is different from nerves that suffer from prolonged ischemia, which has a longer recovery period. (**Staal et al., 1999; Wilgis, 1971**)

Another study that evaluated the rapidity of sensory loss when artificial ischemia is induced was done by Gilliatt and Wilson. (**Gilliatt & Wilson, 1954**) In a study of 40 patients with peripheral nerve and dorsal root lesions affecting the hand, compared with 50 control subjects, they found that the controls had sensory loss within an average of 14 minutes. The patients with peripheral nerve lesions developed sensory loss usually within 5 to 10 minutes of the onset of ischemia.

There are specific characteristics that one can look for in the tourniquet test in different areas. For example, in carpal tunnel syndrome the paresthesia that develops is of median nerve distribution, whereas in normal subjects the tourniquet test causes ulnar or diffuse paresthesia. (**Nitz & Dobner, 1989; Gilliatt & Wilson, 1953**) Patients evaluated for unilateral disturbance are best studied by applying the tourniquet test to the suspected abnormal limb and then the normal one, comparing the time and amount of pain and paresthesia.

There are orthopedic tests designed for evaluating the vascular system at the various areas where blood flow restriction may occur. These include maneuvers such as Adson's and Wright's tests for thoracic outlet syndrome, which will be discussed when the anatomical area is considered in the following chapters.

Several types of non-invasive instrumentation are available to provide specific data regarding the circulatory system. By themselves they are incapable of making a diagnosis, but they are significant diagnostic aids. Usually, the best approaches are non-invasive because they lack the potential side effects present in angiography. (**Benya et al., 1989; Kopell, 1980**) With non-invasive equipment, a diagnosis of patency or occlusion can usually be made; if the vessels are patent, the potential hazards of angiography can be avoided.

Plethysmography

In general, plethysmography relates to measuring blood flow volume, usually at the distal extremities. There are many types of plethysmographs available. Some are transducers that convert the mechanical pulsation of blood circulation to electricity. The mechanism may be a strain gauge directly contacting the skin or using a conductive medium that is hydraulic or pneumatic to couple with the transducer. (**Higashi & Yoshizumi, 2003**) Other plethysmographs are primarily electrical, measuring the impedance and inductance. In a class by itself is the photoelectric cell. The latter shines a light into the capillary bed that is reflected back in various intensities, depending on the bed's engorgement. The reflected light is picked up by a photoelectric cell. All modern plethysmographs display the electrical current on an oscilloscope or on a strip recorder. The latter has the advantage of making a continuous recording while the patient is maneuvered into provocative test positions. When one can look at the recorded strip rather than remembering what was on the oscilloscope, comparisons are easier to make.

The plethysmograph can be used to simply record the pulse wave in the various digits on a comparative basis. The pulse wave can also be recorded as the patient goes through various orthopedic maneuvers, such as Adson's and Wright's tests. This gives an improved evaluation of blood flow change over simply monitoring the pulse at the radial artery by the examiner's digital sensation. The provocative tests and their use with the plethysmograph are discussed with the various syndromes. Current interpretation of many of these tests, especially at the thoracic outlet, is modified over that of the past.

Another test done with plethysmography is called reactive hyperemia, which helps determine the patency of the blood vessels and whether the nervous system reacts to the body's physiologic needs. First, a recording is made of the peripheral circulation in the resting state; then the vessel is occluded proximal to the area being evaluated. This is accomplished with a blood pressure cuff pumped up above the systolic blood pressure. Complete occlusion is indicated by eliminating the pulse wave display on the plethysmograph. The restriction of blood flow is maintained for 3–5 minutes to produce ischemia. Blockage of the blood flow during this time is not harmful, but it must be pointed out that over 30 minutes of complete occlusion causes microscopic changes in human muscle. (**Wilgis, 1971**) The blood flow is then released while a recording is made of the pulse wave. In a normal reactive hyperemia test, the pulse wave will increase over the resting state. (**McGrath et al., 1980**) This is a normal body reaction to increase blood flow into the area to eliminate the ischemia resulting from the restriction. This activity is moderated by the nervous system. A positive test is one in which there is no increased blood flow after the restriction; in some cases, the blood flow may even be reduced from that of the resting state. A positive test indicates a disturbance in the nervous system's regulation of blood flow or an occlusion of the blood vessel that will not allow adequate increased flow. This test is valuable as a before and after treatment evaluation when a diagnosis of peripheral nerve entrapment has been made as the primary cause of ischemia. After the therapeutic effort to return the nerve to normal, there should be improvement in the reactive hyperemia test. Improvement of the reactive hyperemia test is often seen immediately after applied kinesiology treatment to release a peripheral nerve entrapment. In chronic conditions, there may be a delay of one or two weeks before the improvement is seen, even though correct therapy has been applied. This may be due to the fibers responsible for improving the circulation needing time for their own recovery before they can control circulation normally.

The control of the vascular bed can also be evaluated by a plethysmograph recording made at room temperature, after which the patient's extremity is placed in ice water for 3 minutes. The blood flow volume is then recorded after removal of the hand from the bath and should return to near-normal within 5 minutes; if it does not, there is evidence of vasospasm. Foster's test is similar, but rather than blood flow volume as an indicator, the extremity temperature is taken before and after the ice stress. **(Wilgis, 1980)**

Disturbed circulation observed by the plethysmograph must be differentially diagnosed. It has been shown that radiculopathy treated by a chiropractic vertebral adjustment can be improved with no additional treatment. **(Figar & Krausova, 1965; Figar et al., 1967)** Budgell & Sato have also demonstrated that regional blood flow can be affected by cervical manipulation aimed at the correction of vertebral subluxations. **(Budgell & Sato, 1997)**

Doppler

The Doppler instrument — an ultrasonic generator and flow detector — is named after the Doppler effect. The transducer emits ultrasound waves through piezoelectric crystals that are transmitted through the skin directly to the blood vessel. They are reflected back to the probe, which has a receiving unit. Ultrasound waves that strike moving tissue are reflected back differently from those that strike stationary tissue. The greater the intensity of movement, the greater the change. The waves that are reflected back are amplified and demonstrated by audible sound or by demonstration of the pulse waveform on an oscilloscope, or on a strip recorder. Arterial blood flow has an alternating high-pitched sound as the blood pulsates in the vessel, while venous blood flow has a slow, waving sound much like that of the wind. Doppler evaluation can be used to locate areas of occlusion in the artery or vein. Various methods are used in Doppler evaluation to locate an occlusion. The method is not limited to the extremities; it is also used in studying disturbances in cranial circulation.

Blood pressure in the lower extremity can be accurately taken by Doppler testing. A standard blood pressure cuff of the proper size for the limb is applied at the ankle, below the knee, or above the knee. The pulse is usually monitored at the posterior tibial artery by the Doppler. This means of accurately obtaining the arm and ankle systolic blood pressures gives the ability to compare them. This is called the ankle-to-arm index. Pressure at the ankle should be at least equal to, but preferably higher than, that of the arm. When there is a diminished ankle-to-arm index, it indicates that there is some type of vascular disturbance capable of producing symptoms. Less than .9 is considered abnormal; .9–.7 is consistent with intermittent claudication; .4–.3 is consistent with resting pain; below .3, gangrene may develop.

X-ray

X-ray is used for evaluating the circulatory system and for peripheral nerve entrapment syndromes. Only a brief description of its use is within the scope of this text.

Angiography is the injection of a contrast medium into the circulatory system to study its anatomy, areas of occlusion within the vessels, and impinging structures such as neoplasms and tortuous routes. The procedure is not without its hazards. There may be a reaction to the iodides used in the contrast medium or complications at the site of arterial puncture. It is primarily an anatomical study providing no information about the dynamic state of circulation. It is not recommended for routine or repetitive use. **(Kopell, 1980)**

X-ray is more routinely used to evaluate for bony spurs, congenital anomalies, and trauma that can impinge upon a nerve or occlude a vessel. X-ray is also valuable in locating neoplasms and other pathologies that may be involved with the condition.

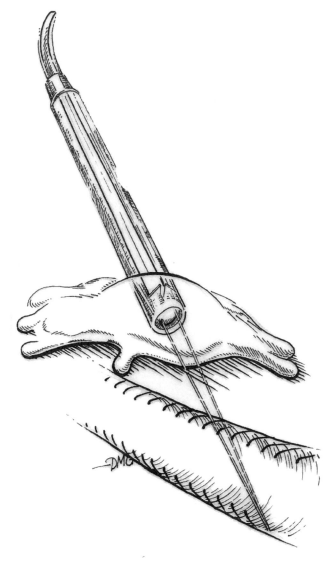

Schematic of Doppler

References

- Adson AW. Cervical ribs: Symptoms, differential diagnosis and indications for section of the scalenus anticus muscle. J Int Coll Surgeons. 1951;16(5):546-559.
- Aguayo AJ. Neuropathy due to compression and entrapment. In: Peripheral Neuropathy, Vol I. Eds. Dyck PJ, Thomas PK, Lambert EH. W.B. Saunders Co.:Philadelphia; 1975.
- Akuthota A, Herring SA (Eds). Nerve and Vascular Injuries in Sports Medicine. Springer: Dordrecht; 2009.
- Alis GP. Medial epicondylitis: a case study. Collected Papers International College of Applied Kinesiology, 2003-2004;1:3-4.
- American Medical Association: Guides to the Evaluation of Permanent Impairment, 5th Ed. 2001:510.
- Amadio PC. Discussion on carpal tunnel syndrome and vitamin B6. (article by Kasdan & Janes) Plastic Reconstruct Surg. 1987;79.
- Asbury AK. Ischemic disorders of peripheral nerve. In: Handbook of Clinical Neurology, Vol 8. Ed. Vinken PJ, Bruyn GW. Elsevier Pub Co.: New York; 1970.
- Barral JP, Croibier A. Manual therapy for the peripheral nerves. Churchill-Livingstone: Edinburgh; 2007.
- Barral JP. Manual thermal diagnosis. Eastland Press, Inc.: Seattle; 1996.
- Bastron JA. Neuropathy in diseases of the thyroid. In: Peripheral Neuropathy, Vol II. Ed Dyck PJ, Thomas PK, Lambert EH. W.B. Saunders Co.: Philadelphia; 1975.
- Benya R, Quintana J, Brundage B. Adverse reactions to indocyanine green: a case report and a review of the literature. Cathet Cardiovasc Diagn. 1989;17(4):231-3.
- Blaich RM. Manual and machine muscle testing before and after correction of 'respiratory faults. In: Selected Papers of the International College of Applied Kinesiology. Ed Markham BC. ICAK USA: Detroit, MI;1981.
- Blaich RM,. Mendenhall EI. Manual muscle testing and Cybex machine muscle testing, a search for a correlation. In: Selected Papers of the International College of Applied Kinesiology (1984). Ed. Schmitt WH, Jr. ICAK USA: Park City, UT; 1984.
- Blum CL, Cuthbert S. Cranial Therapeutic Care: Is There any Evidence? Chiropractic & Osteopathy 2006;14:10.
- Breig A. Adverse Mechanical Tension in the Central Nervous System. John Wiley & Sons: New York; 1978.
- Breig A, Tumbull I, Hassler, O. Effects of mechanical stresses on the spinal cord in cervical spondylosis. A study of fresh cadaver material. J Neurosurg 1966;25:45-56.
- Bronston LJ, Larson KM. A new approach to upper rib fixation mobilization. Chiropr Tech. 1990;2(2):39-44.
- Browning JE. Distractive manipulative protocols in treating the mechanically induced pelvic pain and organic dysfunction patient. Chiropractic Tech 1995;7(1):1-11.
- Brumagne S, Cordo P, Lysens R, Verschueren S, Swinnen S. The role of paraspinal muscle spindles in lumbosacral position sense in individuals with and without low back pain. Spine. 2000;25(8):989-994.
- Budgell B, Sato A. The cervical subluxation and regional blood flow. J Manipulative Physiol Ther. 1997;20(2):103-107.
- Burstein EM. Raynaud's phenomenon secondary to scleroderma: a case study. Collected Papers International College of Applied Kinesiology. 1989-1990;1:17-19.
- Butler DS. The Sensitive Nervous System. Noigroup: Adelaide; 2000.
- Butler DS. Mobilisation of the Nervous System. Churchill-Livingstone: Melbourne, 1991.
- Chaitow L, et al. Naturopathic Physical Medicine: Theory and Practice for Manual Therapists and Naturopaths. Churchill Livingstone: Edinburgh; 2008.
- Chaitow L. Cranial Manipulation: Theory and Practice. Churchill-Livingstone; 2005.
- Chaitow L. Soft-Tissue Manipulation. Healing Arts Press: Rochester, VT; 1987.
- Charash R. The human factor: Tools need not harm the hands that use them. Industrial Design Mag; 1982.
- Chien AJ, Jamadar DA, Jacobson JA, Hayes CW, Louis DS. Sonography and MR imaging of posterior interosseous nerve syndrome with surgical correlation. AJR. 2003;181:219–221.
- Christiansen JA, Meyer JJ. Altered metabolic enzyme activities in fast and slow twitch muscles due to induced sciatic neuropathy in the rat. J Manip Physiol Ther. 1987;10(5):227-231.
- Cohen JS. Overdose: The case against the drug companies. Prescription drugs, side effects, and your health. Putnam: New York; 2001.
- Conable K, Corneal J, Hambrick T, Marquina N, Zhang J. Investigation of Methods and Styles of Manual Muscle Testing by AK Practitioners. J Chiropr Med. 2005;4(1):1-10.
- Coq JO, Xerri C. Tactile impoverishment and sensorimotor restriction deteriorate the forepaw cutaneous map in the primary somatosensory cortex of adult rats. Exp Brain Res. 1999;129(4):518-531.
- Cramer GD, Darby SA. Basic and Clinical Anatomy of the Spine, Spinal Cord, and ANS. Mosby-Yearbook: St. Louis; 1995.
- Crymble B. Brachial neuralgia and the carpal tunnel syndrome. Br Med J. 1968;3(5616):470-471.
- Cseuz KA, et al. Long—term results of operation for carpal tunnel syndrome. Mayo Clin Proc. 1966;41(4):232-241.
- Cuthbert S, Rosner A, McDowall D. Association of manual muscle tests and mechanical neck pain: Results from a prospective pilot study. J Bodyw Mov Ther. 2011;15(2):192-200.
- Cuthbert S, Rosner A. Applied Kinesiology Management of Long-Term Head Pain Following Automotive Injuries: A Case Report. Chiropr J Aust. 2010;40:109-16.
- Cuthbert SC, Barras M. Developmental delay syndromes: psychometric testing before and after chiropractic treatment of 157 children. J Manipulative Physiol Ther. 2009;32(8):660-9.
- Cuthbert SC, Goodheart GJ Jr. On the reliability and validity of manual muscle testing: a literature review. Chiropr Osteopat. 2007;15:4.
- Cuthbert S, Blum C. Symptomatic Arnold-Chiari malformation and cranial nerve dysfunction: a case study of applied kinesiology cranial evaluation and treatment. J Manipulative Physiol Ther. 2005;28(4):e1-6.
- Cuthbert S. The piriformis muscle and the genitor-urinary system: The anatomy of the muscle-organ-gland correlation. Int J AK and Kinesio Med. 2003;15.
- Cyriax J. Fibrositis. Br Med J. 1948;2(4569):251-5.
- Dahlin LB, Nordborg C, Lundborg G. Morphologic changes in nerve cell bodies induced by experimental graded nerve compression. Exp Neurol. 1987;95(3):611-21.

- Dahlin LB, Sjöstrand J, McLean WG. Graded inhibition of retrograde axonal transport by compression of rabbit vagus nerve. J Neurol Sci. 1986;76(2-3):221-30.
- Dahlin LB, Rydevik B, McLean WG, Sjöstrand J. Changes in fast axonal transport during experimental nerve compression at low pressures. Exp Neurol. 1984;84(1):29-36.
- Daniels L, Worthingham K: Muscle Testing – Techniques of Manual Examination, 7th Edition. Philadelphia, PA: W.B. Saunders Co.; 2002.
- Danneels LA, Vanderstraeten GG, Cambier DC, et al. CT imaging of trunk muscles in chronic low back pain patients and healthy control subjects. Eur Spine J. 2009;9(4):266-272.
- Dauphine DB. Case report of applied kinesiology treatment and allergic cutaneous vasculitis. Collected Papers International College of Applied Kinesiology. 1993-1994;1:217-221.
- Dawson DM, Hallett M, Millender LH. Entrapment Neuropathies. Little, Brown & Co: Boston; 1983.
- DeJarnette MB. Collected Writings, 1929-1984. www.soto-usa.com; http://www.sorsi.com/products-services.
- Denny-Brown D, Brenner C. The effect of percussion of nerve. J Neurol Neurosurg Psychiatry. 1944;7(3-4):76-95.
- Devor M, Seltzer Z. Pathophysiology of damaged nerves in relation to chronic pain. In: The Textbook of Pain. Eds. Wall PD, Melzack R. Churchill Livingstone: Edinburgh;1999.
- Drugs.com www. Literature produced by the pharmaceutical companies on drug side-effects.
- Dyck PJ, Thomas PK. Peripheral neuropathy, Vol 2, 3rd Ed. WB Saunders: Philadelphia; 1993:513-567.
- Edwards FO. Pyriformis syndrome. Acad Applied Osteopath Yr Bk; 1962.
- Einaudi G. Research on vasomotor disorders of the upper extremities present in patients with cervicalgia & cervicobrachialgia; changes induced by static variations of the cervical spine. Reumatismo. 1959;11(3):173-8.
- Evans RC. Illustrated Orthopedic Physical Assessment, 3rd Ed. Mosby; 2008.
- Fernandez-de-las-Penas C, Albert-Sanchis JC, Buil M, et al. Cross-sectional area of cervical multifidus muscle in females with chronic bilateral neck pain compared to controls. J Orthop Sports Phys Ther. 2008;38(4):175-180.
- Figar S, Krausova L. A plethysmographic study of the effects of chiropractic treatment in vertebrogenic syndromes. Acta Universitatis Carolinae. Supp 21; 1965.
- Figar S, Krausova L, Lewit K. Plethysmographic examination following treatment of vertebrogenic disorders by manipulations. (German translation), Acta Neuroveg. 1967;29.
- Fullerton PM. The effect of ischaemia on nerve conduction in the carpal tunnel syndrome. J Neurol Neurosurg Psychiatry. 1963;26:385-397.
- Gaist D, Jeppesen U, Andersen M, García Rodríguez LA, Hallas J, Sindrup SH. Statins and risk of polyneuropathy: a case-control study. Neurology. 2002;58(9):1333-7.
- Gardner WJ. Trigeminal neuralgia. Clin Neurosurg. 1967;15.
- Garten H. Lehrbuch Applied Kinesiology Muskelfunktion-Dysfunktion-Therapie. (in German) Urban & Fischer: Munich; 2004.
- Gasser HS, Erlanger J. The role of fiber size in the establishment of a nerve block by pressure or cocaine. Am J Physiol. 1929;88.
- Gelberman RH, Hergenroeder PT, Hargens AR, Lundborg GN, Akeson WH. The carpal tunnel syndrome: A study of carpal canal pressures. J Bone Joint Surg. 1981;63(3):380-383.
- Gerz W. Lehrbuch der Applied Kinesiology (AK) in der naturheilkundlichen Praxis. (in German) AKSE-Verlag: Munchen; 2001.
- Giggey K, Tepe R. A pilot study to determine the effects of a supine sacroiliac orthopedic blocking procedure on cervical spine extensor isometric strength. J Chirop Med. 2009;8(2):56-61.
- Gilliatt RW, Wilson TG. A pneumatic–tourniquet test in the carpal—tunnel syndrome. Lancet. 1953;265(6786):595-597.
- Gilliatt R., Wilson TG. Ischaemic sensory loss in patients with peripheral nerve lesions. J Neurol Neurosurg Psychiatry. 1954;17(2):104-114.
- Goodheart GJ, Jr. The carpal tunnel syndrome. Chiro Econ. 1967;10(1).
- Goodheart GJ, Jr. Tarsal tunnel syndrome. Chiro Econ. 1971;13(5).
- Goodheart GJ, Jr. Sacroiliac and ilio sacral problems, Part 2. Chiro Econ. 1972;15(1).
- Goodheart GJ, Jr. Applied Kinesiology Research Manuals. Applied Kinesiology Systems, Inc: Grosse Point Farms, MI; 1964-1998.
- Goodman CE. Unusual nerve injuries in recreational activities. Am J Sports Med. 1983;11(4):224-227.
- Granit R, Skoglund CR. Facilitation, inhibition and depression at the ;artificial synapse' formed by the cut end of a mammalian nerve. J Physiol. 1945;103(4):435-48.
- Greenawalt MH. Feet and the dynamic science of chiropractic. Chiro Econ. 1988;30(6).
- Guyton AC, Hall JE. Textbook of Medical Physiology, 11th ed. WB Saunders Co: Philadelphia; 2005.
- Halasi T, Kynsburg A, Tállay A, Berkes I. Changes in joint position sense after surgically treated chronic lateral ankle instability. Br J Sports Med. 2005;39(11):818-24.
- Haldeman S, Meyer BJ. The effect of constriction on the conduction of the action potential in the sciatic nerve. S Afr Med J. 1970;44(31):903-906.
- Hammer W. Functional Soft-Tissue Examination and Treatment by Manual Methods, 3rd Ed. Jones and Bartlett Publishers, Inc.; 2007.
- Harms-Ringdahl K, Ed.. Muscle Strength. Edinburgh: Churchill Livingstone; 1993.
- Heidrich JM. Case history: Dupuytren's contracture and cervical disc. Collected Papers International College of Applied Kinesiology, 1994-1995;1:11-13.
- Heidrich JM. Atypical fibular head subluxation in a case of lower leg neuritis. Collected Papers International College of Applied Kinesiology, 1992-1993;1:73-74.
- Helm PA, Nepomuceno C, Crane CR. Tibial nerve dysfunction during pregnancy. South Med J. 1971;64(12):1493-4.
- Hess K, et al. Acute ischaemic neuropathy in the rabbit. J Neurol Sci. 1979;44(1):19-43
- Hides JA, Stokes MJ, Saide M, et al. Evidence of lumbar multifidus muscle wasting ipsilateral to symptoms in patients with acute/subacute low back pain. Spine 1994;19(2):165-172.
- Higashi Y, Yoshizumi M. New methods to evaluate endothelial function: method for assessing endothelial function in humans using a strain-gauge plethysmography: nitric oxide-dependent and -independent vasodilation. J Pharmacol Sci. 2003;93(4):399-404.
- Hirasawa Y, Sakaida K. Sports and peripheral nerve injury. Am J Sports Med. 1983;11(6):420-426.
- Hogg JDW. The efficacy of applied kinesiology protocols in correcting peripheral nerve entrapment associated with carpal tunnel syndrome: An inter-examiner study. Collected Papers International College of Applied Kinesiology. 1994-1995;1:169-173.
- Hong C-Z, Simons DG. Response to treatment for pectoralis minor myofascial pain syndrome after whiplash. J Musculoskeletal Pain. 1993;1(1):89-131.
- Hoppenfeld S. Physical Examination of the Spine and Extremities. Appleton–Century–Crofts: New York; 1976.

- Hurst LC, Weissberg D, Carroll RE. The relationship of the double crush to carpal tunnel syndrome (an analysis of 1,000 cases of carpal tunnel synrome). J Hand Surg. 1985;10(2):202-204.
- Ian'shina EN, Liubchenko PN, Ian'shin NP, Kasatkina LF, Samoĭlov MI. Multiple local involvement of peripheral nerves in workers suffering from occupational hands disorders. Med Tr Prom Ekol. 2009;(2):24-8.
- ICAK USA and ICAK International websites. www.icakusa.com; www.icak.com.
- Idler RS. Persistence of symptoms after surgical release of compressive neuropathies and subsequent management. Orthop Clin North Am. 1996;27(2):409-16.
- Janda V: Muscle Function Testing. Butterworths: London; 1983.
- Johnson RH, Lambia DG, Spalding JMK. Neurocardiology: The interrelationship between dysfunction in the nervous and cardiovascular systems. WB Saunders: London; 1984.
- Juhan D. Job's Body. Station Hill Press: New York; 1987.
- Jull G, Sterling M, Falla D, Treleaven J, O'Leary S. Whiplash, Headache, and Neck Pain. Ch. 4: Alterations in Cervical Muscle Function in Neck Pain. Churchill Livingstone: Edinburgh; 2008:41-58.
- Kendall FP, McCreary EK, Provance PG, Rodgers MM, Romani WA. Muscles: Testing and Function, with Posture and Pain. Williams & Wilkins: Baltimore; 2005.
- Kharrazian D. The role of dermal proprioceptors in reactive muscle patterns. Collected Papers International College of Applied Kinesiology, 2001-2002;1:129-130.
- Kharrazian D. The role of the transverse ligament in suprascapular nerve entrapments. Collected Papers International College of Applied Kinesiology, 2000-2001;1:81-83.
- Kline D, Hudson A. Kline and Hudson's Nerve Injuries: Operative Results for Major Nerve Injuries, Entrapments and Tumors, 2nd Ed. Saunders; 2007.
- Kobesova A. Twenty-year-old pathogenic "active" postsurgical scar: a case study of a patient with persistent right lower quadrant pain. J Manip Physiol Ther. 2007;30(3):234-238.
- Kolosova LI, Nozdrachev AD, Moiseeva AB, Ryabchikova OV, Turchaninova LN. Recovery of mechanoreception at the initial stage of regeneration of injured sciatic nerve in rats in conditions of central axotomy of sensory neurons. Neurosci Behav Physiol. 2004;34(8):817-20.
- Kopell HP, Thompson WAL. Peripheral Entrapment Neuropathies, 2nd Ed. Robert E. Krieger Pub Co: Huntington, NY; 1976.
- Kopell HP. Lower extremity lesions. In: Management of Peripheral Nerve Problems. Eds. Omer GE, Spinner M. W.B. Saunders Co: Philadelphia; 1980.
- Korr IM. The spinal cord as organizer of disease processes: Some preliminary perspectives. J Am Osteopath Assoc. 1976;76(1):35-45.
- Korting GW, Denk R. Differential Diagnosis in Dermatology. WB Saunders Co.: Philadelphia; 1976.
- Langer P. Great Feet for Life: Footcare and Footwear for Healthy Aging. Fairview Press: Minneapolis, MN; 2007.
- Leaf D. Applied Kinesiology Flowchart Manual, 4th Ed. Privately Published: David W. Leaf: Plymouth, MA; 2010.
- Lee D. The Pelvic Girdle: An approach to the examination and treatment of the lumbopelvic-hip region. Churchill Livingstone: Edinburgh; 2004.
- Leinonen V, Kankaanpää M, Luukkonen M, Kansanen M, Hänninen O, Airaksinen O, Taimela S. Lumbar paraspinal muscle function, perception of lumbar position, and postural control in disc herniation-related back pain. Spine. 2003;28(8):842-8.
- Lewit K: Manipulative Therapy in Rehabilitation of the Locomotor System, 3rd ed. London: Butterworths; 1999.
- Liebenson C. Ed: Rehabilitation of the Spine: A Practitioner's Manual, 2nd ed. Lippincott, Williams & Wilkins: Philadelphia; 2007.
- Loeser JD. Pain due to nerve injury. Spine. 1985;10(3):232-235.
- Logan AL. The Foot and Ankle: Clinical Applications. Gaithersburg, MD: Aspen Publishers, Inc; 1995.
- Logan AL. The Knee: Clinical Applications. Gaithersburg, MD: Aspen Publishers, Inc; 1994.
- Lombardini R, Marchesi S, Collebrusco L, Vaudo G, Pasqualini L, Ciuffetti G, Brozzetti M, Lupattelli G, Mannarino E. The use of osteopathic manipulative treatment as adjuvant therapy in patients with peripheral arterial disease. Man Ther. 2009;14(4):439-43.
- Lundborg G. Nerve injury and repair. Churchill Livingstone: Edinburgh; 1988.
- Lundborg G. Intraneural microcirculation. Orthop Clin North Am. 1988;19(1):1-12
- Lundborg G. Structure and function of the intraneural microvessels as related to trauma, edema formation, and nerve function. J Bone Joint Surg Am. 1975;57(7):938-948..
- Lusskin R. Peripheral neuropathies affecting the foot: Traumatic, ischemic, and compressive disorders. In: Disorders of the Foot, Vol 2. Ed. Jahss MH. W.B. Saunders Co: Philadelphia; 1982.
- Luttges MW, Stodieck LS, Beel JA. Postinjury changes in the biomechanics of nerves and roots in mice. J Manip Physiol Ther. 1986;9(2):89-98.
- Maffetone P. Fix Your Feet. The Lyons Press: Guilford, CT; 2003.
- Maffetone P. Complementary Sports Medicine: Balancing traditional and nontraditional treatments. Human Kinetics: Champaign, IL; 1999.
- Maitland G. Maitland's vertebral manipulation, 6th Ed. Butterworth-Heinemann: Oxford; 2001.
- Marmor L. Bowler's thumb. J Trauma. 1966;6(2):282-4.
- McGrath MA, Verhaeghe RH, Shepherd JT. The physiology of limb blood flow. In: Peripheral Vascular Diseases, 5th ed. Eds. Juergens J, Spittel JA, Fairbairn J, II. W.B. Saunders Co: Philadelphia; 1980.
- Mc Knight ME, DeBoer KF. Preliminary study of blood pressure changes in normotensive subjects undergoing chiropractic care. J Manip Physiol Ther. 1988;11(4):261-266.
- Mense S, Simons DG. Muscle Pain: Understanding Its Nature, Diagnosis, and Treatment. Philadelphia, PA: Lippincott Williams & Wilkins; 2001:131-157.
- Moberg E. Objective methods for determining the functional value of sensibility in the hand. J Bone Joint Surg. 1958;40B(3):454-76.
- Moller AR. The cranial nerve vascular compression syndrome. Rev Treat Acta Neurochir. 1991;113(1-2):18-23.
- Moncayo R, Moncayo H. Evaluation of Applied Kinesiology meridian techniques by means of surface electromyography (sEMG): demonstration of the regulatory influence of antique acupuncture points. Chin Med. 2009;4:9.
- Mondelli M, de Stefano R, Rossi S, Aretini A, Romano C. Sympathetic skin response in primary Raynaud's phenomenon. Clin Auton Res. 2009;19(6):355-62.
- Mooney V, Pozos R, Vleeming A, Gulick J, Swenski D. Exercise treatment for sacroiliac pain. Orthopedics. 2001;24(1):29-32.
- Muir JM, Vernon H. Complex regional pain syndrome and chiropractic. J Manipulative Physiol Ther. 2000;23(7):490-7.
- Murphy EL. Observations on craft palsies. Trans Ass Industr Med Offrs. 1955;5/1.
- Myers JB, Lephart SM. Sensorimotor deficits contributing to glenohumeral instability. Clin Orthop. 2002;400:98-104.
- Myers TW. Anatomy Trains: Myofascial Meridians for Manual and Movement Therapists. Churchill-Livingstone; 2001.
- Myers R, Powell H. Endoneurial fluid pressure in peripheral neuropathies. In: Hargens A (ed). Tissue fluid pressure and composition. Williams & Wilkins, Baltimore;1981.

- National Research Council, Institute of Medicine. Musculoskeletal Disorders and the Workplace. National Academy Press: Washington, DC; 2001.
- Neary D, Ochoa J, Gilliatt RW. Sub—clinical entrapment neuropathy in man. J Neurol Sci. 1975;24(3):283-298.
- Nicholas JA, Strizak AM, Veras G. A study of thigh muscle weakness in different pathological states of the lower extremity. Am J Sports Med. 1976;4(6):241-248.
- Nitz AJ, Dobner JJ. Upper extremity tourniquet effects in carpal tunnel release. J Hand Surg Am. 1989;14(3):499-504.
- Ochoa J. Nerve fiber pathology in acute and chronic compression. In: Management of Peripheral Nerve Problems. Ed. Omer GE, Spinner M. W.B. Saunders Co: Philadelphia; 1980.
- O'Riain S. New and simple test of nerve function in hand. Br Med J. 1973;3(5881):615-616.
- Page P, Frank CC, Lardner R. Assessment and treatment of muscle imbalance: The Janda Approach. Human Kinetics: Champaign, IL; 2010.
- Passatore M, Grassi C, Filippi GM. Sympathetically-induced development of tension in jaw muscles: the possible contraction of intrafusal muscle fibres. Pflugers Arch. 1985;405(4):297-304.
- Patten J. Neurological Differential Diagnosis, 2nd Ed. Springer-Verlag: Berlin; 1995.
- Phalen GS. The carpal–tunnel syndrome — clinical evaluation of 598 hands. Clin Orthop Relat Res. 1972;83:29-40.
- Pham K, Gupta R. Understanding the mechanisms of entrapment neuropathies. Review article. Neurosurg. Focus. 2009;26(2):E7.
- Punnett L, et al. Soft tissue disorders in the upper limbs of female garment workers. Scand J Work Environ Health. 1985;11(6):417-425.
- Punnett L, Keyserling WM. Exposure to ergonomic stressors in the garment industry: Application and critique of job–site work analysis methods. Ergonomics. 1987;30(7):1099-116
- Racinais S, Bringard A, Puchaux K, Noakes TD, Perrey S. Modulation in voluntary neural drive in relation to muscle soreness. Eur J Appl Physiol. 2008;102(4):439-46.
- Ramsak I. AK diagnosis and nosode therapy. Medical Journal for Applied Kinesiology. 1997;1:12-14.
- Retzlaff EW, et al. The piriformis muscle syndrome. J Am Osteopath Assoc.. 1974;73(10):799-807.
- Ritts GD, Wood MB, Linscheid RL. Radial tunnel syndrome — a ten-year surgical experience. Clin Orthop. 1987;219:201-205.
- Rob CG, Standeven A. Arterial occlusion complicating thoracic outlet compression syndrome. Br Med J. 1958;2(5098):709-712.
- Robinson DR. Pyriformis syndrome in relation to sciatic pain. Am J Surg. 1947;73(3):355-358.
- Russell SM. Examination of Peripheral Nerve Injuries: An Anatomical Approach. Thieme Medical Publishers, Inc.: New York; 2006.
- Rydevik B, McLean WG, Sjöstrand J, Lundborg G. Blockage of axonal transport induced by acute, graded compression of the rabbit vagus nerve. J Neurol Neurosurg Psychiatry. 1980;43(8):690-8.
- Saal JA, et al. The pseudoradicular syndrome — lower extremity peripheral nerve entrapment masquerading as lumbar radiculopathy. Spine. 1988;13(8):926-930.
- Saratsiotis J, Myriokefalitakis E. Diagnosis and treatment of posterior interosseous nerve syndrome using soft tissue manipulation therapy: a case study. J Bodyw Mov Ther. 2010;14(4):397-402.
- Robertson C, Saratsiotis J. A review of compressive ulnar neuropathy at the elbow. J Manipulative Physiol Ther. 2005;28(5):345.
- Schmitt WH, Cuthbert SC. Common errors and clinical guidelines for manual muscle testing: The "arm test" and other inaccurate procedures. Chiropr Osteopat. 2008;16:16.
- Schmitt WH, Jr. Muscle testing as functional neurology differentiating functional upper motorneuron and functional lower motorneuron problems. 1986 Selected Papers of the International College of Applied Kinesiology. ICAK: Shawnee Mission, KS; 1986.
- Schmitt WH. But what if there's no water in the hose? Collected Papers International College of Applied Kinesiology. 1986;Winter:125-144.
- Selander D, Månsson LG, Karlsson L, Svanvik J. Adrenergic vasoconstriction in peripheral nerves of the rabbit. Anesthesiology. 1985;62(1):6-10.
- Sharma L. Proprioceptive impairment in knee osteoarthritis. Rheum Dis Clin N Am. 1999; 25(2):299-314.
- Simons D, Travell J, Simons L. Myofascial pain and dysfunction: the trigger point manual: Upper half of body. Vol. 1, 2nd Ed. Williams & Wilkins: Baltimore; 1998.
- Skaggs CD, Winchester BA, Vianin M, Prather H. A manual therapy and exercise approach to meralgia paresthetica in pregnancy: a case report. J Chiropr Med. 2006;5(3):92-6.
- Smith MW, Marcus PS, Wurtz LD. Orthopedic issues in pregnancy. Obstet Gynecol Surv. 2008;63(2):103-11.
- Spaans F. Occupational nerve lesions. In: Handbook of Clinical Neurology, Vol 7. Eds. Vinken PJ, Bruyn GW. American Elsevier Pub Co: New York; 1970.
- Spinner M, Spencer PS. Nerve compression lesions of the upper extermity — a clinical and experimental review. Clin Orthop. 1974;104.
- Staal A, van Gijn J, Spaans, F. Mononeuropathies: Examination, Diagnosis and Treatment. WB Saunders: London; 1999.
- Steiner C, Staubs C, Ganon M, Buhlinger C.Piriformis syndrome: pathogenesis, diagnosis, and treatment. J Am Osteopath Assoc. 1987;87(4):318-323.
- Sternschein MJ, et al. Causalgia. Arch Phys Med Rehabil. 1975;56(2):58-63.
- Stux G, Pomeranz B. Acupuncture: Textbook and Atlas. Springer-Verlag: Berlin; 1987.
- Sunderland S. Nerves and Nerve Injuries, 2nd Ed. Churchill Livingstone: Melbourne; 1978.
- Sunderland S. Traumatic injuries of peripheral nerves. I — Simple compression injuries of the radial nerve. Brain. 1945;68:243-99.
- Sutherland WG. Contributions of Thought: The Collected Writings of William Garner Sutherland, D.O. Rudra Press: Portland, OR; 1998.
- Swanson S, Macias LH, Smith AA. Treatment of bowler's neuroma with digital nerve translocation. Hand (NY). 2009;4(3):323-6.
- TePoorten BA. The piriformis muscle. JAOA. 1969;69(2):150-160.
- Toth C. Peripheral nerve injuries attributable to sport and recreation. Neurol Clin. 2008;26(1):89-113.
- Travell JG, Simons DG. Myofascial Pain and Dysfunction: The Trigger Point Manual: The Upper Extremities. Williams & Wilkins: Baltimore; 1983.
- Travell JG, Simons DG. Myofascial Pain and Dysfunction: The Trigger Point Manual: The Lower Extremities. Williams & Wilkins: Baltimore; 1992.
- Triano JJ, Luttges MW. Subtle, intermittent mechanical irritation of sciatic nerves in mice. J Manip Physiol Ther. 1980;3(2).
- Upton ARM, McComas AJ. The double crush in nerve–entrapment syndromes. Lancet. 2(7825):359-362;1973.
- Veves A, Giurini JM, LoGerfo FW. The Diabetic Foot: Medical and Surgical Management. Humana Press: Totowa, NJ; 2002.
- Walsh MJ, Polus BI. The frequency of positive common spinal clinical examination findings in a sample of premenstrual syndrome sufferers. J Manipulative Physiol Ther 1999;22(4):216-220.
- Walther DS. Applied Kinesiology, Volume I — Basic Procedures and Muscle Testing (Pueblo, CO: Systems DC, 1981).

- Walther DS. Upper extremity peripheral nerve entrapment. Selected Papers of the International College of Applied Kinesiology, ed W.H. Schmitt, Jr. ICAKUSA: Detroit, MI; 1982.
- Walther DS. Applied Kinesiology, Volume II — Head, Neck, and Jaw Pain and Dysfunction — The Stomatognathic System. Systems DC: Pueblo, CO; 1983.
- Walther DS. Applied Kinesiology — Synopsis, 2nd Ed. International College of Applied Kinesiology: Shawnee Mission, KS; 2000.
- Watanabe N, Polus B. A single mechanical impulse to the neck: Does it influence autonomic regulation of cardiovascular function? Chiropr J Aust. 2007;37(2):42-48.
- Werder SF. Managing polypharmacy: Walking the fine line between help and harm. Current Psychiatry. 2003;2(2):1.
- Wetz HH, Hentschel J, Drerup B, Kiesel L, Osada N, Veltmann U. Changes in shape and size of the foot during pregnancy. Orthopade. 2006;35(11):1124, 1126-30.
- Wilgis EFS. Observations on the effects of tourniquet ischemia. J Bone Joint Surg. 1971;53(7):1343-1346.
- Wilgis EFS. Special diagnostic studies. In: Management of Peripheral Nerve Problems. Eds. Omer GE, Spinner M. W.B. Saunders Co: Philadelphia; 1980.
- Wilkins RH, Brody IA. Tinel's sign. Arch Neurol. 1971;24(6).
- Wilson JL. Adrenal Fatigue: The 21st Century Stress Syndrome. Smart Publications; 2002.
- Wright HM, Sympathetic activity in facilitated segments: vasomotor studies. J Am Osteop Assoc. 1955;54(5):273-276.
- Wright HM. The origin and manifestations of local vasomotor disturbances and their clinical significance. J Am Osteop Assoc. 1956;56(4):217-224.
- Zhao C, Ettema AM, Berglund LJ, An KN, Amadio PC. Gliding resistance of flexor tendon associated with carpal tunnel pressure: a biomechanical cadaver study. J Orthop Res. 2011;29(1):58-61.

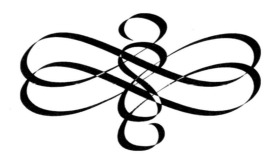

> "The sciatic nerve innervates the knee flexors and all the muscles below the knee. Therefore, complete palsy of the sciatic nerve would lead to marked instability of the foot and to severe impairment of gait because of knee and ankle impairment."
> – (Dawson et al., 1983)
>
> "Sometimes patients will have symptoms but will show no evidence of I.V.F. narrowing - WHY?"
> – George J. Goodheart, Jr.

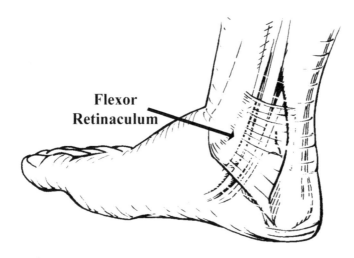

Flexor Retinaculum

CHAPTER THREE
Peripheral Nerve Entrapment Lower Extremity

Extraspinal peripheral nerve entrapment of the lower extremity is often misdiagnosed as lumbar or lumbosacral root compression, or a disturbance of the autonomic nervous system. **(Akuthota & Herring, 2009; Russell, 2006; Ruch, 2001; Staal et al., 1999)** Most nerves that go to the extremities have autonomic and afferent fibers, the median and the sciatic nerves particularly. The frequency of radiculopathy and the primary attention that a manipulative physician gives to the spine may preclude consideration of the more distal origins of nerve dysfunction. Further, many physicians — especially those who have been in practice for several years — may not have received adequate education about peripheral nerve entrapment in their initial training.

This chapter emphasizes differential diagnosis of lower extremity peripheral nerve entrapment, with particular attention to the level of involvement and local correction. The condition may be a double crush combination of spinal involvement and a more distal peripheral nerve entrapment. If only the spinal column is corrected, which includes dural tension technique, and the peripheral entrapment is ignored, the patient will improve but the condition will not be completely corrected.

Lateral Femoral Cutaneous Nerve

The lateral femoral cutaneous nerve arises from the 2nd and 3rd lumbar nerves. It is formed in the psoas muscle (**Moore et al., 2009**) and emerges from its lateral border to cross the iliacus and exit the pelvis. The most common point considered for possible entrapment is as the nerve passes between the two slips of the inguinal ligament's lateral attachment to the anterior superior iliac spine, where it exits the pelvis. The nerve is tightly bordered by the tendinous fibers of the inguinal ligament at this point and makes a right-handed bend to change direction from a horizontal course in the pelvis to a more vertical course in the lateral and anterolateral thigh. The lower slip of the inguinal ligament also gives origin to some sartorius fibers. (**Staal et al., 1999; Keegan & Holyoke, 1962**) The nerve may pass in front of or through the sartorius into the thigh. (**Gray's Anatomy, 2004**)

Shortly after it leaves the abdomen, the nerve divides into an anterior and a posterior branch. The anterior branch supplies the skin of the anterior and lateral parts of the thigh to the knee. The posterior branch supplies the skin on the lateral surface of the thigh, from the greater trochanter to about the middle of the thigh. The lateral femoral cutaneous nerve is strictly sensory, giving no motor supply.

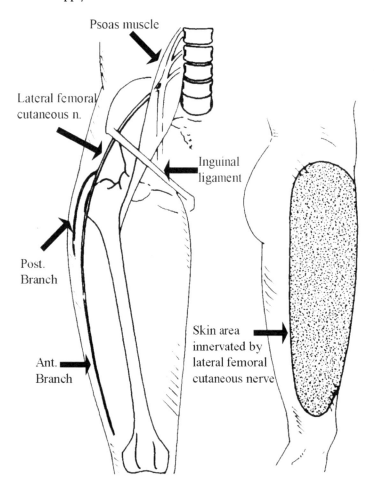

Lateral femoral cutaneous nerve.

As the nerve reaches a point just medial to the anterior superior iliac spine where it enters the thigh, it changes its course from nearly horizontal to vertical. This angulation is increased by extension and lessened by flexion of the thigh. In addition to this major point of possible entrapment, other locations may be where the nerve emerges from the psoas muscle, (**Skaggs et al., 2006**) the passage underneath the fascia lata, and the area where the nerve leaves the fascia. (**Staal et al., 1999; Biemond, 1970**) After the lateral femoral cutaneous nerve exits the pelvis at the opening of the inguinal ligament, it is held down as it pierces the fascia lata; thus, movement between the thigh and pelvis can stretch the nerve and increase entrapment at the opening of the inguinal ligament. (**Kopell & Thompson, 1976**)

Entrapment is prone to occur in obesity, with a lax abdominal wall, and in pregnancy. (**Harney & Patijn, 2007; Staal et al., 1999; Stites, 1986**) Diabetics have an increased susceptibility to compression neuropathy, which is especially apparent in this nerve. (**Veves et al., 2002; Staal et al., 1999; Aguayo, 1975**)

Nerve entrapment causes pain called meralgia paresthetica (Barnhardt-Roth syndrome) in the anterolateral thigh. (**Harney & Patijn, 2007; Pearce, 2006**) Interestingly, Sigmund Freud reported that he and one of his sons suffered from the condition. (**Freud, 1895**) The condition affects men more than women due to possible occupational considerations and can also be bilateral in some 25 percent of cases. (**Sunderland, 1978**) It most commonly develops from the nerve's fascial attachment in the thigh pulling the nerve tightly against the opening at the lateral end of the inguinal ligament, (**Staal et al., 1999; Kopell & Thompson, 1976, 1960**) which usually relates to thigh obesity and/or a lateral shift in trunk posture. (**Skaggs et al., 2006; Kopell, 1980**) Generally, meralgia paresthetica develops without prior trauma.

Symptoms consist of increased (**Staal et al., 1999; Aguayo, 1975**) or decreased (**Harney & Patijn, 2007; Kopell & Thompson, 1960**) sensitivity. Rubbing of clothing and other cutaneous stimuli causes a characteristic burning pain on the outer side of the thigh. Pelvic and hip motion, such as walking and running, aggravates the pain. In general, those with meralgia paresthetica have poor posture; consequently, the postural muscles are overactive with prolonged standing, which also aggravates pain. The postural deficiency most common with meralgia paresthetica is lumbar hyperlordosis and a protuberant abdomen. (**Russell, 2006; Skaggs et al., 2006; Massey, 1977; Bellis, 1977**) Applied kinesiology examination usually finds weak abdominal and gluteus maximus muscles failing to provide support to prevent anterior rotation of the pelvis, with subsequent hyperlumbarlordosis. Aggravating the condition will probably be hypertonic lumbar extensor muscles and shortening of the iliopsoas. Sitting and recumbent positions relieve the pain; however, sitting with one leg crossed over the other — especially the ankle on the knee — can exacerbate the pain. Postural shifts by a short leg, whether physiological or anatomical, may be a factor in the condition. Other factors include the muscle fibers from the internal oblique and transverses abdominus muscles originating from the inguinal ligament, or the external oblique muscle inserting into it. (**Keegan & Holyoke, 1962**) Additional causes have been reported, including bodybuilding, (**Szewczyk et al., 1994**) falling asleep in siddha yoga position, (**Mattio et al., 1992**) seat-belt (**Beresford, 1971**) and pocket watch trauma, (**Mack, 1968**) and misplaced injections. (**Ecker & Woltman, 1938**)

Meralgia paresthetica is presented more often to chiropractors than is generally recognized. (**Arcadi, 1996**) In 215 consecutive examinations of patients in a chiropractic office, 12 cases of meralgia paresthetica were diagnosed. The method of diagnosis was "…standard orthopedic and neurologic testing procedures that evaluate the lateral femoral cutaneous nerve territory for superficial tactile sensation, superficial pain, sensitivity to vibration, sensitivity to temperature, and temperature gradient studies." (**Kadel & Godbey, 1983**) The patients had been aware of their condition from two days to 12 years. Five of the 12 had received treatment by the medical profession, with an unsatisfactory diagnosis or results.

The condition should be differentially diagnosed from conditions such as 2nd and 3rd lumbar nerve root compression, appendicitis, spinal cord tumor, colon cancer, and trochanteric or iliopsoas bursitis. (**Staal et al., 1999; Stites, 1986**) Disturbance at the lumbar level is usually associated with diminished or absent patellar reflex and weakness of the quadriceps muscles. Because the lateral femoral cutaneous nerve is purely sensory, there is never weakness or reflex change when disturbance is limited to its entrapment. (**Staal et al., 1999; Aguayo, 1975**) In femoral neuropathy or L2, 3 root lesions, sensory changes usually spread out to extend more anteromedially than in entrapments of the lateral femoral cutaneous nerve where the sensory change is limited to its dermatome. (**Harney & Patijn, 2007; Stites, 1986**)

Staal (**Staal, 1970**) considers meralgia paresthetica rare and suggests that "…one should look carefully for vertebral or disc lesions, intrapelvic anomalies, old operation scars, compression from outside due to clothing or chronic microtrauma: in short for all kinds of local pathology along the course of the nerve, from its beginning until distal to the anterior superior iliac spine. In the majority of the cases no obvious cause will however be found even after thorough investigation." Failure to find a cause for meralgia paresthetica is not consistent with applied kinesiology findings. This may be due to a more functional evaluation of spinal, pelvic, and muscular function with AK methods.

There will usually be pain below and slightly medial to the anterior superior iliac spine, close to the inguinal ligament attachment. To determine the attachment of the inguinal ligament, palpate along the ligament to the bone. A positive indication that meralgia paresthetica is present when deep pressure in this area causes radiation of pain in the skin supplied by the lateral femoral cutaneous nerve. Another source of pain at the anterior superior iliac spine is the origin of the sartorius, especially when there is a category II pelvic involvement. Since skeletal muscular imbalance is often present in meralgia paresthetica, sartorius pain and a category II may be present in combination with lateral femoral cutaneous nerve entrapment. To differentiate sartorius pain, palpate its origin at the anterior superior iliac spine and the upper half of the notch below it. From the sartorius origin, palpate down the sartorius tendon into the muscle, observing for pain; this indicates muscle involvement rather than, or in addition to, nerve entrapment. Usually the muscle will also test weak.

Low back derangement may cause pain in the greater trochanteric region or in the tensor fascia lata. It is distinguished from meralgia paresthetica by the absence of sensory alteration in the skin, and lack of pain on digital pressure to the nerve. (**Skaggs et al., 2006**)

Entrapment of the superior gluteal nerve that supplies the gluteus medius and minimus and tensor fascia lata, as well as the greater trochanteric region and a portion of the hip joint, can cause what has been termed false meralgia paresthetica. (**Staal et al., 1999; Kopell & Thompson, 1976**) In this condition there is pain over the gluteus medius and minimus that radiates down the lateral aspect of the thigh to the knee. Lack of cutaneous sensory findings and no tenderness of the lateral femoral cutaneous nerve differentiate this from true meralgia paresthetica. Structural correction of the spine and pelvis typically corrects the condition.

Body language that first indicates that entrapment of the lateral femoral cutaneous nerve and meralgia paresthetica may be present is the location of the pain, and whether it is relieved in seated or recumbent positions and exacerbated by standing, walking, and hip extension. Hip extension applied as a provocative test for meralgia paresthetica will aggravate the paresthesia and discomfort. (**Russell, 2006; Casscells, 1978**) A therapeutic test that is sometimes done is a local anesthetic injection at the point of probable entrapment. (**Williams & Trzil, 1991; Bellis, 1977**) Williams and Trzil (**Williams & Trzil, 1991**) report on an impressive series of 277 patients for whom only 24 cases required surgery. In a surgical study for the relief of meralgia paresthetica there were findings of constrictive fascial bands around the lateral femoral cutaneous nerve in 19 of 21 cases, indicating that soft-tissue disturbances of the mechanical interface with the nerve the most common etiology. (**Edelson & Stevens, 1994**)

Correction of entrapment of the lateral femoral cutaneous nerve is usually readily accomplished using applied kinesiology techniques. Even without the effective procedures of applied kinesiology, exercises directed toward improvement of muscles that support the pelvis — especially the abdominals — have been shown to be effective in treating this condition. (**Skaggs et al., 2006**) Surgical procedures are usually not done as a method of treatment. (**Aguayo, 1975**) Correction of spinal subluxations has been indicated as a method of treatment. (**Skaggs et al., 2006; Ferezy, 1989; Dunn, 1975**) Obviously, the optimal approach is to evaluate all aspects of the condition and correct any dysfunction that is present.

There is usually a pelvic category I or II and weak abdominal and gluteus maximus muscles. Total postural balance should be evaluated and corrected, as well as modular interaction of PRYT, cloacal synchronization, and dural tension. (**Walther, 2000, 1981**) These techniques restore organization to the muscles and postural balance, and are often important in improving range of motion. A painful myofascial trigger point (MTrP) may be found just medial to the anterior superior iliac spine. (**Biemond, 1970**) Muscle stretch reaction of the hip flexors differentiates this from radiating pain due to pressure on the entrapped nerve. The trigger point is treated with the usual AK methods of percussion, trigger point pressure release, or stretch and spray methods. **(Leaf, 2010; Cuthbert, 2002; Walther, 2000)** A recently published systematic review of the literature on the chiropractic management of myofascial trigger points and myofascial pain syndromes **(Vernon & Schneider, 2009)** reviewed 112 publications and came to the recommendation that moderately strong evidence supports some manual therapies (manipulation and ischaemic pressure) for immediate pain relief for myofascial trigger points. According to Leibenson **(Liebenson, 2007)** the combination of muscular inhibition, joint dysfunction and trigger point activity is the key peripheral component leading to functional pathology of the motor system. In AK, the presence of myofascial trigger points can be identified using the muscle stretch procedure that produces detectible changes in muscle strength on MMT. **(Cuthbert, 2002; Mense & Simons, 2001; Walther, 2000; Goodheart, 1998; Travell & Simons, 1992)** After these techniques have been applied, evaluate hip extension range of motion; if limited, evaluate the hip flexors for muscle stretch reaction and apply either trigger point pressure release, percussion, or stretch and spray technique to obtain improved range of motion. Frequently the percussion and/or trigger point pressure release technique is the appropriate one. This appears to release the fascial pull on the lateral femoral cutaneous nerve that is causing the entrapment at its pelvic outlet. Stretching exercises designed to stretch the iliopsoas and rectus femoris muscles are usually contraindicated since they will cause additional irritation to the entrapped nerve. Generalized body organization or local muscle treatment is usually satisfactory for obtaining effective correction.

Medical approaches have included ultrasound at L2 and where the nerve leaves the pelvis, and high voltage electrogalvanic stimulation. "The medical approach, after attempting to relieve external stress to the nerve, would include mild analgesia or local injection of xylocaine or corticosteroids." **(Stites, 1985)** These procedures are usually not necessary in an applied kinesiology practice. Kopell and Thompson **(Kopell & Thompson, 1960)** have found that stretching exercises to relax the tensor fascia lata and shoe lifts are ineffective. They do, however, state that most often the condition can be corrected without neurolysis.

Obturator Nerve

The obturator nerve arises from the 2nd, 3rd, and 4th lumbar nerves. After forming in the substance of the psoas major, **(Moore et al., 2009)** it passes down the pelvis immediately anterior to the sacroiliac joint. It traverses the obturator canal with the obturator vessels and is the closest to the bone of all of the structures passing through the canal. It divides into an anterior and a posterior branch. The anterior branch supplies the hip joint, adductor longus, gracilis, and usually the adductor brevis. It innervates the skin at the medial portion of the thigh. The posterior branch innervates the obturator externus, adductor magnus, and adductor brevis when not supplied by the anterior branch. There is an articular branch for the knee joint.

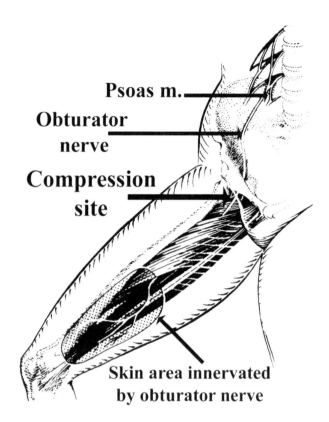

Obturator nerve.

Entrapment is at the osteofibrous tunnel of the obturator canal. The tunnel is formed by the obturator muscles covered by the obturator groove of the pubic bone. **(Barral & Croibier, 2007; Kopell & Thompson, 1960)** It may be aggravated by psoas muscle hypertonicity. The obturator nerve is in the upper part of the canal, and the obturator artery is directly beneath it.

Pain from entrapment is located in the groin; it radiates down into the medial aspect of the thigh, which is the obturator nerve's cutaneous distribution. It is typically not relieved by rest. There will probably be adductor muscle weakness. With injury to the obturator nerve, the legs cannot be crossed easily.

Entrapment may be the result of complications of genital or urinary surgery. The edema of osteitis pubis (inflammation of the pubic bone), which often follows surgery, can compress the obturator canal and be the major factor. Kopell and Thompson **(Kopell & Thompson, 1960)** consider that the severe pain of osteitis pubis is from this cause.

Obturator hernia can be a cause of entrapment. Increased pain radiation localized to obturator nerve distribution that develops with abdominal pressure is a positive sign of hernia. **(Staal et al., 1999; Kopell & Thompson, 1976; Staal, 1970)**

The obturator nerve can be injured during difficult labor. Pressure from the fetal head or from forceps may be the injuring factor. **(Aguayo, 1975)** The ileocecal valve dysfunction found in applied kinesiology examination can produce a lymphadenitis in the region that may irritate the obturator nerve. These pseudo-appendix conditions can be effectively resolved with applied kinesiology methods aimed at the ileocecal valve. (**Walther, 2000**)

The obturator nerve is formed within the psoas muscle; consequently, hip motions that stretch or contract the psoas may aggravate the pain. The psoas should be evaluated for muscle stretch reaction, which is most easily done by moving the patient close to the edge of the table and dropping the leg over the edge with extension and abduction. Immediately move the leg into position to complete the psoas muscle test. The psoas is also often involved with strain/counterstrain, in which a maximum contraction for three seconds causes a previously strong muscle to weaken. (**Walther, 2000; Chaitow, 2002**) Treat appropriately and balance the pelvis, lumbar spine, and other postural muscles. Isolated weakness of the adductor muscles, with or without sensory changes, is a strong indication of obturator nerve injury. However because the nerve roots making up the obturator nerve also contain fibers to the femoral nerve, particular attention should be paid to the manual muscle test finding of the quadriceps muscle as well as to the patellar reflex; both should be normal in an isolated obturator neuropathy.

Patients who fail to respond to conservative treatment are subjected to an intrapelvic section of the nerve. An extrapelvic section obviously leaves the entrapment neuropathy intact. (**Skaggs et al., 2006; Kopell & Thompson, 1976**)

Femoral Nerve

The femoral nerve arises from L2, 3, and 4 and is the longest and largest nerve of the lumbar plexus; it runs from the 2nd lumbar vertebra to the big toe. The nerve is formed in and passes through the psoas muscle; (**Moore et al., 2009**) it continues between the psoas and iliacus muscles and travels under the inguinal ligament. (**Biemond, 1970**) After leaving the pelvis, the nerve runs in the psoas fascia (**Mozes et al., 1975**) and comes in close proximity to the femoral head. It is separated by tough, unyielding tissues, including a thin slip of the iliacus muscle, reflected head of the vastus intermedius tendon, psoas tendon, and the hip joint capsule. These structures provide minimal cushioning. Anteriorly, the nerve is protected only by the skin, making it vulnerable to trauma; (**Moore et al., 2009; Kopell, 1980**) this can cause a subperineural hematoma that may need to be surgically removed. (**Kopell & Thompson, 1976**)

Lumbar Plexus

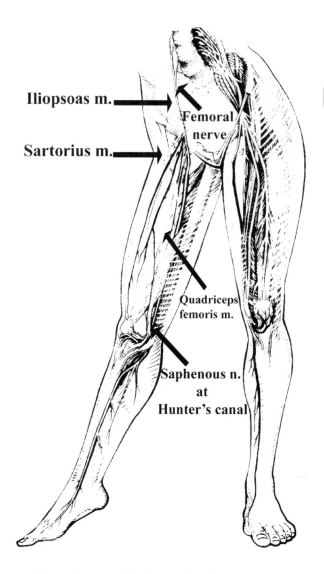

Femoral nerve and saphenous branch.

Muscles supplied in the abdomen are the psoas major and minor and the iliacus; in the thigh they are the sartorius, pectineus, and four parts of the quadriceps group. An articular branch supplies the knee.

There are cutaneous branches supplying the skin of the medial and anterior surface of the thigh, medial surface of the leg, and — finally — the medial aspect of the foot to the metatarsophalangeal articulation of the big toe. The saphenous nerve is the largest cutaneous branch of the femoral nerve and is its termination. It originates near the inguinal ligament and descends into the adductor or subsartorius canal (Hunter's canal) to run along with the superficial femoral artery. It emerges from the canal approximately 10 cm proximal to the medial femoral condyle. In the leg the nerve is in close proximity to the saphenous vein. The name "saphenous" derives from the Arabic word for visible. (**Mozes et al., 1975**)

Entrapment of the femoral nerve may develop in the pelvis or at the inguinal region. According to Aguayo, (**Aguayo, 1975**) it may be associated with inguinal hernia or adhesions from careless surgery. Johnson and Montgomery (**Johnson & Montgomery, 1958**) state, "The most common cause of femoral neuropathy is involvement of the femoral nerve in the course of abdominopelvic surgery." It may also develop as a result of retroperitoneal hematoma, a complication of hemophilia, or in patients receiving anticoagulants. (**Staal et al., 1999; Aguayo, 1975**)

In femoral neuropathy, there is weakness of the quadriceps muscles and diminished or absent patellar reflex. Complete paralysis of the femoral nerve is indicated by three conditions. (**Russell, 2006; Biemond, 1970**) (1) There is motor loss of the quadriceps muscle. In the event of a high lesion, the iliopsoas function is also lost. (2) There is hypoesthesia (seldom anesthesia) of the anterior and medial aspects of the thigh, following down the tibial aspect of the legs and sometimes extending to the medial margin of the foot. (3) The quadriceps reflex is absent.

Femoral neuropathy can be caused by severe stretching of the nerve, as in accidents that cause extreme extension and abduction of the thigh or by abduction at surgery. (**Russell, 2006; Biemond, 1970**) Surgical accidents frequently follow gynecologic procedures. In more severe cases, the patient's leg collapses under her when first trying to ambulate following surgery. In a less severe condition, the patient feels weakness when attempting to climb stairs, walk up a hill, or with general activity. It is easier to walk backward than forward because the leg can support in a stance stage with accessory muscles; when hip flexion is attempted, the leg tends to collapse. If bracing is not provided, a pronounced genu recurvatum will develop in long-standing paralysis. Even when there is no paralysis of the quadriceps muscles but muscle weakness as identified by applied kinesiology, the patient will stand with the knee hyperextended.

Radiculopathy, neoplasms, abscesses, and other pathology must be ruled out. The metabolic problem often identified with femoral neuropathy is diabetes. Usually diabetic neuropathy is more widespread, but it can be localized to the femoral nerve. (**Veves et al., 2002; Kopell, 1980; Aguayo, 1975**) Possibly the reason the femoral nerve is so vulnerable to diabetic neuropathy is the poor blood supply to the intra-abdominal portion.

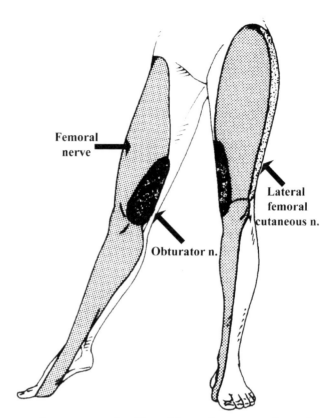

Cutaneous supply of lateral femoral cutaneous, femoral, and obturator nerves.

It should be remembered that there is a direct relationship between the ovaries and the femoral nerve. Pathophysiological consequences of spinal dysfunction on the segmentally related visceral organs have been described. (**Leach, 2004**) Lumbar and sacral nerve root compression as the result of lumbar and sacral articular dysfunction and degeneration has been identified as a potential cause of pelvic pain and organic dysfunction (PPOD), a term coined by Browning. (**Browning, 2009, 1988**) The impact of spinal subluxations upon sexual function may be important. (**Rome, 2010, 2009; Nansel & Slazak, 1995**)

Because the femoral nerve is formed in the psoas muscle, its hypertonicity can cause entrapment. Compression of the femoral nerve often causes pain radiation into the groin, which is increased by extension of the thigh. Sensory impairment is over the anteromedial aspect of the thigh, occasionally extending into the leg.

Time is a factor in recovering from surgical complications. Johnson and Montgomery (**Johnson & Montgomery, 1958**) indicate there is generally spontaneous recovery in three to six months, but if there is failure of some degree of recovery within three months, surgical intervention to explore and repair the nerve is indicated.

Femoral nerve involvement of a functional nature can be successfully treated using AK diagnostic and treatment methods, especially when treatment causes an immediate improvement of quadriceps muscle function. Even in the absence of frank radiculopathy, spinal and pelvic manipulation can improve femoral nerve involvement. Several cases successfully treated by such methods both with and without the advantage of AK methods have been reported. **(Picard, 2008; Skaggs et al., 2006; Kaufman, 1997; Kadel et al., 1982)** The addition of evaluation and treatment of psoas muscle weakness or hypertonicity is of primary importance, along with spinal and pelvic function. There is sometimes a tendency to concentrate examination and treatment on the side of nerve involvement. Remember to always evaluate the contralateral psoas and other muscles that may be involved. Hypertonicity of the ipsilateral psoas is frequently secondary to contralateral psoas weakness.

The saphenous nerve distal to its exit from the adductor (subsartorius) canal is superficial and innervates the medial thigh, calf, and foot. **(Russell, 2006; Worth, 1984)** Sensory symptoms of saphenous distribution are rare when the lesion involves the main trunk of the femoral nerve. **(Aguayo, 1975)** The most reliable evidence of saphenous nerve entrapment is point tenderness in which the nerve emerges through the fascia from Hunter's canal. **(Staal et al., 1999; Worth, 1984; Kopell & Thompson, 1960)** This is about 10 cm proximal to the medial condyle of the femur. At this level, palpate posteriorly across the vastus medialis to the edge of the sartorius, the point where the nerve emerges. Since the saphenous nerve is wholly sensory, there will be no muscle weakness that is dependent on its individual entrapment.

Pain is usually present at the knee when there is entrapment of the saphenous nerve. The pain at the medial knee is aggravated by walking, standing, and quadriceps exercise. The tenderness at the medial knee must be differentiated from the tenderness that is present with a weak sartorius and/or gracilis and a category II pelvic fault. Weakness of the sartorius on manual muscle testing is not adequate differential diagnosis, because the sartorius receives its nerve supply from the femoral nerve.

Misdiagnosis of saphenous nerve entrapment is common. The increase of pain on walking may lead to the diagnosis of intermittent claudication with symptoms of fatigue and heaviness of the leg. **(Russell, 2006; Staal et al., 1999)** Saphenous nerve entrapment can simulate hip pain. One patient is described who had a fracture of the femur neck and was operated on for the condition. She suffered pain attributed to hip joint pathology for a full year following the operation. A second operation to free the saphenous nerve from its entrapment point resulted in complete relief from the pain. Likewise, irritation of the infrapatellar branch of the saphenous nerve may cause pain simulating arthritis of the knee joint **(Mozes et al., 1975)** or other knee dysfunction.

Entrapment of the saphenous nerve is often a sequel to knee surgery. In a study by Worth et al. **(Worth, 1984)** of fifteen entrapments, ten followed knee surgery. Their surgical treatment is neurolysis or neurectomy. Conservatively, repeated injections of Novocain are sometimes effective on a lasting basis. **(Mozes et al., 1975)** This also gives a definitive diagnosis. **(Staal et al., 1999; Kopell & Thompson, 1960)** Other iatrogenic causes of saphenous nerve disturbances are operations for varicose veins or harvesting of the saphenous vein for arterial grafting. **(Staal et al., 1999)**

Applied kinesiology techniques are effective in many cases. There are usually many spinal and pelvic factors to correct including dural tension release. Also, examination and treatment, if necessary, to local muscular imbalance of the hip and knee with particular attention given to fascial release techniques are important.

Piriformis Syndrome

The sciatic nerve arises from L4, 5, and S1, 2, and 3. It is the largest nerve in the body, about 2 cm in diameter. In reality, the sciatic nerve is two nerves, the tibial and peroneal, contained within the same epineurium. In some cases the nerves fail to combine in this manner. **(Prakash et al., 2010; Moore, 2009)** The tibial division originates in the sacral plexus from the ventral divisions of L4, 5, and S1, 2, and 3, and the peroneal from the dorsal divisions of L4, 5, and S1 and 2.

The sciatic nerve usually passes out of the pelvis through the greater sciatic foramen below the piriformis. Here it is flattened or in an elliptical form. Beneath the

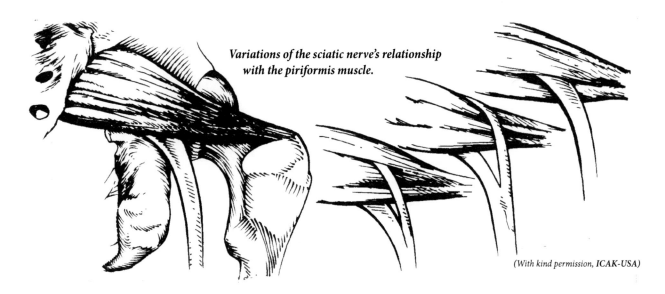

Variations of the sciatic nerve's relationship with the piriformis muscle.

(With kind permission, ICAK-USA)

nerve is the obturator internus muscle. The nerve may divide at the piriformis muscle; when it does, it is usually the peroneal trunk that deviates. (**Prakash et al., 2010**) The nerve then passes under the gluteus maximus muscle midway between the greater trochanter of the femur and the ischial tuberosity. It continues to course down the thigh to divide in the lower one-third into the tibial and peroneal nerves.

In a study of 240 cadavers, Beaton and Anson (**Beaton & Anson, 1938**) found 10% variation from the usual nerve passing beneath the piriformis muscle. In 9.2%, the nerve divided; one portion passed through the muscle and the other above or below it. In .8%, the undivided nerve passed through the belly of the piriformis muscle. Travell & Simons (**Travell & Simons, 1992**) review many cadaveric studies on the varying courses of the two divisions of the sciatic nerve, and they list similar proportions to Beaton & Anson.

It is in its relationship with the piriformis muscle that sciatic entrapment neuropathy can develop distal to radiculopathy. If the sciatic nerve passes under the piriformis muscle, it exits the pelvis crossing the greater sciatic notch, where the nerve is flattened and adjacent to the sharp edge of the bone. (**Prakash et al., 2010; Kopell & Thompson, 1960**) When the piriformis is in spasm or shortened, it can cause sciatic compression and perhaps ischemia, especially if a portion of the nerve goes through the muscle. (**Travell & Travell, 1946**)

Travell & Simons (**Travell & Simons, 1992**) also note that the inferior gluteal nerve that innervates the gluteus maximus muscle penetrated the piriformis muscle in 15% of 112 patients, making it vulnerable as well to piriformis entrapment. The inferior gluteal artery and vein cross the sciatic trunk under the belly of the piriformis. Contraction of the piriformis may conceivably produce sustained congestion in the vein and limit circulation from the artery. This may explain the tenderness in the entire piriformis muscle in addition to sciatic pain. It may also cause an ischemia of the nerve contributing to the pain. (**Cox, 1999; Freiberg & Vinke, 1934**) Kendall et al (**Kendall et al., 1993**) report that either a contracted or a stretched piriformis can contribute to sciatic pain.

The piriformis syndrome may be caused by subtle, moderate, or severe trauma. (**Chaitow & DeLany, 2002; Cox, 1999; TePorteen, 1969; Edwards, 1962**) Acute trauma usually relates with a sudden twist of the hip, which is either activated or counteracted by the piriformis. The trauma may be such that it injures the fascia and/or tendons of the rotator muscles. Fascial trauma may cause a disrelation between the muscle and fascial function that requires the fascial release technique of applied kinesiology. (**Walther, 2000, 1981**) The ends of the taut bands of the piriformis muscle are likely to create enthesopathies; for this reason stretching the muscle may further inflame the attachments thereby creating weakness on subsequent manual muscle testing. Immediately after the trauma the muscles become hypertonic but, as the stress continues, the tendons develop enthesopathy. (**Palesy, 1997; Lund et al., 1991**) The enthesopathy causes inhibited firing of the muscles leading eventually to atrophy and eventually fatty infiltration. (**Hodges & Moseley, 2003**) Experimentally induced pain has been shown to change muscle activity, including reflex activity and muscle spindle sensitivity.

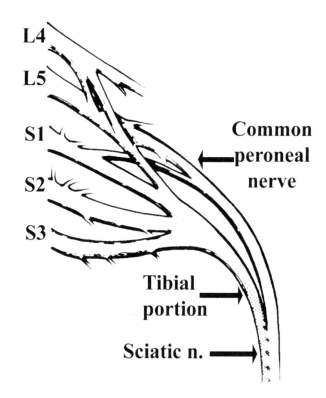

Sacral plexus.

(**Zedka et al., 1999**) The neuromuscular spindle cell or Golgi tendon organ may be injured, causing the muscle to test weak on manual muscle testing. It should be remembered that physical trauma can directly affect receptor axons (articular receptors, muscle spindles and other proprioceptors), (**Lederman, 1997**) and that muscle spindles possess extreme sensitivity, where the muscle spindle reacts to a pull of only 1 gram and a stretch of $1/1000^{th}$ millimeter. (**Korr, 1997, 1979**) Acute trauma often relates with sudden starts or stops, as in playing tennis, or from falls and other direct injuries to the pelvis. A sit-down type fall is often described. (**Robinson, 1947**) Twists that develop in the pelvis, such as entering or leaving an automobile on one leg, can cause strain to the muscle. Basically any type of trauma to the pelvis or buttocks may be responsible for piriformis imbalance.

More subtle etiology of piriformis imbalance can develop as a result of the patient assuming positions for prolonged periods. Usually the problem posture is in a seated position. Sitting cross-legged, especially with the ankle crossed over the opposite knee, shortens the piriformis. When quickly brought out of the position, the nervous system may not have adequate time to adapt. A muscular syndrome, as described by Jones (**Jones, 1981**) in his strain/counterstrain technique, may develop. Sitting on the feet or a foot provides a similar etiology. When driving an automobile, one may put the leg in a twisted position to contact the accelerator with the right foot, or have the left foot in a position to cause strain to the piriformis. Waddell (**Waddell, 1998**) warns that driving and sitting improperly (ergonomic stress positions) for long periods are major causes of back pain. Ergonomic seats, good posture, frequent periods of movement, and optimal tire pressure and suspension systems are important factors for reducing the jarring forces inherent in driving. Imbalance of the piriformis in all these conditions probably relates with muscle dysfunction that can be treated with the strain/counterstrain technique. This type of dysfunction can also develop from rapid trauma. Brugger (**Brugger, 1960**) suggests a postural exercise that eases the stresses of sitting and contribute to neck and back pain. Exercise like this and ergonomic advise should be used for patients whose jobs require prolonged sitting.

Piriformis muscle dysfunction often develops after obstetric or urologic procedures in which the patient is in stirrups, especially when under general anesthesia. (**TePorteen, 1969**) Coital positions are also responsible for piriformis strain. The patient frequently does not relate this to the physician. Goodheart (**Goodheart, 1992**) observes that backache from sexual intercourse probably relates to disruption of the craniosacral primary respiratory action of the pelvis and spine. On inspiration the pubes should drop slightly inferiorly and raise superiorly on expiration. If during the sexual act there is a pelvic thrust by the male during inspiration by the female, the pelvic respiratory action may be disturbed. This disturbance will likely involve the piriformis muscle and its function.

Structural disorganization anywhere in the body, especially in the pelvis and hips, can be the causative factor of piriformis syndrome. Lumbar hyperlordosis with anterior pelvic rotation and hip flexion tightens the rotator group of muscles over the nerve and draws the nerve trunk more tightly against the rim of bone. (**Kendall et al., 1993; Kopell & Thompson, 1960**)

Careful differential diagnosis must be done in sciatic nerve conditions. There is a close symptomatic association in piriformis syndrome, radiculopathy, and sacroiliac subluxations and fixations. (**Travell & Simons, 1992**) There may be a combination of conditions, or symptoms of one being created by another condition.

Piriformis entrapment of the sciatic nerve is not accepted as a common syndrome by all. Dawson et al. (**Dawson et al., 1983**) take the view of surgeons, stating, "It should be mentioned at the outset that sciatic entrapment neuropathy has not been fully accepted as a legitimate syndrome, and in spite of occasional cases it remains difficult to substantiate." It appears that their concern is for total paralysis as a result of sciatic entrapment. They continue, "The sciatic nerve innervates the knee flexors and all the muscles below the knee. Therefore, complete palsy of the sciatic nerve would lead to marked instability of the foot and to severe impairment of gait because of knee and ankle impairment." The entrapment found by AK examination is more subtle, leading to pain and muscle dysfunction that readily respond to the various types of treatment used in applied kinesiology. Piriformis entrapment may also complicate sciatic involvement by contributing to a double crush of the nerve.

Travell & Simons and Steiner et al. (**Travell & Simons, 1992; Steiner et al., 1987**) support the concurrency concept of radiculopathy and piriformis syndrome. These conditions may appear almost identical. There is a consistent absence of true neurologic findings in the piriformis syndrome. Back pain is common to both. Even when there are specific indications of a herniated disc, one should examine for piriformis involvement.

Dawson et al. (**Dawson et al., 1983**) indicate the following in differentiating piriformis syndrome from radiculitis due to a lumbar disc. "Pain on coughing, sneezing, or laughing is a sure indicator of epidural disease and should be inquired for in every case of this type. Most patients with sciatic pain due to lumbar disc disease have monoradicular symptoms. Therefore, the pain generally radiates to the lateral side of the foot and the small toe (S1), to the dorsum of the foot (L5), or to the medial part of the calf (L4). Many patients also have proximal sciatic pain in the posterior thigh and the buttocks, but this has less anatomic or localizing significance." Piriformis syndrome must also be differentiated from lumbar spinal stenosis.

In addition to other differential diagnosis in sciatic-type pain, particular attention must be given to the correlation of the piriformis and sacroiliac syndromes. Vleeming et al (**Vleeming et al., 1995**) and Poole (**Poole, 1985**) assert that many conditions diagnosed as sacroiliac or lumbosacral involvement are, in reality, piriformis syndromes. Vleeming et al. observe that sacroiliac joint pain may not be a local problem "but symptomatic of a failed load transfer system" of the muscles that attach to the sacroiliac joint. Many other muscles are involved in the stability of the sacroiliac joints, including the gluteus medius, multifidus, biceps femoris, transverse and obdominal obliques, gluteus maximus, and latissimus dorsi, among others. (**Harrison et al., 1997**) Various tests of the sacroiliac designed to stretch the ligaments, such as Gaenslen's test, iliac maneuvers designed to stretch the anterior and posterior ligaments, such as Hibbs' test, and others should be analyzed to determine the structure involved in addition to the sacroiliac. Physicians from many disciplines appreciate that maneuvers that move nerves and receptors often reproduce radiating pain. This symptom reproduction has important implications for the diagnosis and management of radiating pain syndromes. Applied kinesiology examination reveals close association between piriformis dysfunction and sacroiliac subluxations and fixations. The concurrent presence of both conditions is common. (**Knutson, 2004; Corwin, 1987**)

The piriformis muscle is important in stabilizing sacral function. Sacral subluxations are attributed to its imbalance. (**Boyajian-O'Neill et al., 2008; Chaitow & DeLany, 2002; Kendall et al., 1993; Travell & Simons, 1992; Retzlaff, 1974; Goodheart, 1972**) A sacroiliac fixation may develop with piriformis hypertonicity and shortening. On the other hand, piriformis spasm can be present as a vain effort of the piriformis to correct pelvic distortion.

Skillern (**Skillern, 1944**) explains the mechanism of piriformis spasm as coming from peripheral irritation caused by minor strain of the sacroiliac joint. Fourteen cadavers were studied by Freiberg and Vinke (**Freiberg & Vinke, 1934**) to establish the relationship of the sacroiliac joints with the sacral and lumbar plexus and the sciatic nerve to the piriformis. There is a very intimate relationship of the branches of the plexus with the blood vessels. They are so closely interwoven that it is difficult to separate them. Some symptoms of piriformis syndrome may occur from local inflammation and congestion caused by the muscular compression of small nerves and vessels – including the pudendal nerve and blood vessels, which emerge at the medial inferior border of the piriformis muscle. (**Moore, 2009**)

Pain associated with the piriformis syndrome is in the area of the femoral head and greater trochanter (**Travell & Simons, 1992; Khoe, 1975**) and may radiate into the inguinal area. There is local tenderness over the piriformis muscle and tendon. The pain radiating into the hip often causes difficulty in walking, and it may radiate down the leg in the form of sciatic neuralgia. There may also be

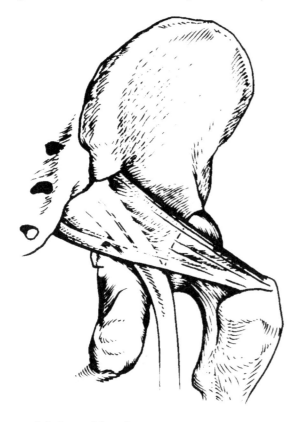

Sciatic notch location.

pain at the lateral knee behind the head of the fibula, due to involvement of the common peroneal nerve. Pain is often relieved by a change in position or walking. There is difficulty in sitting, standing, or lying comfortably. One may find the patient walking around the examination room rather than sitting and waiting for the physician. The patient often gives a poor description of the pain and has a difficult time localizing its origin. The symptoms are vague in the area of the posterolateral hip and down the tensor fascia lata muscle. When the physician palpates into the piriformis, the patient usually quickly recognizes that is the source of the pain.

The pudendal nerve leaves the pelvis between the piriformis and coccygeus muscles. It supplies much of the external genitalia and related peroneal musculature in both males and females. When the pudendal nerve and blood vessels are involved in piriformis muscle entrapment, there may be a lack of genital sensation, pain, incontinence, or impotence. (**Cuthbert & Rosner, 2012; Browning, 2009; Retzlaff, 1974**)

The first phase of examination includes palpation of the piriformis muscle's insertion and around the greater trochanter for tenderness. From this insertion proceed toward the sciatic notch to palpate the piriformis muscle mass. The sciatic notch is located approximately 2" lateral and 1" caudal to the posterior superior iliac spine. (**Lee, 2004; Corwin, 1987**)

In sciatic neuralgia and with other signs of piriformis syndrome, observe for an antalgic position of external thigh rotation and slight hip flexion, which indicates a possible piriformis syndrome. This position takes tension off the piriformis and may thus reduce irritation on the sciatic nerve. (**Chaitow & DeLany, 2002; Freiberg & Vinke, 1934**) This body language position is called the "piriformis sign". (**Foster, 2002; TePoorten, 1969**)

Internally rotating the thigh stretches the piriformis and usually increases the irritation on the sciatic nerve, causing pain radiation. Evaluate the effect of stretching the piriformis with the patient prone, using the lower leg as a lever to internally rotate the thigh. What is often referred to as "tenderness at the sciatic notch" is usually tenderness of the piriformis belly. (**Freiberg & Vinke, 1934**) The Beatty test is useful for diagnosis of the syndrome too. The patient lies on the uninvolved side, and then lifts and holds the superior knee approximately 6 inches off the examination table. If sciatic symptoms are recreated, the test is positive for piriformis syndrome.

The above descriptions are of stretching or shortening the piriformis muscle passively. When the patient actively contracts the piriformis, such as in a manual muscle test, there is external thigh rotation and tightening of the piriformis. This may cause the same type of pain as passively stretching it.

Pain and dysfunction from piriformis nerve entrapment must be differentiated from sciatic involvement at a higher level. The straight leg raise and Lasegue tests can aid in this determination. (**Morris, 2006; Robinson, 1947**) The piriformis muscle is put on stretch after only a few degrees of leg raising. Differentially the straight leg-raising test is usually positive for a pathological disc rather than the first stage of Lasegue's sign before the knee is extended, although both may be present. In a piriformis syndrome, pain is caused by hip movement and stretching of the muscle rather than sciatic nerve stretch. Freiberg and Vinke (**Freiberg & Vinke, 1934**) evaluated straight leg raising on cadavers. When the thigh reached approximately 25–50° of flexion, there was tightening of the sacrotuberous ligament and the piriformis muscle. They point out that in many patients with sciatic pain, straight leg raising is limited at only a few degrees of hip flexion. This is long before the sciatic nerve is stretched by the maneuver, indicating entrapment below the vertebral level.

There will be tenderness of the sciatic nerve in a piriformis syndrome; however, it is often difficult to elicit because of the bulk of tissue covering the structure. When a straight leg raise test elicits pain, lower the leg slightly, just below the point of pain. Internal rotation in this position stretches the piriformis and will increase the pain. There will be relief by passive external rotation. (**Morris, 2006**)

The superior gluteal nerve supplies the gluteus medius and minimus, and the inferior gluteal nerve supplies the gluteus maximus. These nerves originate from the sacral plexus with the sciatic nerve, and travel with it in its short intrapelvic course. Both gluteal nerves leave the pelvis through the greater sciatic foramen, but the superior nerve leaves above the piriformis and the inferior one below it. **(Gray's Anatomy, 2004)**

There is controversy over involvement of the gluteal muscles in a piriformis syndrome. Some indicate that if there is weakness of the gluteal muscles along with others of sciatic distribution, there is indication of spinal or intrapelvic entrapment, such as tumors. When the lesion is due to piriformis entrapment, the gluteal muscles will not be involved; however, there is involvement of the hamstrings and all muscles of the lower leg. **(Boyajian-O'Neill et al., 2008; Aguayo, 1975; Kopell & Thompson, 1960)**

The other side of the question is that with chronicity there may be gluteal atrophy on the affected side. **(Staal et al., 1999; Robinson, 1947)** Because the superior gluteal nerve leaves the pelvis above the piriformis and the inferior nerve below it, it seems that the gluteus medius and minimus will be spared and the gluteus maximus will be vulnerable in a piriformis syndrome. In manual muscle testing one observes a varied involvement of the gluteus medius, minimus, and maximus. In any event, when the piriformis muscle is responsible for gluteal weakness, they strengthen after the piriformis muscle is treated. This seems to indicate that there is a varying morphology of the superior and inferior gluteal nerves in relationship to the piriformis.

Patients with a vascular neuropathy, particularly a diabetic one, may acquire a piriformis syndrome. The symptoms are relatively acute in onset, developing over several days. **(Veves et al., 2002; Dawson et al., 1983)** Pain develops in intermittent claudication with a specific amount of walking, and is eased upon standing, **(MacNab, 1977)** which is opposite that of typical piriformis syndrome. The pattern of pain with activity in claudication is consistent, and inconsistent in neuromuscular, osseous conditions. **(Greenfield & Andersen, 1982)**

In summary, the piriformis syndrome should be differentially diagnosed from iliotibial band contracture, **(Ober, 1987)** lumbar, lumbosacral, and sacroiliac disturbances, hip joint lesion, various types of arthritis, prostate conditions, female organ dysfunction, cystitis, and epidural and pelvic tumors that may be found with CT scan. **(Boyajian-O'Neill, 2008; Cohen et al., 1986)**

The use of specific manual muscle tests can facilitate the diagnosis of the syndrome, and differentiate it from other disturbances to the sciatic nerve. In patients with piriformis syndrome, manual muscle test results will be normal for muscles proximal to the piriformis muscle and abnormal for muscles distal to it.

Specific treatment of the piriformis syndrome is directed toward the piriformis muscles. In addition, total body function must be considered and any pelvic and spinal problems corrected, as well as modular interaction of the body and gait problems, including foot dysfunction. Applied kinesiology could be defined as diagnosis and treatment of disorders of the body within a conceptual basis of the inter-dependency and continuity of all the tissues of the body. For patients with piriformis syndromes, TePoorten **(TePoorten, 1969)** recorded decreased range of motion of the T10-T11 segments, tissue texture changes at T3-T4, pain and decreased range of motion of the contralateral C2 segment, and ipsilateral occipital-atlantal lesion. TePoorten's observations are consistent with the experience of applied kinesiologists.

The piriformis muscles should be evaluated bilaterally for hyper- or hypoactivity. Early in applied kinesiology Goodheart proposed that much hyperactivity or muscle spasm is secondary to weakness of the antagonist muscle(s). This observation has been confirmed by the work of Lund **(Lund et al, 1991)**, Wall et al., **(1988)** and Mense and Simons **(2001)**. Mense and Simons suggest that the recognition of the muscle weakness caused by myofascial trigger points is often a critical step in the restoration of normal function. In their view other muscles suffer from compensatory overload due to the inhibition created by the myofascial trigger points in the inhibited muscles. Travell and Simons also state that "weakness is generally characteristic of muscles with active myofascial trigger points". **(Travell & Simons, 1983)** When tightness or spasm of a piriformis muscle is present and there is weakness on the opposite side, it is advantageous to correct the weak side first. This is done with the usual subluxation and fixation correction techniques, reflex techniques, cranial-sacral primary respiratory correction, meridian balancing, and sometimes nutritional corrections as used in applied kinesiology.

Berry and Retzlaff **(Berry & Retzlaff, 1978)** describe a method to treat a hypertonic muscle by giving attention to the hypotonic or weak muscle on the opposite side, based on Sherrington's law of reciprocal innervation. **(Denny-Brown, 1979)** When a muscle on one side of the body contracts, its homologue on the opposite side relaxes. Retzlaff and Fontaine **(Retzlaff & Fontaine, 1960)** have described a neuronal mechanism that provides simultaneous excitation of the neurons to one muscle (or group) and inhibition to the contralateral muscle (or group).

These basic principles are used when indirectly treating the hypertonic piriformis syndrome by improving function or stimulating the contralateral (non-affected) piriformis muscle. In addition, the piriformis reciprocal innervation technique (Berry method) **(Retzlaff, 1974)** can be applied. The patient is supine, with knees and hips flexed. The physician stands on the side opposite involvement. The patient abducts the knee of the uninvolved side against the physician. This contracts the non-involved piriformis to send inhibitory impulses to the hyperactive piriformis. The physician monitors the relaxation of the involved piriformis by palpating it, while the patient continues to isometrically contract the non-involved side. The procedure is repeated several times, with a short rest period in between. With each period of isometric contraction by the patient there is usually greater relaxation of the affected piriformis.

There are several applied kinesiology techniques applicable to the piriformis syndrome. The involved side is first tested to determine if the muscle is strong in the clear, and it usually is. If there is weakness on manual muscle testing, treat the muscle with the five factors of the IVF. **(Walther, 2000, 1981)** When the muscle tests strong in the clear, it is further evaluated for muscle stretch reaction by stretching the muscle and then immediately testing to determine if the previously strong muscle weakens. If so, percussion, trigger point pressure release, or stretch and spray methods are indicated. Most often fascial release technique is effective. (**Leaf, 2010; Cuthbert, 2002; Walther, 2000**) Whichever technique is used, internal thigh rotation should be increased.

Further evaluation is done by means of the strain/counterstrain technique. This treatment is often needed when no acute trauma is known to have caused the condition. The neuromuscular condition develops as a result of the individual maintaining a strained postural position of the piriformis for a prolonged period and then quickly moving out of the position. Jones (**Jones, 1981**) relates this to the nervous system's inability to quickly adapt to the change in muscle position.

To test for the need for applying the strain/counterstrain technique, have the patient maximally contract the piriformis, externally rotating the thigh. The contraction is held for three seconds, immediately after which the previously strong muscle is tested for weakening; if it does, strain/counterstrain technique is indicated. This is most easily accomplished with the patient prone. The hip and knee of the involved side are flexed, with the knee resting on the examination table. The tender point in the piriformis is usually located close to the greater trochanter. The thigh is passively rotated externally until the position is found that minimizes the tenderness in the piriformis. Some spinal extension and rotation may be required to relieve the tender point. When the point is found, have the patient maximally exhale while you simultaneously spread the tissue over the tender point with a two–finger contact. Hold this for approximately thirty seconds, after which the patient's involved limb and spine are slowly and passively brought back to a neutral position.

Another method of strain/counterstrain treatment is with the patient prone near the edge of the treatment table so the involved leg drops off the table with hip and knee flexion. The physician supports the patient's leg at the knee and passively directs external thigh rotation and abduction to relieve the tender point. The same exhalation and two–point finger spread over the tender point is done, after which the leg is passively brought back to neutral. Optimal relief of tenderness may not be accomplished in this position because of inability to extend and rotate the spine. Successful treatment by strain/counterstrain technique is indicated when there is no weakening after the patient maximally contracts the piriformis for three seconds.

There are several pressure techniques for relaxing a piriformis muscle. Chaitow & DeLany (**Chaitow & DeLany, 2002**) and Edwards (**Edwards, 1962**) recommend that heavy pressure be applied to the belly of the muscle by the thumbs, thenar eminence, or elbow pressure. The patient lies on the unaffected side, rotated slightly toward the examination table, with slight hip and knee flexion. The physician applies a heavy (40–60 lbs) pressure for approximately ten seconds. The pressure is sustained, not a massaging movement. This is repeated eight to twelve times. The pressure should be heavy enough to stretch fibrous tissue, yet not so heavy that it produces further inflammation and additional fibrositis. Pressure in the piriformis is painful but it does not accentuate the pain the patient is experiencing; rather, it tends to block it. After treatment the patient typically feels better, but the pain will return somewhat within a few hours. Repeated treatments eliminate the condition. People with good muscle physiology and in good health respond more rapidly than those with sedentary habits and poor general health.

A stretching technique described by TePoorten (**TePoorten, 1969**) is also done with the patient side–lying on the unaffected side. The knees are flexed to 90°, with very slight hip flexion. The physician applies pressure to the piriformis with his or her elbow. The contact is just posteromedial to the greater trochanter. The thigh is moved into internal rotation by using the leg as a lever. Simultaneously pressure is maintained on the piriformis while the muscle is thus stretched for one or two minutes and repeated two or three times.

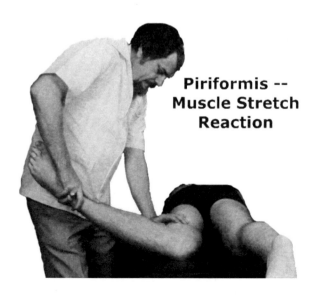

Piriformis muscle stretch reaction

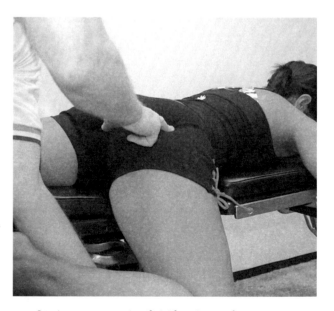

Strain-counterstrain of piriformis muscle

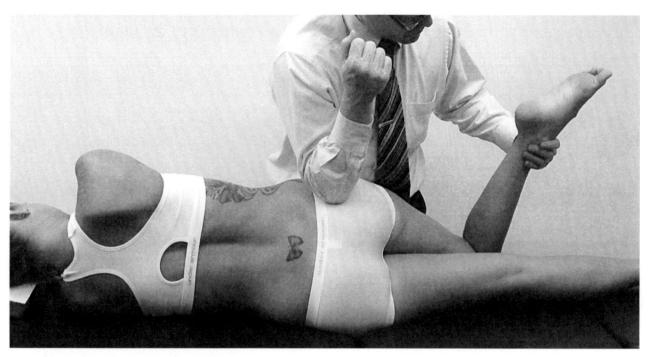

TePoorten's piriformis stretching technique

Another stretching activity described by TePoorten (**TePoorten, 1969**) and Retzlaff et al. (**Retzlaff, 1974**) is to have the supine patient flex his knee and hip to approximate the thigh on the abdomen. Use the leg as a lever to rotate the thigh while flexing and extending the hip and knee. When flexing, rotate the thigh externally; when extending, rotate the thigh internally. A following step is to hyperflex the hip and knee so the thigh approximates the abdomen. The physician uses his shoulder on the patient's knee to adduct it, with passive pressure at the patient's ankle to internally rotate the thigh. Throughout the first and second procedures the patient resists the motion, providing a type of isometric contraction.

Quite often the applied kinesiology approaches previously mentioned will correct the condition. Additional help may be obtained by the stretching and pressure procedures. An additional approach recommended by Poole (**Poole, 1985**) that might occasionally be beneficial is treatment of the piriformis spasm done rectally. A finger is inserted into the rectum with the patient side-lying, with the painful side up and the hips and knees slightly flexed. From contact of the coccyx, the flexor surface of the finger moves laterally to touch the anterior surface of the levator ani and coccygeus muscle. The fibers of the piriformis muscle are felt posterior to the sacrospinous ligament and stroked in a lateral motion. There may be extremely sensitive trigger points which, according to Nimmo, (**Nimmo, 1985**) cause the muscle spasm.

Whatever therapeutic approach is used, there should be an increase of passive internal thigh rotation. The standing posture should improve, with reduction of the excessive external leg rotation on the involved side. There should be balance of the piriformis muscle test bilaterally.

Finally, other therapeutic approaches have been advocated. Procaine hydrochloride injection into the piriformis muscle and sacroiliac joint has been recommended by Skillern. In 1947, Robinson (**Robinson, 1947**) recommended section of the piriformis muscle. This was the original publication about the piriformis syndrome reporting about two patients. Although he felt no disability resulted from the surgery, no follow-up reports of this approach have been found. Surgery for this condition is still advocated by some orthopedists (**Staal et al., 1999**) The piriformis syndrome responds well to the conservative approaches described herein.

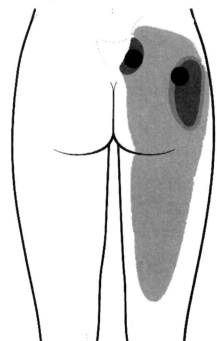

Piriformis MTrP Referred Pain Zone

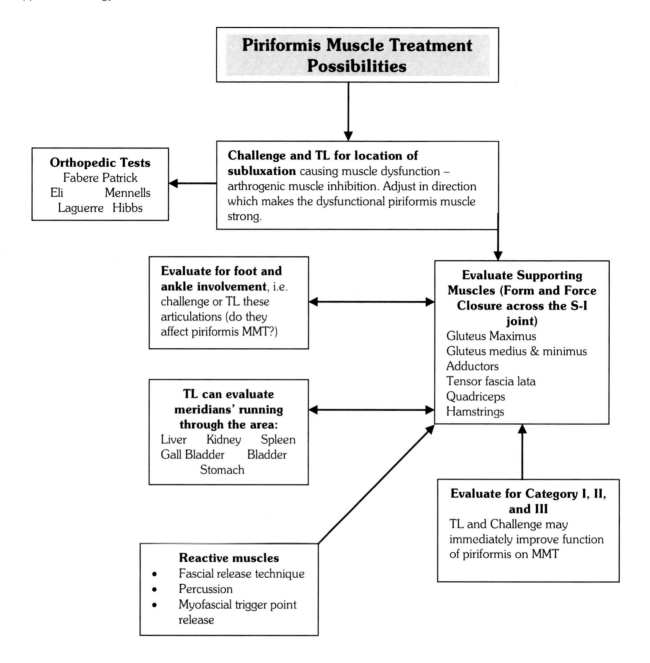

Peroneal Nerves

The sciatic nerve divides in the lower thigh into the common peroneal and tibial nerves. The common peroneal — the lateral branch — is smaller than the tibial nerve by about one-half. It is derived from the dorsal branches of the ventral rami of the 4th and 5th lumbar and the 1st and 2nd sacral nerves. After passing behind the head of the fibula, close to the tendon of the biceps femoris, it divides into the superficial and deep peroneal nerves. The peroneal nerve, with its superficial and deep branches, supplies the dorsiflexors and everters of the foot, the skin on the lateral portion of the leg and dorsal portion of the foot, a portion of the ankle joint, and the tarsal joints.

Common Peroneal

The common peroneal nerve courses down the lateral aspect of the popliteal fossa, emerging between the biceps femoris tendon and the lateral head of the gastrocnemius. It pierces the fascia to lie against the fibular neck and lies directly upon the periosteum where it is particularly exposed to compression, stretch or other trauma. The nerve is covered by unusually thick fascia as it winds around the lateral surface of the fibular neck, deep to the origin of the peroneus longus. This fascia may be stretched tightly over the nerve to form an osteofibrous tunnel in which the nerve runs for about half an inch. **(Russell, 2006; Marwah, 1964)** There may sometimes be an opening in the origin of the peroneus longus muscle through which the nerve passes. **(Kopell & Thompson, 1976)**

Contributing to the tunnel are the two heads of the peroneus longus muscle that form a bridge under which the common peroneal nerve courses. The bridge is made up of the dense fascia overlying the two heads. The superficial head attaches to the head of the fibula and adjacent tibia. The deep head attaches to the neck of the fibula below the nerve. This has been called the fibular tunnel, where entrapment has been reported. **(Butler, 2000; Schweitzer et al., 1997; Lusskin, 1982; Kopell & Thompson, 1976; Fettweis, 1968)**

Palpation of the common peroneal nerve can be accomplished by beginning at the apex of the popliteal fossa and following the medial side of the biceps tendon distally and laterally to the posterior head of the fibula. **(Gray's Anatomy, 2004)** Here the nerve can be palpated against the bone. At this point 60% of the nerve is connective tissue and is quite flattened here and will not roll under the fingers as it will at the posterior knee.

Peroneal nerve entrapment usually involves the common peroneal. The extent of the impairments depends upon whether the entrapment involves the fascicles of the deep or the superficial peroneal nerve, or both, and also on the duration of the compression. The common peroneal is vulnerable to compression where it is fixed and angulated at the fibular head and neck. **(Sunderland, 1953)** The etiology relates to its superficial position and lying directly on the bone under the sharp crescentic arch of the tendinous origin of the peroneus longus. The disturbance may result from direct trauma, plaster cast compression, external pressure, fracture of the fibula, or pressure from a neoplasm. Indirectly, osteoarthritis of the knee or platform fracture of the tibia may be involved.

Entrapment of the common peroneal may be due to traction on it, especially from the superficial peroneal nerve. With inversion and plantar flexion of the foot, the nerve is pulled taut against the fascial edge, close to the fibular head. **(Butler, 2000)**

Common peroneal nerve entrapment can develop at the level of the fibular head from habitually sitting with one leg crossed over the other. **(Staal et al., 1999)** The nerve is vulnerable to compression in this position because it is relatively fixed, and flexion of the knee exerts abnormal tension on it. In this position the nerve is also compressed against the neck of the fibula. Both the pressure and the traction can be increased in positions of crouching, squatting, or kneeling. This is particularly problematic if the same position is repeatedly required in the person's occupation. People most vulnerable to this compression are those who are tall, long-legged, and asthenic. **(Nagler & Rangell, 1947)** External pressure caused by casts and tight, high boots may also cause compression of the common peroneal nerve. **(Staal et al., 1999; Mitra et al., 1995; Kopell & Thompson, 1976)** The neurologist Staal **(Staal et al., 1999)** observes that for patients who are bed-ridden or spend many days in hospital "…the common peroneal nerves at the head of the fibula should be protected by soft padding around the head of the fibula. It often proves difficult to convince residents and nurses of the rationale of these measures." Also, for patients with a leg in a plaster case that reaches as far as the head of the fibula or higher, "…the common peroneal nerve should be protected, preferably by an appropriate window in the plaster cast. Such precautions may avert not only pressure palsies but lawsuits. It makes sense to instruct every patient with a leg plaster to check every day if the strength of the toe extensors is still normal; if there is weakness the plaster should be opened."

The initiating injury of common peroneal nerve entrapment may be a sprained ankle. There may be a continued tendency for the ankle to turn over due to muscle weakness. **(Palmieri-Smith et al., 2009; McVey et al., 2005; Sidey, 1969)** Muscular imbalance and structural distortion applicable to applied kinesiology examination and treatment may result from trauma or have an insidious origin. Often attention is focused on the ankle weakness as a supposed ligamentous laxity, yet such laxity is demonstrable in only a few cases. This condition, if allowed to progress, could develop into a complete foot drop. Foot drop is a deceptively simple term for a potentially complex disorder. Foot drop is most commonly associated with intervertebral disc syndromes and radiculopathy, but also with peripheral

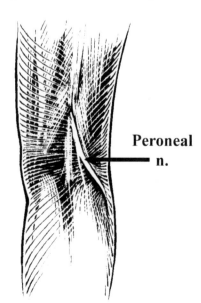

Common peroneal nerve.

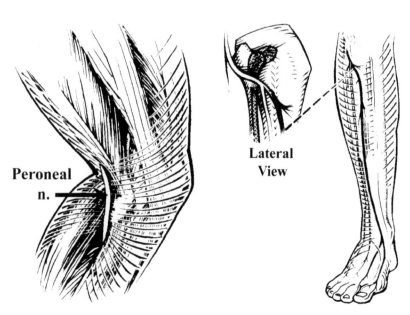

Common peroneal nerve adjacent to fibular neck.

nerve entrapments, neuropathies, stroke, drug toxicities, diabetes mellitus, multiple sclerosis, peripheral arterial disease or hyperthyroidism, myofascial trigger points in the dorsiflexor muscles of the ankle, among others. (**Brief et al., 2009**) Walking with a more or less complete foot drop leads to the typical stepping gait in which there will be compensatory overaction of the hip and knee flexors to gain clearance for the foot during the swing phase. With moderate foot drop the foot may still land on the heel but immediately afterwards the foot dorsiflexors give way, producing a flat-footed sound. With mild weakness the patient may appear to walk normally to an inexperienced examiner, but the patient is not able to walk on his heels or will swing the foot with less clearance from the floor than on the normal side.

Symptoms of common peroneal entrapment may be both motor and sensory, or either one individually. The most common symptom is pain, which may be severe and disabling, but paresthesia and numbness can also occur. Nocturnal cramping is common, and occasionally there is subjective coldness below the knee and in the foot. (**Staal et al., 1999; Sidey, 1969**) Usually the condition is unilateral, but occasionally it can be bilateral. (**Moller & Kadin, 1987**)

By way of the superficial and deep peroneal nerves the common peroneal supplies the dorsiflexors and everters of the foot, the skin on the lateral portion of the leg and dorsal portion of the foot, a portion of the ankle joint, and the mid-tarsal joints. (**Moore et al., 2009**) When entrapment is at the level of the common peroneal, symptoms and objective tests will be present in the distribution of both the superficial and deep peroneal nerves. Travell and Simons (**Travell & Simons, 1992**) note that when symptoms of neural entrapment of the peroneal nerves are present, the peroneus longus muscle must be evaluated and treated due to its ability to compress neural structures. The ankle jerk is typically normal with disorders of the peroneal nerve. If this reflex is decreased in combination with inhibition of the foot dorsiflexor muscles on manual muscle testing, the disturbance is more likely to be proximal, in the course of the sciatic nerve or in combination with the L5 and S1 nerve roots. A diminished ankle jerk indicates involvement of the S1 nerve root or sciatic nerve. Because sensory deficits in lumbosacral nerve root lesions are vaguely demarcated, they are not very helpful in distinguishing between entrapments of the peroneal nerve and the L5 nerve root. EMG studies are helpful in making this differential diagnosis. (**Staal et al., 1999**) CT or MRI scans may well show a disc protrusion, but such abnormalities are very common in asymptomatic people. (**Jensen et al., 1994**)

Entrapment of the common peroneal nerve may be associated with various types of athletic activities. (**Mitra et al., 1995**) Turco (**Turco, 1987**) comments on factors that are predisposing to this type of involvement. "The peroneal nerve winds around the neck of the fibula and is closely adherent to the bone. The biceps tendon and lateral collateral ligament are lax in flexion, hence stability of the fibular head is correspondingly less in the flexed knee position. Slight rotary and axial motion of the head of the fibula occurs with ankle motion. In some individuals the head of the fibula is more prominent, hypermobile, and laterally positioned. Runners with generalized joint laxity, hypermobility of the proximal tibiofibular joint, and hyperextension of the knees appear predisposed to develop peroneal neuritis associated with overuse." Some athletes with peroneal neuritis have a genu varum that predisposes the nerve to tension. Symptoms may develop only after running or other exercise. In this case, the patient should be examined before and after exercise. (**Leach et al., 1989**)

Under normal conditions, the common peroneal nerve at the neck of the fibula is not sensitive to palpation or to gentle percussion. When there is entrapment, it generally shows a local tenderness. The major symptoms may be summarized as follows: pain, paresis, paresthesia and numbness, and nocturnal cramp. (**Russell, 2006; Sidey, 1969**)

Stretching the nerve usually increases pressure on it at the fibular head entrapment site. Diagnostic stretch may be done by straight leg raise, which may cause proximal and distal pain in the presence of common peroneal entrapment. Inversion and plantar flexion of the foot may increase pain from the nerve being stretched taut against the fascial edge near the fibular head. Pain may not be localized to the distribution of the peroneal nerve. There may also be antidromic impulses causing sciatic nerve pain. (**Mense & Simons, 2001; Kopell & Thompson, 1976**)

There may be post–traumatic autonomic dystrophy (causalgia) in association with neuropathy of the peroneal nerve. When observing x-rays look for osteoporosis, which is associated with causalgia. (**Mense & Simons, 2001; Kopell & Thompson, 1976**) Peroneal nerve irritation may also be associated with a popliteal (Baker's) cyst, peroneal nerve cyst, compression by a sesamoid bone in the gastrocnemius muscle, or a bunching of fibers from an avulsed or ruptured peroneus longus muscle. (**Travell & Simons, 1992**)

Kopell and Thompson (**Kopell & Thompson, 1960**) find that this condition is often misdiagnosed as subtalar traumatic synovitis or continuing subluxation of the talus in the ankle mortise under inversion stress. They recommend treatment with a lateral sole wedge to relax the peroneal musculature and nerve.

Ironically, muscles supplied by the deep and superficial peroneal nerves often test weak with applied kinesiology initially; then after the "weakness" is corrected with the five factors of the IVF, they test positive to the muscle stretch reaction, indicating a shortened muscle. This often occurs with the peroneus longus and brevis. Delay in the peroneal muscle's onset times during contraction has been observed when this muscle is injected with an irritant. (**Richie, 2001**) Generally, the timing of activation of synergistic muscle groups during co-contraction and reciprocal activation of muscles are also affected in injury and pain, and this usually produces muscular inhibition. (**Lederman, 2010**)

The most common finding causing weakness of the muscles is a subluxation of the fibular head on the tibia. In the presence of a weak peroneus longus and brevis, one can challenge the proximal fibula to find the optimal vector that creates strength in the peroneus longus and brevis. The fibular head is usually posterior, and it needs to be adjusted in a generally anterior direction. Take care to determine the exact direction by vector challenge. It is important when evaluating for fibular head dysfunction that the distal articulation of the fibula be carefully examined as well at the ankle joint.

Heidrich (**Heidrich, 1993**) offers a case-report on a

50-year-old physically active male with left leg neuralgia of 3 months duration due to a fibular head subluxation. Symptoms began as a burning sensation that extended from the lateral knee to the dorsum of the foot. Analgesics, anti-inflammatories, physical therapy, and orthopedic treatments provided little relief and the patient was on Percocet® to control pain. Lumbar MRI was negative for discopathy. EMG and nerve conduction velocity tests were positive for left deep peroneal radiculopathy. Orthopedic testing of the knee was unremarkable, with a left genu valgus present. Palpatory pain was elicited at the anterosuperior border of the fibular head. Manual muscle testing showed weakness of the left sartorius, tensor fascia lata, popliteus, anterior tibialis, and peroneus tertius muscles. An AK posterior to anterior challenge of the fibular head produced weakness in a previously strong indicator muscle, indicating an anterior displacement. This displacement was manipulated and fascial release was performed on the tensor fascia lata muscle. Within 9 days there was resolution of the deep peroneal nerve symptomatology.

When the peroneus longus and brevis test strong in the clear, evaluate for muscle stretch reaction. When positive, the treatment usually effective is percussion, trigger point pressure release, or stretch and spray methods. (**Leaf, 2010; Cuthbert, 2002; Walther, 2000**) Following treatment for muscle stretch reaction, there will usually be an increased range of motion in foot inversion combined with plantar flexion. There should also be reduction of pain caused by stretching the nerve with inversion and plantar flexion. All muscles of the foot, leg, and knee should be evaluated and corrected as appropriate. There is frequently extended pronation in this condition.

Superficial Peroneal Nerve

The superficial peroneal (musculocutaneous) nerve branches from the common peroneal deep to the peroneus longus. It supplies the peroneus longus and brevis muscles. In the distal lateral portion of the leg, the nerve divides into two branches before or after piercing the deep fascia. It is at the point of passage through the fascia that entrapment neuropathy may occur. (**Chaitow & DeLany, 2002**)

The medial branch of the superficial peroneal supplies the skin of the medial side of the 1st toe, and the skin between the 2nd and 3rd toes. The lateral branch has dorsal digital branches that supply the skin between the 3rd through 5th toes and the skin over the lateral side of the ankle.

The nerve is strictly sensory from the deep fascial point on; consequently, entrapment at that area can only result in pain and altered sensation. There will be no associated muscle weakness. Digital pressure and palpation of the nerve will probably increase the pain. Antidromic impulses may cause pain in the proximal leg.

Inversion or plantar flexion trauma stretches the superficial peroneal nerve as it passes through the fascia, or at the nerve's anchor at the fibular neck. This type of trauma often causes a weakness of the peroneus longus and brevis. (**Leach et al., 1989; Kopell & Thompson, 1976**) Applied kinesiology challenge will usually identify a fibular head subluxation relating with common peroneal nerve entrapment. This seems to perpetuate the weakness of the peroneus longus and brevis, making the individual subject to the recurrent trauma of "twisting an ankle," usually in the direction of inversion. When the fibular head is adjusted, the peroneus longus and brevis will usually immediately test strong. Effective treatment of this articulation upon global muscle and joint dysfunctions have been recorded. (**Zatterstrom et al., 1994**)

Metatarsalgia of the 2nd and 3rd metatarsals may develop as a result of poor function of the peroneus longus muscle. Part of its insertion is to the base of the 1st metatarsal. Weakness causes a failure of the 1st metatarsal to flex during the last phase of stance, shifting weight to the 2nd and 3rd metatarsals. (**Dananberg, 2007; Lusskin, 1982**)

When there is structural imbalance, the fascia lata may be tight and, because of its continuity with the deep fascia of the leg, cause tension on the portion of the fascia through which the superficial peroneal nerve passes. Here again we see the continuity of fascia throughout the body. Kopell and Thompson (**Kopell & Thompson, 1976**) point out that fasciotomy of the fascia lata has been done in the past with some success in reducing irritation of the superficial peroneal nerve. They do not recommend a revival of the procedure; they just note the mechanism. Staal et al. (**Staal et al., 1999**) recommend surgery for the peroneal nerve only in three instances: 1) penetrating trauma, 2) with local mass lesions, and 3) with compartment syndromes. A tight fascia lata can be identified by Ober's test. (**Ober, 1987**) The tensor fascia lata should be tested for muscle stretch reaction and treated with percussion, trigger point pressure release, or stretch and spray methods if positive. (**Leaf, 2010; Cuthbert, 2002; Walther, 2000**) This will often eliminate a positive Ober's test.

The dorsal cutaneous nerves of the superficial peroneal nerve can be compressed by tight shoes. This is especially a problem with running shoes (**Abshire, 2010; Maffetone, 2010; Krissoff & Ferris, 1979**) and with ski boots. (**Whitesides, 1982**) This is corrected by a larger shoe, a shoe that laces over a wider portion of the foot, or one that

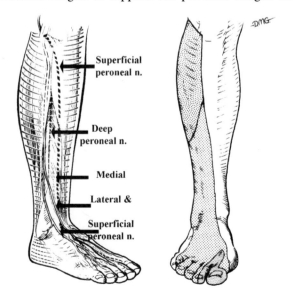

Superficial and deep peroneal nerves

Dermatomes of common superficial and deep peroneal nerves

has fewer eyelets (three or four) that do not cross over the area of entrapment. One can also skip the eyelets over the area of entrapment when lacing the shoes.

Deep Peroneal Nerve

The deep peroneal (anterior tibial) nerve begins at the bifurcation of the common peroneal nerve between the fibula and the proximal part of the peroneus longus muscle. It passes down the leg between the extensor digitorum longus and the tibialis anterior to the inferior extensor retinaculum.

In the leg the deep peroneal nerve supplies muscular branches to the tibialis anterior, extensor hallucis longus, extensor digitorum longus, and peroneus tertius. Deficiency of these muscles is caused by entrapment at the common peroneal nerve or higher. There is rarely an involvement of the deep peroneal nerve itself to affect these muscles. The deep peroneal nerve also has an articular branch in the leg to the ankle joint.

Entrapment of the deep peroneal nerve may occur in the anterior tarsal tunnel. This is a deep passage on the dorsum of the foot covered by the inferior extensor retinaculum, which extends from the lateral to the medial malleolus. After passing through the tunnel, the nerve divides into medial and lateral branches. The medial branch supplies the skin between the 1st and 2nd toes and the 1st dorsal interosseous muscle. The lateral branch innervates the extensor digitorum brevis and the lateral tarsal joints. **(Moore et al., 2009)**

The nerve on the dorsum of the foot, especially over the tarsal bones, is subject to trauma that can be caused by a striking blow or ill-fitting shoes. Trauma that stretches the nerves, such as severe plantar flexion and/or inversion, can also cause neuropathy locally in the foot or further up in the peroneal nerve. An increase of pain is present with stretching of the nerve by forced plantar flexion of the foot and toes. This is not specifically a lesion of the peroneal nerve in the foot, because stretching applies traction in which the common peroneal nerve is anchored to the fibular neck. When only the deep peroneal nerve is involved, there will be dorsiflexion weakness but strength on eversion. In addition, there will be sensory loss between the 1st and 2nd toes. Pressure over the area of suspected neuropathy should cause both local pain and reproduction of the radiated pain.

Tibial Nerve

The tibial nerve is the larger terminal division of the sciatic nerve. Its source consists of the 4th and 5th lumbar and 1st, 2nd, and 3rd sacral nerves. As it descends the leg it is well-protected by muscle. **(Moore et al., 2009; Russell, 2006)** It descends through the popliteal fossa and continues in the leg with the posterior tibial vessels to enter the tarsal tunnel. The nerve splits into the medial and lateral plantar nerves within the tarsal tunnel 93% of the time and proximal to it the other 7%. It also gives off the calcaneal nerve, the origin of which varies. **(Andreasen Struijk et al., 2010)** Prior to the tarsal tunnel

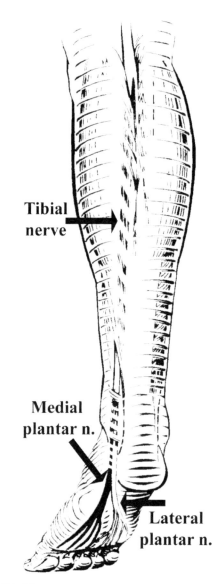

Tibial Nerve.

there is minimal or no incidence of peripheral nerve entrapment because of the excellent protection provided the nerve by the muscles.

When there is entrapment of the tibial nerve in the leg, its source is often a deep posterior compartment syndrome. **(Lusskin, 1982)** Compartment syndromes are associated with muscle activity in which the contents of the compartment enlarge past the closed compartment's ability to yield. An acute compartment syndrome can be a surgical emergency; it must be differentially diagnosed from other conditions.

Tarsal Tunnel Syndrome

The first description in the literature of the tarsal tunnel syndrome was in 1962, when two operative cases were reported. **(Lam, 1967)** Lam indicates that in the early investigation of the carpal tunnel syndrome, advanced cases showed marked motor and sensory disturbance; at operation, pathological changes of the median nerve were evident. Tarsal tunnel syndrome is also analogous to the carpal tunnel because the tunnel contains tendons and blood vessels together with the tibial nerve. Lam states in his paper of 1967, "Nowadays the carpal tunnel syndrome is sufficiently well recognized to ensure that most cases are treated before this stage is reached. The same pattern of events may evolve in the case of the tarsal tunnel syndrome." Unfortunately, there are still a large number of people with unrecognized entrapment at the tarsal tunnel. Many physicians overlook the tarsal tunnel syndrome. (**Hudes, 2010**) Alshami et al. (**Alshami et al., 2007**) and Keck (**Keck, 1962**) note that tarsal tunnel syndrome is frequently diagnosed as acute foot strain or plantar fascitis. In applied kinesiology experience, many patients who have previously been unsuccessfully treated for foot dysfunction are found to have functional tarsal tunnel syndromes. This allows the plantar intrinsic muscles to become weak and causes increased development of extended pronation affecting total body function. (**Mondelli et al., 2004**)

The term "functional tarsal tunnel syndrome" was used above to differentiate from frank pathological entrapment of the tibial nerve, which may require surgical decompression. The functional tarsal tunnel syndrome highlighted in this discussion is the type of nerve entrapment that causes dysfunction, yet will respond and return to normal with conservative treatment. Since Goodheart's (**Goodheart, 1971**) introduction of tarsal tunnel treatment into applied kinesiology, many have been treating it successfully. (**Hudes, 2010; McDowall, 2004; Hambrick, 2001; Cuba et al., 1995**) With an increased awareness of this type of tarsal tunnel syndrome we may yet treat these conditions before marked motor and sensory disturbance with pathologic change develops.

Anatomy

The tarsal tunnel is a superficially located osseous tunnel behind and below the medial malleolus and covered by the flexor retinaculum (laciniate ligament) with the bones making up the base of the tunnel. The tibial nerve passes through this osteofibrous passageway with the tendons of the tibialis posterior, flexor digitorum longus, and flexor hallucis longus muscles, each within its own synovial sheath. The other components of the neurovascular bundle are the posterior tibial artery and vein. (**Gray's Anatomy, 2004; Goodgold et al., 1965**)

The flexor retinaculum extends between the malleolus and the medial side of the calcaneus but has several deep fibrous septa that blend with the periosteum of the calcaneus. The neurovascular bundle in the tarsal tunnel is often attached to some of these septa, rendering itself more liable to minor degrees of traction on movement of the foot. (**Henricson & Westlin, 1984**)

Vascular supply to the nerve may have a bearing on its susceptibility to compression. As in the carpal tunnel syndrome, the median nerve in the wrist and the posterior tibial nerves have better arterial supply than the ulnar and lateral tibial nerves. The ulnar and lateral tibial nerves rarely have "spontaneous" compression symptoms, though they run through osteofibrous tunnels. The median and tibial nerves have a much more common incidence of "spontaneous" compression. Their ample arterial blood supply may make them more susceptible to the effects of localized vascular insufficiency. Lam (**Lam, 1967**) states, "When exploring the posterior tibial nerve one is struck by the density of the areolar tissue binding the structures under the retinaculum, and by their relative lack of mobility — compared with the extreme mobility and lack of adherence of the median nerve in the carpal tunnel. Hence even slight degrees of compression, possibly caused by oedema following minor strains, may produce vascular insufficiency locally and render a nerve lesion more likely."

It is suggested by Mense and Simons, and Butler (**Mense & Simons, 2001; Butler, 2000**) that sensory symptoms in nerve compression syndromes are partially due to arterial insufficiency. More slowly occurring motor paralysis is thought to be due to later structural changes produced within the nerve, and the paralysis is less likely to benefit from decompression. Manipulative treatment is known to have an effect on the diameter of arteries and veins. (**Rome, 2010, 2009**) It is important therefore to make the diagnosis and treat the patient before the onset of demonstrable motor involvement resulting from ischemia. Fluid flow through a blood vessel is strongly affected by small changes in the diameter of the vessel, demonstrating how even a small swelling or compression around the neural tissues will severely reduce the flow of blood and lymph in the area. (**Zink, 1977**)

In Lam's (**Lam, 1967**) early series of ten cases surgically treated, one case had enlarged tortuous veins within the tarsal tunnel. There was no other demonstrable pathology in any of the cases. This points out the necessity in an applied kinesiology examination of determining whether actual paralysis develops in the muscle from nerve impingement, or whether there is functional weakness as observed in the manual muscle test. This is readily determined by challenge and therapy localization. If one is unable to return muscle function to normal, as observed by manual muscle testing, surgical intervention may be necessary. Hoskins et al. (**Hoskins et al., 2006**) and Denny-Brown and Brenner (**Denny-Brown & Brenner, 1944**) have shown that when peripheral nerve entrapment is treated in its early stages, it is reversible.

Extended pronation (**See chapter 4**) is a factor in nearly all cases of tarsal tunnel syndrome treated by applied kinesiologists. (**Leaf, 2010; Goodheart, 1998**) Because of the posterior movement of the calcaneus on the talus during extended pronation, the flexor retinaculum is stretched. In extended pronation there is intermittent compression, then release, then compression, then release of the laciniate ligament on the posterior tibial nerve. (**Rosson et al., 2009; Carrel & Davidson, 1975**)

Symptoms

Tarsal tunnel nerve entrapment can be discovered by an astute examiner when the patient has no complaint of foot or ankle dysfunction. As discussed in **Chapter 4**, foot dysfunction plays a major role in total body organization.

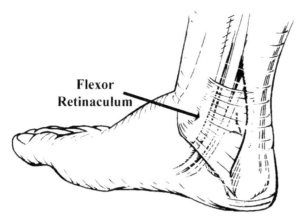
Tarsal tunnel - medial & posteromedial views

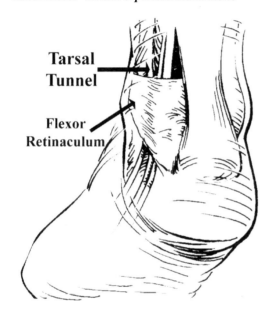

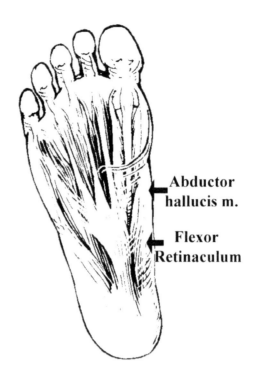
Plantar view

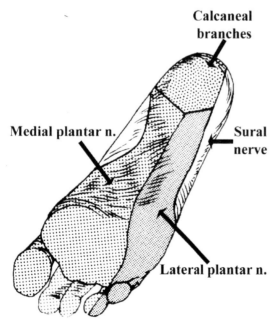
Dermatomes of plantar surface of foot

Recognizing an asymptomatic tarsal tunnel entrapment is just as important as examining and finding the cause of the painful foot.

Patients' complaints of tarsal tunnel syndrome include burning pain and paresthesia in the toes and along the sole of the foot. This may cause "burning feet" or the "restless legs syndrome." **(Staal et al., 1999)** One study reports that in 43% of cases the pain is more severe at night. **(Mondelli et al., 2004)** Moving the limb, getting out of bed, or hanging the limb over the edge of the bed may provide relief. **(Langer, 2007)** Transitory nerve ischemia or compression may be relieved by massage or walking. **(Barral & Croibier, 2007; Edwards et al., 1969)**

Proximal pain, tingling and numbness may radiate up the leg from a tarsal tunnel syndrome and is seen in approximately 30% of cases and is called the Valleix phenomenon. **(Lau et al., 1999)** It may simulate a disc problem, peripheral vascular disease, or neuritis. Sometimes the pain is attributed to an existing condition, such as diabetes or peripheral vascular disease, when the problem is really an unassociated tarsal tunnel syndrome. This error frequently occurs with older patients. This is why tarsal tunnel syndrome may be under-diagnosed because it can be difficult to differentially diagnose it from other conditions of the foot. For these reasons, specific as compared to group manual muscle tests are an important advantage in the diagnosis of specific articular-muscular impairments. **(Leaf, 2010; Walther, 2000; Goodheart, 1998; Kendall et al., 1993)**

Many aspects of an examination may appear normal when there is entrapment at the tunnel. There may be normal dorsalis pedis and posterior tibial pulses. Skin color, hair distribution, and capillary circulation may also appear normal. **(Keck, 1962)**

With chronicity the plantar muscles of the foot may be atrophied, giving an appearance of a high arch. There is indication in the literature that sensory deficit develops prior to motor deficit; (**Lusskin, 1982**) this does not agree with applied kinesiology findings nor with more recent neurophysiological research. When muscles are painful, fatigued, or injured, there is an inhibition of muscle strength, timing, and a decrease in their endurance. (**Racinais et al., 2008; Mense & Simons, 2001**) In any exercise inducing muscle damage, decreased neural drive to the muscles is thought to be an attempt of the neuromuscular system to protect the muscle-tendon unit from additional damage. (**Nicol et al., 2006; Strojnik & Komi, 2000**) For this reason muscular imbalance is thought to be a primary reflection of the functional state of the neuron; this is particularly true because of the extreme sensitivity of the muscle spindle cells, where a muscle spindle reacts to a pull of only 1 gram and a stretch of 1/1000th millimeter. (**Korr, 1997, 1979**) This makes the muscular system and the manual muscle test an extremely sensitive organ and tool.

One often finds severe plantar muscle atrophy with no sensory deficit or hyperesthesia. (**Mondelli et al., 2004**) The patient with tarsal tunnel entrapment will have weak plantar muscles, and the long toe flexors will probably be strong. This, of course, is because the long muscles receive their nerve supply prior to the tarsal tunnel, and the intrinsic muscles after the point of entrapment. With this muscle imbalance, the patient will have hyperextension at the metatarsophalangeal articulations and hyperflexion at the interphalangeal articulations, giving a claw–like appearance to the toes, which is often called "claw toes." (**Lau et al., 1999**) This is because the long muscles insert into the distal phalanx, and the intrinsic muscles into the intermediate phalanx. When the individual stands or walks, one can see the distal phalanx gripping the substrate with elevation of the proximal phalanges. There will often be calluses on the distal ends of the toes.

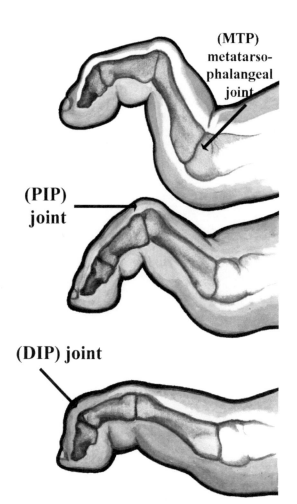

(With kind permission, ICAK-USA)

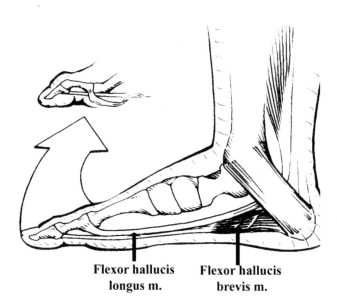

Insertion of muscles to the toes.

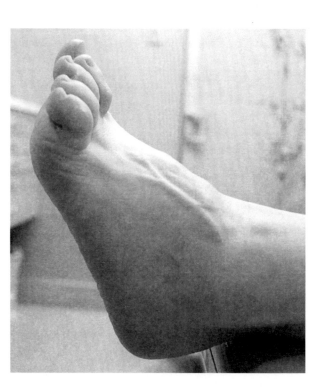

Claw Toes

Etiology

Often there is no known etiology for functional tarsal tunnel syndrome. Space-occupying lesions such as ganglions or tumors are rare. **(Taguchi et al., 1987)** Even in a surgical series of sixteen cases reviewed by Edwards et al., **(Edwards et al., 1969)** eight of the cases were "spontaneous" entrapment, that is, no space-occupying lesion was found. Five were post-traumatic fibrosis due to fracture, three had accessory or hypertrophied abductor hallucis muscle, and one was due to tenosynovitis.

Tenosynovitis within the tarsal tunnel can develop in runners, **(Abshire, 2010)** from trauma, **(Kopell & Thompson, 1976)** or from infectious processes. **(Kopell, 1980)** The latter, in effect, causes a space-occupying lesion to produce pressure on the nerve. There will be crepitation at the tunnel and severe pain on digital pressure over the tendons.

Circulation problems may be responsible for tarsal tunnel syndrome. Peripheral nerves in a diabetic are more easily affected by compression force than are normal nerves. **(Veves et al., 2002)** Venous engorgement of the tunnel can develop as a result of proximal venous obstruction or valvular deficiency. **(Goodgold et al., 1965)** Some of the pain present with an obvious venous stasis or thrombophlebitis may be caused by encroachment on the posterior tibial nerve rather than pressure from the vein distension alone. **(Alshami et al., 2008)**

Any condition that creates fluid retention such as the ileocecal valve syndrome can be the final factor that causes entrapment at the tarsal tunnel. Helm et al. **(Helm et al., 1971)** studied 164 pregnant women to determine if there is an increased incidence of tarsal tunnel syndrome during pregnancy. They found 56.1% had abnormal nerve conduction through the tarsal tunnel. Twenty of the subjects were studied six weeks postpartum. Thirteen of these had abnormal conduction studies of the tibial nerve during the course of pregnancy, and twelve (92.4%) reverted to normal values at postpartum follow-up. Of the subjects with abnormal studies, 69.5% had some complaint of leg cramps, burning of feet, tingling, numbness or pain during the pregnancy.

Tarsal tunnel syndrome must be differentially diagnosed from interdigital neuritis, dropped metatarsal heads, plantar calluses, arch strain, various types of arthritis, tenosynovitis, peripheral neuritis, peripheral vascular disease, and various causes of sciatic pain. **(Staal et al., 1999; Kopell & Thompson, 1976; Aguayo, 1975; Lam, 1967; Goodgold et al., 1965)** Antidromic impulses can cause tenderness along the entire sciatic nerve. The pain may simulate root pain of spinal origin. **(Russell, 2006; Staal et al., 1999; Ricciardi-Pollini et al., 1985)** Moloney **(Moloney, 1964)** reports on a case where three spinal surgeries were done without permanent improvement of pain until the tarsal tunnel was diagnosed and operated, which eliminated the painful condition. Patients who have a systemic propensity toward peripheral entrapment may have a history of problems in other areas such as a previous carpal tunnel syndrome. **(Mondelli & Cioni, 1998; McGill, 1964)**

Examination

When there is frank entrapment of the tibial nerve in the tarsal tunnel, the Tinel sign is sometimes present over the tunnel or the medial arch. One study reports that Tinel's sign is positive in only 67% of cases. **(Mondelli et al., 2004)** The Valleix phenomenon or nerve trunk tenderness may be present proximal or distal to the area of entrapment. **(Lau & Daniels, 1999)** Palpation over the retinaculum may reveal tenderness or a small fusiform swelling of the nerve. **(Barral &**

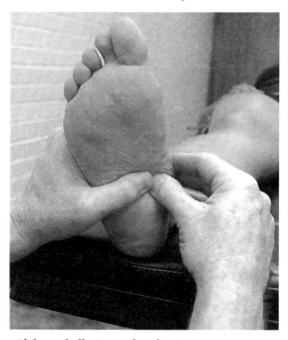

Abductor hallucis muscle palpation

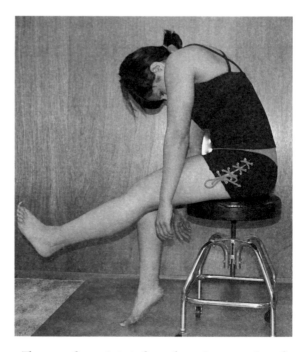

The neurodynamic test places the entire neuraxis under tension. The tibial & peroneal nerves' tension is felt at the head of the fibula.

Croibier, 2007) Hypo- or hyperesthesia may develop in the tibial nerve branches (medial and lateral plantar nerves), or the calcaneal branches. Sensory disturbance should only be in this distribution, and it should not affect the dorsum of the foot except over the distal phalanges of the toes. There may also be loss of two–point discrimination. Pressure on the calf of the leg created by inflating a sphygmomanometer cuff may reproduce symptoms on the affected side more quickly than in a normal foot. (**Butler, 1991; Cipriano, 1985; Henricson & Westlin, 1984**)

There must be careful differentiation in cases of leg pain. In a series of thirteen patients, seven of whom were operated and four being treated conservatively but expected to go to operation, Mann (**Mann, 1974**) found over half the patients had pain radiating up the medial side of the calf but not past the knee. In one patient the sharp pain in the plantar aspect of the foot could be reproduced by straight leg raising and dorsiflexion of the patient's foot.

Maneuvering the heel into a valgus position narrows the tarsal tunnel and may increase pain; on the other hand, maneuvering the heel into a varus position may reduce pain. (**Butler, 2000; Staal et al., 1999; Kopell & Thompson, 1976**) Neurodynamic testing (**Butler, 2000, 1991**) employs this type of nerve tension testing. When a neurodynamic tension test is positive (i.e. pain or other sensations result from one or another element of the test – for example, the initial position alone, or with 'sensitizing' additions) it indicates that there exists abnormal mechanical tension (AMT) somewhere in the continuous nervous system, but not that this is necessarily at the site of reported pain. For more detail of this assessment and treatment approach, the books by Butler (**Butler, 2000, 1991**) are recommended.

Patients may not complain of muscle weakness in a tarsal tunnel syndrome; (**Aguayo, 1975**) however, muscle deficiency will be found if carefully examined for by inspection, palpation, and specific muscle testing and AK sensorimotor challenges. The abductor hallucis located along the medial longitudinal arch is inspected and palpated for atrophy, along with the other intrinsic muscles of the foot. The ability to flex the toes at the metatarsophalangeal articulation is evaluated by muscle testing. There is no weakness of toe extensor muscles.

The flexor digitorum longus and brevis and the flexor hallucis longus and brevis should be evaluated. The long muscles will have improved function over the short ones. Challenge, usually directed to the calcaneus, may improve the function of the intrinsic muscles. There will be positive therapy localization over the tarsal tunnel at the area of entrapment, and probably at the subtalar articulation, making an immediate and non-invasive diagnosis of this disorder readily available to the clinician.

Electrodiagnosis should be done to determine nerve conduction in cases that are unresponsive to conservative care; however, there may be false–positive findings from these studies. Gatens and Saeed (**Gatens & Saeed, 1982**) studied the adductor hallucis, extensor digitorum brevis, 1st dorsal interossei, and abductor digiti minimi in seventy individuals with asymptomatic feet. They found that 38.6% had at least one of the four muscles examined showing abnormal potentials. They concluded that using abnormalities in intrinsic foot muscles by needle EMG as a diagnostic criterion could be misleading. An interesting question regarding studies like these is what an applied kinesiology examination would have found in the individuals with positive tests. Functional problems are often found in asymptomatic feet. (**Abshire, 2010; Maffetone, 2010**) It is possible in studies like these that the positive findings were in feet that were functionally inadequate but asymptomatic. If the muscles of the foot and ankle are dysfunctioning, the lower legs, knees, thighs, hips, lower back, shoulders, neck and head are also disturbed. (**Maffetone, 2010; Dananberg, 2007; Chaitow & DeLany, 2002; Walther, 2000; Travell & Simons, 1992**)

Treatment

Most cases of tarsal tunnel syndrome respond well to the conservative approach of applied kinesiology. If there are neoplasms or other space–occupying lesions that are irreversible with conservative care, surgical intervention will be necessary.

Most often accompanying and frequently the precipitating factor for tarsal tunnel syndrome is extended foot pronation. The first effort toward correction is to examine for and correct extended pronation, which includes any subluxations of the foot and other foot dysfunction. Intrinsic muscles of the foot may not be corrected until specific adjustments are made for the tarsal tunnel because their nerve supply is being interfered with. (**See Chapter 4**) One of the most common forms of muscle inhibition comes from joint subluxations, commonly called arthrogenic weakness. Even non-noxious stimulation of the joints of the foot and ankle can elicit strong inhibition of the muscles of the foot and ankle. (**McVey et al., 2005**) There are few better options for patients with joint subluxations of the foot and ankle affecting the tarsal tunnel than specific manipulative joint correction. Non-manipulative treatments for the arthrogenic weakness of muscles will be less effective and more time-consuming when joint disturbances are present.

After correcting pronation, challenge the calcaneus in its relationship to the talus. The calcaneus will usually be subluxated posterolaterally, with its posterior surface somewhat superior. There are several methods for adjusting the calcaneus; it can be done with the patient supine or prone.

With the patient supine, the physician stands at the patient's feet facing the foot to be corrected. To adjust the calcaneus, the physician grasps the posterior superior surface of the bone solidly with his or her hand. The left

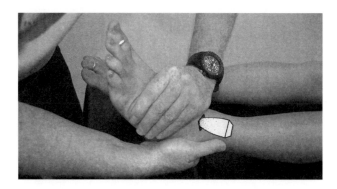

Calcaneus adjustment

Pain from calcaneus malposition & foot pronation.

hand makes a broad contact across the dorsal surface of the foot. A broad contact is important to avoid creating a subluxation with this stabilizing and controlling hand. The correction is an extension thrust directed toward moving the posterior surface of the calcaneus in a generally inferior anterior direction. The vector of force is determined by challenge. It usually requires moving the inferior portion of the calcaneus medially. It is often necessary to have the patient hold the side of the examination table to prevent slipping, especially if the table is covered with tissue paper. There is often an audible release of the calcaneus; however, it is not necessary for an adequate correction.

Another method is to flex the prone patient's knee to 45°. The physician stands on the side to be corrected. To correct the right calcaneus, the physician's right hand cradles the superior posterior aspect of the calcaneus between his thumb and forefinger. The other hand cradles the dorsum of the foot. The thrust is directed to the calcaneus as previously determined by challenge. Usually the thrust is directed to move the posterior inferior aspect of the calcaneus in an anterior, inferior, and medial direction.

Successful adjustment of the calcaneus and pronation correction are indicated by relief in the painful area inferior to the medial malleolus. Goodheart **(Goodheart, 1971)** states that the disappearance of this diagnostic feature is essential. It may be necessary to adjust the calcaneus several times to completely eliminate this pain.

Leaf **(Leaf, 2010)** points out that the navicular bone is a key in the correction of tarsal tunnel syndrome and other disorders of the foot. The navicular is controlled by the tibialis posterior and when the navicular drops inferiorly from tibialis posterior inhibition, the alignment and dynamics of the rest of the foot is disturbed. As the navicular descends it fails to support the talus and thereby causes the calcaneus to shift posteriorly. The posterior calcaneus that results produces increased pressure on the flexor retinaculum. The initiating disturbance in this sequence is the inferior navicular and it must be corrected.

When tarsal tunnel syndrome is associated with tendinitis, as from jogging, it is treated with ice. **(Redwood & Cleveland, 2003)** Correction of any foot, leg or other problems associated with the running pattern should be done to prevent recurrence.

The intrinsic plantar muscles should be individually evaluated. They often require treatment to the muscle proprioceptors, or need origin-insertion technique, or fascial release. The muscles are often exquisitely tender in the area requiring treatment. It may be of value to use the percussor or vibrator unit for treatment. **(IMPAC, 2011)**

It is very important to follow up with proper treatment of the extended pronation. Failure to adequately stretch the triceps surae if necessary, adjust subluxations or fixations, or to provide proper footwear will result in failure to permanently correcting the tarsal tunnel syndrome.

When conservative treatment fails, surgical release is often effective **(Rosson et al., 2009; Murphy & Baxter, 1986)** if proper diagnosis has been made. Surgical release may find a small tag or tags of tissue **(Lam, 1962)** often secondary to trauma. **(Edwards et al., 1969)** At the point of entrapment there may be enlargement of the tibial nerve, **(Keck, 1962)** with poor nerve conduction with electrical stimulation past the point of constriction. In a group of sixteen patients who required surgical decompression, Edwards et al. **(Edwards et al., 1969)** found that "Local injection of cortisone produced only transient, if any, improvement. Shoe modifications provided no improvement and arch supports always increased the severity of symptoms." Burning feet from tarsal tunnel entrapment do not improve with vitamin treatment. **(Babbage, 1965)** When there is a sensation of burning feet without the nerve entrapment, they often respond to vitamin B-complex. **(Cuthbert, 2007; Patrick, 2007)** Goodheart recommends vitamin B complex for burning feet; and burning feet accompanied by swelling in hot water needs thiamine. **(Goodheart, 1991)**

Plantar Nerves

The posterior tibial nerve branches into the medial and lateral plantar nerves and the calcaneal nerve in or around the tarsal tunnel. The medial and lateral plantar nerves arise within the tarsal tunnel 93% of the time and proximal to it the other 7%. The origin of the calcaneal nerve varies. **(Moore et al., 2009)** A portion of the calcaneal branch usually pierces the laciniate ligament. The other portion, which is the sensory supply to the skin inferior to the calcaneus, traverses the tunnel. Where the abductor hallucis muscle originates on the calcaneus, there are two openings through which the plantar nerves pass. The calcaneonavicular ligament is above the medial plantar nerve. Above the lateral plantar nerve is the origin of the quadratus plantae muscle on the calcaneus. These two structures form superior boundaries for the foramen and are locations for entrapment neuropathies. Excessive pronation stresses the nerve in the opening, bringing the superior boundaries down against the nerves. **(Langer, 2007; Lau & Daniels, 1999)** Since the medial and lateral plantar nerves go through individual openings, each can be subject to isolated entrapment neuropathy. **(Rosson et al., 2009; Staal et al., 1999)** Thus entrapment can be of an individual plantar nerve, of all three nerves at the tarsal tunnel, or at a higher level.

The calcaneal nerve is purely sensory; the medial and lateral plantar nerves are mixed. **(Moore et al., 2009)** They correlate with the median and ulnar nerves of the hand. The medial plantar nerve innervates the three-and-one-half medial digits, and the lateral plantar nerve the one-and-one-half lateral digits. The lateral nerve passes under the

Intrinsic plantar muscles.

flexor digitorum brevis and supplies the interosseous muscles and the adductor hallucis. The medial nerve innervates the abductor hallucis, flexor digitorum brevis, flexor hallucis brevis, and lumbrical muscles. The medial and lateral plantar nerves terminate as the interdigital nerves.

Extended pronation tightens the medial and plantar nerves against the calcaneonavicular ligament or the fibrous-edged openings of the abductor hallucis muscle. (**Kopell & Thompson, 1960**) In addition, extended pronation is conducive to narrowing of the tarsal tunnel. In both cases, the extended pronation must be corrected to obtain relief with conservative care. The exact location of entrapment is academic unless it is the type requiring surgical release; otherwise treatment is similar for either level of entrapment.

Pain from entrapment neuropathy of the plantar nerves is in the distribution of the nerves. There is tenderness at the point of nerve entrapment. This is located one finger width below and three fingers anterior to the medial malleolus in an adult. Pressure at this location should cause radiation to the forefoot in a painful condition. (**Russell, 2006; Cailliet, 1997**)

If there is sensory disturbance of the skin of the heel the entrapment site is higher, probably involving the tarsal tunnel. (**Hamilton, 1985**) If the heel is not involved in a tarsal tunnel entrapment, it is due to the calcaneal branch arising proximal to the tunnel. (**Staal et al., 1999**)

The more common cause of heel pain is plantar fasciitis, which must be differentiated from entrapment of the calcaneal branch of the tibial nerve. In plantar fasciitis there is tenderness of the fascia and its attachment to the calcaneus. In nerve entrapment the tenderness is over the medial anterior part of the heel pad, the anterior calcaneal branch of the tibial nerve, and the origin of the abductor hallucis muscle. X-rays of the calcaneus are negative. Stretching of the adductor hallucis and contraction of the flexors increase the pain; it decreases when the muscles relax. (**Travell & Simons, 1992**) Lusskin (**Lusskin, 1982**) discounts the frequency of heel pain being due to plantar spur, placing more interest on nerve entrapment. Applied kinesiology experience concurs with this, finding that functional peripheral nerve entrapment, foot strain, calcodynia (painful heel), and irritation of the plantar fascia are all part of the complex of extended pronation and general foot strain. (**See Chapter 4**) However, there can be frank neuropathy accompanying the more common structural strain cause of pain. (**Russell, 2006**)

Kopell and Thompson (**Kopell & Thompson, 1976**) describe interesting "...cases of a burning, causalgic type of foot pain." Characteristics of the condition are severe pain on standing or walking (to the point of almost complete disability), moderate hallux rigidus, atrophy of medial foot muscles, anteromedial plantar navicular pain, medial first toe hypoesthesia, and a groove on the first cuneiform observed on x-ray. The groove results from the tibialis anterior tendon wearing down the bone. Overactivity of the tibialis anterior was found to be a natural response of muscle activity to counteract extended pronation. With the extended pronation, entrapment of the plantar nerves developed with the subsequent symptomatic pattern. If conservative treatment was ineffective, they found it necessary to do a neurolysis. "The nerve was usually found to lie within a tough, constricted, fibrous sheath with barely enough room for its passage."

Entrapment of the medial plantar nerve can develop as a result of extended pronation while running for long distances. Rask (**Rask, 1978**) emphasizes the role of extended pronation in running problems by stating, "A fibromuscular tunnel for the nerve is formed by the abductor hallucis muscle and eversion of the foot brings the medial plantar nerve into stretch against the tunnel whose superior surface is the navicular tarsal bone." Entrapment has been found to occur to the medial plantar nerve just distal to the tarsal tunnel, at the entrance of the fibromuscular tunnel behind the navicular tuberosity. (**Oh & Lee, 1987**) It is at this point that there will consistently be tenderness on digital palpation when this condition is present. Often there is a burning heel pain associated with the condition. Rask (**Rask, 1978**) relates this to reflex excitation of the medial calcaneal branches that have their origin higher in the posterior tibial nerve. He applies the term "jogger's foot" to this condition.

It is probable that many applied kinesiologists have unknowingly effectively treated entrapment neuropathy of the plantar nerves. The therapeutic approach to two commonly treated conditions is similar to the therapeutic approach for the nerve entrapment. First is acute foot strain where pain is attributed to strain on the calcaneonavicular ligament and plantar aponeurosis. The second is painful heel, which is attributed to strain on the origin of the intrinsic plantar muscles at the calcaneus and the plantar aponeurosis attachment to the calcaneus. Both of these conditions require correction of foot subluxations and extended pronation, as well as returning all muscles to normal function using applied kinesiology techniques. If there is neuropathy of the plantar nerves as they pass under the calcaneonavicular ligament and the quadratus plantae muscle, an arch support may aggravate the condition. (**Kopell, 1980**) If it is impossible to maintain subluxation corrections, that is, the patient loses the correction as soon as he walks, it may be necessary to temporarily use tape support to the foot. The tape can supply a wider support than the arch support, preventing direct pressure on the area of nerve entrapment. As the condition is corrected, the patient may no longer need arch supports to maintain the correction, or he will be able to tolerate them if needed.

The medial and lateral plantar nerves terminate as the

interdigital nerves. They pass over the transverse metatarsal ligament on the plantar surface, and then angle to reach the dorsum of the foot. If the toes are hyperextended at the metatarsophalangeal articulation, the nerves may become irritated and cause pain between the toes. There is increased pain on digital pressure between the toes as opposed to over the metatarsal heads, as in metatarsalgia. This very often happens as a result of wearing high-heeled shoes, or sitting on one's haunches with the toes extended. **(Cailliet, 1997)**

Plantar Interdigital Neuralgia (Morton's Neuroma)

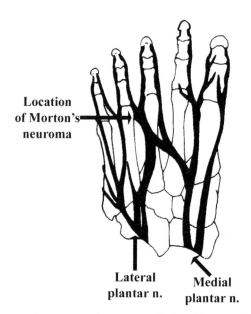

Anastomosis between medial and lateral plantar nerves.

Plantar interdigital neuralgia (Morton's neuroma) is a type of peripheral nerve entrapment not regularly included in an overview of the subject. **(Langer, 2007)** Often misdiagnosed, it may be mistaken for other causes of pain in the forefoot **(Staal et al., 1999)** or for other causes of metatarsalgia.

Plantar interdigital neuralgia was originally described by Thomas Morton in 1876. The condition has been known as Morton's toe, Morton's neuroma, Morton's neuritis, Morton's disease, Morton's affection of the foot, metatarsal neuralgia, and plantar interdigital neuroma. Care should be taken to avoid confusing it with the condition described by Dudley Morton in 1927, which is the 1st metatarsal insufficiency syndrome discussed in **Chapter 4**. The term "plantar interdigital neuralgia" or "neuritis" is chosen here because there is evidence that a true neuroma is not often present in this condition. **(Staal et al., 1999; Viladot 1982; Lusskin, 1982)**

Morton's original treatment for the condition was excision of the 4th metatarsophalangeal joint. In 1891 Bradford **(Bradford, 1891)** produced a paper on "Morton's Affection of the Foot" in which he stated, "It is somewhat singular that an affection which is not infrequent, in these days of thorough investigation of all ailments, should have attracted but little attention either in the researches of surgeons or of neurologists. The cases are usually classed among the ill-defined hysterical or nervous affections, and not thoroughly investigated...." It is interesting that with the passage of time there is still not complete agreement about the etiopathology and optimal treatment for the majority of these cases. Some look at the condition as a neuroma **(Betts, 1940)** with the recommendation of surgical excision, **(Turek, 1984)** while others take the position of microtraumatism to the nerve that walking brings about **(Viladot, 1982)** and treat the condition, in most cases, conservatively. **(Govendor et al., 2007; Brantingham et al., 1994)**

Anatomy

Plantar interdigital neuralgia is a lesion of one of the interdigital nerves of the foot, which arise from the medial and plantar nerves, at the point where these course between the heads of the metatarsal bones, just before they divide into two digital nerves. These lesions are found mostly between the 3rd and 4th metatarsal bones, particularly in women. **(Staal et al., 1999)** The 2nd cleft is occasionally involved. The condition has been attributed to the anatomical characteristics of the nerve in the involved individual.

Walther points out **(Walther personal communication, 2007)** that Viladot **(Viladot, 1982)** describes an anastomosis between the medial and lateral plantar nerves that is not illustrated in most general anatomy texts. In this case there is an enlargement of the nerve at the common site of Morton's neuroma. He proposes that this may "...explain why it is precisely this nerve that is the most sensitive to the microtraumisms that it receives from walking, especially when it is caught between the metatarsal heads." Turek **(Turek, 1984)** disagrees with this, stating that the digital nerve between the 3rd and 4th toes "...lies in relation to the plantar surface of the transverse ligament and therefore cannot be compressed between the metatarsal heads."

Symptomatic Pattern

The description of plantar interdigital neuroma has been mostly consistent, from Morton's **(Morton, 1876)** early description to the present. In the next major description, Bradford **(Bradford, 1891)** stated, "The symptoms can be removed by kicking off the boot and squeezing the forepart of the foot, especially if the heads of the middle metatarsals are at the same time pressed upwards by the finger. When this is done, a kind of grating is again felt together with a sharp twinge of pain, but almost immediately afterward the pain entirely ceases."

Betts **(Betts, 1940)** subscribes to surgically removing a neuroma. His description parallels many aspects of the early descriptions. A synopsis of his well-known paper is that neuritis of the 4th digital nerve is caused by a neuroma. It is initially characterized by an attack of severe pain while walking with shoes on. Relief is obtained by removal of the shoe, compression of the foot, and flexion of the toes. Upon examination, deep pressure over the site of the nerve does not produce acute pain, only tenderness. This correlates with relief being obtained by compression of the foot, and indicates that stretching of the nerve rather than compression initiates the attacks. The sharp pain of the acute attack is often followed by numbness and tingling. The acute pain may be diffuse, which also applies to the numbness and tingling after the attacks. A dulling sensation is frequently present adjacent to the sides of the 3rd and 4th toes. Passive movement of the foot and toes in various positions does not usually cause any pain. The condition usually occurs in what appears to be a generally healthy foot, perhaps with slight clawing of the toes. It does not occur in the weak, flat foot with poor circulation and calluses under the heads of the middle metatarsals. **(Betts, 1940)**

The fact that squeezing the metatarsals together relieves the acute attack, and that digital pressure over the site of disturbance does not elicit acute pain, differentiates this condition from intermetatarsal bursitis. Not all authors agree with this. Krissoff and Ferris **(Krissoff & Ferris, 1979)** state that transverse compression of the metatarsal heads causes pain. They further recommend confirming the diagnosis with lidocaine (xylocaine) ingestion to determine if it eliminates the pain. It is difficult to see how either of these approaches differentiates the condition from intermetatarsal bursitis.

The pain is not always localized. It can cause a radiation that is sciatic-like, radiating from the outer side of the foot into the calf and posterior thigh. **(Cohen, 1952)**

Many authors concur that this is not a true neoplasm; **(Subotnick, 1991; Viladot, 1982; May, 1976)** rather, it is a traumatic neuritis with proliferation of connective tissue elements.

Some surgeons move immediately to surgery following positive diagnosis of Morton's neuritis, **(Keh et al., 1992; Turek, 1984; Betts, 1940)** while others prefer a conservative approach, reserving surgery for unresponding cases. **(Maffetone, 2010; Logan, 1995; Subotnick, 1991; Viladot, 1982; Whitesides, 1982; Wickstrom & Williams, 1970)** Additionally, a subsequent neuritis may form after surgery and the permanent numbness of the toes may be very troublesome to the patient. In any event, there a re times when the patient is a poor surgical risk and conservative treatment is the preferred choice, regardless of the surgeon's usual preference. This is usually when there is poor vascular supply to the extremity. **(Veves et al., 2002)**

Treatment

Once plantar interdigital neuralgia has been diagnosed, further evaluation of the rest of the foot and factors influencing it must be made. Frequently extended pronation or some other factor increases the strain throughout the foot. Correcting remote factors enhances the opportunity for successful conservative care. Particular attention should be given to gait and foot function.

Evaluate the patient closely for subluxations or fixations throughout the foot. Make certain there is adequate range of motion between the metatarsal heads by evaluating with a scissors-type motion palpation. Use the rehabilitation exercises and procedures discussed in **Chapter 4**, and particularly evaluate footwear for tightness in the toe-box. If the patient has to buy new shoes to meet your requirements, be certain that the toe-box is wide enough, because often as corrections are obtained the forefoot widens considerably. **(Leaf, 2006)**

Specific padding designed to spread the metatarsal heads can be applied for plantar interdigital neuritis. To locate the position for the pad, press with the index finger between the metatarsals just behind or at the heads into the affected interspace. Continue changing the pressure and vector until the location is found that maximally spreads the toes. **(Langer, 2007; Ball & Afheldt, 2002)** A triangular foam adhesive rubber pad approximately 1/4" thick, 3/8" to 1/2" wide, and 3/4" long is applied to that location by the pad's adhesive and additionally held in place by tape. A small pad can be placed between the toes to additionally provide spreading action. It should be attached to only one of the toes.

Each patient must be evaluated to determine what structurally sound portion of the forefoot may be used to help transfer weight away from the involved area. Often an L-shaped pad **(described under foot support in Chapter 4)** can be applied to the 1st metatarsal to transfer some of the weight of the metatarsal arch to that ray. Determining the exact pad location to separate the metatarsals and to shift weight to another area is often a time-consuming process. Adequate examination time should be allowed, or the patient should be scheduled at the end of the day so the procedure is not rushed. When pads are applied, it is important to evaluate the individual with applied kinesiology techniques to determine that the padding is not creating a neurologic insult that causes muscles to weaken when standing or walking.

Its important to note that placing orthotics in shoes may result in the shoe not fitting properly, usually too tightly; this is one of the most common causes of plantar interdigital neuritis in the first place. In this case the shoe must be modified or a different shoe used altogether. Unless this is done, many people with orthotics supports have ill-fitting shoes. **(Maffetone, 2003)** The same AK procedures are used here as when evaluating an individual in the various weight-bearing modes.

Determining whether a correction in the foot is effective and whether the correction endures is one of the major advantages of using applied kinesiology as an adjunct to the diagnostic approach for peripheral nerve entrapments in the feet. For example when subluxations and muscular or ligament injuries in the feet have been corrected but the problem returns immediately when the patient walks, there is an indication that something is wrong in the individuals gait mechanism which may include the feet or muscles of ambulation as well as the torquing patterns within the body. Without the weight-bearing evaluation of manual muscle testing in AK, how would a physician who had just corrected a foot dysfunction recognize if there was a persisting disturbance (unless the patient complains of pain immediately upon walking)?

With the complexity of symptoms on display in the typical patient with lower extremity dysfunction, including pain and dysfunctional soft tissues, joints, etc., where would it be most appropriate to evaluate causes and initiate treatment? The manual muscle test identifies the dysfunctional tissues and the process of therapy localization and challenge allows for the identification of the precise articular correction that will change that finding with a rapidity and certainty that palpation alone cannot provide.

References

- Abshire D. Natural Running: the simple path to stronger, healthier running. VELO Press: Boulder; 2010.
- Aguayo AJ. Neuropathy due to compression and entrapment. In: Peripheral Neuropathy, Vol I. Eds. Dyck PJ,. Thomas PK, Lambert EH. W.B. Saunders Co: Philadelphia; 1975.
- Akuthota A, Herring SA (Eds). Nerve and Vascular Injuries in Sports Medicine. Springer: Dordrecht; 2009.
- Alshami AM, Souvlis T, Coppieters MW. A review of plantar heel pain of neural origin: differential diagnosis and management. Man Ther. 2008;13(2):103-11.
- Alshami A, Babri A, Souvlis T, Coppieters M. Biomechanical evaluation of two clinical tests for plantar heel pain: the Dorsiflexion-Eversion test for Tarsal Tunnel Syndrome and the Windlass test for plantar fasciitis. Foot & Ankle Intern. 2007; 28(4):499–505.
- Andreasen Struijk LN, Birn H, Teglbjaerg PS, Haase J, Struijk JJ. Size and separability of the calcaneal and the medial and lateral plantar nerves in the distal tibial nerve. Anat Sci Int. 2010;85(1):13-22.
- Arcadi VC. Lower back pain in pregnancy: chiropractic treatment and results of 50 cases. Collected Papers International College of Applied Kinesiology, 1995-1996;1:55-57
- Babbage NF. Burning foot syndrome: Bilateral occurrence with radical cure. Med J Aust. 1965;1(21):764-765.
- Ball KA, Afheldt MJ. Evolution of foot orthotics--part I: coherent theory or coherent practice? J Manipulative Physiol Ther. 2002;25(2):116-124.
- Barral JP, Croibier A. Manual therapy for the peripheral nerves. Churchill-Livingstone: Edinburgh; 2007.
- Beaton LE, Anson BJ. The sciatic nerve and the piriformis muscle: Their interrelation a possible cause of coccygodynia. J Bone Joint Surg. 1938;20(3).
- Bellis CJ. Meralgia paresthetica. (Letter to the Editor) JAMA. 1977;238(6):482.
- Beresford HR. Meralgia paresthetica after seat-belt trauma. J Trauma. 1971;11(7):629-30.
- Berry AH, Retzlaff EW. Reciprocal innervation of the piriformis muscles. JAOA. 1978;77.
- Betts LO. Morton's metatarsalgia: Neuritis of the fourth digital nerve. Med J Aust. 1940;April 13.
- Biemond A. Femoral neuropathy. In: Handbook of Clinical Neurology, Vol 8. Eds. Vinken PJ, Bruyn GW. American Elsevier Pub Co: New York; 1970.
- Boyajian-O'Neill LA, McClain RL, Coleman MK, Thomas PP. Diagnosis and management of piriformis syndrome: an osteopathic approach. JAOA. 2008;108(11):657-664.
- Bradford EH. Metatarsalgia neuralgia, or 'Morton's affection of the foot'. Boston Med Surg J. 1891;CXXV(3).
- Brantingham JW, Hubka MJ, Snyder WR. Chiropractic management of Morton's metatarsalgia (Morton's Neuroma): A review of 29 patients. Chiropr Tech. 1994;6(2):61-66.
- Brief JM, Brief R, Ergas E, Brief LP, Brief AA. Peroneal nerve injury with foot drop complicating ankle sprain--a series of four cases with review of the literature. Bull NYU Hosp Jt Dis. 2009;67(4):374-7.
- Browning JE. Pelvic Pain and Organic Dysfunction: A new solution to chronic pelvic pain and the disturbances of bladder, bowel, gynecologic and sexual function that accompany it. Outskirts Press, Inc.: Denver, CO; 2009.
- Browning JE. Chiropractic distractive decompression in the treatment of pelvic pain and organic dysfunction in patients with evidence of lower sacral nerve root compression. J Manipulative Physiol Ther. 1988;11(5):426-32.
- Brugger A. Pseudoradikulare syndrome. Acta Rheumatologica. 1960;18:1.
- Butler DS. The Sensitive Nervous System. Noigroup: Adelaide; 2000.
- Butler DS. Mobilisation of the Nervous System. Churchill-Livingstone: Melbourne, 1991.
- Cailliet R. Foot and Ankle Pain, 3rd Ed. F.A. Davis Co: Philadelphia; 1997.
- Carrel JM, Davidson DM. Nerve compression syndromes of the foot and ankle: A comprehensive review of symptoms, etiology and diagnosis utilizing nerve conduct. J Am Podiatry Assoc. 1975;65(4):322-341.
- Casscells W, et al. Meralgia paresthetica due to unusual compression (Letter to the Editor). N Engl J Med. 1978;298(5):285.
- Chaitow L, DeLany JW. Clinical Application of Neuromuscular Techniques: Volume 2, The Lower Body. Elsevier: London; 2002.
- Chaitow L, DeLany JW. Clinical Application of Neuromuscular Techniques, Vol.1: The Upper Body. Churchill Livingstone: Edinburgh; 2000:150-151.
- Cipriano JJ. Photographic Manual of Regional Orthopaedic Tests. Williams & Wilkins: Baltimore; 1985.
- Cohen BA, et al. CT evaluation of the greater sciatic foramen in patients with sciatica. AJNR. 1986;7(2):337-342.
- Cohen HH. Morton's metatarsalgia (interdigital neuroma). Bull Hosp Joint Dis. 1952;13(1):206-211.
- Corwin JM. Piriformis syndrome in the athlete. ACA J Chiro. 1987;24(1).
- Cox JM. Low Back Pain: Mechanism, diagnosis and treatment, 6th Ed. Williams & Wilkins: Baltimore; 1999.
- Cuba M, Klinginsmith D, Mavissakalian S. Tarsal tunnel syndrome and the chiropractic adjustment: a case report. Journal J Am Chiropr Assoc. 1995;32(2):77-78.
- Cuthbert S, Rosner AL. Conservative Management of Urinary Incontinence using Applied Kinesiology: A Retrospective Case Series Report. J Chiro Med. 2012.
- Cuthbert S. Restless legs syndrome: a case series report. Collected Papers International College of Applied Kinesiology, 2006-2007;1:45-54.
- Cuthbert S. Applied Kinesiology and the Myofascia. Int J AK and Kinesio Med, 2002;13-14.
- Dananberg H. Lower back pain as a gait-related repetitive motion injury. In: Movement, Stability and Low Back Pain. Eds: Vleeming et al. New York, NY: Churchill Livingstone;2007:253-264.
- Dawson DM, Hallett M, Millender LH. Entrapment Neuropathies. Little, Brown & Co: Boston; 1983.
- Denny—Brown D, Ed. Selected Writings of Sir Charles Sherrington. Oxford Univ Press: New York; 1979.

References

- Abshire D. Natural Running: the simple path to stronger, healthier running. VELO Press: Boulder; 2010.
- Aguayo AJ. Neuropathy due to compression and entrapment. In: Peripheral Neuropathy, Vol I. Eds. Dyck PJ,. Thomas PK, Lambert EH. W.B. Saunders Co: Philadelphia; 1975.
- Akuthota A, Herring SA (Eds). Nerve and Vascular Injuries in Sports Medicine. Springer: Dordrecht; 2009.
- Alshami AM, Souvlis T, Coppieters MW. A review of plantar heel pain of neural origin: differential diagnosis and management. Man Ther. 2008;13(2):103-11.
- Alshami A, Babri A, Souvlis T, Coppieters M. Biomechanical evaluation of two clinical tests for plantar heel pain: the Dorsiflexion-Eversion test for Tarsal Tunnel Syndrome and the Windlass test for plantar fasciitis. Foot & Ankle Intern. 2007; 28(4):499–505.
- Andreasen Struijk LN, Birn H, Teglbjaerg PS, Haase J, Struijk JJ. Size and separability of the calcaneal and the medial and lateral plantar nerves in the distal tibial nerve. Anat Sci Int. 2010;85(1):13-22.
- Arcadi VC. Lower back pain in pregnancy: chiropractic treatment and results of 50 cases. Collected Papers International College of Applied Kinesiology, 1995-1996;1:55-57
- Babbage NF. Burning foot syndrome: Bilateral occurrence with radical cure. Med J Aust. 1965;1(21):764-765.
- Ball KA, Afheldt MJ. Evolution of foot orthotics--part I: coherent theory or coherent practice? J Manipulative Physiol Ther. 2002;25(2):116-124.
- Barral JP, Croibier A. Manual therapy for the peripheral nerves. Churchill-Livingstone: Edinburgh; 2007.
- Beaton LE, Anson BJ. The sciatic nerve and the piriformis muscle: Their interrelation a possible cause of coccygodynia. J Bone Joint Surg. 1938;20(3).
- Bellis CJ. Meralgia paresthetica. (Letter to the Editor) JAMA. 1977;238(6):482.
- Beresford HR. Meralgia paresthetica after seat-belt trauma. J Trauma. 1971;11(7):629-30.
- Berry AH, Retzlaff EW. Reciprocal innervation of the piriformis muscles. JAOA. 1978;77.
- Betts LO. Morton's metatarsalgia: Neuritis of the fourth digital nerve. Med J Aust. 1940;April 13.
- Biemond A. Femoral neuropathy. In: Handbook of Clinical Neurology, Vol 8. Eds. Vinken PJ, Bruyn GW. American Elsevier Pub Co: New York; 1970.
- Boyajian-O'Neill LA, McClain RL, Coleman MK, Thomas PP. Diagnosis and management of piriformis syndrome: an osteopathic approach. JAOA. 2008;108(11):657-664.
- Bradford EH. Metatarsalgia neuralgia, or 'Morton's affection of the foot'. Boston Med Surg J. 1891;CXXV(3).
- Brantingham JW, Hubka MJ, Snyder WR. Chiropractic management of Morton's metatarsalgia (Morton's Neuroma): A review of 29 patients. Chiropr Tech. 1994;6(2):61-66.
- Brief JM, Brief R, Ergas E, Brief LP, Brief AA. Peroneal nerve injury with foot drop complicating ankle sprain--a series of four cases with review of the literature. Bull NYU Hosp Jt Dis. 2009;67(4):374-7.
- Browning JE. Pelvic Pain and Organic Dysfunction: A new solution to chronic pelvic pain and the disturbances of bladder, bowel, gynecologic and sexual function that accompany it. Outskirts Press, Inc.: Denver, CO; 2009.
- Browning JE. Chiropractic distractive decompression in the treatment of pelvic pain and organic dysfunction in patients with evidence of lower sacral nerve root compression. J Manipulative Physiol Ther. 1988;11(5):426-32.
- Brugger A. Pseudoradikulare syndrome. Acta Rheumatologica. 1960;18:1.
- Butler DS. The Sensitive Nervous System. Noigroup: Adelaide; 2000.
- Butler DS. Mobilisation of the Nervous System. Churchill-Livingstone: Melbourne, 1991.
- Cailliet R. Foot and Ankle Pain, 3rd Ed. F.A. Davis Co: Philadelphia; 1997.
- Carrel JM, Davidson DM. Nerve compression syndromes of the foot and ankle: A comprehensive review of symptoms, etiology and diagnosis utilizing nerve conduct. J Am Podiatry Assoc. 1975;65(4):322-341.
- Casscells W, et al. Meralgia paresthetica due to unusual compression (Letter to the Editor). N Engl J Med. 1978;298(5):285.
- Chaitow L, DeLany JW. Clinical Application of Neuromuscular Techniques: Volume 2, The Lower Body. Elsevier: London; 2002.
- Chaitow L, DeLany JW. Clinical Application of Neuromuscular Techniques, Vol.1: The Upper Body. Churchill Livingstone: Edinburgh; 2000:150-151.
- Cipriano JJ. Photographic Manual of Regional Orthopaedic Tests. Williams & Wilkins: Baltimore; 1985.
- Cohen BA, et al. CT evaluation of the greater sciatic foramen in patients with sciatica. AJNR. 1986;7(2):337-342.
- Cohen HH. Morton's metatarsalgia (interdigital neuroma). Bull Hosp Joint Dis. 1952;13(1):206-211.
- Corwin JM. Piriformis syndrome in the athlete. ACA J Chiro. 1987;24(1).
- Cox JM. Low Back Pain: Mechanism, diagnosis and treatment, 6th Ed. Williams & Wilkins: Baltimore; 1999.
- Cuba M, Klinginsmith D, Mavissakalian S. Tarsal tunnel syndrome and the chiropractic adjustment: a case report. Journal J Am Chiropr Assoc. 1995;32(2):77-78.
- Cuthbert S, Rosner AL. Conservative Management of Urinary Incontinence using Applied Kinesiology: A Retrospective Case Series Report. J Chiro Med. 2012.
- Cuthbert S. Restless legs syndrome: a case series report. Collected Papers International College of Applied Kinesiology, 2006-2007;1:45-54.
- Cuthbert S. Applied Kinesiology and the Myofascia. Int J AK and Kinesio Med, 2002;13-14.
- Dananberg H. Lower back pain as a gait-related repetitive motion injury. In: Movement, Stability and Low Back Pain. Eds: Vleeming et al. New York, NY: Churchill Livingstone;2007:253-264.
- Dawson DM, Hallett M, Millender LH. Entrapment Neuropathies. Little, Brown & Co: Boston; 1983.
- Denny—Brown D, Ed. Selected Writings of Sir Charles Sherrington. Oxford Univ Press: New York; 1979.

- Leach RA. The chiropractic theories. A textbook of scientific research, 4th Ed. Lippincott, Williams & Wilkins: Philadelphia; 2004.
- Leach RE, Purnell MB, Saito A. Peroneal nerve entrapment in runners. Am J Sports Med. 1989;17(2).
- Leaf D. Applied Kinesiology Flowchart Manual, 4th Ed. Privately Published: David W. Leaf: Plymouth, MA; 2010.
- Leaf D. Effectiveness of applied kinesiology procedures on foot size. Collected Papers International College of Applied Kinesiology, 2005-2006;1:99-100.
- Lederman E. Neuromuscular rehabilitation in manual and physical therapies: principles and practice. Churchill Livingstone: Edinburgh; 2010.
- Lederman E. Fundamentals of Manual Therapy: Physiology, Neurology and Psychology. Churchill Livingstone: New York; 1997.
- Lee D. The Pelvic Girdle. Churchill Livingstone: Edinburgh; 2004.
- Liebenson C. Ed: Rehabilitation of the Spine: A Practitioner's Manual, 2nd ed. Lippincott, Williams & Wilkins: Philadelphia; 2007.
- Logan AL. The Foot and Ankle: Clinical Applications. Gaithersburg, MD: Aspen Publishers, Inc; 1995.
- Lund JP, Donga R, Widmer CG, Stohler CS. The pain-adaptation model: a discussion of the relationship between chronic musculoskeletal pain and motor activity. Can J Physiol Pharmacol. 1991;69(5):683-94.
- Lusskin R. Peripheral neuropathies affecting the foot: Traumatic, ischemic, and compressive disorders. In: Disorders of the Foot, Vol 2. Ed. Jahss MH. W.B. Saunders Co: Philadelphia; 1982.
- Mack GJ. Watchpocket meralgia paresthetica. IMS Ind Med Surg. 1968;37(10):778-9.
- MacNab I. Backache. Williams & Wilkins: Baltimore; 1977.
- Maffetone P. The Big Book of Endurance Training and Racing. Skyhorse Publishing: New York, NY; 2010.
- Maffetone P. Fix your feet. The Lyons Press: Guilford, CT; 2003.
- Mann RA. Tarsal tunnel syndrome. Orthop Clin North Am. 1974;5(1):109-15.
- Marwah V. Compression of the lateral popliteal (common peroneal) nerve. Lancet;1964;2(7374):1367-9.
- Massey EW. Meralgia paraesthetica, an unusual case. JAMA. 1977;237(11):1125-6.
- Mattio TG, Nishida T, Minieka MM. Lotus neuropathy: report of a case. Neurology. 1992;42(8):1636.
- May VR, Jr. The enigma of Morton's neuroma. In: Foot Science. Ed. Bateman JE. W.B. Saunders Co: Philadelphia, 1976.
- McDowall D. Fix foot problems without orthotics. Int J AK and Kinesio Med. 2004;18.
- McGill DA. Tarsal tunnel syndrome. Proc R Soc Med. 1964;57.
- McVey ED, Palmieri RM, Docherty CL, Zinder SM, Ingersoll CD. Arthrogenic muscle inhibition in the leg muscles of subjects exhibiting functional ankle instability. Foot Ankle Int. 2005;26(12):1055-61.
- Mense S, Simons DG. Muscle Pain: Understanding Its Nature, Diagnosis, and Treatment. Philadelphia, PA: Lippincott Williams & Wilkins. 2001:275-277.
- Milgram JE. Office measures for relief of the painful foot. J Bone Joint Surg. 1964;46A(5).
- Milgram JE. Design and use of pads and strappings for office relief of the painful foot. In: Symposium on the Foot and Ankle. Eds. Kiene RH, Johnson KA. The C.V. Mosby Co: St. Louis; 1983.
- Mitra A, Stern JD, Perrotta VJ, Moyer RA. Peroneal nerve entrapment in athletes. Ann Plast Surg. 1995;35(4):366-368.
- Moller BN, Kadin S. Entrapment of the common peroneal nerve. Am J Sports Med. 1987;15(1):90-1.
- Moloney S. Tarsal tunnel syndrome pain abated by operation to relieve pressure. CA Med. 1964;101:378-9.
- Mondelli M, Morana P, Padua L. An electrophysiological severity scale in tarsal tunnel syndrome. Acta Neurol Scand. 2004;109(4):284-9.
- Mondelli M, Cioni R. Electrophysiological evidence of a relationship between idiopathic carpal and tarsal tunnel syndromes. Neurophysiol Clin. 1998;28(5):391-7.
- Moore KL, Dalley AF, Agur AMR. Clinically Oriented Anatomy, 6th Ed. Lippincott Williams & Wilkins; 2009.
- Morris CE. Low back syndromes: integrated clinical management. McGraw-Hill: New York; 2006.
- Morton TG. A peculiar and painful affection of the fourth metatarso-phalangeal articulation. Am Med Sci. 1876;71.
- Mozes M, Ouaknine G, Nathan H. Saphenous nerve entrapment simulating vascular disorder. Surgery. 1975;77(2).
- Murphy PC, Baxter DE. Nerve entrapment of the foot and ankle in runners. Foot & Ankle. 1985;4(4):753-63.
- Nagler SH, Rangell L. Peroneal palsy caused by crossing legs. JAMA. 1947;133(11):755-61.
- Nansel D, Slazak M. Somatic dysfunction and the phenomenon of visceral disease stimulation: a probably explanation for the apparent effectiveness of somatic therapy in patients presumed to be suffering from true visceral disease. J Manipul Physiol Ther. 1995;18(6):379-397.
- Nicol C, Avela J, Komi PV. The stretch-shortening cycle: a model to study naturally occurring neuromuscular fatigue. Sports Med. 2006;36:977–999.
- Nimmo R. Opinion referred to by P.B. Poole in "Sacroiliac, lumbosacral or piriformis. Ortho Briefs — Am Col Chiro Orthop. 1985;7(1).
- Ober FR. Back strain and sciatica. Clin Orthop. 1987;219:4-7.
- Oh SJ, Lee KW. Medial plantar neuropathy. Neurology. 1987;37(8):1408-10.
- Palesy PD. Tendon and ligament insertions – a possible source of musculoskeletal pain. J Craniomandibular Practice. 1997;15:194-202.
- Palmieri-Smith RM, Hopkins JT, Brown TN. Peroneal activation deficits in persons with functional ankle instability. Am J Sports Med. 2009;37(5):982-8.
- Patrick LR. Restless legs syndrome: pathophysiology and the role of iron and folate. Altern Med Rev. 2007;12(2):101-12.
- Pearce JM. Meralgia paraesthetica (Bernhardt-Roth syndrome). J Neurol Neurosurg Psychiatry. 2006;77(1):84.
- Picard L. Management of an anatomical short leg following L4-L5 disc surgery: a case study. Collected Papers International College of Applied Kinesiology, 2007-2008:5-6.
- Poole PB. Sacroiliac, lumbosacral or piriformis. Ortho Briefs — Am Coll Chiro Orthop. 1985;7(1).
- Prakash, Bhardwaj AK, Devi MN, Sridevi NS, Rao PK, Singh G. Sciatic nerve division: a cadaver study in the Indian population and review of the literature. Singapore Med J. 2010;51(9):721-3.
- Racinais S, Bringard A, Puchaux K, Noakes TD, Perrey S. Modulation in voluntary neural drive in relation to muscle soreness. Eur J Appl Physiol. 2008;102(4):439-46.
- Rask MR. Medial plantar neurapraxia (jogger's foot). Clin Orthop. 1978;134:193-5.
- Redwood D, Cleveland CS III. Fundamentals of Chiropractic. Mosby: St. Louis; 2003.

- Retzlaff E, Fontaine J. Reciprocal inhibition as indicated by a differential staining reaction. Science. 1960;131:104-5.
- Retzlaff EW et al. The piriformis muscle syndrome. JAOA. 1974;73(10):799-807.
- Ricciardi–Pollini PT, Moneta MR, Falez F. The tarsal tunnel syndrome: A report of eight cases. Foot & Ankle. 1985;6(3):146-9.
- Richie DH, Jr. Functional instability of the ankle and the role of neuromuscular control: a comprehensive review. J Foot Ankle Surg. 2001;40(4):240-251.
- Robinson DR. Pyriformis syndrome in relation to sciatic pain. Am J Surg. 1947;73(3):355-8.
- Rome PL. Neurovertebral influence upon autonomic nervous system: some of the somato-autonomic evidence to date. Chiropr J Aust. 2009;39:2-17.
- Rome PL. Neurovertebral influence on visceral and ANS function: some of the evidence to date, Part 2: Somatovisceral. Chiropr J Aust. 2010;40:9-33.
- Rosson GD, Larson AR, Williams EH, Dellon AL. Tibial nerve decompression in patients with tarsal tunnel syndrome: pressures in the tarsal, medial plantar, and lateral plantar tunnels. Plast Reconstr Surg. 2009;124(4):1202-10.
- Ruch WJ. Autonomic neuroanatomy of the vertebral subluxation complex. In: Somatovisceral aspects of chiropractic: An evidence-based approach. Ed. Masarsky & Masarsky. Churchill Livingstone: New York; 2001.
- Russell SM. Examination of Peripheral Nerve Injuries: An Anatomical Approach. Thieme Medical Publishers, Inc.: New York; 2006.
- Schweitzer ME, Eid ME, Deely D, Wapner K, Hecht P. Using MR imaging to differentiate peroneal splits from other peroneal disorders. Am J Radiol 1997;168:129-33.
- Sidey JD. Weak Ankles. A study of common peroneal entrapment neuropathy. Br Med J. 1969;3(5671):623-6.
- Skaggs CD, Winchester BA, Vianin M, Prather H. A manual therapy and exercise approach to meralgia paresthetica in pregnancy: a case report. J Chiropr Med. 2006;5(3):92-6.
- Skillern PG. The relief of painful thigh stump and sciatica. JAMA. 1944;126.
- Staal A, van Gijn J, Spaans F. Mononeuropathies: Examination, Diagnosis and Treatment. WB Saunders: London; 1999.
- Staal A. The entrapment neuropathies. In: Handbook of Clinical Neurology, Vol 7, Ed. Vinken PJ & Bruyn GW. American Elsevier Pub Co.: New York; 1970.
- Steiner C, et al. Piriformis syndrome: Pathogenesis, diagnosis, and treatment. JAOA. 1987;87(4):318-23.
- Stites JS. Meralgia paresthetica: A case report. Research Forum. 1986;2(2).
- Strojnik V, Komi PV. Fatigue after submaximal intensive stretchshortening cycle exercise. Med Sci Sports Exerc. 2000;32(7):1314–1319.
- Subotnick S. Sports & Exercise Injuries: Conventional, homeopathic & alternative treatments. North Atlantic Books: Berkeley, CA; 1991.
- Sunderland S. Nerves and Nerve Injuries, 2nd Ed. Churchill-Livingstone: New York; 1978.
- Sunderland S. The relative susceptibility to injury of the medial and lateral popliteal divisions of the sciatic nerve. Br J Surg. 1953;41(167):300-2.
- Szewczyk J, Hoffmann M, Kabelis J. Meralgia paraesthetica in a body-builder. Sportverletz Sportschaden. 1994;8(1):43-5.
- Taguchi Y, et al. The tarsal tunnel syndrome. Clin Orthop. 1987;217:247-52.
- TePoorten BA. The piriformis muscle. JAOA. 1969(2):150-60.
- Thomas PK, Fullerton PM. Nerve fibre size in the carpal tunnel syndrome. J Neurol Neurosurg Psychiatry. 1963;26:520-7.
- Travell JG, Simons DG. Myofascial Pain and Dysfunction: The Trigger Point Manual. Baltimore, MD: Williams & Wilkins, 1992.
- Travell J, Travell W. Therapy of low back pain by manipulation and of referred pain in the lower extremity by procaine infiltration. Arch Phys Med Rehabil. 1946;27:537-547.
- Turco VJ. Commentary on article by B.N. Moller & S. Kadin: Entrapment of the common peroneal nerve. Am J Sports Med. 1987;15(1).
- Turek SL. Orthopaedics — Principles and Their Application, Vol 2, 4th Ed. J.B. Lippincott: Philadelphia; 1984.
- Vernon H, Schneider M. Chiropractic management of myofascial trigger points and myofascial pain syndrome: a systematic review of the literature. J Manipulative Physiol Ther. 2009;32(1):14-24.
- Veves A, Giurini JM, LoGerfo FW. The Diabetic Foot: Medical and Surgical Management. Humana Press: Totowa, NJ; 2002.
- Viladot A. The metatarsals. In: Disorders of the Foot, Vol 1. Ed. Jahss MH. W.B. Saunders Co: Philadelphia; 1982.
- Vleeming A, Pool-Goudzwaard AL, Stoeckart R, van Wingerden JP, Snijders CJ. The posterior layer of the thoracolumbar fascia. Its function in load transfer from spine to legs. Spine (Phila Pa 1976). 1995 Apr 1;20(7):753-8.
- Waddell G. The back pain revolution. Churchill Livingstone: Edinburgh; 1998.
- Wall PD, Coderre TJ, Stern Y, Wiesenfeld-Hallin Z. Slow changes in the flexion reflex of the rat following arthritis or tenotomy. Brain Res. 1988;447(2):215-22.
- Walther DS. Applied Kinesiology, Volume I — Basic Procedures and Muscle Testing. Systems DC: Pueblo, CO; 1981.
- Walther DS. Applied Kinesiology — Synopsis, 2nd Ed. ICAK USA: Shawnee Mission, KS; 2000.
- Whitesides TE. Compartment syndromes. In: Disorders of the Foot, Vol 2, Ed. Jahss MH. W.B. Saunders Co: Philadelphia; 1982.
- Wickstrom J, Williams RA. Shoe corrections and orthopaedic foot supports. Clin Orthop. 1970;70:30-42.
- Williams PH, Trzil KP. Management of meralgia paresthetica. J Neurosurg. 1991;74(1):76-80.
- Williams PL, Warwick R, Eds. Gray's Anatomy, 36th Br Ed. W.B. Saunders Co: Philadelphia; 1980.
- Worth RM, et al. Saphenous nerve entrapment — a cause of medial knee pain. Am J Sports Med. 1984;12(1):80-1.
- Zatterstrom R, Friden T, Lindstrand A, Moritz U. The effect of physiotherapy on standing balance in chronic anterior cruciate ligament insufficiency. Am J Sports Med. 1994; 22(4):531-6.
- Zedka M, Prochazka A, Knight B, Gillard D, Gauthier M. Voluntary and reflex control of human back muscles during induced pain. J Physiol. 1999;520 Pt 2:591-604.
- Zink JG. Respiratory and circulatory care: the conceptual model. Osteopathic Annals. 1977;5(3):108-112.

> "The foot is the generator for the brain which is the battery."
> – George J. Goodheart, Jr.
>
> "85 percent of the US population will someday seek medical care for foot pain."
> – American Podiatric Medical Association (2010)
>
> "…our walking is admittedly merely a constantly prevented falling…"
> – Arthur Schopenhauer

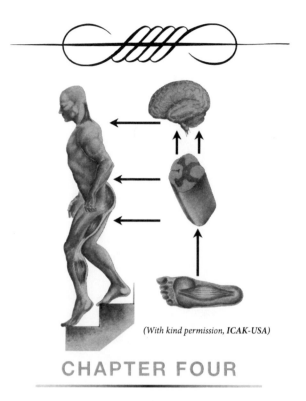

(With kind permission, ICAK-USA)

CHAPTER FOUR

Foot and Ankle

The foot is often considered an odd-looking and unglamorous appendage with the sole purpose of bearing one's weight. It usually does not attract a physician's attention unless a patient complains of foot pain. In reality, foot dysfunction is often the cause of remote health problems that respond poorly or not at all to treatment until the dysfunction is corrected.

The American Podiatric Medical Association says that 85 percent of the US population will someday seek medical care for foot pain. **http://www.apma.org/**

The importance of evaluating all patients' feet is emphasized by a case (SC) treated when he first joined in practice with (DSW). "Peggy's" chief complaint was intractable left shoulder pain that had failed to respond to any physician's therapeutic efforts. Cortisone injections, manipulation of the shoulder and neck, immobilization of the shoulder, and various types of muscle relaxants and analgesics had all failed to bring anything but temporary relief. During consultation Peggy informed SC that she had been unable to sleep except for very brief periods for about two months. Examination of the shoulder with applied kinesiology techniques proved difficult. Range of motion was extremely limited, with almost no ability to abduct the shoulder. Direct testing of the shoulder muscles was impossible because of pain and limited motion. My therapeutic attempts consisted of local muscle treatment, various reflexes, acupuncture, spinal and pelvic corrections, cranial manipulations, and several forms of electrotherapy, all to no avail. At one time Peggy did indicate that she was able to get two hours of sleep a night as a result of these treatments. She stated that the only way she could obtain relief was to stand next to a wall and press her shoulder hard against the wall by leaning on it.

After numerous treatments with minimal positive results, Peggy became the type of patient a doctor hates to see come in the office. It got to the point that when I would see her name as the next patient the thought would surface, "Do I have to go in and see her again?" Just before I went in with her I described to DSW the problems I had been having with Peggy. DSW suggested, "You ought to evaluate her feet." To make matters worse, as I walked into the treatment room it seemed empty with no patient on the examination table. Glancing around the room, I found Peggy leaning against a wall, pressing her shoulder hard against it.

After discussing this case with DSW and having already spent more time evaluating and treating Peggy's shoulder and spine than is usually allotted for a single treatment in our office, I asked Peggy to move her arm into abduction with minimal improvements and some pain. I then asked Peggy to remove her shoes and socks. I found the most gnarled set of feet I have ever seen, and immediately wondered how she could get them into shoes and still walk on them.

The results of examining Peggy's feet were absolutely amazing! After finding subluxations in the feet by challenge and making the corrections, the pain in the shoulder was reduced. After correcting her feet to the maximum her shoulder muscles could be tested without pain, and a large percentage of them tested normal. This enabled me to evaluate and correct the remaining shoulder dysfunction. Unfortunately, as soon as she walked she lost all corrections in her feet, and the shoulder pain returned. Again her feet were corrected and solidly taped for support.

The rest of Peggy's story is uneventful. The shoulder pain was gone after several more treatments, most of which were directed toward the feet. Apparently the reason she was able to obtain relief in her shoulder by leaning against a wall was that the position changed the structural relationship of her feet.

Since that time, a firm rule has been practiced in our office: No new patient is examined with their shoes and socks on. Some in chiropractic recognize the need to examine and correct the feet; **(Brantingham et al., 2009)** more often they are overlooked, even though there have been many efforts to encourage their routine examination. **(Ferrari, 2007; Dananberg, 2007; Keating, 2002; Greenawalt, 1980)** It is important to at least screen the foot as a possible contributing factor to remote health problems.

The type of problem not generally overlooked by the physician is foot symptoms in the form of pain, corns, calluses, or bunions, or an ankle injury. Most of these will be considered here; however, they are not the major problem areas. These conditions are generally looked into because they produce specific complaints for which the patient requests attention. The area of most concern is that of aberrant neurologic ramifications that the symptomatic or asymptomatic foot is capable of producing. In some types of primary foot dysfunction, the patient's complaint is remote from the foot, and the foot is asymptomatic. It is often the secondary conditions that receive attention, possibly for years, without any permanent help. Treatment may be directed to the shoulder, tension headache, low back pain, and on and on — and the basic underlying cause of the problem is not found until the patient possibly develops a symptomatic condition in the foot, or comes in contact with a doctor knowledgeable about the many ramifications of foot dysfunction.

Routine applied kinesiology evaluation of the foot cannot be overemphasized. Yes, the foot is often considered an unglamorous appendage, primarily for holding the body's weight. It seems unglamorous because it is often ugly, distorted with dysfunction; it may have obvious circulatory problems, corns, calluses, in-grown or broken nails. All these are evidence of foot dysfunction. A general rule is that a pretty foot is a healthy, properly functioning foot.

Leonardo da Vinci (1450-1519) regarded anatomic-artistic studies as not only having to do with muscles and the skeleton, but as being part of every other aspect of anatomy and physiology, as well as a polemic with the great anatomists of the past – from Hippocrates to Galen, Avicenna, and Mondino. For instance when Leonardo dissected a bear, he also focused his attention on the anatomy of the leg and foot which he drew.

Comparative anatomy reveals significant alteration from the hominoids to man for the conversion to bipedal stance. **(Lovejoy, 2007, 1988; Burke et al., 1982)** Bipedalism mostly occurred for hominoids because it was essential for species survival. We moved down from the trees because the climate cooled and there were no longer enough trees or sufficient food in the trees that remained to sustain our ancestors there. **(Gould, 2000)** The development of spinal curves and change in pelvic shape allow for a conservation of energy as man's trunk balances over the pelvis and legs. The spinal engine theory **(Gracovetsky, 1989)** proposes that the spine improves its efficacy of motion by using the spine and pelvis to propel the legs forward by capturing the ground reaction force to derotate the spinal segments with each step of the gait cycle. Converting the ground reaction force as potential energy in the viscoelastic tissues of the lower limb and spine into kinetic energy as the spine derotates makes the mechanics of the foot an essential factor in optimal locomotion. **(Dananberg, 2007)** In man, the gluteus maximus has become the largest of the gluteal muscles, while in apes it is much smaller; the gluteus medius in apes is larger than that in man. In apes, the ischium is long, giving the hamstring muscles leverage for great strength. In man the lever is short, which is consistent with the speed required of the hamstrings in walking and running.

Leonardo's anatomical study of the bear

In apes, the foot has grasping capability with divergence of the great toe. Man has lost the ability of the great toe to diverge, with the other four digital rays being

Leonardo — Bear foot anatomical study

aligned toward the great toe. The load lines of the foot in man project from the calcaneus through the 1st and 5th digital rays. This creates a tripod, comprising the heel, hallux, and small toe, with weight bearing primarily at the heel and across the entire width of the distal metatarsals. **(Cavanagh, 1987)** This is important in the bipedal stance of man. Relating to man's weight passing through a coronal plane to the firmly planted tripodal feet, Tobias **(Burke et al., 1982)** states, "His back and front are nearly evenly balanced on either side of this plane. Seen in this light, the upright posture is a precariously balanced state. If our body were merely a nerveless framework, with atonic muscles, it could be thrown off-balance by a push or a gust of wind." He goes on to state, "Long ago, Schopenhauer pondering man's individuality said that our walking is admittedly nothing but a constantly prevented falling." Bipedalism also made us much slower runners than when we were quadrupeds, and many of the big predators can run several times faster than we can. **(Johanson, 1981)**

The foot has many homologues to the hand, yet often when they are compared the foot comes out second-best because of the hand's ability to grip, its dexterity, and its apparent beauty in comparison to the foot. From an evolutionary point of view, several million years spent sitting in trees helped bipedalism emerge because it freed up the prehominoid hand, which allowed the gradual evolution of the opposable thumb. This led to a leap forward in hominoid survivability and versatility in every environment the hominoid lived. **(Napier, 1993)**

When the foot's functional architecture is understood however, its functionality should at least equal that of the hand and perhaps surpass it, especially in human evolution. Bordelon **(Bordelon, 1987)** cites a change in concept indicating the importance of man's foot being different from primates. He cites a *National Geographic* (November 1985) article stating that "…recently discovered fossils have revolutionized our concepts of the human past. Our earliest, most distinguished characteristics were not large brains, language or tool making, but the ability to habitually walk upright." With this ability man's hands were freed to allow for their development. Hand use may have then caused the need for greater brain development. **(Blechschmidt, 2004; Napier, 1993)**

The 26 bones of the foot and its soft tissue comprise a marvel of engineering design. With 200,000 nerve endings, 19 major muscles, 33 joint centers and 17 ligaments, it may be that 6 million years of evolution created the perfect foot. When one considers the stress of walking, running, and jumping that must be accommodated and dissipated by the foot, it's a wonder that anyone has healthy feet. The normal stresses the foot must adapt to are compounded by poor quality or improperly designed and fitted shoes.

The foot has gone through many changes to adapt to human bipedal upright posture. There are three basic ways the human foot differs from that of other primates. (1) In man the 1st ray is not markedly divergent; thus it is normally not useful as a grasping appendage, as is the hand. (2) The human foot has both longitudinal and transverse arches. (3) In certain functions the foot is a rigid structure with strong ligamentous support, yet it can become supple to adapt to the contour of the ground. These factors are true of the normal foot, which is best adapted to weight bearing. Many poorly functioning feet revert back to the simian or prehensile foot, with loss of the arches, hypermobility, and medial deviation and flexibility of the 1st ray.

British trained osteopath Hugh Milne remarks in his excellent book *The Heart of Listening: A Visionary Approach to Craniosacral Work*, that a former agent told him that the C.I.A. teaches its novice agents to watch a suspect's body language, especially their feet. **(Milne, 1995)** Milne notes that since watching the feet and toes of patients under duress, "it is amazing what tales a toe can tell you. The person says one thing, the toe says another. Watch, always watch."

The human foot normally develops its rigidity and lack of flexibility during normal use and development as a weight-bearing appendage. Its marvelous ability to adapt can be observed in those born without arms who use their feet from early childhood to grasp objects. Many have developed the ability to write and paint by holding a pencil or brush between the toes, to eat by similarly holding a spoon or fork between the toes, and even to drive automobiles by using the feet to steer and shift gears. **(BBC News, 2002)**

If the feet are so marvelous, why do so many people have poor foot function? There is a tendency to attribute foot dysfunction to congenital conditions, or failure of foot design to meet the requirements of weight bearing and the general pounding they receive on a daily basis. Stress to the feet has certainly increased with the recent popularity of running and jogging, which is now shifting to walking for exercise. From 1970 to '79 the "bible" of the running crowd, *Runner's World*, increased its circulation from 3,000 to 500,000. **(Clement & Taunton, 1981)** Running USA released a portion of the results from their 2009 *State of the Sport* series. **(Running USA, 2008)** In 2008 there was an

18% increase in the total running population (35,904,000 runners) compared to 2007, and a 15% increase in the estimated number of trail runners in 2008 (4,857,000). The mean age of road race finishers in 2008 was 36.3. If you ran at least 100 days in 2008, you were one of 14.9 million who did likewise. And despite the recession, running shoes (39.9 million pairs) and dollars spent on them ($2.3 billion) exceeded those of the previous year. Many people are unaware of foot dysfunction. It may be manifested as end-of-the-day fatigue. When activity increases, as in jogging, it may present a serious problem. (**Abshire, 2010; Marshall, 1978**)

General Examination and Body Language

One should recognize that much foot dysfunction is locally asymptomatic, or the patient fails to discuss the problem because "I have normal aching in my feet after being on them all day." In addition, women especially recognize that their foot problems result from improper shoes, but with style-conscious thinking they fail to discuss the matter with their physician. It is therefore necessary to be capable of reading the body language of foot dysfunction.

Body language may be recognized in the pattern of symptoms, by visual observation of the foot, by palpation, and/or by specialized tests. Dananberg, Maffetone and Cailliet (**Dananberg, 2007; Maffetone, 2003; Cailliet, 1997**) point out that there are three ways in which the foot functions: (1) in willed, free movements, (2) as a static support mechanism subject to gravitational forces and postural changes, and (3) as a dynamic mechanism for moving the body. Keeping these three functions in mind, we will direct our examination toward each aspect.

Willed, free movements are present when the foot is non-weight-bearing, such as on an examination table. The foot is not generally used as a freely movable appendage, yet analysis of the foot in manual muscle testing is usually done in this manner. It is necessary to consider the other two aspects — a static support mechanism and a dynamic mechanism for moving the body — since many foot dysfunctions are only observable under these circumstances. The physician who examines the foot and other weight-bearing mechanisms only when the patient is non-weight-bearing is destined to overlook much dysfunction.

With the thought in mind of non-weight bearing, static weight bearing, and dynamic weight bearing, body language of foot involvement will be described as (1) symptomatic, which is primarily the patient's complaint recorded during consultation, (2) the foot's appearance as observed by the examiner, and (3) shoe quality and shoe wear. The examination is divided into three sections: (1) non-weight bearing, which includes structural balance of the foot, palpatory findings, and applied kinesiology's shock absorber and sensorimotor challenge tests, (2) static weight bearing, which evaluates postural structural balance and strain (muscle tests will be used to examine for remote dysfunction with various foot positions), and (3) the dynamic examination, which takes walking, running, and treadmill activities into consideration.

Most foot dysfunction is functional, such as that discussed in this chapter. (**Abshire, 2010; McDowall, 2004; Maffetone, 2003**) As with any health problem, however, it is necessary to thoroughly consider all possibilities. Neuropathy, which may be due to diabetes, Hansen's disease, tabes dorsalis, yaws, syringomyelia, vascular occlusion of vasa nervorum, or direct trauma, can cause foot deformities simulating those of dysfunction. (**Staal et al., 1999**) Foot deformity may result in the loss of intrinsic musculature, causing clawing or hammering of the toes, cavus feet, and/or depressed metatarsal heads simulating functional problems. There may also be varus or valgus deformities as a result of extrinsic muscle imbalance. Breakdown of the foot can result from a Charcot joint. (**McCormack & Leith, 1998; Cailliet, 1997; Wagner, 1983**)

Symptomatic

It should be determined during consultation whether the patient's symptoms become worse with standing, running, and other weight-bearing activities, regardless of the chief complaint. Many low back conditions are not optimally treated because the fault is diagnosed primarily within the lumbar spine. It may be hyperlordosis leading to facet syndrome, but the question "Why is the hyperlordosis present?" must be asked and answered. As noted later, if the positive support reaction is not functioning properly, extensor muscles of the pelvis may be giving inadequate support, which allows anterior pelvic rotation and increases the lumbar lordosis. In this case, the primary condition is foot dysfunction even though the feet are asymptomatic. Treatment directed to the facet syndrome may well give relief, but if the basic underlying cause of the condition is not corrected, it will only be temporary. One of the documented reasons for facet syndromes and lumbar hyperlordosis is inhibition of the gluteus maximus muscles. (**Travell & Simons, 1992**) One of the advantages of the applied kinesiology approach to manual muscle testing is that the gluteus maximus muscles can be tested in both the prone, supine, and standing positions – particularly after applied kinesiology challenge procedures have been applied to the foot. Immediate strengthening or weakening of the gluteus maximus muscles after such a challenge provides immediate, non-invasive evidence that foot dysfunction is related to the gluteus maximus muscle inhibition and related low back and pelvic problems.

Additional questions should be asked to further categorize any increase of symptoms with weight bearing. Some patients develop symptoms on static standing, with no problem when walking and moving about; on the other hand, some have no problem from static weight bearing, but they experience significant symptoms when walking or running. Structural strain and disorganization may not develop until the gait mechanism is brought into play.

Knowledge of disturbance from the weight-bearing mechanism can be enhanced by determining if the patient feels better on arising in the morning, and the symptoms develop or increase with weight-bearing activity

AK Foot and Ankle Examination

Body language is observed by the health problem and history, e.g., worse when on feet for a long time, running, eases with sitting. Muscular imbalances observed standing and in supine position. Corns, calluses and fat pads easily observed during static examination. Hammer toes indicate weak intrinsic foot muscles.

Foot Pronation
Talus subluxation
Also:
Tarsal, metatarsal dysfunctions

Ankle support muscle impairments:
- Peroneus tertius
- Peroneus longus, brevis

Anterior and posterior tibial

Tarsal Tunnel Syndrome
Compare for weak flexor hallucis brevis and strong flexor hallucis longus. Correct foot pronation first.

Challenge calcaneus and adjust. Calcaneus will usually be in a posterior position.

Golgi tendon organ or muscle spindle cell treatment is frequently needed, especially for intrinsic muscles of the foot. Raw bone-concentrate and/or Folic Acid-B12 for myofascial trigger points if nutritional support if corrections do not hold.

Weight bearing that causes weakness indicates need for further evaluation and support. Evaluation of knees, pelvis, and other structures above may be needed.

Tape Support
Figure 8 Support
Anchor on leg. Tape extends under scaphoid to return onto leg in figure 8 pattern.

Calcaneus Anchor
First application around & behind calcaneus, followed by a support to scaphoid with 2 or 3 strips of tape.

Frequent cause of recurring cranial, sacral, pelvic, respiratory, shoulder and neck involvements.

Meridian involvement through area
Liver   Kidney   Stomach
Gallbladder   Bladder   Spleen

Neck, shoulder girdle, and shoulder dysfunction are frequently caused or perpetuated by foot pronation. These specific interactions dynamically assessed with AK MMT.

Bladder Involvement
Infection or irritation and chronic pelvic pain can cause reflex muscle weakness, perpetuating foot and ankle problems. These problems can cause recurrent or chronic bladder involvement: viscero-somatic interactions.

Circulatory disturbances of any nature can be complicated by foot/ankle dysfunction.

Calcaneal spurs from traction on plantar aponeurosis.

To further evaluate the holding capability of foot and ankle corrections, have the patient walk, jump, run, or repeatedly activate the involved muscles. Re-examine immediately to determine return of any factors which were corrected. If return, there is indication to test for:
- Gait Mechanism
- Microavulsions of muscles involved
- Shoes the patient wears
- Consider orthotic supports

throughout the day. The weight-bearing mechanism is especially indicted if relief is obtained by sitting or lying down at the end of the workday.

Foot dysfunction is the cause of recurrent neurologic disorganization (switching) in many individuals. (**Walther, 1983**) The afferent supply from the foot is instrumental in much organization of body movement and postural regulation. Many studies have been done in other areas of the body showing that improper stimulation to proprioceptors disturbs the proper temporal pattern of muscle contraction, i.e., agonist and antagonist muscles contract at the same time, or there is contraction or inhibition not in keeping with the joint movement taking place at that time. (**Gardner-Morse & Stokes, 1998; Graven-Nielsen et al., 1997; Bard et al., 1992; Munro, 1972**) It is now well established that the feet are very important for proprioception as well as for posture and balance. Applied kinesiology clinical evidence certainly indicates this. Normal facilitation and inhibition of muscles are often demonstrated in applied kinesiology by testing the shoulder flexors and extensors with an individual in various gait positions. (**Walther 2000, 1983, 1981; Goodheart, 1982**) When an individual is in a simulated gait position with the right leg forward and the left trailing, the left shoulder extensors test weak by manual muscle tests although they were strong in a normal stance. This is expected in order for the arm to swing forward easily during that phase of gait. However, if one places pencils under the 1st and 5th metatarsals, causing an artificially produced dropped metatarsal arch, the normal facilitation and inhibition of the shoulder muscles with various gait positions become unpredictable. It is not known how they will test, but in nearly all cases the results are different from the predictable norm.

All patients who exhibit recurrent neurologic disorganization, known as "switching" in applied kinesiology, should be evaluated for foot and gait problems. Improper stimulation to the proprioceptors of the joints, skin, and muscles of the feet and legs can cause a bombardment of improper afferent impulses to the neuronal pools. This is especially true if there is a facilitated state at a spinal segment from a vertebral subluxation. (**Korr, 1976, 1975**) The facilitated spinal neurons are initiated by afferent impulses arising from nociceptors or visceral receptors and

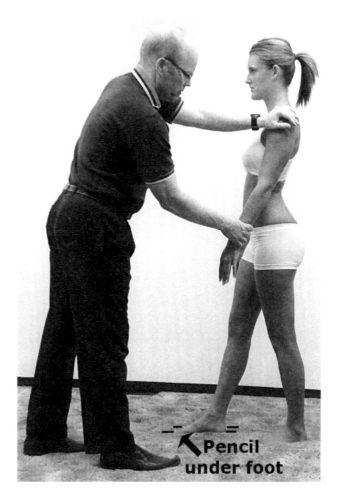

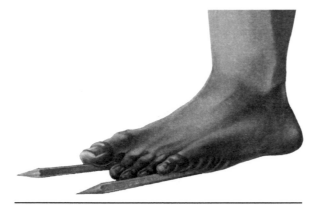

Pencil under 1st & 5th metatarsals

are then transmitted to the dorsal horn of the spinal cord, where they synapse with interconnecting neurons. These stimuli are then transmitted to motor and sympathetic efferents, resulting in changes in the physical tissues, such as skeletal muscle, skin and blood vessels.

Areas of facilitation can be exacerbated by stressors of many types which place increased stress upon many aspects of the individual, primarily the triad of health in applied kinesiology – physical, chemical, and psychological. Korr's concept of facilitation fits directly into the applied kinesiology contextual model for understanding pain and dysfunction. It also provides a model for understanding the influence of a wide gamut of stressors (postural, nutritional, emotional, etc.) on the health of individuals.

General structural strain throughout the body gives indication to evaluate the weight-bearing mechanism. Rolf and Myers (**Myers, 2001; Rolf, 1977**) and many others have pointed out the continuity of myofascia throughout the body, and how distortion in one area can transmit strain to remote body parts through the fascia. When there is structural strain, such as that present when the positive support reaction is not functioning adequately, the most superior aspect of the fascia — at the crown of the head — is tense. Tenderness to digital pressure posterior to the bregma indicates probable fascia tension. Applied kinesiology clinical evidence indicates that the pull of the fascia around the skull can jam sutures. Especially important is the sagittal suture which, when jammed, may cause weak abdominal muscles. (**Walther, 2000, 1983**)

The weak abdominal muscles allow the pelvis to rotate anteriorly, compounding the increased AP spinal curves from poor function of the positive support mechanism. It is possible that the abdominal muscle weakness correlates with a sagittal suture cranial fault because the abdominal ptosis causes a generalized traction on the fascia, which ultimately results in tightening of the superficial and deep fascia of the epicranium. This could provide a vicious circle of cranial etiology that disturbs the function in such a manner that the cranial faults cause additional muscle weakness. Another observation is the important role the abdominal muscles play in pelvic support, **(Hodges & Richardson, 1996)** especially in category I pelvic faults. The category I pelvic fault is especially important in the cranial-sacral primary respiratory mechanism. Correcting the sutural fault will very likely be only temporary unless other structural corrections are made, because the tense fascia will cause the suture to jam again.

Recurrent cranial faults are body language to examine the weight-bearing and gait mechanisms. There are many reasons for this. Ferguson **(Ferguson, 1991)** suggests that the powerful muscles attaching directly to the cranial bones are capable of generating sufficient pressure or forces upon the skull to produce the flexibility and palpable motion at the cranial sutures which may exist for this purpose. Just as it is imprudent to diagnose the status of the pelvic joints without knowledge of the status of the dynamic muscles which attach to the bones of the pelvis (such as the gluteus maximus, hamstrings, piriformis, quadratus lumborum, or psoas), so must evaluation and treatment of cranial dysfunctions demand the evaluation of those muscles which attach to the skull. Page **(Page, 1952)** documents the continuity of the fascia from the abdominal area to the dura mater. The continuity of the fascia is emphasized by Wright and Brady **(Wright & Brady, 1958)** in their statement, "The deep cervical fascia is so continuous with that of the head that the two should be considered as one." In addition, the deep cervical fascia "…forms the pectoral fascia which in turn blends with the fascial covering of the rectus abdominis." Chaitow **(Chaitow, 2005)** observed: "Pick, **(1999)** in his landmark text on cranial sutures, inexplicably fails to mention the profound potential impact of muscular attachments that frequently overlie and traverse sutures." Chaitow counsels that treatment of dysfunctional muscles attaching to the cranial and facial sutures are essential prior to any attempt at treating the osseous structures themselves.

It is clinically obvious that many times when there is a "rigid" skull on applied kinesiology examination, it relates with postural strain often associated with foot dysfunction. One can often feel greater rigidity of the skull when challenging an individual in the standing weight-bearing position than when sitting, supine, or prone. Before lasting corrections can be obtained in the stomatognathic system, the structural strain from weight-bearing dysfunction must be corrected.

The applied kinesiology approach has shown that because of the connection in the fascial system, a change in any part of the body may create a disorder elsewhere. An anterior cruciate ligament injury can produce changes in the masseter, anterior temporalis, posterior cervicals, upper and lower trapezius and sternocleidomastoid muscles. **(Tecco et al., 2006)** Dvorak & Dvorak **(Dvorak & Dvorak, 1990)** injected a saline solution into the transverse process of C7 while using electromyography and observed muscle contractions in zones distal from the spinal myomere where the injection was made. An increase in active mouth opening and a decrease in myofascial trigger point sensitivity in the masseter muscle were observed in response to the stretch of the hamstring muscles, demonstrating a functional relationship between the masticatory and hamstring muscles. **(Fernández-de-las-Peñas et al., 2006)**

During normal walking, there is alternate facilitation and inhibition of the sternocleidomastoid and upper trapezius muscles to keep the head pointed forward as the shoulder girdle rotates with the gait; otherwise, one's head would rotate with the shoulder girdle. The organization of these muscles relates to the input from the foot and ankle proprioceptors. **(Dananberg, 2007)** Since the muscles insert into the skull, an improper temporal pattern can cause strain in the cranium and possibly create cranial faults. Tecco et al. **(Tecco et al., 2010)** have also shown there are detectable interrelationships between occlusion and locomotion. It is not uncommon in applied kinesiology practice to observe loss of cranial fault correction as soon as the patient walks or runs. The correction is usually maintained after a foot, gait, or some other modular problem is corrected.

As noted, shoulder motion harmony depends on proper inhibition and facilitation of the shoulder muscles, which are integrated with proper proprioception from the foot, leg, and pelvis during walking. **(Geyer & Herr, 2010)** All patients with shoulder problems should have screening tests of the foot mechanism early in the evaluation. With recurrent or resistant shoulder problems, a thorough evaluation of the feet, legs, and pelvis must be done.

Afferent supply from the feet is also instrumental in the organization of the sacrospinalis and other large muscles of the spine. **(Kavounoudias et al., 2001)** Disorganization of these muscles can cause spinal or pelvic subluxations to recur as soon as the patient walks, even though they were adequately corrected.

Proper muscle activity is important in supporting the ankle and arch during walking and running. Disorganization and possible muscle weakness in the foot and ankle are often indicated by frequent twisting of the ankle. Again, unfortunately, the patient will not necessarily volunteer this information because it is often considered a normal occurrence, or "I just have weak ankles, and nothing can be done about it."

During consultation the presence of symptoms may be volunteered by the patient, but more often they are observed or dug out by the physician. One must often ask leading questions that can be analyzed by the physician's knowledge of body integration.

The patient can often provide a clue about what is taking place by comparing the two extremities. Ask the patient if one extremity seems to feel different from the other when running or walking. Is there recurrent injury to only one side? Does the patient seem to favor one side in walking or running? Is there more difficulty turning in one direction than in another? Answers to these questions may lead an applied kinesiologist to think of the conditions that cause unilateral dysfunction, such as a lateral atlas subluxation or cranial faults; in addition, can modular factors or equilibrium proprioceptive function be at fault?

Appearance of the foot

It cannot be overemphasized that a patient should remove his shoes and socks for the initial examination. Most chronic foot involvements have characteristic body language that can be readily observed during examination. Foot problems that are not of a long-standing nature may not have revealing visual body language, but dysfunction can easily be determined by examination procedures that include palpation, range of motion, muscle testing, and applied kinesiology therapy localization and challenge. (**Logan, 1995**)

Goodheart has often said the body language does not lie. Clinical experience finds this to be true. When there is no apparent correlation between body language and clinical findings, it pays to persist in an effort to find a correlation. Take, for example, the situation described above where an individual gains relief after getting home and relaxing. If the weight-bearing and other factors are not found positive, persist with your evaluation by asking the patient about his working conditions. The type of physical activity may not be reproduced in examination for gait, etc. Situations have been discovered where an individual stands on one leg and pushes a lever with the other leg all day long. Simulating that activity in the office reveals a reactive muscle condition which, when corrected, eliminates the problem. In another case, it may be the work boots an individual wears. We have had people bringing in their work shoes and found them to be severely run over, with broken down counters and heel wear. Simply standing in the work boots reveals many positive findings using applied kinesiology. In some cases, correction of a low back problem is as simple as having the patient purchase a new pair of boots.

In general, a pretty foot is a normally functioning one. If a pretty foot is dysfunctioning, it probably has not done so for long. Dysfunction causes the foot to distort or appear strained, and ultimately become gnarled. (**Langer, 2007**)

Visual observation will reveal wear points on the skin. Corns and calluses on the tops and sides of the toes and foot usually indicate improper shoe fit. In the average individual, calluses on the sole of the foot usually indicates improper weight distribution of the foot, (**Maffetone, 2010, 2003; Langer, 2007**) probably relating with subluxations. They may also be due to improper footwear. Often there will be calluses under the distal heads of the metatarsals. They may be limited to only one metatarsal, usually the 2nd or 5th, or there may be calluses across all the metatarsal heads. The callus, of course, is nature's effort at building resistance to the abnormal weight distribution and is the natural response of skin to external pressure applied against the underlying bony surface. Neurovascular corns are very tender and painful and should be evaluated by a podiatric specialist. Distance runners may have normal calluses; they will not only be concentrated at abnormal points of wear or pressure.

There may be a subcutaneous fat pad along with abnormal calluses, indicating an additional body effort to protect the structure. In this case, the fat pad usually develops prior to a callus; thus an increased fat pad without a callus generally indicates the condition is not as chronic. In some chronic conditions there may be a fat pad without a callus, especially in individuals who lead a sedentary life with minimal weight-bearing activities.

What appears to be an increase in the fat pad may be localized edema from the tissue trauma of improper function. Also, the fat pad may move anteriorly, discussed with forefoot dysfunction.

When adequate correction is obtained, calluses that have been present for many years may slowly disappear. This amazes the patient because of the long-term necessity of having to have the calluses removed by abrasion or application of a keratolytic agent.

Observation of the skin often provides clues to circulation deficiency. The circulatory problem might be the patient's complaint during consultation, but more often it is found by the physician during examination. Poor circulation can be caused or aggravated by foot dysfunction, or it may be coincidentally present with a foot condition. Venostasis is observed as cyanosis of the nail beds. Further evaluation can be obtained if the doctor blanches the skin by digital pressure and records the time it takes for blood to return. (**Langer, 2007; Greenawalt, 1982**) Vascular tests such as the evaluation of pulses and for deep vein thrombosis have been described by Petty & Moore. (**Petty & Moore, 1998**) The entire foot and ankle, as well as the lower extremity, may show evidence of plethora. If the plethora is localized in the foot only, there may be compression of the posterior tibial vein in the tarsal tunnel.

The foot should be examined for temperature variances, edema, dependent cyanosis, and blanching of the elevated foot. An ischemic foot will blanch when elevated and flush when dependent. Skin will have a thin, elastic appearance and lack hair growth on the dorsum of the proximal toes. Arterial deficiency causes the foot to become cyanotic on dependency and blanch with elevation. (**Veves et al., 2002; Cailliet, 1997**)

Venous insufficiency is indicated by pitting edema, which is absent or decreased after a night's rest. Varicosities enhance the diagnosis, but edema must always create suspicion of a systemic condition. Good muscular function of the lower extremity enhances venous and lymphatic drainage. Contraction of neighboring muscles compresses lymph vessels, moving lymph in the directions determined by their valves. Manual or mechanical lymphatic reflex techniques may be effective ways to increase lymph removal from stagnant or swollen tissue. (**Chaitow & DeLany, 2008; Goodheart, 1998; Chaitow, 1987**) Arterial supply can be evaluated at the dorsalis pedis between the 1st and 2nd metatarsals, and at the posterior tibial artery behind the medial malleolus.

Weak, brittle toenails may indicate circulatory disturbance. Ingrown toenails may indicate short shoes or socks. Stretch socks are particularly problematic for some individuals. A dark lesion under the nail may be melanoma. (**Langer, 2007; Kalivas, 1983**)

Shoe wear

Evidence of disturbed function of the foot and ankle is often indicated by a patient's shoes. (**Abshire, 2010; Langer, 2007**) The heel should wear on the posterolateral aspect. When the wear deviates more laterally, pronation is indicated. Sometimes excessive sole wear underneath the metatarsal heads can be observed, but it is not a constant finding with dropped metatarsals. The counter and vamp of the shoe should remain balanced over the heel and sole. Running over of the shoe in any direction indicates foot problems or poorly constructed shoes. Toe wear is indicative of weakness of the dorsiflexor muscles. Runners' shoe wear varies considerably from that of the average patient.

The patient should be questioned regarding his work,

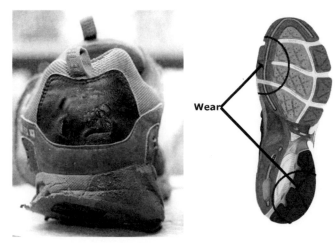

Shoe wear with extended pronation.

dress, sport, and any other type of shoes worn. Often the shoes worn to the doctor's office are not those in which the patient spends most of his time. Questioning the patient about whether symptoms develop when certain types of shoes are worn may elicit no positive response at the time; however, it plants a seed for the patient's future evaluation and may later evoke a positive response. For example, correcting upper cervical subluxations and balancing the cervical musculature can effectively reduce a patient's headache for two to three days, only to have it return and need correcting again. It is not uncommon for the patient to report that the headache returned after he wore a pair of Western boots over the weekend. Evaluate the possible connection by having the patient bring the boots in question to the office. Correct the cervical subluxation and have the patient walk in the shoes usually worn during the week. If the subluxation does not return, have the patient walk in the boots. If the subluxations and muscular imbalance return, either the boots or the patient's feet are indicted as the cause of the recurrence; probably both are at fault.

Meridian system influenced by foot/ankle dysfunction

The meridian system is covered in the text *Applied Kinesiology Essentials*. However, it should be mentioned as part of the body language associated with foot and ankle problems. There are six meridians that flow through the ankle and foot. The normal flow of Chi (the energy of acupuncture meridians) may be interfered with by foot and/or ankle problems such as trauma or structural strain. Correction of a subluxation, fixation, or fascial or muscular strain may immediately improve energy flow in a meridian.

When there is interference with a meridian's energy flow, it may cause a problem associated with that meridian, such as a gallbladder attack when the energy flow in the gallbladder meridian is reduced. If the patient has had no previous disposition toward gallbladder involvement and develops an attack within a day or two after ankle or foot trauma, the meridian system should be evaluated. Appropriate treatment to the gallbladder meridian at the location of injury may produce gratifying results within minutes. **(O'Connor & Bensky, 1981)**

There are occasions when a gallbladder attack is so severe that hospitalization is required if results are not obtained rather quickly. Returning the gallbladder meridian to normal by simply stimulating the meridian through the area of trauma at the ankle may alleviate the attack within minutes. If the physician does not have this specialized knowledge, the patient may have to be hospitalized and have gallbladder studies done; often the diagnosis will be a "sluggish gallbladder." Treatment usually consists of dietary recommendations and probably medication, which is credited for the improvement of the condition. On the other hand, the gallbladder dysfunction would probably have been self-limiting as the ankle healed and the meridian spontaneously returned to normal function.

Other meridians running through the ankle and foot are the stomach, bladder, spleen, liver, and kidney. When considering the foot and ankle for possible meridian involvement, remember that not only the meridians traversing the foot and ankle may be involved; the other six meridians can also be involved as a result of the mother-child effect, five-element law, or other pathways of energy exchange within the meridian system.

Costa and Araujo **(Costa & Araujo, 2008)** demonstrated one of the approaches applied kinesiology that has used for decades in evaluating the meridian system. They showed that stimulation of the sedation point for the Bladder meridian (acupuncture point stomach-36) induced decreased strength in the tibialis anterior muscle as measured by electromyography. According to applied kinesiology, the tibialis anterior muscle corresponds to the Bladder meridian.

Moncayo & Moncayo have written extensively about AK and the meridian system and have been able to demonstrate the usefulness of the working principles of AK in relation to Traditional Chinese Medicine. **(Moncayo & Moncayo, 2009)** Garten, Walther, Dale, Corneal and Dick, and Larson have demonstrated this relationship as well. **(Garten, 2002; Walther, 2000; Dale, 1993; Corneal & Dick, 1987; Larson, 1985)** (The AK approach to traditional Chinese medicine will be substantially updated in teh subsequent volume of this series, *Applied Kinesiology Essentials*.)

Non-Weight Bearing Examination

Foot-ankle-leg balance.
When the patient is prone or supine and the foot and ankle are hanging over the table end, the stabilized structure becomes the leg; the foot may deviate into structural distortion, which may not be observed when standing.

Abnormal flexion of the toes can be of three types. 1) "Mallet toe" is flexion deformity of the distal interphalangeal articulation. 2) "Hammertoe" is flexion deformity of the proximal interphalangeal articulation with extension of the metatarsophalangeal articulation. 3) "Claw toe" is flexion deformity of both interphalangeal articulations with extension of the metatarsophalangeal articulation. **(Kwon et al., 2009)**

The most commonly observed deviation of the toes is claw toes. The claw-like appearance to the toes is often to the point that weight bearing on the toes is on the distal end of the distal phalanges. This position may be due to anomalous insertion of the extensor tendons, **(Kwon et al., 2009; Sgarlato et al., 1969)** for which surgical procedures are available. **(Kwon et al., 2009; Sgarlato, 1970)** Surgery in advanced cases may not even restrict marathon runners.

A more common cause of claw toes is relative weakness of the intrinsic compared to the extrinsic flexor muscles. **(Kwon et al., 2009; Ramamurti, 1979)** The extrinsic flexor muscles are the flexor digitorum longus and flexor hallucis longus. They insert into the distal phalanx and receive their nerve supply from the tibial nerve prior to the tarsal tunnel. The major intrinsic flexors are the flexor digitorum brevis and flexor hallucis brevis. The flexor

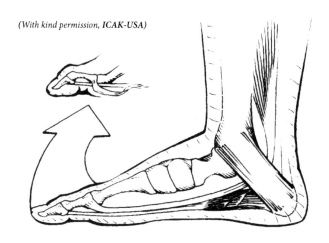

(With kind permission, ICAK-USA)

Insertion of the muscles into the toes. (See Tarsal Tunnel Syndrome, Chapter 3)

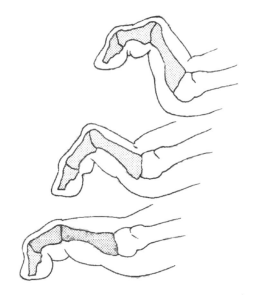

Toe deformities.

digitorum brevis inserts into the sides of the shaft of the intermediate phalanx, and the flexor hallucis brevis inserts into the sides of the base of the proximal phalanx of the hallux. Nerve supply to both muscles is the medial plantar nerve, a division of the tibial nerve at the tarsal tunnel. **(Gray's Anatomy, 2004; Hamilton, 1985)** Peripheral nerve entrapment at the tarsal tunnel may deprive the intrinsic muscles of normal nerve supply, causing the muscular imbalance that allows the interphalangeal articulation to rise, while the distal phalanx is pulled into flexion by the strong extrinsic muscles. Because of the imbalance between the longus and brevis muscles of the foot the distal toes curl inferior and posterior creating claw toes. Tarsal tunnel syndrome and foot pronation almost always precedes the phenomenon of claw toes.

With poor nerve supply to the intrinsic muscles, there is often plantar muscle atrophy. This is often mistaken by both the patient and the physician as a high arch. In reality,

the arch is usually broken down, receiving very poor support while walking and running. Generally, when there is extension-flexion deformity of the toes there is also an elevation of the extensor tendons, giving a strained look to the foot. This apparently fits the observation in applied kinesiology that when there is muscle weakness, antagonist muscles contract.

Because the abductor hallucis muscle also receives nerve supply from the medial plantar nerve, hallux valgus may develop along with the peripheral nerve entrapment at the tarsal tunnel. Other causes of hallux valgus are poorly fitting shoes or socks, and congenital variations of bone and of the insertion of the abductor hallucis.

Non-weight-bearing balance of the muscles that cross the ankle can be observed with the patient supine. The feet should hang over the end of the table far enough so that the edge of the table does not interfere with the foot position.

An imaginary line down the anterior ridge of the tibia should extend into the second toe. When it extends lateral to the second toe, the turgor of the lateral ankle stabilizing muscles — the peroneus tertius, longus, and brevis — is less than the medial ankle stabilizing muscles. This indicates either weakness of the peroneus group or hypertonicity of the tibialis anterior and posterior. This is more common than when the line projects medial to the second toe, indicating less turgor of the tibialis muscles. Again, it could be caused by hypertonicity of the peroneus group or weakness of the tibialis muscles.

When observing lateral or medial deviation of the foot, care must be taken that it is properly related to the anterior tibial ridge. There should be neutral thigh rotation observed by the position of the patella. Some individuals have increased muscle or adipose tissue on the lateral aspect of the leg. In that instance, a line drawn down the center of the leg does not correspond with the tibial ridge.

Balance between the dorsiflexor and plantar flexor muscles can be observed by the amount of plantar flexion of the relaxed foot hanging over the end of the table. The peroneus tertius and tibialis anterior tendons

Foot and Ankle

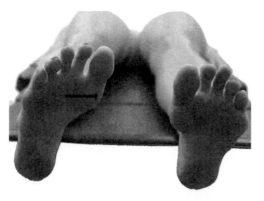

Peroneus longus and brevis weakness.

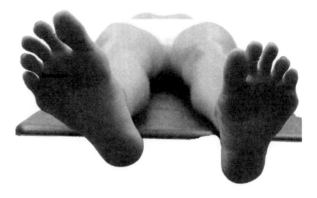

Tibialis posterior weakness.

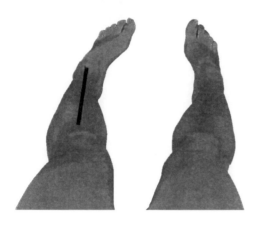

Tibialis posterior weakness

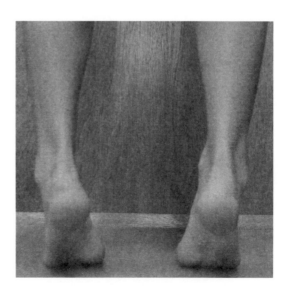

Indicates strain of left posterior tibialis muscle

~ 105 ~

course anterior to the malleoli and are thus dorsiflexors. The peroneus longus and brevis and the tibialis posterior tendons course behind the malleoli and are plantar flexors; thus, if the relaxed foot is adducted with considerable plantar flexion, the peroneus group has less turgor than the tibialis group, and the peroneus tertius less turgor than the peroneus longus and brevis. This only takes into account the peroneus and tibialis groups. When considering dorsiflexion and plantar flexion, the length and strength of the soleus and gastrocnemius must be taken into account, as well as the flexor and extensor digitorum longus muscles. (Length of the gastrocnemius and soleus will be discussed further under "Hypermobile Flatfoot" and "Pronation".)

Palpation

A major tool in examination of the foot and ankle is palpation. Palpatory findings rapidly give the physician considerable knowledge. Observe for subluxated articulations, indications of congenital anomalies, muscles that are strained or hypertonic, and evidence of muscle atrophy. Palpatory findings may give evidence of pathologic or systemic conditions, such as arthritis, edema, circulatory disturbances, or neuropathy. Palpatory findings will be discussed with the conditions to which they relate.

Shock absorber test

The shock absorber test (**Zampagni et al., 2009; Walther 2000, 1981**) is unique to applied kinesiology. It appears to evaluate joint integrity to determine if the nervous system can rapidly recover from stimulation to joint proprioceptors applied as a quick, shocking force. Clinical evidence reveals that when the joint is functioning normally, there will be normal function of muscles associated with it (or remote from it) immediately after the shock is applied. When there is joint fixation or subluxation, a previously strong muscle will test weak for many seconds up to several minutes following the shock.

Muscles to test when applying the shock absorber test to the foot are those of the foot or leg involved with the joints being tested, and muscles of gait that have associated function with the foot, like the tensor fascia lata. (**Zampagni et al., 2009**) The psoas muscle is often used as an indicator muscle when doing the shock absorber test on the foot or ankle.

In addition to being a valuable screening test for the physician's examination, the shock absorber test can often be used to educate the patient about his condition. For example, if the patient is being examined for a shoulder condition and the examination is leading to the basic underlying cause being the foot and/or gait mechanisms, one can test shoulder muscles before and after the shock absorber test. If foot dysfunction disturbs normal activity of the shoulder muscles, it will be revealed in the shoulder muscle tests following the shock to the foot. One may find relatively normal tests of the shoulder muscles in conjunction with a shoulder problem; following the shock absorber test several muscles may test weak, and there may be pain in the shoulder upon testing the muscles when none was present in the clear. These findings demonstrate to the patient in a clear and dramatic way the relationship of foot dysfunction with the shoulder complaint. As noted before, the patient often has no complaint about his feet, yet the basic underlying cause of his chief complaint — in this case the shoulder — resides in foot dysfunction.

Because of the numerous articulations in the foot, the shock absorber test must be applied with many vectors to analyze most of the joints. Varying vectors of force into the distal metatarsal heads with glancing blows are usually effective in testing for dropped metatarsals; however, challenge — discussed later — is a better test. The transverse tarsal arch is shocked with vectors completely encircling it, that is, from the plantar surface to the lateral, dorsal, and medial surfaces, ending on the plantar surface. Sharp blows are applied continuously as the transverse tarsal bones are circled. Solid blows to the entire plantar surface of the foot test many of the articulations, including the ankle and perhaps even the knee

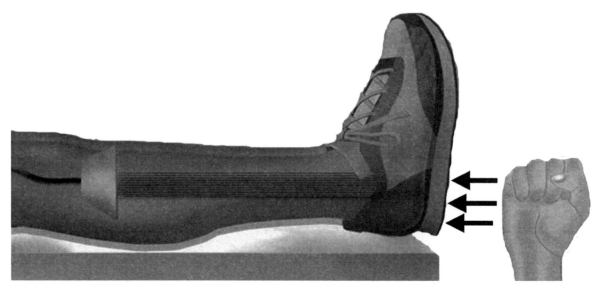

Shock Absorber Test

and hip. Medial, lateral, and posterior blows to the calcaneus further evaluate the ankle mortise, subtalar articulation, and the talar and calcaneal articulations with the midfoot.

The shock absorber test is designed to evaluate for joint dysfunction only; however, it is possible that shock to dysfunctioning neuromuscular spindle cells, Golgi tendon organs, or cutaneous proprioceptors may cause a muscle to test weak. It is also possible that joint dysfunction may not be discovered by this test because of the many articulations and vectors required to stimulate them.

Recent research from the Rizzoli Orthopedics Institute in Bologna specifically investigated the AK shock absorber test and showed that altered signals originating in the subtalar joint of the foot induce inhibition when the tensor fascia lata muscle is tested experimentally in soccer players with ankle injuries and muscle imbalances, eventually altering stability of the knee. **(Zampagni et al., 2009)**

The tensor fascia lata muscle, in particular, plays an important role in maintaining stability of the body and of the knee joint. If this muscle control should become impaired, functional disturbance of the lower extremity might occur and thus represent a risk factor for both the knee and ankle joints. The experimental team evaluated whether the shock absorber test would trigger differences in the force activation pattern of the tensor fascia lata muscle in athletes with ankle imbalance. In particular, the maximal isometric strength of the tensor fascia lata muscle was evaluated to determine whether soccer players who suffer from ankle imbalances experienced weakness in this muscle after the shock absorber test, comparing the unstable and stable ankles from groups of 15 and 14 male soccer players, respectively.

Using load cells to detect forces on the subject's ankle (electromyography), the experimenter applied manual percussion under the subtalar joint of all subjects. Before the shock absorber test was applied, the soccer players who experienced ankle imbalance showed no differences from the control (balanced ankle group) in the duration of tensor fascia lata muscle resistance to the force imposed by the operator during the manual muscle test. After the shock absorber test, however, the ankle imbalance group now displayed a unique and significant decrease in the duration of muscle resistance to the force applied by the operator. No such differences were seen in the uninjured side. **(Zampagni et al., 2009)**

This finding suggests that there is a change in the activation of the tensor fascia lata muscle in subjects with ankle imbalance, and that the AK shock absorber test could dependably diagnose this impairment. In practical terms, this observation suggested that ankle imbalance could be a risk factor for tensor fascia lata muscle weaknesses and ultimately instability of the knee.

It was suggested by these researchers that broader use of the shock absorber test might prevent further injury to the lower extremities exposed to external shocks, which could conceivably be experienced during everyday walking. When the shock absorber test is positive, a whole kinetic chain of events may occur including misalignment of the knees, anterior tilt of the pelvis, upper crossed syndrome, and forward head posture. **(Dananberg, 2007; Rothbart, 2006)**

The shock absorber test is a good screening examination for joint dysfunction when put into proper perspective in the total examination. It should be remembered that it does not find most muscular and cutaneous receptor dysfunction; there may be joint dysfunction in a weight-bearing position that is not discovered in the non-weight-bearing shock absorber test.

Static Weight-Bearing Examination

Static postural examination of the foot includes visual inspection, palpation of the foot and remote structures, and remote muscle testing while standing with the foot in various positions. A weight-bearing examination should be done with the patient barefooted and also wearing shoes. Examination of the effect of various types of shoes and orthopedic support to the feet will be discussed later.

Helbing's sign

Helbing's sign.

In the structurally sound ankle and foot, a line bisecting the calcaneus is in alignment with one bisecting the lower one-third of the leg when both feet are resting on the ground in bipedal stance. **(Evans, 2008)** Stated another way, a line bisecting the lower one-third of the leg should enter into the midline of the Achilles tendon and continue to project down through the midline of the calcaneus without deviation. If the line deviates laterally, it is called Helbing's sign. In the pronated foot, the line turns laterally as it extends to the midline of the calcaneus. There should be an approximately equal amount of medial and lateral malleolus viewed on each side of the Achilles tendon when observation is made directly posterior to the foot.

Knee and leg position

When viewed from the anterior, the patella should be centered. With pronation the knee medially rotates. **(Dananberg, 2007; Prior, 1999; Rothbart & Estabrook, 1988)** A vertical line from the center of the patella should drop between the 1st and 2nd toes. **(Cailliet, 1997)** When pronation or internal tibial torsion or internal femoral torsion is present, the line from the patella will drop medial to the 1st and 2nd toes.

An individual with tibial or femoral torsion, as well as primary genu varus or valgus, is very likely to develop secondary involvement of the foot. Treatment, if applicable, should be directed to the primary cause and attention given to the foot dysfunction.

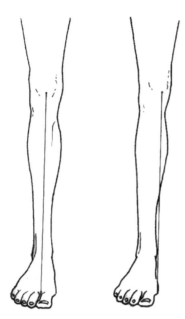
Line from center of the patella should drop between the 1st and 2nd toes.

General postural balance should be related to possible foot dysfunction. As previously noted, a normally functioning positive support reaction is important to facilitation of the extensor muscles for upright posture. Increased spinal curves in the sagittal plane may be due to improper proprioception from foot dysfunction.

Weight-bearing tests

The results of testing muscles remote from the foot will change when weight bearing is causing improper proprioceptive impulses. Muscles that tested strong in the prone or supine position may fail to perform adequately during manual muscle testing when in the static weight-bearing position. Delahunt et al have shown that patients with ankle instability **(Delahunt et al., 2006)** exhibit altered muscle activation and movement patterns in areas remote from the primary physiopathology in the foot. Higher motor system functions usually compensate for functional instability and motor weakness that may result from subluxations, soft-tissue injuries, proprioceptive disturbances and pain. Edgerton and colleagues **(1996)** proposed that decreased muscle recruitment (from an inhibited muscle) can result in increased recruitment from compensating motor neuron pools, possibly leading to further injury. In some cases, basically all muscles tested while weight bearing will fail, while in others only certain muscles do not initially lock on the manual muscle test. When muscle dysfunction is selective, it usually mirrors the patient's symptoms. For example, when foot dysfunction contributes to a shoulder problem, some or all of the shoulder muscles will fail to test normal during a weight-bearing manual muscle test.

Many factors can contribute to the change in muscle function from non-weight bearing to weight bearing. In addition to foot and leg problems, weight bearing in the spinal column can change the vertebral subluxation or fixation status. Various baroreceptors in the vessel systems of the body can be influenced, as well as receptors in the skin, abdominal cavity, equilibrium mechanism, and others. To help differentiate the change of muscle function due to static standing from other causes, have the patient stand on the lateral borders of his feet in the case of pronation. Usually this will cause the muscles that weakened when standing flatfooted to test normal. Standing on the lateral borders of the feet changes stimulation to the foot receptors and also to the receptors throughout the leg, knee, thigh, hip, and pelvis. **(Rothbart, 2006; Prior, 1999)** To more specifically localize the change to the foot, one can place pads under the navicular bone, metatarsals, or other appropriate area of the foot, and then re-test muscles for improvement.

Some muscles will develop or have increased tenderness when the patient with foot dysfunction stands. The origin, insertion, and belly of the muscles should be evaluated. Common muscles having increased tenderness with weight bearing are the upper trapezius, sternocleidomastoid, gluteus maximus, and tibialis posterior. **(Simons et al., 1999)** The tenderness can often be relieved by having the patient stand on the lateral borders of his feet for a few minutes, or stand on pads as previously mentioned. Temporary taping, discussed later, may also relieve strain. The physician can hold various bones of the foot in different positions, such as elevating the navicular bone while palpating the tibialis posterior or gluteus maximus for reduction of tenderness. A two-person examination team can do the same with muscles that are more distant from the foot, such as the sternocleidomastoid and upper trapezius.

The effect of pronation on circulation can be observed in a general manner with a finger pressure test. **(Greenawalt, 1982)** With the patient in a neutral standing position, observe for localized redness of the medial dorsal portion of the foot. Apply finger pressure to different areas, expressing blood from the area. Upon releasing the pressure, observe the length of time required for the color to return. Have the patient externally rotate the leg while the foot remains in place. This takes pressure off the pronated foot. Repeat the blanching test and observe for reduced time for normal color to return.

Many subtle factors regarding the patient's stance can be observed when he is not aware that a postural evaluation is being made. The patient may subconsciously move into a more comfortable position, or one that in all probability does not adversely stimulate the proprioceptors. This is often seen, especially in children who have foot pronation; they stand on the outer borders of their feet. The tests described previously explain why this is a more comfortable position. Many additional observations can be made by keeping in mind that the body subconsciously puts itself in a position for better function or relief.

In foot pronation there is excessive tension on the plantar fascia. To evaluate plantar fascia tension, have the patient stand with two blocks under each foot. One block is put under the calcaneus and the other under the metatarsal arch. Palpate the tension of the plantar muscles and their fascia when the patient is in neutral stance on the blocks. Then have the patient rotate the knees externally, changing the stance to stand on the lateral foot border. Again palpate the muscle and fascia. There should be little tension change if the ligaments and bones are supporting the arch as they should be in static weight bearing. **(Basmajian & Stecko, 1963)**

Dynamic Examination

The weight-bearing examination up to this point has dealt only with static weight bearing. As the examination progresses into walking and running, many other factors come into play, such as the equilibrium proprioceptors, gait mechanism, dural tension, body module organization (PRYT technique), reactive muscles, and others discussed thoroughly elsewhere. **(Leaf, 2010; Frost, 2002; Walther, 2000; Maffetone, 1999; Goodheart, 1998-1964)**

It serves here to put into perspective the addition of walking and running when specifically evaluating for weight-bearing influence on and by foot function. Occasionally, adverse influence of foot dysfunction is not recognized until after the patient walks. Also, there may not be loss of foot and ankle correction until the patient walks. It is important to have the patient walk in a figure-eight pattern or in a circle, first in one direction and then in the opposite one, because turning in either direction may stress the foot in a manner different from walking in a straight line.

The importance of this is illustrated by a patient being treated for phlebitis. Her major etiologic factor was foot dysfunction. On initial examination it was recognized that severe pronation and foot strain were present and contributing to leg strain and the phlebitis. The patient easily recognized the foot problem, but she was hesitant to believe that it might be contributing to the phlebitis. Foot corrections and attention to the supporting musculature were accomplished, but the patient lost the corrections immediately upon walking. Spinal Pelvic Stabilizers® from Foot Levelers, Inc., **(Foot Levelers, Inc.)** and adequate shoes were prescribed. The patient refused the Spinal Pelvic Stabilizers®, stating that eight to ten orthotics had been prescribed and fitted over the years and none helped her chronic foot strain. (She had spent well over a thousand dollars on them.) In light of this, it was necessary to provide tape support to the foot and ankle. She did agree to purchase and wear proper shoes.

The current episode of phlebitis had lasted approximately three months before my examination and had not responded to standard medical treatment. Her condition was so severe that she could walk no farther than across her living room. With correction of her foot problem and other chiropractic and AK treatment, the condition significantly improved in one week; within three weeks there was no evidence of phlebitis. During the three weeks, most of the structural corrections were maintained. The foot and ankle were re-taped as needed. At the end of the three weeks, an attempt was made to eliminate the tape support; within a week, the phlebitis began returning. She then agreed to be fitted for Spinal Pelvic Stabilizers® which, with proper shoes, allowed the corrections to be maintained.

Approximately nine months passed with no recurrence of phlebitis, and the patient was relieved of the chronic foot and leg pain she had previously experienced. Then she mentioned on a routine maintenance office visit that she was beginning to develop leg pain again. Examination revealed foot subluxations and numerous muscle weaknesses of the leg and foot. It was curious that the corrections, which had held so well, had been lost. She was questioned about wearing her Spinal Pelvic Stabilizers® on a regular basis; she stated that she did. Corrections were made, and she was asked to walk; the corrections were not lost. The patient was again scheduled for a routine one-month examination. Within two weeks she made an appointment because the leg pain was worsening. Examination revealed similar findings; corrections were made, and she was able to walk without losing the corrections. Again she was scheduled for re-examination in one month. In two weeks she again returned to the office with worsening of the leg pain. Examination revealed the same dysfunction; again correction was not lost after walking. She began to make appointments more frequently until she was being seen twice a week. Each time she came in with foot and leg dysfunction; each time it was corrected and the corrections were not lost upon walking. By this time two-and-a-half months had passed since the recurrence of symptoms, and frank phlebitis was now present. She was back to the point of not being able to walk farther than across her living room without stopping because of pain. Her days were again spent with the leg elevated and heat applied. She was ready to return to the previous doctor for medication, even though his therapy had failed.

Because of her discouragement, SCC spent more time talking with her on this occasion, completing the record and discussing with her who she could be referred to. When I finished she walked ahead of me to leave the treatment room; she made a sharp turn, and I observed a major foot roll within her shoe. Calling her back, I re-examined her; all the corrections just made had been lost. My previous test was for her to walk in a straight line, stop, turn around, and walk in a straight line. In other words, she was not walking and turning at the same time. Under those conditions no corrections were lost. When she turned while walking, her foot would roll and the corrections were lost.

I asked her when she started wearing the shoes she had on, and it turned out to be just prior to the time her symptoms began to return. The purchase of a new pair of shoes of the same type that I had originally prescribed allowed the corrections to be maintained, and her recovery from the phlebitis was uneventful. Her condition was followed for several years; there was no recurrence of the phlebitis, although she previously had been having several episodes per year.

Since that experience, it has been standard procedure to always test a patient walking in either a figure-eight pattern or around a circle, first in one direction and then in the other, so that both directions of turning are incorporated into the walking pattern. The figure eight-pattern is optimal.

Examination proceeds from dynamic evaluation of the foot to correlating it with the patient's gait. There are several AK methods that find functional conditions otherwise missed. Corrective treatment is usually easily applied and produces lasting results.

It must be remembered that whether you consider the foot as an architectural wonder or a pair of stumps to be ignored, it is an integral component of whole-body function with its influences being felt both locally and throughout the body. As this case demonstrated, the health and integrity of the foot should be a high priority in most therapeutic interventions.

Foot and Ankle Anatomy and Physiology

Divisions and Motions of the Feet

The foot is divided into three functional segments: the hindfoot, the midfoot, and the forefoot. The hindfoot is composed of the talus and calcaneus. The talus is the foot's only bony connection with the rest of the body. It is unique in that it has no direct muscular attachment. The heel of the foot is composed of the calcaneus. The talus and calcaneus articulate with the cuboid and navicular, which is the hindfoot's connection to the midfoot. This joint complex is called Chopart's joint. In addition to the cuboid and navicular, the midfoot is composed of the three cuneiform bones.

The mechanics between the talus, calcaneus, navicular, and cuboid bones allow the two important foot movements: inversion/eversion and pronation/supination. The flexibility of these movements allows the foot to accommodate to the shape of the substrate, and yet lock together to become a solid lever of great strength during certain phases of gait. **(Langer, 2007; Dananberg, 2007; De Wit, 2000; Hamilton & Ziemer, 1983)**

The forefoot is composed of 5 metatarsals and 14 phalanges, making a total of 26 bones in the foot. The articulations of the cuboid and cuneiform bones with the metatarsal bones are collectively called Lisfranc's joint. The anterior pillars of the longitudinal arches are composed of the forefoot, which is active in the propulsive stage of gait.

Classification of the forefoot

The forefoot is classified according to the relationship of the length of the 5 rays or toes and the relationship of the length of the metatarsals. These are the digit and metatarsal formulae, respectively. Viladot **(Viladot, 1982)** reports the percentage of individuals with each type.

Classifying the forefoot according to the length of the toes, there are three types: 1) big toe shorter than the 2nd toe, 22%, Greek foot; 2) big toe longer than the 2nd, 69%, Egyptian foot; and 3) big toe same length as 2nd toe, 9%, squared foot.

Metatarsals are classified according to three types. 1) The 1st metatarsal is equal in length to the 2nd, termed index plus–minus, 28%. 2) The 1st metatarsal is shorter than the 2nd, termed index minus, 56%. 3) The 1st metatarsal is longer than the 2nd, which is termed index plus, 16%.

The digit formula can be combined with the metatarsal formula. The Egyptian-type foot with index–minus metatarsals is the most frequent type.

The Greek-type foot with index plus-minus (short big toe and 1st metatarsal equal to 2nd) is ideal. **(Hoppenfeld, 1982)** It rarely has pathological alterations. The metatarsal index–minus predisposes an individual to hallux valgus, and the metatarsal index–plus favors hallux rigidus in men and sesamoiditis in women. The Greek-type foot with index–plus or index–plus–minus adapts best to modern footwear.

Motions, positions, and fixed structural positions of the foot

Terms used to describe motions, positions, and fixed structural positions vary somewhat among authors. **(Gray's Anatomy, 2004; Kleiger, 1956)** Levangie & Norkin **(Levangie & Norkin, 2001)** point out in reference to the terminology issue for foot motions, "Terms used in research and published literature should carefully be defined to impart the most and clearest information."

The description of terms given here is that generally used. In any event, the reader can correlate with information provided by other authors. **(Table 1)**

Adduction. Adduction, which takes place in the transverse plane, is movement of the forefoot toward the midline of the body. An adducted foot has the forefoot deviated toward the midline of the body. The term "adductus" refers to a fixed structural position of a part of the foot, as if it were adducted.

Abduction. Abduction is motion in the transverse plane in which the forefoot moves away from the midline of the body. An abducted foot occurs when the forefoot has moved away from the midline in the transverse plane. "Abductus" means a fixed structural position of the forefoot, as if it were abducted.

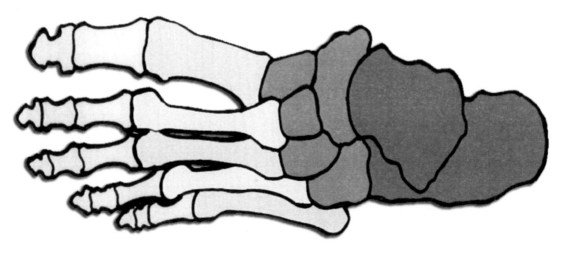

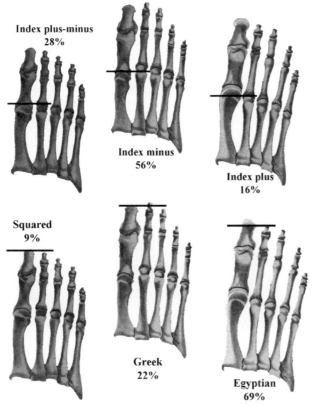

There is different use of the terms "abduction" and "adduction" when referring to active movement of a part, such as the hallux, than that presented for the forefoot. Abduction of the hallux, as with activity of the abductor hallucis, is movement away from the midline of the foot; adduction of the hallux, as accomplished by the adductor hallucis, is movement toward the midline of the foot. **(Gray's Anatomy, 2004; Turek, 1984)**

Inversion. Inversion is often used synonymously with supination and occurs in the frontal plane. It is motion of the foot so that the plantar aspect of the foot, or a portion of it, tilts to face toward the midline of the body. Inverted refers to the foot, or a portion of it, in the position of inversion. When the foot, or a portion of it, is fixed in an inverted position, it is termed "varus."

Eversion. Eversion is often used synonymously with pronation and is foot motion in the frontal plane in which the plantar surface of the foot faces away from the midline of the body. When the foot, or a portion of it, is in eversion, it is termed "everted." If the everted position of the foot, or a portion of it, is fixed, it is termed "valgus."

Dorsiflexion. When the foot, or a portion of it, moves in the sagittal plane so that the forefoot moves toward the tibia, it is termed "dorsiflexion." Note that all dorsiflexion does not take place at the ankle joint. A position of dorsiflexion is termed "dorsiflexed." A fixed limitation of dorsiflexion at the ankle joint is called "ankle equinus" or "talipes equinus." Less than 10° of ankle dorsiflexion is considered talipes equinus.

Plantar flexion. Movement of the foot, or a portion of it, in the sagittal plane so that the forefoot moves away from the tibia is termed "plantar flexion." Occasionally in the literature one will find the term "plantar extension," which is synonymous with plantar flexion. Plantar flexed means that the foot, or a portion of it, is in plantar flexion. A fixed limitation of plantar flexion of the foot at the ankle joint is called "ankle calcaneus" or "talipes calcaneus."

Pronation. Pronation is simultaneous abduction, eversion, and dorsiflexion. The term "pronated" refers to the position into which the foot moves with pronation.

Supination. Supination is simultaneous adduction, inversion, and plantar flexion movement of the foot. "Supinated" is the position of the foot as a result of supination.

Foot Osteology

The foot consists of 26 bones that can be divided into the tarsus, metatarsus, and phalanges. Almost ¼ of the body's bones are in the feet. The tarsus consists of the bones making up the posterior half of the foot — the talus, calcaneus, cuboid, navicular, and 1st medial, 2nd

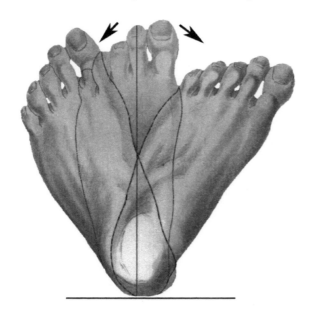

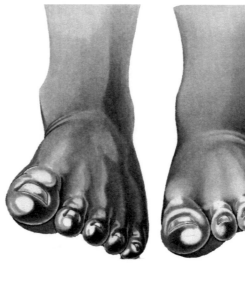

Inversion   *Eversion*

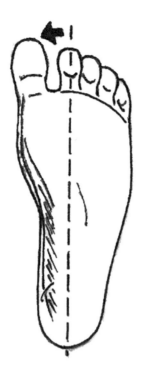

Hallux abduction

Talus (astragalus). The talus is the only articulation between the foot and leg. It is unique because it is the only bone in the body that does not have direct muscular attachment. The body of the talus is basically cuboid. The superior surface is a trochlear surface for articulation with the tibia. The medial surface articulates with the medial malleolus of the tibia, and the lateral surface articulates with the fibula. The medial and lateral surfaces form a wedge, with the widest portion at the anterior. The articulation of the body of the talus with the tibia and fibula is called the ankle mortise; it will be discussed later with the ankle joint. The ankle mortise is designed to handle high levels of force. Compressive forces transmitted across this joint during gait may reach five times the body weight, while tangential shear forces (the result of internal and external rotational forces associated with the body moving over the foot) may reach 80% of the body weight. **(Gray's Anatomy, 2004)** The fibula has little weight-bearing responsibility. Levangie & Norkin state "no more than 10% of the weight that comes through the femur is transmitted through the fibula." **(Levangie & Norkin, 2001)**

The inferior or distal surface of the body has two or three surfaces that articulate with the calcaneus, called the subtalar or talocalcaneal articulations. The larger surface is posterior and is concave to reciprocate with the convex surface of the calcaneus. These surfaces allow multiaxial movement between the two bones, including inversion and more limited

intermediate, and 3rd lateral cuneiform. The metatarsus consists of 5 bones, and there are 14 phalanges, 3 for each toe except the first, which has 2 phalanges.

A digital ray is a digit of the hand or foot consisting of the metacarpal or metatarsal and its corresponding phalanges. The entire unit or the great toe, then, is referred to as the 1st ray, 2nd toe the 2nd ray, and so on.

For convenience sake anatomists divide the foot into three main sections: the *hindfoot*, the *midfoot*, and *forefoot*.

Hindfoot

The hindfoot consists of only two bones, the calcaneus and talus. The calcaneus is the major contact of the posterior foot with the substrate, and the talus is the foot's only contact with the leg.

Calcaneus (os calcis). The calcaneus is the largest bone of the foot. It forms the heel of the foot and transmits the hindfoot's weight to the substrate. The posterior surface projects posteriorly beyond the bones of the leg to provide an effective lever for the insertion of the calf muscles. The superior surface has two or three facets that articulate with the talus. The posterior articular surface is the largest. Anteriorly there are one or two articular surfaces to reciprocate with the corresponding surfaces of the talus.

The sustentaculum tali arises at the anterior superior medial surface of the calcaneus. Its superior surface is the medial articular surface for the talus mentioned above. The inferior surface of the sustentaculum tali is grooved for the tendon of the flexor hallucis longus. The entire anterior surface of the calcaneus is a concavoconvex facet for articulation with the cuboid bone.

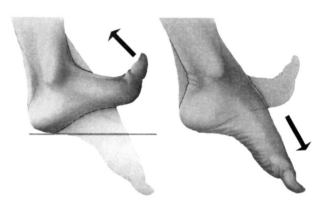

Dorsiflexion   *Plantar flexion*

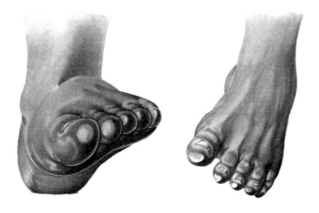

Pronation   *Supination*

MOTION	POSITION	FIXED	COMBINATION
Abduction	Abducted	Abductus	
Eversion	Everted	Valgus	Pronation – Pronated
Dorsiflexion	Dorsiflexed		
Adduction	Adducted	Adductus	
Inversion	Inverted	Varus	Supination – Supinated
Plantar flexion	Plantar flexed		

Table 1
Summary of foot and ankle positions and motions

Superior view of left calcaneus

eversion, adduction and abduction. Combinations of these movements contribute to pronation and supination.

These motions of the foot are not limited to the subtalar articulation; they include movement at the transverse tarsal articulations as well.

The posterior articular surface is separated from the anterior one by a groove called the sulcus talus. The calcaneus has a similar groove called the sulcus calcanei. When the two are combined, the tarsal canal is formed between the two bones. The largest end of the canal is called the sinus tarsi, and it opens laterally. It can be palpated in front of the fibular malleolus when the foot is markedly inverted. From the sinus tarsi the canal angles slightly posteriorly and medially to open just behind the sustentaculum tali of the calcaneus.

Anterior to the sulcus talus is the anterior calcaneal articulation, which may be singular or double. The anterior facets are convex on the talus to reciprocate with the concave articular surfaces of the calcaneus. This is exactly opposite the posterior articulation of the talus and calcaneus.

The talocalcaneal articulations are also called the subtalar articulations. The major motion at the subtalar articulation is inversion and eversion. With the foot in free motion, the calcaneus moves on the talus, whereas when the foot is planted on the floor, the talus moves on the calcaneus.

Projecting anteriorly and medially from the body is the neck of the talus, which joins the head of the talus to the body. The head of the talus is a large oval surface for articulation with the navicular bone.

Midfoot

The midfoot consists of five tarsal bones — the navicular, cuboid, and three cuneiforms. These bones are wedge-shaped, being wider dorsally and narrower plantarly. When articulated they form longitudinal and transverse curves, with the concavities facing plantarly. The bases of all five metatarsal bones are also wedge-shaped, forming a medial-to-lateral arch. (**McDougall, 2009; Levangie & Norkin, 2001; Cailliet, 1997**) The hindfoot and midfoot are joined by the transverse tarsal articulation (Chopart's joint). The head of the talus fits into the posterior concave surface of the navicular bone. Inversion and eversion are allowed at this articulation; there is a significant gliding action because the articular surface of the head of the talus is larger than that of the navicular.

The rest of the transverse tarsal articulation is the calcaneocuboid joint. More stability is gained in the foot if the talonavicular and calcaneocuboid articulations are not parallel. This causes joint restriction for better foot stability. In pronated feet the articulations are usually parallel.

Navicular. The navicular bone (scaphoid) provides three facets for articulation with the cuneiform bones on its anterior surface. The lateral surface usually does not articulate with the cuboid; however, it sometimes does present an articulating surface. Its posterior concave surface articulates with the head of the talus. The navicular bone often requires manipulation. When it is subluxated with the talus, the deep articulation of the head of the talus into the navicular must be considered when designing a line of drive for correction. For this reason, and because of the varying articulating surfaces, careful challenge is necessary to find the optimal vector of force for effective adjustment of the bone.

Cuboid. The cuboid bone derives its name from its cuboidal shape. It is lateral to and longer than the navicular bone. The posterior surface is almost flat and articulates with the calcaneus. The anterior surface is divided into two facets for articulation with the 4th and 5th metatarsal bases. The proximal medial surface is adjacent to, but only occasionally articulates with, the navicular. The distal medial surface has a flat facet for articulation with the lateral 3rd cuneiform.

Cuneiform bones. The three cuneiform bones are called medial, intermediate, and lateral or 1st, 2nd, and 3rd, respectively. They are wedge-shaped, being wider dorsally and narrower plantarly. The 1st cuneiform is the largest. The

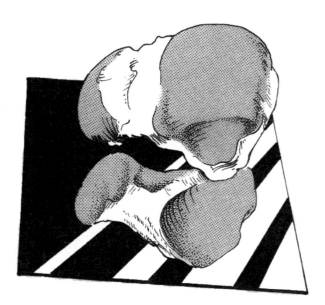

Left talus (With kind permission, ICAK-USA)

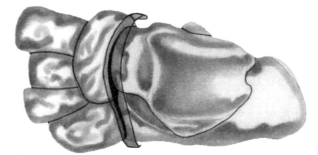

Foot integrity is improved when the articulations of Chopart's joint are not in a straight line.

posterior surface articulates with the most medial and largest of the three facets on the anterior surface of the navicular. The inferior or plantar surface is wider than the wedge portion of the bone, forming the base of the wedge. The lateral surface is concave, articulating with the 2nd cuneiform.

The 2nd cuneiform is the smallest of the three. Specifically it is shorter than the 1st and 3rd cuneiform bones, thus forming a groove for articulation of these three bones with the 2nd metatarsal base. The lateral surface is a smooth facet for articulation with the 3rd cuneiform bone.

The 3rd cuneiform is intermediate in size between the 1st and 2nd cuneiform bones. The posterior surface articulates with the lateral facet of the navicular surface. The proximal surface articulates with the 2nd cuneiform, and the distal one with the lateral surface of the base of the 2nd metatarsal. The lateral surface of the 3rd cuneiform also has two articular surfaces. The proximal one articulates with the cuboid, and the distal with the medial side of the base of the 4th metatarsal bone. The anterior articular surface of the 3rd cuneiform mates with the 3rd metatarsal bone.

Forefoot

Metatarsal bones. The metatarsal bones have common characteristics of a base, which is proximal and wedge-shaped to articulate with the tarsal bones. The base has facets on its sides to articulate with contiguous metatarsal bones. The body is the shaft, which gradually grows smaller as it progresses distally. The head is the distal portion extremity, which presents a convex surface to articulate with the proximal phalanx. The plantar surface of the head is grooved for the flexor tendons.

The 1st metatarsal is the thickest, and usually the shortest, of the metatarsal bones. Differing from the other metatarsals, the base as a rule has no articular facet on its sides. The proximal portion of the base of the 1st metatarsal is kidney-shaped to articulate with the 1st cuneiform. It has dorsal and plantar flexion and rotation, as well as a gliding action on the surface of the 1st cuneiform. The head is large and has two grooved facets on the plantar surface to articulate with sesamoid bones, which are incorporated into the tendons of the flexor hallucis brevis. They bear body weight and act as a fulcrum for the tendon. This arrangement is important for action of the medial longitudinal arch, which will be discussed later.

The 2nd metatarsal bone is usually the longest of the metatarsals. Its base has four articular surfaces. Proximally it articulates with the 2nd cuneiform. The medial surface of the base articulates with the 1st cuneiform. There are two lateral articular surfaces that articulate with the 3rd cuneiform and 3rd metatarsal bones.

The 3rd metatarsal bone articulates proximally with the 3rd cuneiform. The medial and lateral surfaces of the base articulate with the 2nd and 4th metatarsals, respectively.

The 4th and 5th metatarsal bones are progressively smaller. The base of the 4th metatarsal articulates with the cuboid. The lateral and medial surfaces of the base articulate with the 3rd and 5th metatarsals, respectively. The 5th metatarsal base articulates with the cuboid, and medially with the 4th metatarsal. It has a tuberosity on its lateral side, into which the tendon of the peroneus brevis inserts. The peroneus tertius inserts into the medial part of the dorsal surface of the 5th metatarsal bone.

The wedge shape of the mid-tarsal bones, along with the wedge shape of the base of the metatarsal bones, forms longitudinal and transverse curves with the concavities facing plantarly, thus forming a medial-to-lateral arch frequently called the transverse arch. The proximal, medial, and lateral portions of the 3rd and 4th metatarsals, obliquely shaped, permit a progressive rotary motion of the 3rd on the 2nd, 4th on the 3rd, and 5th on the 4th. This motion increases the transverse arch and thus "cups" the sole of the foot.

Phalanges. There are three phalanges for each ray, with the exception of the first where there are two. The great toe has the flexor hallucis longus attached to the distal phalanx on the plantar surface, and on the dorsal surface the extensor hallucis longus. The proximal phalanx of the great toe has attachment on its dorsal surface for the extensor digitorum brevis. At the medial side of the base there is attachment for the abductor hallucis. The plantar surface of the proximal metatarsal has two attachments for the flexor hallucis brevis on the lateral and medial aspects. There is also attachment on the lateral plantar surface of the base for the adductor hallucis.

The proximal or 1st phalangeal row of the 2nd through 5th rays has proximally a base that is concave and a head that presents a trochlear surface for articulation with the

Foot and Ankle

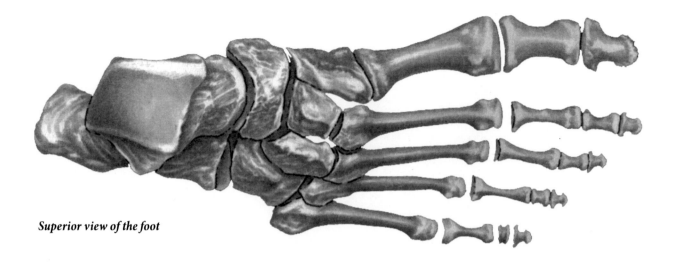
Superior view of the foot

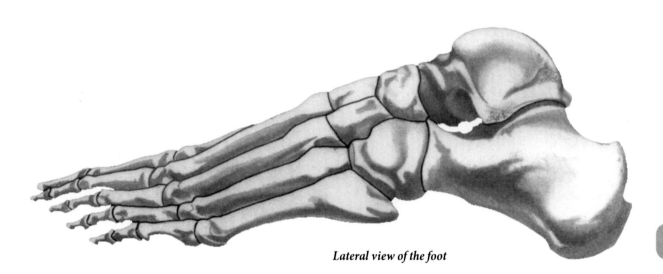
Lateral view of the foot

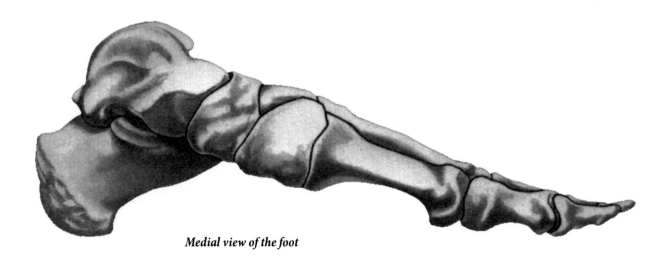
Medial view of the foot

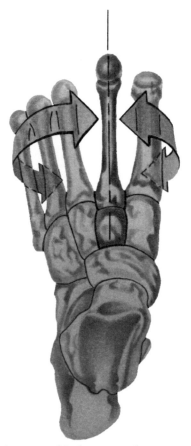

Rotation around 2nd metararsal

middle phalanx. The 1st tendon of the extensor digitorum brevis inserts into the dorsal surface of the 1st metatarsal bone; the 2nd through 4th tendons attach to the lateral sides of the extensor digitorum longus tendons.

The middle or second row of the 2nd through 5th toes has characteristics similar to the proximal row, but they are shorter and broader. The flexor digitorum brevis inserts on the plantar surface at the base, and the extensor digitorum brevis on the dorsal surface.

The distal row is called the ungual phalanges. Each presents a base to articulate with the trochlear surface of the head of the middle phalanx, and ends in a point. The distal phalanx of the 2nd through 5th toes has an attachment point on the plantar surface for the flexor digitorum longus, and on the dorsal surface for the extensor digitorum longus.

There are many additional, smaller intrinsic muscles of the metatarsals and phalanges that are extremely important in proprioceptive function of the foot and its relationship with the positive support mechanism.

Accessory bones. Accessory bones of the foot are supernumerary bones. They appear by age 10 and are fully formed by age 20. (**Espinosa et al., 2010; Jaffe & Laitman, 1982**) They follow fairly standard types and may cause significant problems or be unnoticed.

The accessory navicular bone can give the appearance of a flat foot. (**Hamilton, 1985**) There are three types of accessory navicular bones. (**Coskun et al., 2009; Sgarlato et al., 1969**) Type 1 is a sesamoid in the tibialis posterior tendon. Approximately 30% of accessory naviculars are this type. Type 2 accessory naviculars are united to the navicular by a cartilaginous synchondrosis measuring 1-3 mm. This is the type that creates the major problem of accessory naviculars. Type 3 is an accessory navicular united to the parent navicular by a bony ridge, producing a cornuate navicular.

Another common accessory bone is the ununited lateral tuberosity of the posterior aspect of the talus, which is called the os trigonum tarsi. It is present in 3-8% of the population. It sometimes remains attached to the talus as the trigonal process. It is rarely symptomatic, but it can cause symptoms with certain types of stress.

The foot and the ankle together possess some thirty-three joints. The bones and joints provide a solid foundation and leverage for the muscles to move the body. More than one hundred ligaments are part of the foot-ankle system and they stabilize and help the body sense the position and functional state of the bones of the foot during movement.

Foot Mechanics – The Close Examination

The American Medical Athletic Association reports that every year 37 to 50 percent of runners suffer injuries severe enough to reduce or stop their training or cause them to seek medical care. (**Wilk et al., 2009**) With about 36 million runners in the United States, this means that 14 to 18 million runners are getting hurt every year.

Abshire, Maffetone, and Robbins and Hanna (**Abshire, 2010; Maffetone, 2003; Robbins & Hanna, 1987**) view the foot with a different perspective than most. Rather than being a functional unit that adapts poorly to the stresses put upon it and needs arch supports and supportive shoes to prevent injury, they find the properly conditioned bare foot is more injury resistant. Their investigation originated from recognizing that barefoot runners in international competitions have minimal running-related injuries. They developed the hypothesis that the bare foot contacting uneven surfaces of the substrate is stimulated to react with the shock absorbing mechanism of the arches and muscular action of the foot. The feet of a shoe-wearing population are insulated from the uneven stimulating surface of the substrate, causing an impaired somatosensory feedback.

To test this hypothesis, Robbins and Hanna compared a shoe-wearing group — who shifted to a minimum of one hour per day of barefoot activity, and were encouraged to do their running and walking in this manner — with a control group who continued wearing shoes. In the barefoot group there was a significant ($p<0.05$) shortening of the median longitudinal arch, indicating an elevation of that structure. They attribute this to activation of the muscles by increased somatosensory feedback from foot contact with the substrate. The foot, along with the hand, has "…an extremely high density of neuroreceptors which respond to small discrete displacements, directionally applied force (shearing force), and low intensity repetitive force (vibration). These mechanoreceptors have a relatively low threshold. There is another group of receptors that respond to similar forces, but with such high thresholds that they only respond to mechanical stimuli intense enough to induce tissue damage (nociceptors). These two groups of receptors are responsible for the perception of pressure and pain.

"The data suggest that, in our subjects, their normal footwear prior to the experiment did not produce the sensation necessary to induce protective adaptations inherent with barefoot weight-bearing activity. It is obvious that footwear with soft flexible polymer foam and a relatively inflexible rubber sole does not allow discrete skin deflections from irregular surfaces. Foam, in addition, has vibration-dampening properties. It has been elegantly demonstrated that there is diminished shearing force in the barefoot state. (**Pollard, 1983**) The shearing force which is applied by the plantar skin when barefoot is transferred to the shoe lacings and counter of the shoes when shod."

Robbins and Hanna (**Robbins & Hanna, 1987**) give the example of almost complete joint obliteration in Charcot joints because of damage resulting from lack of somatosensory feedback as a result of neurological damage of tertiary syphilis. "The runner, like the neuropathic syphilitic, damages his or her lower extremities due to lack of somatosensory feedback-mediated protective behavior."

"The arch support, which is present in all running footwear, would interfere with the downward deflection of the medial arch on loading. Furthermore, the use of orthotics, or structures that are fitted to the mould of the soft tissues of the foot, could cause similar difficulty. Such designs occur when an engineer looks at the foot as an inflexible lever that is delicate and thus requires packaging. Various myths persist about foot behavior due to poor understanding of its biology." (**Robbins & Hanna, 1987**) It is now accepted that proprioceptive awareness of movement in the ankle is heightened when the person is barefoot than when wearing shoes. (**Waddington & Adams, 2003**) Sensory information coming from proprioceptors in the muscles and joints and from the plantar surface of the foot is essential for postural control.

A growing body of research on "functional footwear" has occurred simultaneously with the development of new athletic shoes such as Vibram Fivefingers, Masai Barefoot Technology, FitFlop, Nike Free's, Vivo Barefoot and Newton among others. This research has shown that when shod and unshod populations are compared, unshod populations have a lower prevalence for many of the most common running injuries including ankle sprain, plantar fasciitis, iliotibial band syndrome, patellar pain, back pain, shin splints, and Achilles tendinopathy, (**Warburton, 2001**) and the "functional footwear" populations have better arch development (**Mauch et al., 2008; Rao & Joseph, 1992**).

Researchers have found impairments in muscle strength, endurance, and activation times in the hip abductors for subjects with ankle instability and injury. (**Zampagni et al., 2009; Beckman & Buchanan, 1995; Bullock-Saxton et al., 1994; Nicholas et al., 1976**) Functional limitations and muscle weaknesses are not uncommon for patients with a history of chronic ankle sprains, and up to 70% of individuals involved in basketball with an initial ankle sprain could suffer from injury recurrence. (**Yeung et al., 1994**) Two studies revealed that the gluteus maximus muscle was significantly delayed in activation in subjects following an ankle sprain injury compared to control groups. (**Bullock-Saxton, 1994; Bullock-Saxton et al., 1994**)

A major reason that the manual muscle test for the muscles of the foot and ankle should be added to the standard diagnostic methods used and taught in the manipulative professions, including podiatry, is that patients with foot and ankle disorders demonstrate joint instability, ligament strain, and muscle inhibition. (**Maffetone, 2010; Logan, 1995; Yeung et al., 1994**) Applied kinesiology MMT adds significant information to foot and ankle examination because it detects and specifically treats these muscle impairments that either cause or perpetuate extremity and whole body dysfunctions. (**Goodheart, 1998**)

The foot evolved some 270 million years ago whereas shoes have been on human feet for a few thousand years at most. (**Haines, 2000**) Foot dysfunction and its relation to total body function are better understood today, and effective methods of examination have been developed, within applied kinesiology and without (**Maffetone, 2010; Logan, 1995**) Bahler (**Bahler, 1986**) cites a quotation by Georg Hohmann, "...the old master of orthopaedics," from his book *Fuss und Bein*. "The human foot is one of nature's works of art and as such, it has not yet been fully recognized and explained. It will require a great deal of scientific investigation before this structure is fully understood." Partially answering this need is the use of computed tomography and magnetic imaging to diagnose pathology and trauma in the foot. (**Tourne et al., 2010; Sartoris & Resnick, 1987**) Far less attention has been given to functional aspects that influence performance. The relationship of the foot to remote body function is still understood by too few in general health care. For the most part doctors in every field are inadequately prepared in the anatomy and physiology of the foot; consequently, they tend to overlook it. This accentuates the basic fact that we cannot recognize what we do not know. Increased attention by surgeons has been directed to athletes' feet. Sammarco (**Sammarco, 1986**) wrote, "Unfortunately [foot and ankle injuries] were often relegated to the trainer or paraprofessional since such conditions and injuries were felt to compromise performance level only to a minor degree. As more information has become available, however, it is obvious that such injuries tend to affect the performance much more than expected."

Another notable recognition of the foot's importance to total body organization is by Dananberg (**Dananberg, 2007**) and Bordelon. (**Bordelon, 1987**) Dananberg notes that a specific condition of the foot (*functional hallux limitus*, discussed later), "because of its asymptomatic nature and remote location, has hidden itself as an etiological source of postural degeneration. Functional hallux limitus is a unifying concept in understanding the relationship between foot mechanics and postural form. Identifying and treating this can have a profound influence on the chronic lower back pain patient." Bordelon notes that "since the function of the foot is to set the stage for the entire gait pattern and since the pattern of gait and the function of the foot are what set the stage for any activity of bipedalism whether it be walking, running, or jumping and since the entire function of the ankle, hip, knee, and back depends upon the shape of the foot, its function, and its position, perhaps we should consider the shape and function of the foot in evaluating any abnormality of the lower extremity and back." In applied kinesiology we could add: Consider the function of the foot in any health problem.

Goodheart, Walther, Chaitow and DeLany, Liebenson, Steindler (**Goodheart, 1998-1964; Walther, 2000; Chaitow & DeLany, 2002; Liebenson, 2007; Steindler, 1955**) and many others have written extensively on the closed kinematic chain of the body. When the foot is in contact with the ground, the foot, leg, thigh, and pelvis make up a modified closed kinematic chain. Imbalance in any part of the chain will cause change in function of the remote portions of the chain; thus extended pronation puts torsion into the leg, thigh, and pelvis, which would not ordinarily be present. Because foot malfunctions lead to instability during gait, compensation patterns emerge that have body-wide implications. Simons et al (**Simons et al., 1999**) report that a cascade of myofascial conditions are likely to emerge in the patient with disturbances in foot structure and function, including pain in the low back, thigh, knee, and foot.

The manual muscle test as used in AK makes the diagnosis of these specific interactions between the joints and muscles of the foot and remote structures and muscles throughout the body far easier. The *visual diagnosis* of a specific joint or muscle impairment in the foot and simultaneously its relationship to a specific joint or muscle impairment in the hip, shoulder, neck or jaw is difficult. (**Lederman, 2010**) The different elements within the chain of events occurring during any particular movement that a patient exhibits in front of the examiner occur within a fraction of a second; far too rapidly to be accessed individually in the absence of laboratory tools. Therefore, what is actually observed by the examiner who depends upon visual diagnosis of these muscle-joint interactions is the grand total of how rapidly and smoothly a person's global posture can change between two activities – it is almost impossible to make a diagnosis of a specific muscle or joint dysfunction on this basis.

The AK manual muscle test permits a specific challenge to a specific muscle or joint in the foot to be immediately followed by another specific test to a distant joint or muscle, thereby making evident to both the physician and the patient the dynamic interactions going on between two distant structures. Additionally if a specific dysfunction is suspected by visualization or palpation, then the AK sensorimotor challenge to this articulation or muscle can be tested in relationship to the remote muscle dysfunction.

Many think of the sacrum as the foundation of the spine, but as Gillet and Liekens (**Gillet & Liekens, 1981**) point out the ischia are the base when sitting and the feet when standing. We see that many in the chiropractic profession's history have emphasized the foot's role in spinal function. (**Keating, 2002; Greenawalt, 1980**) In 1954 Janse (**Janse, 1976**) recognized faulty body mechanics from serial distortions as he described the possible scenario resulting from foot pronation or "falling of the medial longitudinal arch." As the foot pronates the talus rotates, carrying the tibia into medial rotation that extends up to the femur, moving the greater trochanter anteriorly and laterally and stretching the piriformis by a windlass mechanism. Because of improper piriformis support, a subluxation of the sacrum may result in an anterior and inferior position. To compensate, the gluteus maximus muscle contracts to resist the forward and downward disposition of the pelvis. Secondary to the gluteus maximus contraction, the innominates become subluxated. As a result of the sacral anterior inferior subluxation, the 5th lumbar is "mobile" and, according to Lovett's Law, will gravitate and rotate toward that side; thus the beginning of structural scoliosis is established. This has been a synoptic review of a description originally presented by Janse in the May, 1954, issue of the *National Chiropractic Association Journal*. Other early descriptions of spinal problems resulting from the feet include hyperlordosis (**Harrison, 1964**) and sciatica. (**Goodheart, 1967**)

Lewin (**Lewin, 1959**) states, "Flat feet and a weak back are often found in the same child and adult." Patients will sometimes lose a vertebral subluxation correction as soon as they walk, often before they reach the reception area to pay for the office visit.

The importance of foot function on the spinal column is shown in a study (n=72) by Ceffa et al. (**Ceffa, 1982**) in Italy. The feet and spine were studied with thermography on individuals who complained of back pain and were under chiropractic care. In this study, 12 had flatfeet, 41 "hollow" feet, and 47 load alterations revealed by thermography. The study was instituted because of "...significant reoccurrence, after a successful manipulative treatment, of muscular

Visual diagnosis of specific joint-muscle interactions is very difficult.

pains and defense contractures, in the same area or in other segments of the spine…." The patients were treated with orthotics and examined several times and at various time intervals. The imbalanced thermographic patterns in the lumbar and thoracic regions were balanced or greatly improved, as was the thermographic pattern of the feet. Comparison of the subjective and objective results of this study showed 21 patients pain-free with normalized thermographic pictures; 34 had been pain-free over one year, with significant improvement in the thermographic pictures; 14 had less frequent and less severe pain, with improved thermographic pictures; and 3 had no improvement, with no improvement of the thermographic pictures.

In 1954 Janse (**Janse, 1976**) summarized the body's structural integration and the mid-20th-century chiropractic conceptual approach by stating "…faulty body mechanics is usually a consequence of a serial distortion rather than a single local lesion. We mean thereby that the problem of postural and mechanical pathology is the result of distortions that may begin in the foot or feet, extend up into the leg or legs, then into the knee or knees. From there it may ascend into the hips and sacroiliacs, reflecting onto the vulnerable lumbosacral articulations and eventually the spine to the occiput."

Early in applied kinesiology, Goodheart (**Goodheart, 1967**) associated psoas muscle dysfunction and sciatic neuralgia with excessive foot pronation. He dramatically demonstrated the immediate effects of improving body function and reducing pain by eliminating adverse stimulation to the foot proprioceptors by simply having the patient stand on the lateral borders of his feet to take the strain off the medial longitudinal arch. This can easily be demonstrated when some muscles test weak when standing but are strong when the patient is supine or prone. The muscles will immediately test strong when the patient stands on the foot's lateral border, and weaken with usual weight bearing. Foot dysfunction is not the only reason muscles may become weak when standing, but it is the most frequent cause.

Monte Greenawalt, the founder of the orthotics company Foot Levelers that has been a staple within the chiropractic profession for more than five decades, quoted Goodheart's research on the feet regularly in his books and published research papers. (**Greenawalt, 1980**)

An individual who has had a hyperextension/hyperflexion cervical sprain/strain, or the so-called "whiplash" accident, may have their condition complicated by extended foot pronation. This, of course, is probably not part of the original traumatic condition; however, it can be a perpetuating factor of the cervical problem. This results from the role the foot proprioceptors play during walking on the facilitation and inhibition of the head rotating muscles, such as the sternocleidomastoid. (**Meyer et al., 2004**) Because of this disorganization, the insertion of these muscles at the mastoid processes may be exquisitely tender. Travell & Simons (**Travell & Simons, 1992**) have noted that myofascial trigger points from as far away as the soleus muscle can refer pain and muscular dysfunction into the temporomandibular region. Simply having the patient stand on the lateral borders of his feet for one or two minutes often reduces or eliminates the pain at the mastoid processes.

Because of the intricate connections in the neuromuscular system, a change in any part of the body can disturb function elsewhere in a distant part. An example of this would be that an anterior cruciate ligament injury has been shown to generate changes in the posterior cervical, upper and lower trapezius, sternocleidomastoid, anterior temporalis and masseter muscles. (**Tecco et al., 2006**) There is no easy way to visually assess the specific muscle impairments seen in the neck after a knee injury; but the use of the AK challenge test to the knee, followed immediately by specific AK MMT to the muscles of the neck, can make each of these specific cervical muscle impairments produced by an anterior cruciate ligament injury evident to the examiner. (**Sprieser, 2003; Duffy, 1999; Schmidhofer, 1997; Zatkin, 1990; Raffelock, 1987**)

Another area that might be influenced by foot dysfunction is the stomatognathic system. (**Cuccia, 2011**) Cranial faults can be created or perpetuated by disorganized function of the sternocleidomastoid muscles as they pull on the mastoid processes during gait, (**Walther, 1983**) or by the fascial pull that may develop with postural sagging. This is a primary cause of neurologic disorganization because of the stomatognathic system's close integration with the equilibrium proprioceptors. Hicks notes that talocalcaneonavicular restriction reduces our ability to stand on one foot. (**Hicks, 1953**) Richie (**Richie, 2001**) notes that balance and postural control of the ankle appear to be diminished after a lateral ankle sprain and this can be restored through treatment; this improvement in balance is mediated through central nervous system mechanisms. One might question what is major; does the inflexibility of the foot or the improper stimulation of the foot proprioceptors contribute more to the body's inability to maintain proper orientation in space?

Unilateral hyperpronation and the resultant "short leg" causing pelvic obliquity have been mentioned as factors in scoliosis development. (**Vleeming et al., 1997; Cailliet, 1997; Silverman, 1986; Botte, 1981; Janse, 1976**) Possibly a greater contribution to the problem is the foot–related neurologic disorganization found by applied kinesiology examination. A major component of almost every idiopathic scoliosis case is neurologic disorganization. (**Walther, 2000**)

Dananberg has shown that symptoms associated with foot dysfunction include low back pain, tibialis posterior dysfunction, and anterior knee pain. (**Dananberg, 2007**) Dananberg and others have also shown that functional hallux limitus (*FHL, discussed later*) to be a remote, often hidden source of postural degeneration and pain. Functional hallux limitus involves limitation in dorsiflexion of the 1st metatarsal-phalangeal joint during walking, despite normal function of this joint when non-weight bearing. Mattson, Ferrari, and Dananberg have each shown that improvements in foot function, mobility and strength resulted in marked improvements for patients with chronic nonspecific low back pain. (**Mattson, 2008; Ferrari, 2007; Dananberg & Guiliano, 1999**)

The work of Gracovetsky (**Gracovetsky, 1989**) (the spinal engine theory), developed in the mid-1980s (and published in 1989 in his book of the same name), proposes that the spine optimizes its efficiency of motion in the gravitational field by using the spine to propel the legs forward by capturing the ground reaction force to decouple the spinal segments with each step of the gait cycle. Storing the ground reaction force as potential energy (similar to the Hicks windlass mechanism which stores elastic energy through the plantar fascia during the swing phase of gait) in the viscoelastic tissues of the lower extremity and spine, and then expressing that potential energy as kinetic energy as the spine propels forward. Disturbances then in the feet may alter the ground reaction force described by Gracovetsky and thereby impair the force transmission potentials of the spinal engine through the rest of the body.

In addition to foot dysfunction adversely affecting spinal function, the converse is applicable. Interplay between the spine and feet are demonstrated by Gillet and Liekens. **(Gillet & Liekens, 1981)** They examined the spine but made no corrections to it; only foot corrections were made. "In 86% of the cases, there were varying changes (small to important) in the spine…either immediately or slowly." In another portion of the study, the height of the arch was periodically measured while spinal corrections were administered. "In 37% of these cases there was definite change in the feet as the spinal fixations were eliminated."

General association may be made between different areas of the foot with levels of the spine. Gillet and Liekens **(Gillet & Liekens, 1981)** observed that the toes relate with the upper cervical vertebrae and the metatarsals to the rest of the cervical spine, while the joints between the tarsals and metatarsals correspond to the dorsal spine. Intrametatarsal fixations relate with fixations between the ribs and the spine, and between the ribs themselves. Fixations in the midfoot — that is, cuneiform–navicular and cuboid with the calcaneus — are associated with the lumbar region.

In an applied kinesiology practice, most foot correction is for chronic problems. When acute trauma is present, one must not only provide the proper type of treatment for the local injury, but also be aware of remote conditions that might develop. The effects of trauma to the foot may cause the afferent nerve supply to cause disorganization, adversely affecting remote areas as well as the problems that develop with disuse during rehabilitation. Nicholas and Marino **(Nicholas & Marino, 1987)** state, "…*the more distal the injury site, the greater the total weakness of the affected limb.* Thus, distal injuries produce more weakness to the entire limb than do proximal ones." When foot and ankle trauma causes continued problems, there is significant weakness of the hip adductors and abductors when compared to the uninvolved contralateral side. "Immobilization of the foot and ankle triggers a cascade of negative effects on skeletal muscle tissue, including muscle atrophy, adaptive shortening of muscle, and periarticular tissue (contracture), a decrease in muscle activity through a diminished motor discharge frequency, and/or recruitment and synchronization, a shift from a tonic to a more phasic pattern of firing, loss of myofibrillar and sarcomal protein, and diminished blood flow. It is well established that muscles atrophy when not used either because of immobilization or disuse."

Rothbart and Estabrook **(Rothbart & Estabrook, 1988)** found a high correlation between excessive pronation, static pelvic abnormalities, and chondromalacia patellae, with 96% of the patients (n=97) in their study showing excessive pronation and low back pain. Treatment was based on a combination of chiropractic and podiatric therapy with a 6-month follow-up. Analysis of the success in this tandem approach was promising.

Rothbart and Estabrook suggest a model that asymmetrical pronation patterns (one arch dropping more than the other) initiates a forward and downward rotation within the sacroiliac joint. Entrapment of the sciatic nerve then occurs between the piriformis muscle and sacrospinous ligament. They suggest that paresis is then observed clinically with weakness, numbness, and eventually paralysis of the affected limb. They also proposed a model for chondromalacia explaining the pathomechanical events associated with oblique tracking patellar syndrome. They suggest that excessive pronation is the causative factor directing asynchronous rotation between the shin and femur. This forces the patella out of its normal tracking groove, which in turn generates erosion between the inferior margin of the patella and femoral epicondyles. In patients with symptomatic and asymptomatic patellar pain, assessment for muscle strength impairments will be essential to restoring the normal "track" or "path" of the patella through any particular movement. Any aberrations in the recruitment or coordination of these sequential movements of the patella (generated by the muscles crossing the knee) will be signaled to the central nervous system. In patients with knee or patella pain, the CNS will seek to inhibit this inappropriate movement by weakness, stiffness or pain. **(Kapandji, 2010)** This reorganization of movement control is a protective strategy which serves to alleviate some of the stresses imposed on the damaged tissues of the knee.

When the feet are subjected to a foundational change, such as in heel height, muscular balance is changed by the nervous system. With the negative heel of the "earth shoe," activity in the erector spinae and soleus decreases, while it increases in the tibialis anterior, rectus femoris and biceps femoris muscles. **(Soderberg & Staves, 1977)** Barefoot running and running in "functional footwear" like Vibram Fivefinger shoes results in more of midfoot or forefoot strike, whereas running in traditional shoes results in more heel strike. **(Squadrone & Gallozzi, 2009)** In the first month of running with Vibram Fivefingers the author (SC) discovered more 2nd metatarsal head pain than when running in normal gym shoes. However as Squadrone & Gallozzi note, barefoot running creates a pressure loading spike around the 2nd metatarsal head; almost identical to running in Vibram Fivefingers. High heels increase the activity of the gastrocnemius and peroneus longus muscles during quiet standing. **(Csapo et al., 2010; Basmajian & Bentzon, 1954)** Normally there are only bursts of muscle activity seen on electromyography to maintain balance. This activity change indicates a failure to conserve energy, probably resulting in excessive fatigue over that which would ordinarily develop during quiet standing.

Applied kinesiology uses numerous reflexes that effectively improve muscle function when the reflex is active and therapeutically stimulated. It has long been observed that reflex stimulation results are not long-lasting if there is a primary factor causing the weakness. For example, if foot dysfunction is causing disorganization to the sternocleidomastoid muscles, as previously mentioned, they will usually test weak with manual muscle testing. The neurolymphatic reflexes for the muscles will probably be active; stimulating the reflexes will usually improve the muscles' function, providing a strong test. The primary effect of foot dysfunction is evidenced by the muscles again testing weak immediately after the patient walks. The effect of muscle dysfunction causing active associated reflexes may also be seen in neurovascular reflexes and stress receptors becoming active secondary to muscle dysfunction from another cause. When the foot is properly corrected, the active secondary reflexes will usually become inactive spontaneously. Applied kinesiology techniques are seldom employed in isolation but as part of a process designed to restore maximal pain free movement of articulations, restoration of postural balance, systemic functionality and facilitation of the self-regulatory mechanisms of the body.

One reason the cause of a patient's musculoskeletal pain is so enigmatic is that an adequate examination to cover the most common causes of neuromusculoskeletal pain requires skills characteristic of as many as 5 disciplines. The clinician may be required to examine for muscle imbalance in the kinesiological sense, neurologic function, myofascial trigger points, somatization in the mind-body sense, and articular dysfunction. Such a complete examination is indicated for patients with chronic musculoskeletal pain who have seen many specialists without finding a satisfactory answer to the cause of their pain.

The examination for functional muscle imbalance is part of the applied kinesiologist's training. The examination detects weak muscles, inhibited muscles, compensatory movement patterns, antalgic movement patterns and muscles recruited in an abnormal sequence. (**Walther, 2000**) The manual muscle test and AK diagnostic tests help to identify which muscle or muscles are dysfunctional and where this trouble is coming from. The weakness of muscles in the distribution of the motor nerve must by distinguished from the dysfunctional patterns of weakness induced by micro-evulsions, enthesopathies, myofascial trigger points, fear of movement, acupuncture meridian problems, cranial-sacral problems and other strains related to functional muscle groups, regardless of innervation. An effective examination requires the development of adequate manual muscle testing skills so that specific muscles can be isolated during the testing and knowledge of every muscle's origin and insertion and myofascial attachments, anatomy and physiology. The details of this examination also vary from muscle to muscle and are not yet routinely taught in many chiropractic and medical training programs. With the development and expansion of applied kinesiology, more and more chiropractors, physical therapists, some physiatrists, osteopathic physicians, dentists, and medical practitioners of other clinical specialties have subsequently learned these important skills. (**ICAK, 2012**)

It appears that there are considerable neurologic ramifications to foot dysfunction that are not charted in the scientific literature. Science scoffs at foot reflexology, yet numerous books have been written on the subject and there are thousands who attest to its effectiveness. It is not our purpose to defend or question the effectiveness or mechanism of the method. It is of interest to observe some parallels that exist between the muscle-organ/gland associations of applied kinesiology and foot reflexology. The most common reflexes indicated in reflexology books for the sinuses are at the tips of the toes and the metatarsal heads. (**Kunz & Kunz, 2005; Berkson, 1977; Carter, 1969**)

Stimulating these points affects the sternocleidomastoid muscle test, which is the muscle applied kinesiology associates with the sinuses. If the muscles test weak, the stimulation will usually strengthen the muscle, at least temporarily. If the muscles test strong in the clear, the reflex stimulation will temporarily weaken the ipsilateral sternocleidomastoid muscle in the normally functioning individual. The ipsilateral sternocleidomastoid muscle is the one normally inhibited during the last portion of the stance phase of gait.

A lateral cuboid subluxation has been associated with tensor fascia lata weakness. The neurologic reason for the muscle inhibition is discussed later under foot reflexes and reactions, and has recently been investigated by Zampagni et al. (**Zampagni et al., 2009**) showing that the applied kinesiology shock absorber test to the foot produces specific and measurable inhibition of the tensor fascia lata muscle. For now it is interesting to observe another correlation between foot reflexology and applied kinesiology. The colon reflex is established in foot reflexology over the cuboid bone; applied kinesiology finds the colon to be associated with the tensor fascia lata muscle. The same type of normal and abnormal relationship for the sternocleidomastoid muscle, foot reflex, and sinuses is applicable for the colon.

There is also a relationship between the Gillet and Liekens' (**Gillet & Liekens, 1981**) spine–foot association, mentioned earlier, and foot reflexology.

Foot reflexology is an area needing thorough physiologic investigation. (**Figure next page**) How accurate foot reflexology is with regard to actual health problems is unknown. Foot and hand massage is useful as an effective nursing intervention in controlling postoperative pain, (**Degirmen et al., 2010**) and foot reflex stimulation is beneficial for insomnia. (**Gong et al., 2009**) It is observed that the reflexes will show positive therapy localization when associated with muscle weakness; their stimulation changes muscle function, at least temporarily. Usually there is some underlying foot dysfunction that needs to be corrected, or the positive reflex returns shortly after the patient walks. It may be that foot reflexologists are, by good luck, sometimes correcting foot subluxations, fixations, or muscle dysfunction and obtaining permanent corrections.

There are remote orthopedic conditions caused by foot dysfunction, many of which are generally recognized. Entrapment neuropathies at the tarsal tunnel can cause sciatic pain. Knee problems are commonly caused by extended foot pronation and other foot dysfunction. Reactive muscle weakness, caused by one of the foot muscles, can be responsible for knee, back, neck, or shoulder pain. (**Marshall et al., 2009; Prior, 1999; Rothbart & Estabrook, 1988**)

With today's knowledge we see that foot dysfunction can influence the total body more than just the spine and its nervous system relationship, as described by Janse (**Janse, 1976**) and many others. (**Goodheart, 1998; Walther 2000; Dananberg, 2007; Liebenson, 2007**) Some of the health problems caused by foot dysfunction are well described in the scientific literature; others, such as interfering with the energy flow over the meridian system or causing remote neurolymphatic reflex activity, are not. To put the possible ramifications into proper perspective requires a wide area of interest in the cause of health problems. When first studying this integration, it can certainly be puzzling. Sometimes leading to the confusion is disorganization from improper stimulation to proprioceptors in the foot and leg as a result of extended pronation or some other dysfunction. This may send signals not in keeping with the body's needs; response to the afferent system causes confusion within the nervous system — a neurologic disorganization or, as has been known in applied kinesiology, "switching." In any event, the primary cause of many health problems that originate in the feet is overlooked by doctors, resulting in symptoms being treated rather than the primary cause.

Foot function and its effect on remote structure and function is an excellent example of the body's integration, and how the entire body must be evaluated in order to find the basic underlying cause for many health problems. The major way the foot influences other areas is by way of the nervous system. Proprioceptors in the foot and ankle send organizing impulses throughout the body. The role of these functional sensory receptors is to inform the central nervous system about ground reaction forces when the body stands and moves. When observing the different types of reflexes within the foot, it becomes apparent that orthopedic conditions in the form of stress and strain develop as a result of improper foot function. **(Cuccia, 2011; Abshire, 2010; Maffetone, 2003; Chaitow & DeLany, 2002; Hill, 1995; Robbins & Hanna, 1987)** Additionally the foot may be suffering an adaptive response to pathomechanical alterations in other parts of the body, including the stomatognathic and oculomotor systems. **(Berthoz, 2000)** Rothbart has demonstrated that foot pronation causes problems in the knees. **(Rothbart & Estabrook, 1988)** Rothbart **(Rothbart, 2008)** also demonstrates a fascinating correlation between foot pronation, innominate rotation and vertical facial dimensions, theorizing an ascending foot-pelvic-cranial model to explain these findings.

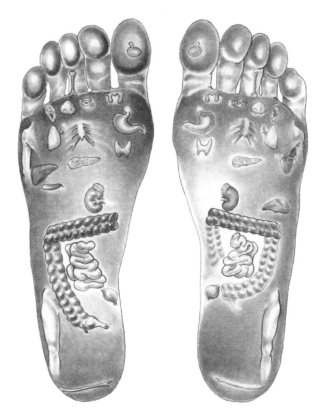

Foot reflexes.

"Abnormal foot pronation (inward, forward, and downward rotation) displaces the innominates anteriorly (forward) and downward, with the more anteriorly rotated innominate corresponding to the more pronated foot; 2) anterior rotation of the innominates draws the temporal bones into anterior (internal) rotation, with the more anteriorly rotated temporal bone being ipsilateral to the more anteriorly rotated innominate bone; 3) the more anteriorly rotated temporal bone is linked to an ipsilateral inferior cant of the sphenoid and superior cant of the maxilla, resulting in a relative loss of vertical facial dimensions; and 4) the relative loss of vertical facial dimensions is on the same side as the more pronated foot." **(Rothbart, 2008)**

Because of the wide-ranging effects of foot dysfunction on the total body, it is important to consider its possibility early in an examination. It is generally estimated that at least 90% of our population has significant foot biomechanical variation from the ideal; **(American**

FOOT	GILLET & LIEKENS	FOOT RELEXOLOGY
Toes	Upper cervical	Cervicals
Metatarsals	Rest of cervicals	Cervicals
Tarsometatarsal	Thoracic spine	Thoracic spine
Intrametatarsal	Rib-spine fixation & fixation between ribs	
Midfoot fixation	Lumbar spine	Lumbar spine
Chopart's joint	Lumbar spine	Sacrum
Calcaneus		Coccyx & sacrum

Foot reflexes (based on empirical reports).

Podiatric Medical Association, 2010; Langer, 2007; Subotnick, 1991, 1975) and approximately 25,000 people sprain an ankle every day in the United States. A study in Spain found that less than 3% of those over sixty had "normal" feet. (Langer, 2007) The majority can be understood and improved by applied kinesiology examination and treatment. When there is dysfunction, especially when it adversely affects the nervous system, "Take out the foot factor first." (Dananberg, 2007; Maffetone, 2003; Aronow & Solomone-Aronow, 1985)

As with all other body structure, a physician must have a thorough knowledge of foot anatomy and function in order to properly examine and treat many problems. Sometimes it is necessary to use orthotic support to the feet, but as more knowledge is obtained in applied kinesiology, the frequency and length of use are getting shorter and overall functional correction is getting better. It is hoped the information presented here will arouse interest in the study of foot anatomy and physiology to better understand problems affecting the feet. Foot dysfunction can literally cause health problems throughout the body.

Ligaments of the Foot

Ligaments are far from merely being restrictive structures that are strategically placed to support and stabilize joints, while maintaining normal tracking during movement. In reality they are sensory organs that provide proprioceptive input to the CNS, as well as having vital reflexive influences on associated muscles, and thereby ligaments become major elements in the stabilization of joints.

It was suggested as far back as the turn of the century that ligamento-muscular reflexes exist from sensory receptors in ligaments to muscles that modify the load imposed on the ligament and joint. Goodheart first discussed the law of the ligaments in 1973. (Goodheart, 1973) Goodheart found that pressure applied to the ends of ligaments towards the belly of the ligament tightens it. The opposite force will elongate the ligament.

It has been shown that ligamento-muscular reflexes exist in most of the lower extremity joints. (Solomonow et al., 1998; Freeman & Wyke, 1967) The ligaments associated with each extremity are richly endowed with afferents that produce reflex activation of the many muscles associated with the extremity's movement. The muscles therefore are a major component in maintaining the stability of the extremity's ligaments, bursae and capsules. (Solomonow et al., 2009, 1987) Goodheart also suggested that when there is a chronically weak muscle there will usually be a ligament involvement (the ligament is stretched) that provides stability in the same direction as the muscle. Conversely, a stretched ligament will cause a weakness in a muscle that provides stability in the same direction. (Solomonow et al., 1987) The usefulness of the manual muscle test and therapy localization and challenge procedures makes the diagnosis of specific ligament involvement (and even the specific portion of the injured ligament) obvious. For foot and ankle dysfunctions, detection of ligament injuries through these ligamento-muscular reflexes can be specifically assessed with the MMT. (Leaf, 2010; Lever, 2007; Sprieser, 2003; Kharrazian, 2001)

Numerous ligaments tie the bones of the foot together. They are extensive and vary somewhat from individual to individual. Nomenclature for the ligaments varies from authority to authority. Emphasis here will be primarily on the massive ligaments of the arch.

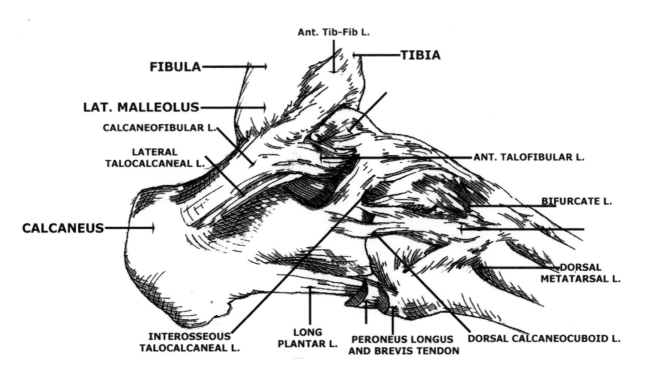

Lateral foot and ankle ligaments

The posterior talocalcaneal articulation has an articular capsule and anterior, posterior, lateral, medial, and interosseous talocalcaneal ligaments. The lateral talocalcaneal ligament attaches, as its name infers, from the lateral process of the talus to the lateral surface of the calcaneus. The medial talocalcaneal ligament connects the medial tubercle of the talus with the posterior sustentaculum tali. It blends with the deltoid ligament, discussed later under the ankle joint. The anterior talocalcaneal ligament connects the anterior and lateral surfaces of the neck of the talus to the superior surface of the calcaneus. The posterior talocalcaneal ligament is short; it connects the lateral tubercle of the talus with the calcaneus.

The major connection between the talus and calcaneus is the interosseous talocalcaneal ligament. It lies somewhat within the sinus tarsi, binding the calcaneus and talus firmly together.

The talocalcaneonavicular articulation often requires adjustment in an applied kinesiology examination. The major portion of the articulation is the head of the talus into the concavity of the posterior surface of the navicular. Contribution to the articulation is made by the anterior articular surface of the calcaneus. The joint has an articular capsule that thickens posteriorly, joining with the interosseous ligament of the talocalcaneal articulation. The talonavicular ligament connects the two bones dorsally.

Five ligaments support the calcaneocuboid articulation: the articular capsule, the calcaneocuboid portion of the bifurcated ligament, the long plantar, and the dorsal and plantar calcaneocuboid ligaments. The bifurcated ligament attaches to the calcaneus on the anterior surface. It divides to attach to the navicular and cuboid bones. The long plantar ligament is the longest of the tarsal ligaments, attaching to the plantar surface of the calcaneus anterior to the tuberosity. Its deep fibers attach to the posterior portion of the plantar surface of the cuboid bone. The superficial fibers are longer than the deep fibers and attach to the bases of the 3rd, 4th, and 5th — and occasionally 2nd — metatarsal bones. The plantar calcaneocuboid ligament is deep to the long plantar ligament. It is short but very strong in its connection between the calcaneus and the cuboid bones. These ligaments are very important in maintaining the lateral longitudinal arch.

Support to the medial longitudinal arch is provided by the plantar calcaneonavicular ligament, sometimes called the "spring" ligament, although there is usually no actual articulation between the calcaneus and the navicular

Plantar ligaments of the foot

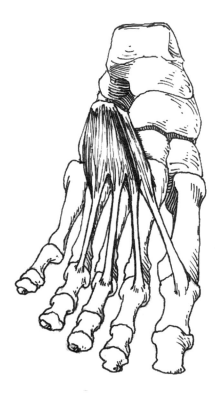

Dorsal intrinsic muscles.

bones. The plantar calcaneonavicular ligament arises from the anterior margin of the sustentaculum tali of the calcaneus, connecting to the plantar surface of the navicular. It supports the head of the talus. Its medial border joins with the anterior part of the deltoid ligament of the ankle joint. In addition, support is given by the medial band of the bifurcated calcaneonavicular ligament.

The American and English versions of *Gray's*

Anatomy differ regarding the histology of the plantar calcaneonavicular ligament. The American version states, "This ligament contains a considerable amount of elastic fibers, so as to give elasticity to the arch and spring to the foot; hence it is sometimes called the 'spring' ligament." **(Gray's Anatomy, 1995 and 1973)** The English version says, "Despite its vernacular name there is no real evidence that the 'spring' ligament is peculiarly resilient." **(Gray's Anatomy, 1980)**

The cuneonavicular articulation consists of the three cuneiform bones and the navicular. There are dorsal and plantar ligaments. The plantar ligaments are supported by slips from the tibialis posterior tendon.

The cuboideonavicular articulation connects the two bones by dorsal, plantar, and interosseous ligaments. This articulation is usually a syndesmosis, but it may be replaced by a synovial joint.

The tarsometatarsal articulations are gliding joints. The 1st metatarsal articulates with the 1st cuneiform. The 2nd metatarsal is given added support by its base articulating with the 2nd cuneiform, and the medial and lateral facets of its base articulating with the 1st and 3rd cuneiform bones, respectively. This forms a groove into which the base of the 2nd metatarsal fits. The 3rd metatarsal articulates with the 3rd cuneiform, the 4th with the 3rd cuneiform and cuboid, and the 5th with the cuboid only.

Giving additional strength to these articulations are the dorsal, plantar, and interosseous ligaments. The plantar ligaments for the 1st and 2nd rays are strongest.

The metatarsophalangeal articulations have plantar and two collateral ligaments. The transverse metatarsal ligament runs across and connects the heads of all the metatarsal bones. The interphalangeal articulations also have plantar and collateral ligaments.

Muscles of the Foot and Lower Leg

The muscles acting on the foot can be divided into extrinsic and intrinsic systems. The extrinsic muscles, in addition to acting on the foot, control its relationship with the leg at the ankle.

Extrinsic muscle system.

The extrinsic muscle system of the lower leg and foot is contained in the five compartments of the leg: the anterior, lateral, and three posterior compartments.

In the anterior compartment are the dorsiflexor and toe extensor muscles: the tibialis anterior, extensor hallucis longus, and the extensor digitorum longus. These are innervated by the deep peroneal nerve.

The peroneal muscles — longus, brevis, and tertius — are in the lateral compartment. The peroneus longus and brevis are innervated by the superficial peroneal nerve, and the tertius by the deep peroneal nerve.

In the superficial posterior compartment are the gastrocnemius, soleus, and plantaris muscles. The deep posterior compartment contains two bipinnate muscles, the flexor digitorum longus and flexor hallucis longus.

Finally, there is the tibialis posterior compartment, which contains only that muscle. **(Gray's Anatomy, 1995; Davey et al., 1984)**

Intrinsic muscle system.

There are two intrinsic muscles on the dorsum of the foot, the extensor digitorum brevis and the extensor hallucis brevis. The extensor digitorum brevis originates from the distal and lateral surfaces of the calcaneus, distal to the groove for the peroneus brevis. It inserts by three tendons into the proximal phalangeal bases at their lateral sides. The extensor hallucis brevis originates medial to the extensor digitorum brevis and inserts into the dorsal surface of the proximal phalanx base of the hallux. Sometimes this muscle is considered a division of the extensor digitorum brevis.

The intrinsic plantar muscles often need applied kinesiology origin and insertion, percussion, or neuromuscular spindle cell treatment to maintain corrections made by manipulation of the foot's articulations. **(Leaf, 2010; Cuthbert, 2002)** These muscles may also need treatment to correct dysfunction of the positive support reaction, discussed later.

There are four layers of intrinsic plantar muscles — the superficial, middle, and two deep layers — all of which are deep to the plantar aponeurosis. Although many of these muscles cannot be easily tested by manual muscle testing, it is necessary to have their anatomy firmly in mind to effectively apply local treatment to them.

The superficial layer contains three muscles, all of which originate from the inferior aspect of the calcaneus. The abductor hallucis inserts with the medial tendon of the flexor hallucis brevis into the medial side of the base of the proximal phalanx of the great toe. The flexor digitorum brevis inserts into the middle phalanges of the 2nd through 5th toes. The abductor digiti minimi inserts into the lateral side of the 5th toe's proximal phalanx base.

The middle layer contains the quadratus plantae and the lumbrical muscles. The quadratus plantae has four tendons that join with the tendons of the flexor digitorum longus to pass through the flexor digitorum brevis tendons and insert on the distal phalanx. The lumbrical muscles originate between the tendons of the flexor digitorum longus, with the exception of the first, which arises from the medial side of the 1st tendon of the flexor digitorum longus. The tendons of these muscles insert on the medial side of the proximal phalanx of the 2nd through 5th toes. Muscle imbalance and specifically weakness of these muscles contributes to "claw toes." **(Kwon et al., 2009; Hamilton, 1985)**

Briefly, the neurophysiology of tarsal tunnel syndrome exists because the brevis muscles of the proximal intrinsic foot muscles receive their nerve supply after the entrapment at the tarsal tunnel. The longus muscles which attach to the distal toes receive their nerve supply before entrapment at the tarsal tunnel. Because of the imbalance between the longus and brevis muscles of the foot the distal toes curl inferiorly and posteriorly thereby creating claw toes. Tarsal tunnel syndrome and foot pronation should be understood to precede the development of claw toes.

The deep layer consists of the flexor hallucis brevis, adductor hallucis, and the flexor digiti minimi brevis. The flexor hallucis brevis originates at the cuboid bone with prolongation of the tibialis posterior tendon. It inserts by two tendons into the medial and lateral sides of the hallux

Applied Kinesiology

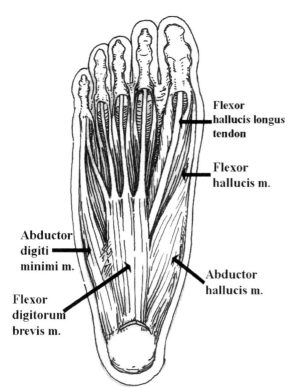

Superficial plantar muscles.

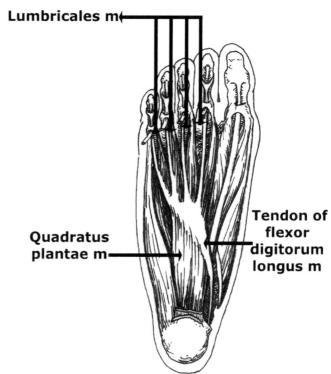

Middle plantar muscles.

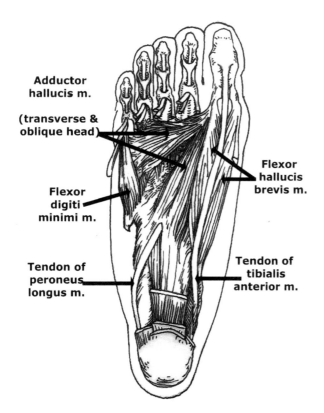

Deep plantar muscles.

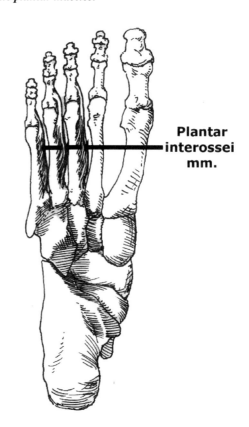

Deepest plantar muscles.

proximal phalanx. Each tendon usually contains a sesamoid bone. The oblique head of the adductor hallucis muscle arises from the bases of the 2nd, 3rd, and 4th metatarsals, and its transverse head from the joint capsules of the 3rd-5th metatarsophalangeal articulations. Its insertion is into the lateral base of the hallux proximal phalanx.

The fourth and deepest layer consists of the four dorsal and three plantar interosseous muscles. "In humans and gorillas, the second ray is the axis of the foot; the dorsal interosseous muscles abduct from that axis and the plantar interosseous muscles adduct to the axis. Thus, the first dorsal interosseous muscle is attached to the medial side of the second toe, and the second, third, and fourth dorsal interosseous muscles attach to the lateral side of their respective toes." (**Hamilton, 1985**) These muscles contract during walking to resist the forefoot's tendency to splay.

Travell and Simons (**Travell & Simons, 1992**) note that myofascial trigger points in the intrinsic plantar muscles are aggravated by the wearing of poorly fitting, tight, or badly designed shoes; ankle and foot injuries, articular dysfunction, conditions that allow the feet to become chilled and systemic conditions that affect the feet. In these cases, the muscle stretch reaction will be present and will guide the clinician toward the solution of this problem in the intrinsic muscles. (**Cuthbert, 2002**)

Arches of the Foot

Lake, (**Lake, 1937**) in discussing the development of the arch for adaptation to the upright posture, speaks of man needing support for his foot behind the axis of the ankle joint in order to keep the center line of gravity within the base. The calcaneus forms a lever for the large calf muscles to balance the weight in the foot. Lake goes on to state, "By Wolff's law, bone would be formed in the proper place and to the extent necessary to resist the stresses it encountered, and the os calcis therefore developed both backwards and downwards to resist the great additional stress thrown upon it in walking." Thus we see great differences in the human foot compared with that of lower primates.

The development and strength of man's arches are not limited to the evolutionary process. The stimulation of the foot from early walking to adult activity is important to the foot's strength and adaptability. Infants have no arch. It has been said that the arch is present in infants but is obscured by a fat pad, (**Morley, 1957**) but this has been shown by pedotopographic x-ray to be untrue. (**Gould, 1989**) The arch begins to develop with walking. With early walking, the arch develops faster in those wearing shoes with support to the medial longitudinal arch than in those without the support but still wearing shoes. At five years of age the arch develops at approximately the same rate with or without an arch support in shoes. (**Gould, 1989**) When the child's feet are subjected to uneven surfaces during early stages of walking, there is less chance of excessive pronation and weak feet developing. (**Lieberman et al., 2010; Rao & Joseph, 1992; Herzmark, 1947**)

The best long-term development is without shoes. A Nigerian study has shown that students wearing shoes most of the time are predisposed to poorly developed arches compared with those who seldom wear them. (**Mauch, 2008; Rao & Joseph, 1992**) This appears to be due to the shoe preventing direct contact with the ground, reducing foot action that would exercise the muscles and plantar fascia. (**Squadrone & Gallozzi, 2009; Didia, 1987**) The feet of people who do not wear shoes are more supple and appear almost flat in the relaxed state, but they become highly arched in action. (**Rasch & Burke, 1978**) Olympic champions and world record holders Abebe Bikila, Tegla Loroupe, and Zola Budd have shown that barefoot running is consistent with superb running performance.

Shoes restrict movement of the feet, reducing the activity and development of the plantar fascia and muscles. They also reduce stimulation to the sensitive mechanoreceptors. (**Richards et al., 2009; Robbins et al., 1989**) Shoes may reduce force to the metatarsals by half. (**Robbins et al., 1988; Collis & Jayson, 1972**) Stimulation to the mechanoreceptors is needed for optimal development of the nerve reflexes and reactions, discussed later. Proprioceptive

Herzmark constructed a playpen pad with an uneven surface and demonstrated better foot development in children.

stimulation coming from the unrestricted sole of the foot may facilitate afferent neural pathways as well as optimizing efferent muscle response and recruitment strategies. As in all sensory-motor complexes, the information going out is only as good as the information coming in. Shoe restriction produces greater demand on the bony architecture with the loss of dynamic activity.

The bony architecture of the human foot is designed for weight-bearing bones to lock together. The importance of proper bony architecture in arch maintenance is emphasized by the fact that a completely paralyzed foot, if once normal, maintains its arch, even when maximum weight is placed upon it. **(Harris & Beath, 1948)** The architectural design of the human arch is a wonder of nature that blends all of the elements of the foot – joints, muscles and ligaments – into a unified system that permits the foot to operate with the best mechanical advantage under the most varied conditions. Any functional or pathological disturbance in this architecture will interfere significantly with the function of the body as it stands as well as with running, walking, and the maintenance of erect posture.

There are three arches in the foot: the medial and lateral longitudinals, and the transverse. They have varying degrees of flexibility, and act as shock absorbers.

In general, an arch or curved structure can be supported in two ways: 1) by being tied together to prevent splaying, and 2) by having a keystone shape of the parts that make up the arch so that they wedge together to support weight from above. The foot uses both methods. **(Kapandji, 2010; Hamilton & Ziemer, 1983)** In addition, muscles have a critical role in arch support under certain conditions.

As study of the foot has progressed over the years, some controversy has developed regarding maintenance of the arches. As we look at the three types of tissue factors in the feet — bony architecture, muscles, ligaments and aponeurosis — we should recognize that there is no single factor responsible for the arch's maintenance. Any one of the three, when deficient, can cause a breakdown of the arches. The proponents of a single approach to the examination and treatment of the feet are right in some instances, but not in all.

The foot's arches have been compared to mechanical structures. This is helpful, but in the final analysis one must bring in the living, dynamic action of function controlled by an active nervous system. Static analysis does not take into consideration the adaptation necessary to meet the onslaught of forces the foot must deal with in daily living.

Only with a thorough combination of actions can the foot provide the shock absorption needed to deal with the daily forces to which it is subjected. The calcaneonavicular ligament has been referred to as the "spring" ligament, which is a misnomer. The concept of the arch having a springing, shock-absorbing effect is based on ligaments having elasticity. Two levers formed as an inverted "V," with a hinge at the apex and the other ends of the levers connected by a spring, would be a simple analogy to the arch of the foot being a spring shock absorbing mechanism. The foot would be a more complex inverted "V," with numerous springs attached to the various bones. **(Lapidus, 1943)**

There is considerable evidence that the arches of the feet are maintained in a static weight-bearing position primarily by ligaments and not muscles. (1) There is minimal activity of the muscles in a neutral standing position, as evidenced by electromyography, except for the control of sway for maintaining balance. **(Landry et al., 2010; Gray's Anatomy, 2004; Gray, 1969; Gresczyk, 1965; Mann & Inman, 1964; Basmajian & Stecko, 1963)** (2) An individual with complete paralysis of the foot and lower leg, such as from poliomyelitis, maintains the full height of the original arch, even though weight bearing is maintained by the use of braces. **(Harris & Beath, 1948)** (3) Nature is a conserver of energy. **(Basmajian, 1961)** If muscle action were necessary to maintain the arch, fatigue would rapidly result from a static stance. Compare a normal static stance with that of standing on the tiptoes for equal lengths of time. The foot is a combination of 1) arches, an example of which is the masonry arch, 2) truss structures like those used in bridges and roofs, and 3) a motor system provided by the muscles for movement. The line of pull for the intrinsic and extrinsic muscles of the foot are essentially in the long arch of the foot and perpendicular to the transverse tarsal joints; for this reason the muscles are important contributors to the muscular support of the arch during gait. **(Gray's Anatomy, 2004)**

A typical masonry arch can stand with great strength, as long as each end of the arch is firmly anchored to the ground. When the arch bears weight, the components wedge against each other, maintaining the arch. This construction is illustrated in the proximal metatarsal and transverse arches of the foot.

Whether standing, walking, or running, the foot is subjected to considerable stress. In mechanical considerations, stress is divided into tension, compression, and shear forces. **(Damm & Waugaman, 1948)** Tensile stress tends to separate particles of the material. Compression stress compacts or brings the particles closer together. Shear stress occurs when the particles tend to slide over one another at a given section, creating a cutting action. Bending stress is a combination of compression and tensile stress.

A simple beam is a combination of tensile and compression strength. When the beam is supported at its two ends and weight is placed on the center of the beam, the force to the lower portion of the beam is tension; from the point of weight to the two support points, it is compressive strength. The area shaded in the illustration produces negligible support to the weight.

A truss is built of materials selected to provide maximum strength with minimum material bulk. To accomplish this, the upper members of the truss are made of materials selected to provide compressive strength, while the lower member is a cable or some other material that provides tensile strength. This provides an example of the medial longitudinal arch of the foot. The bones are excellent providers of compressive strength, while the ligaments provide the tensile strength. Muscles provide the dynamic action of foot function. Proper function and/or structure of all three are necessary to provide optimal support and locomotion to the body.

Bony architecture

The medial longitudinal arch is the highest of the arches. It is formed by the first three rays, the three cuneiforms, the navicular, the talus, and the calcaneus. The 1st metatarsal, its sesamoid bones, and the 2nd and 3rd metatarsals comprise the cuneiforms, and cuboid. The wedge portion of the bones is narrower toward the plantar surface and wider dorsally. When articulated, the bones form longitudinal and transverse arches, with the concavities facing inferiorly. This arch is continued into the metatarsal bones, which are wedge-shaped at their bases.

A basic cause of pronation and flatfoot can be a congenital anomaly of the bones, especially the talus and calcaneus. (**Vukasinović et al., 2009; Harris & Beath, 1948**) In this case, there will be excessive pronation, regardless of the efficiency of the other two aspects — ligaments and aponeurosis, and muscles. Types of bony congenital anomalies that cause pronation will be discussed later under that subject.

Ligaments and aponeurosis

Hicks (**Hicks 1951, 1953, 1954, 1955**) studied the dynamic function of the longitudinal arches of the foot by dissection, x-rays of movement at sequential steps of the dissection, and x-rays of the living foot. He concluded that in the normal foot the arch is maintained by the plantar aponeurosis, with no contribution from the intrinsic or extrinsic muscles. The operative phrase here is "normal foot." It might be added that there must be normal activity of the intrinsic and extrinsic muscles during walking, running, and jumping.

The mechanism that Hicks described is of significant interest in function. Extension of the toes tightens the plantar aponeurosis to elevate the arch. This can be easily observed when one stands with most of the weight on the heel. The toes are easily extended, and it can be observed that the arch rises. When weight is evenly distributed over the full length of the foot, causing compression of the arch, the extension of the toes is limited by tension on the aponeurosis; indeed, one feels tension develop in the plantar aponeurosis.

The influence of toe extension on the plantar aponeurosis results from the attachment of the aponeurosis to the proximal phalanges. As the toe extends, it pulls with it the plantar pad, which is an extension of the aponeurosis. The plantar pad, sliding around the metatarsal head, pulls on the plantar aponeurosis, tightening it as if it were a cable arrangement being pulled around the drum of a windlass. Maximum toe extension shortens the effective length of the aponeurosis by approximately 1 cm. This mechanism exists in all five toes, but it is most marked in the great toe. When

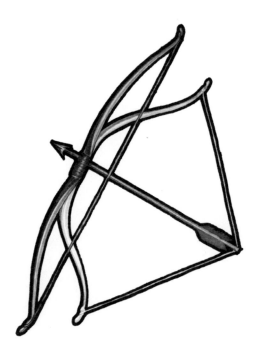

The string of a bow represents tensile strength like the ligaments in the arches.

The stones in a Roman arch wedge together with compressive strength like some of the bones in the arches.

(With kind permission, ICAK-USA)

anterior pillar. The 1st ray contributes to the strong pillar function because of the 1st tarsometatarsal joint's limited range of motion. (**Wanivenhaus & Pretterklieber, 1989**) The tuberosity of the calcaneus is the posterior pillar as viewed from the medial side. The head of the talus is the keystone. (**Kapandji, 2010; Hamilton & Ziemer, 1983**) The arch is slanted in the forefoot, and more steeply arched in the hindfoot.

The 4th and 5th metatarsals make up the anterior pillars of the lateral longitudinal arch, while the tuberosity of the calcaneus comprises the posterior. The apex is the cuboid. The lateral longitudinal arch has minimal height in comparison with the medial. (**Hamilton & Ziermer, 1983**)

The transverse arch is rigid, very similar to a standard Roman arch. It is made up of the wedge-shaped navicular,

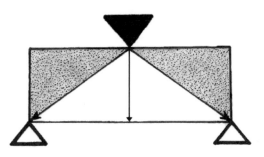

Locations of compressive and tensile strength in a beam.

the hallux is amputated, its role in the windlass is lost; the forefoot weight-bearing shifts laterally throughout gait, and the height of the medial longitudinal arch diminishes. **(Mann et al., 1988)**

This mechanism is important in gait activity. The medial longitudinal arch, controlled by the windlass mechanism, rapidly shortens just before toe-off; it then lengthens at heelstrike, becoming the longest at about the time of toe contact. **(Levangie & Norkin, 2001; Kayano, 1987)** This occurs because during the last portion of the stance phase, toe extension pulls on the plantar aponeurosis, raising the height of the arch. This increases the range and speed of planter flexion over and above that which occurs in the ankle alone. It helps the foot become a solid lever and avoid yielding to increasing forces at the toe-raising phase. This causes the foot to plantar flex and thrust downward, with an additional "flick" on take-off.

In static stance, the mechanism works in reverse. Body weight flattens the arch, which unwinds the windlass mechanism and causes the toes to flex toward the ground. This gives a "gripping" action of the toes on the ground.

Muscle role in support of the arch

There are three ways in which the foot functions: (1) in directed free movements, (2) as a static support mechanism subject to gravitational forces and postural changes, and (3) as a dynamic mechanism for moving the body. **(Chaitow & DeLany, 2002; Houtz & Fischer, 1961)** The foot has limited function as a freely-moving appendage, yet analysis of the foot and manual muscle testing are usually done under this circumstance. Understanding the role of muscles in weight bearing and the dynamic activity of walking and running is

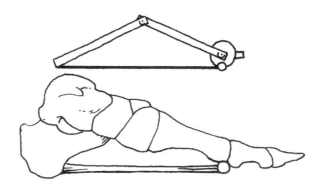

Hicks mechanism without toe extention

(With kind permission, ICAK-USA)

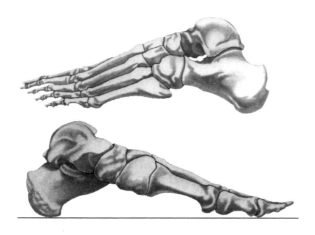

Medial longitudinal arch & Lateral longitudinal arch.

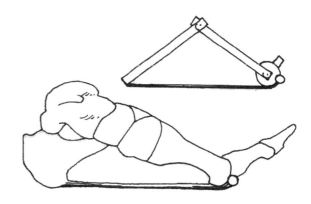

With toe extension the plantar fascia tightens to raise the arch

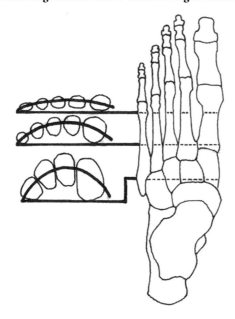

Traverse arch diminishes distally

important in accurately determining areas of dysfunction so that proper treatment can be applied.

From anatomical studies made prior to electromyography, it was concluded that several muscles have a role in arch maintenance in the normal *static* foot. **(Jones, 1941; Kaplan & Kaplan, 1935; Lake, 1937)** Electromyographic studies have modified this conclusion. **(MacConnaill & Basmajian, 1969; Basmajian & Bentzon, 1954; Basmajian & Stecko, 1963; Gray, 1969; Gresczyk, 1965; Mann & Inman, 1964; Smith, 1951)** In gait however both the longitudinally and transversely oriented muscles become active and contribute to support of the arches of the foot. (**Levangie & Norkin, 2001**) Muscles of interest are the tibialis anterior for evaluation of its ability to maintain elevation of the medial longitudinal arch; the tibialis posterior and peroneus longus to determine if they might provide sling support; and the flexor hallucis brevis, abductor hallucis, and flexor digitorum brevis, as all three are intrinsic to the foot and may provide longitudinal bowstrings. (**Thordarson D, et al., 1995**) It is also important to compare the muscles in both static and dynamic action for normal feet and pronated or flatfeet.

Basmajian and Stecko (**Basmajian & Stecko, 1963**) did a fine-wire EMG study of the static foot. Muscles studied were the tibialis anterior and posterior, peroneus longus, flexor hallucis longus, abductor hallucis, and flexor digitorum brevis. The subjects were seated, with the foot supported on an adjustable platform. Weight was applied on the flexed knee in the amount of 100, 200, and 400 pounds by a lever mechanism so arranged that no leg movement would take place during the experiment. One hundred pound weights were chosen to approximate or exceed the normal load on each foot in the upright bipedal stance, and 200 pounds to approximate or exceed the load on the arch in one-legged stance. The 400-pound load was the maximum that could be applied without extreme discomfort at the knee.

Recordings were made with no load, then progressively adding the loads noted previously. The foot platform was set to horizontal and 20° each of dorsiflexion, plantar flexion, inversion, and eversion. With no load there was no muscular activity recorded; with 100 pounds, there was negligible activity or, in a very small percentage of the population, minimal activity. Two hundred pounds caused the muscular activity to increase generally, but only slightly. With 400 pounds of weight, many muscles continued to show no activity. The tibialis anterior had the highest incidence of activity, and the tibialis posterior was the most active in general. Muscle activity varied considerably among the individuals in the various foot positions.

A contradictory EMG needle study by Suzuki (**Suzuki, 1956**) was done on normal subjects positioned in a manner similar to that mentioned earlier. With a load of 60 kg on the flexed knee, the muscles of the foot and leg revealed the following. Contracting against the load were the flexor hallucis longus, tibialis posterior, abductor hallucis, flexor digitorum longus, and the abductor digiti minimi. There was no action in the peroneus longus and brevis, tibialis anterior, and flexor digitorum brevis, which correlated with Basmajian and Stecko's study.

The preponderance of evidence is that muscle activity is not necessary to maintain the *static* arches in the normal foot. (**MacConnaill & Basmajian, 1969; Basmajian & Bentzon, 1954; Basmajian & Stecko, 1963; Gray, 1969; Gresczyk, 1965; Mann & Inman, 1964; Smith, 1951**) Bony architecture and ligaments support the normal static foot. These studies support Basmajian's thought: "Muscles [are] spared when ligaments suffice." (**Basmajian, 1961**) Man stands on his skeleton. Muscle is used only in short bursts to maintain sway control. (**Burke et al., 1982**) When flatfooted subjects are studied however, the findings are significantly different. These are the type of feet that walk into clinician's offices for evaluation and treatment.

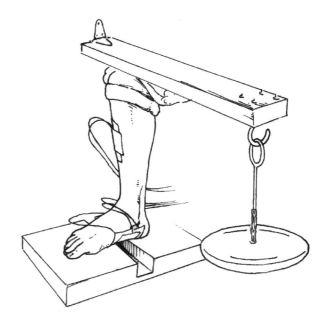

Apparatus of Basmajian and Stecko (1963)

In Gray's study (**Gray, 1969**) of the static foot, subjects were evaluated by EMG while standing for two minutes, having been cautioned not to sway. In the normal group there was marked activity of the soleus and no activity of other muscles. Of the 27 subjects with flatfeet, there was marked activity of the muscles as follows: soleus, 26; tibialis anterior, 23; tibialis posterior, 23; and peroneus longus, 22. Gresczyk (**Gresczyk, 1965**) also found activity in muscles other than the soleus in a flatfooted group (n=27). In no instance was there activity of only the soleus in this group.

When an individual is standing and shifts his weight from one leg to the other, there is a short burst of activity of the posterior leg muscles, the peroneus longus and brevis, and the intrinsic muscles of the foot; there is no activity during the static stance. (**Levangie & Norkin, 2001; Smith, 1954**)

Mann and Inman (**Mann & Inman, 1964**) did an excellent fine-wire EMG study of the intrinsic muscles of the dynamic foot. The following muscles were studied: extensor digitorum brevis, abductor hallucis, flexor hallucis brevis, flexor digitorum brevis, abductor digiti minimi, and the dorsal interosseous muscle between the 3rd and 4th toes. In addition, the extrinsic gastrocnemius and tibialis anterior were considered.

Some of the subjects had asymptomatic bilateral flatfoot, and their muscular patterns varied from subjects with normal feet during level walking. Normal and flatfooted subjects had comparable muscular activity going up or down stairs or slopes, and standing on their toes. When standing on the toes, all intrinsic muscles were active, along with the gastrocnemius. The tibialis anterior muscle was electrically silent. The intrinsic muscles were inactive during quiet standing, except for sporadic bursts.

During walking, there is electrical activity of the intrinsic muscles during the stance phase only. Individuals with normal feet do not have activity in the intrinsic muscles from heelstrike until 40% into the cycle. In flatfeet, the muscular activity begins 10% into the cycle. Likewise, transverse tarsal stabilization begins at 40% into the cycle in the normal subject, and 10% into the cycle in flatfeet. This study shows that the intrinsic muscles of the foot act together as a functional unit. The pronated foot requires greater intrinsic muscle activity to stabilize the transverse tarsal and subtalar joints than does the normal foot.

A balance of muscle turgor is necessary to maintain balance in the arch. When in the weight-bearing position, the triceps surae is normally stretched beyond its resting length. When non-weight bearing, the resting length yields approximately 30° of plantar flexion. When standing, this causes additional tension on the Achilles tendon attachment to the calcaneus, maintaining the calcaneus position. When there is paralysis of the gastrocnemius and soleus, a progressive deformity of the foot occurs. The calcaneus is rotated into the vertical position to cause its posterior aspect to move inferiorly. The paralysis, if combined with a completely paralyzed foot, does not cause deformity. This gives evidence that there is balance between the intrinsic muscles of the feet and the triceps surae to maintain the normal arch. **(Gondin et al., 2004)**

All positions and actions of the feet must be considered in evaluating the muscles' role on arch maintenance. Hamilton, **(Hamilton, 1985)** in a discussion of the surgical anatomy of the foot and ankle, indicates that the tibialis posterior and anterior muscles are important in maintaining the longitudinal arches. Referring to the tibialis posterior, he states, "Clinically the importance of this muscle is seen when its tendon ruptures and a true fallen arch occurs with great rapidity and severity." The tibialis posterior plays an important role in moving the navicular bone slightly on the head of the talus. **(Kapandji, 2010)** By moving the navicular inferiorly and slightly posteriorly, the anterior pillar of the medial longitudinal arch is angled to increase the arch height during the latter portion of the gait's stance phase. In addition, the head of the talus has a different radius from medial-to-lateral as opposed to superior-to-inferior. "Thus as the talonavicular joint is loaded, increasing stability is brought about by the seating of the convex head of the talus into the concave navicular." **(Mann, 1982)**

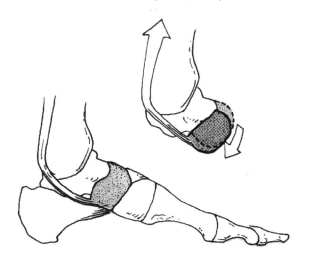

Navicular bone movement with tibialis posterior contraction.

Even muscles remote from the foot and leg may play a role in extended pronation. In addition to supporting the posterior pelvis to decrease lumbar hyperlordosis, the gluteus maximus externally rotates the thigh. With external leg rotation, the arches of the feet rise. This can easily be demonstrated on yourself by contracting the gluteal muscles while standing. Feel the external leg rotation and elevation of your arches. **(Greenawalt, 1988)** As will be seen later, it is apparent that poor foot function can cause an increase in the spinal AP curves; conversely, poor remote muscle function and an increase of the spinal AP curves can cause excessive foot pronation.

Notwithstanding the EMG data indicating that muscles are spared if ligaments suffice, **(Basmajian, 1961)** clinical data suggests that a common form of flatfoot is caused by muscle failure resulting in stretched ligaments and bones in altered positions. When muscles fail to support the arch, the ligaments can be stretched and the medial longitudinal arch of the foot is lost. **(Langer, 2007)** In addition, when excessive pronation is present — from whatever cause — the muscles are abnormally active in an attempt to control the problem.

Strength and flexibility of the musculature spanning the foot is very important. The majority of muscles function eccentrically, not concentrically, during walking. For this reason their strength and flexibility are important. If muscles are weak they are unable to control function and movement; if they are inflexible they will not permit function and movement. Particularly with weakness of the posterior tibialis there will be reduced control of subtalar joint pronation. **(Zwipp et al., 2000)** Subtalar joint pronation may also be compensation for limited dorsiflexion at the ankle joint which encourages abnormal dorsiflexion at the subtalar joint or midtarsal joints **(Donatelli and Wooden, 1996; Bohannon et al., 1989)**.

When examination and correction of foot problems are considered, we will see that there is much that the physician using applied kinesiology can do to correct excessive foot pronation when there are subluxations of the bones of the feet and muscle dysfunction. In some instances, especially those of congenital anomaly of the bones, support or surgery is necessary to effectively treat the foot. Understanding foot function enables one to choose the proper direction of treatment.

Foot weight bearing. Early descriptions indicated that static weight-bearing forces to the foot are those of a tripod. With modern measuring equipment, **(Thordarson et al., 1995; Hughes et al., 1987; Lord, 1981; Soma Staff, 1988)** this has been shown to be incorrect. Weight distribution is in the general form of a triangle, but the forces are not concentrated on the ends of a transverse arch at the metatarsal heads and the calcaneus. Considerable differences are described in the literature regarding the concentration of forces on the static weight-bearing foot. Some indicate that the 1st ray carries twice the weight of each of the others, giving a distribution of 2:1:1:1:1. **(Hamilton, 1985; Viladot, 1973; Manter, 1946)** Other studies **(Luger et al., 1999; Rodgers & Cavanagh, 1989; Cavanagh, 1987)** indicate that the 2nd ray has the highest force. This discrepancy probably relates to the muscle's and nervous system's role in weight distribution, as well as the structure of the arches. In any event, the transverse arch does not extend to the distal metatarsals; thus there is weight bearing across all the metatarsal heads.

Each of the five rays contributes to weight bearing, but not equally. Lying side by side, they are tied together posteriorly and mechanically independent anteriorly, where the level of each ray is independently determined by its amount of flexion and not by any relationship between it and its neighbors. Any metatarsal head projecting below the general level takes a heavier load than the others. Each ray in the neutral weight-bearing foot bears weight by acting variously as a beam and as an arch (or truss). With toe extension there is increased tension on the plantar aponeurosis by the windlass mechanism, causing more arch support and less beam support. If the weight-bearing foot is put on a block where the toes can flex, it relaxes the tension on the plantar aponeurosis. The arch drops somewhat and there is more contribution to weight bearing by the beam function of the rays. (**Hicks, 1955**)

Although the metatarsals are the major weight bearers of the forefoot, it is the toes that "grip" the substrate. The metatarsals are incapable of making this adjusting contact because they lack any tendinous insertions. (**Abshire, 2010; Viladot, 1973**) The great toe also presses against the ground because it has only two phalanges. (**Cailliet, 1997**)

There is great conflict in the literature on weight balance into the foot. The bottom line is *in vivo* weight balance. Levangie and Norkin and Hamilton (**Levangie & Norkin, 2001; Hamilton, 1985**) describe the role of muscle activity on the distribution of forces to the foot. With muscle relaxation, weight loaded onto the knee is distributed 80% into the metatarsals and 20% into the calcaneus. Normally contraction of the triceps surae distributes the weight between the metatarsals and the calcaneus, with greater weight on the hindfoot. In his description the weight distribution into the metatarsals is 2:1:1:1:1, with double the weight being distributed to the 1st metatarsal. In a weight-loaded cadaver foot, weight is distributed 4:1:1:1:1. The 1st ray bears as much as the others combined, rather than half as much as in the normal living foot.

Supraspinal and local neurologic factors play a role in foot protection. Hennig and Cavanagh (**Hennig & Cavanagh, 1987**) studied muscular action of the foot during expected and unexpected falls onto a flatfoot. When a fall is unexpected, there is stronger force to the heel. During an expected fall, there is supination and flexion so the force goes more into the lateral forefoot and midfoot borders, providing protection for the heel. Stimulation to the foot mechanoreceptors must be received by the central nervous system for foot protection. In rheumatoid feet in which there is no neurologic deficit, there are low forces where active arthritis is present. (**Yavuz et al., 2010; Collis & Jayson, 1972**) In diabetic feet, peak pressure is higher at the point of ulcer formation, (**Veves et al., 2002; Rogers et al., 1987**) indicating poor neurologic feedback from diabetic neuropathy. The pressures of the insensitive foot are different and greater than normal when walking. (**Veves et al., 2002; Bauman & Brand, 1963**)

The forefoot support provided by the metatarsals is very irregular. (**Viladot, 1982**) There is consensus that the usual peak pressures during gait are rarely located underneath the 1st ray. (**Chaitow & DeLany, 2002; Levangie & Norkin, 2001; Cavanagh et al., 1987; Viladot, 1982**)

Ankle Anatomy and Function

The articulation between the distal tibia and fibula and the talus is referred to as the ankle mortise or ankle joint (also called the tibiotalar or talotibiofibular or talocrural joint). A high level of integrity is required of this joint for normal foot and leg function. Gray's anatomy (**Gray's Anatomy, 1995**) points out that compressive forces of five times the body weight and tangential shear forces of 80% of the body weight are transmitted across this joint. Stress is put into this joint from excessive foot pronation or abnormal hip or leg rotation. The overall articulation between the three bones is a synovial ginglymus joint. Its structure and ligaments allow a single oblique axis of motion that provides dorsiflexion and plantar flexion of the foot.

The nerve supply to the joint comes from the deep peroneal and tibial nerves while the arterial supply comes from the malleolar rami of the anterior tibial and peroneal arteries.

The distal tibia and fibula make up the proximal portion of the joint. The union between the tibia and fibula is a syndesmosis (a fibrous joint in which relatively opposing surfaces are united by ligaments). Separation of the bones is prevented by the ligaments, primary of which is the crural tibiofibular interosseous ligament. It is continuous with the interosseous membrane, which extends most of the length of the tibia and fibula. This ligament's strength is so great that forces that tend to separate the bones will usually fracture the fibula proximal to the ligament before it will tear. Also supporting the inferior tibia and fibula articulation are the anterior and posterior tibiofibular ligaments and the interosseous membrane. The tibialis posterior muscle has an important role in the physiologic separation of the bones and maintenance of ankle joint integrity. (This will be discussed later.)

The distal aspect of the tibia projects medially to form the medial malleolus. The lateral surface of the medial

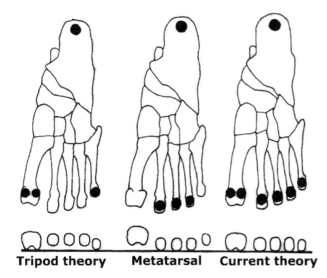

Foot weight bearing theories

malleolus is an articular surface that extends about one-third of the way down the medial talar body. The distal aspect of the fibula comprises the lateral malleolus. Its medial articular surface covers the entire lateral articulating surface of the talar body. Together the bones form a U-shaped articular surface, which is congruent with the articulating surfaces of the talar body.

The ankle joint should be seen as two joints with integrated functional action: the ankle mortise and the subtalar joint. Inman (**Inman, 1976**) showed that during the gait cycle there is more visible medial rotation of the tibia than can be explained by movement solely at the talotibiofibular joint. Inman demonstrated that the increased tibial rotation described resulted from calcaneal eversion about the subtalar axis.

The superior articulating surface of the talus is called the trochlear surface. The trochlea is wider anteriorly than posteriorly, giving it a wedge shape. It is approximately 5 mm wider anteriorly. (**Kapandji, 2010**)

Several ligaments are important in stabilizing and limiting ankle motion. The medial ligament is also called the deltoid collateral ligament. It is a strong ligament that generally attaches proximally at the medial malleolus and spreads with superficial and deep fibers to the navicular anteriorly, the talus, and the calcaneus distally and posteriorly.

The lateral ligament is composed of the anterior talofibular ligament, posterior talofibular ligament, and the calcaneofibular ligament. The lateral ligament, as a whole, is weaker than the medial one; consequently, it is more prone to injury. Freeman reports that 40% of injuries to the lateral ligaments result in functional instability of the ankle. (**Freeman, 1965**) The medial ligament will often avulse the bone rather than tear with trauma.

An *in vitro* study was done by Stormont (**Stormont, 1985**) on the contribution of the ligaments and articular surfaces to the ankle's stability. Evaluation was done in both the loaded and unloaded states for internal and external rotation, inversion, and eversion. In internal rotation, the two primary restraints are the anterior talofibular ligament and the deltoid ligament. In external rotation, the primary restraint is the calcaneofibular ligament. In both internal and external rotation during loading, the articular surface accounts for 30% of the stabilization.

The calcaneofibular ligament is the primary restraint in inversion, and the anterior talofibular ligament is secondary. The deltoid ligament is the primary restraint in eversion. In both inversion and eversion during loading, ligaments do not contribute because 100% of the stabilization is provided by the articular surface. Ankle instability may occur during loading and unloading; it does not occur once the ankle is fully loaded because of the articular surface's resistance to inversion displacement. This helps one understand the importance of the shape of the bones in providing ankle strength to resist the application of forces that may traumatize the ankle. It must be recognized that this study (**Stormont, 1985**) was done *in vitro* by cutting the ligaments. One must consider the effect of muscles in both the loaded and unloaded condition.

Rotation of the leg into the foot, which is fixed on the substrate, may account for clinically symptomatic ankle instability. (**Stormont, 1985**) The ankle is more vulnerable to injury when sports shoes make firm contact with the playing surface. (**Abshire, 2010**)

For optimal function, the ankle mortise must be a congruent articulation throughout its range of motion. An adequate ankle range of motion is a necessary component for many activities such as running, ascending and descending stairs and normal gait (**Brukner and Khan, 2006; Donatelli, 1996**). Since the body of the talus is wedge-shaped, being wider anteriorly, there must be some accommodating mechanism in the tibia and fibula to maintain the articulation's integrity. This is accomplished by changing the distance between the medial and lateral malleoli.

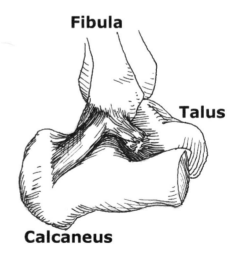

Lateral ankle ligaments.

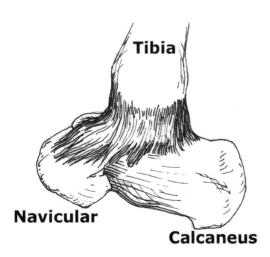

Medial ankle ligaments.

Grath (**Grath, 1960**) studied the widening of the ankle mortise from plantar flexion to dorsiflexion by implanting steel pins into the lateral and medial malleoli of living, locally-anesthetized volunteers. Measurement was accurately taken on the exposed ends of the pins during ankle motion. The mortise width is less on maximal plantar flexion and greater on maximal dorsiflexion. The increase of mortise width may be maximal between plantar flexion and neutral, or neutral and dorsiflexion. It is rare that widening of the mortise fails to occur with dorsiflexion.

It is important that the mortise not widen with weight bearing. Measurement by the above method "… demonstrates that increases in mortise width often fail to occur on loading the ankle with the weight of the body, and that when they do occur they are so insignificant as to defy demonstration with the conventional millimeter gauge." **(Grath, 1960)**

Grath's study definitively documents the adaptation of the tibia and fibula to the wedge shape of the talus body. The adaptation of the mortise to position is of great importance when one examines the ankle for integrity. There is considerable difference in the amount of joint play, depending on whether motion is passive or active.

Jaskoviak **(Jaskoviak, 1983)** indicates that the ankle mortise is weakest during plantar flexion because the narrow posterior wedge shape of the talus causes the joint to loosen. He continues by stating that the mortise is strongest in dorsiflexion because the wide anterior

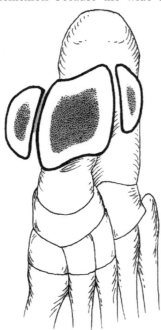

Talus positiong between the malleoli.

wedge of the talus tightens between the medial and lateral malleolus. This is applicable when one examines the ankle mortise by motion palpation when the ankle is non-weight-bearing and movement is passive. When ankle motion is active, especially when weight-bearing, there should be no difference in the articulation's integrity (joint play) in full plantar flexion or dorsiflexion.

Ankle Stability in Active Function

The usual and generally accepted description of human ankle mechanics is based on cadaver studies rather than how the ankle functions *in vivo*. The accepted description of ankle mechanics, based on cadaver studies, is summarized by Cailliet. **(Cailliet, 1997)** The body of the talus is wedge-shaped with the wider portion anterior. As the ankle dorsiflexes, this wider portion comes up between the two malleoli and wedges between them. Plantar flexion of the foot presents the posterior narrower portion of the talus between the malleoli and in this position permits some lateral motion of the talus within the mortise. This mobility creates instability of the joint and places an added burden on its supporting ligaments.

In vivo studies and additional functional consideration show that there is greater functional integrity of the ankle joint than is indicated in cadaver analysis. Weinert et al. **(Weinert et al., 1976)** have shown by high-speed photography of running, followed by cineroentgenography, that during weight bearing the fibula moves inferiorly on the talus to provide greater articular surface to the lateral aspect of the ankle mortise, providing greater stability. This inferior fibular movement is probably a result of the contraction of the peroneus longus and brevis, soleus, and tibialis posterior, all of which have origin or partial origin on the fibula and are active in plantar flexion.

Further dynamic support to the ankle is provided by the arrangement of the tibialis posterior and its action during dorsiflexion and plantar flexion. **(Kapandji, 2010)** The origin of the tibialis posterior is from the lateral part of the posterior surfaces of the tibia and the upper two-thirds of the medial surface of the fibula. Contraction of the muscle pulls the tibia and fibula together because of the muscle's bipinnate arrangement. Since tibialis posterior action is plantar flexion with inversion, the ankle mortise narrows as the narrower portion of the talus body enters the mortise, especially when one is weight-bearing. Rising onto the toes requires strong action of the plantar flexor muscles. The tibialis posterior draws the tibia and fibula together, keeping the ankle mortise tight on the talar body. When the foot is dorsiflexed, the mortise must open to allow the wider portion of the talar body to enter. This occurs because the tibialis posterior is inhibited with dorsiflexion, which releases the approximation of the tibia and fibula.

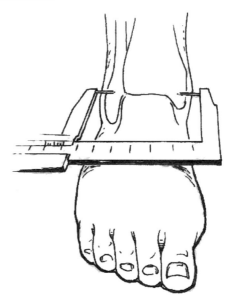

With plantar flexion, the distance between the medial and lateral malleoli narrows.

As the mortise widens to accept the wider portion of the talar body, the fibula moves slightly superiorly and finally rotates medially. **(Kapandji, 2010)** This changes the orientation of the interosseous ligament to horizontal and tightens it, limiting further dorsiflexion. **(Cailliet, 1997)**

The function of the tibialis posterior muscle in adjusting the tibia and fibula to the trochlear surface of the talar body may be augmented by a portion of the soleus. Michael and Holder **(Michael & Holder, 1985)** described an aspect of the soleus muscle that is not in standard anatomy texts. It is a separate and distinct bipennate portion at the anterior surface of the soleus, called the accessory fasciculus muscle. This was a consistent finding in 28 dissections by Michael and Holder. The soleus muscle has a U-shaped origin from the upper one-third of the fibula, the soleus line of the proximal tibia, and the posterior medial third of the tibia. With this origin, it is possible that the bipennate section of the soleus acts to approximate the tibia and fibula in a manner similar to the tibialis posterior, as described by Kapandji. **(Kapandji, 2010)** This would provide additional strengthening of the ankle mortise as the wedge-shaped body of the talus moves with plantar flexion and dorsiflexion.

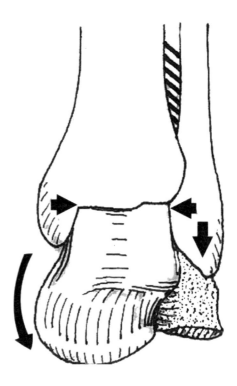

Plantar flexion; fibula movement exaggerated.

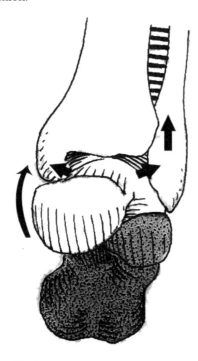

Dorsiflexion

The gastrocnemius and soleus have a greater role in ankle motion than is generally recognized in popular textbooks. The combination of these muscles with the Achilles tendon is called the triceps surae. The term should be quadriceps surae because the soleus functions as medial and lateral divisions. Campbell et al. **(Campbell et al., 1973)** did a fine-wire EMG study on the two heads of the gastrocnemius and the medial and lateral divisions of the soleus. The medial soleus stabilizes the leg on the foot and is a strong mover of the foot. The lateral soleus is not as powerful in these actions; however, it has an ongoing activity stabilizing the leg on the foot. When there is instability of the foot platform, activity of the lateral soleus increases.

The role of the soleus in stabilizing the leg on the foot requires understanding the soleus' contribution to the Achilles tendon. **(O'Brien, 2005)** Its tendon makes up the anterior half of the Achilles tendon, with the gastrocnemius contributing to the posterior half. As the Achilles tendon approaches the calcaneus, it rotates 90° so that the soleus portion inserts on the medial one-third of the calcaneus. The gastrocnemius portion inserts on the lateral two-thirds, causing the lateral portion of the soleus to be an inverter of the calcaneus. **(Michael & Holder, 1985)**

The two heads of the gastrocnemius are inactive until movement begins. **(Campbell et al., 1973)** The gastrocnemius is more sensitive to conditions of length, strength, and rate of contraction. The gastrocnemius restrains the tibia from rotating on the talus during gait as the bodyweight is shifted from the heel to the ball of the foot during the stance phase of the gait cycle. **(Travell & Simons, 1992)** The soleus has a more constant role. The gastrocnemius has its greatest activity when the ankle is plantar flexed, in large contractions, and in the rapid development of tension. The soleus is most active with the ankle in dorsiflexion and in minimal contractions, **(Herman & Bragin, 1967)** usually in controlling postural sway.

Ligament stretch reaction in AK: Preventing recurrence

Of interest is the reason an athlete will sprain an ankle with a turning activity that has been done thousands of times before with no trauma. A possible answer is failure of the soleus, gastrocnemius and tibialis posterior to act in the role of approximating the tibia and fibula during strong plantar flexion. Applied kinesiology

Bipennate fiber arrangement of the tibialis posterior muscle approximates the tibia and fibula during plantar flexion.

recognizes that these muscles are related with the adrenal glands. This is especially of interest in an athlete under the stress of the game but with adrenals incapable of meeting the body's needs. **(Wilson, 2002)** Roberts et al. **(Roberts et al., 1993)** report that overtrained male athletes had significantly increased cortisol levels. Goto et al. **(Goto et al., 2005)** have suggested that exercise-induced metabolic stress is associated with acute responses of growth hormone, epinephrine, and norepinephrine following resistance training. One sees this type of deficiency often in an applied kinesiology practice that focuses on the treatment of athletes. **(Maffetone, 1989)**

The ligament stretch reaction **(Walther, 2000)** is also related to stressed adrenal glands. This is a condition in which previously strong associated muscles test weak immediately after ligaments of the joint are stretched.

Schmitt **(Schmitt, 1977)** using applied kinesiology, designed a clinical study (n=16) to determine the reproducibility and apparent association of the ligament stretch reaction to the adrenal gland. The study consisted of stretching the ligaments of various articulations in the body, and then re-testing muscles associated with the articulation as well as general indicator muscles. The ligament stretch reaction was identified in three separate areas of the body to determine that the reaction was present generally rather than in any one area. After this was accomplished, various factors associated with the adrenal gland in applied kinesiology were evaluated by therapy localization. The reflexes examined were the neurolymphatic, neurovascular, meridian alarm point, and cranial stress receptors. A point -- such as the neurolymphatic -- was therapy localized, and the ligament stretch procedure was repeated, followed by the same muscle tests. If therapy localization to the adrenal reflex point abolished the ligament stretch reaction, there was probable adrenal involvement associated with the mechanical stretch to the ligament. After all reflexes were tested, the individual was asked to chew adrenal concentrate; its effect was evaluated by re-testing for ligament stretch reaction.

Following are the results of the nutritional administration and various reflexes tested in the study of sixteen individuals:

REFLEX POINT TL	ELIMINATED REACTION
Neurolymphatic reflex	16
Neurolymphatic reflex	6
Meridian alarm point (Circulation sex)	8
Cranial stress receptor	8
Adrenal concentrate	10

If therapy localization to the adrenal point listed abolished the positive ligament stretch reaction, it was listed as positive. Note that in 50% or more of the subjects, each reflex point or the nutrition cancelled the positive reaction, with the exception of the neurovascular reflex.

Treatment indicated by the positive tests was initiated. The ligament stretch reaction was removed with these approaches in fourteen of the sixteen patients. Two cases required specialized meridian therapy to abolish the positive ligament stretch reaction. "In all but one case, the symptoms of the patients were generally improved at the next office visit. In most cases, there was no recurrence of the ligament stretch/muscle weakness patterns. In the few where the pattern did recur, fewer reflex areas were found to be involved."

This association between ligament stretch and muscle weakness patterns has been presented in contemporary research by Solomonow; **(Solomonow et al., 2009, 1987)** additionally, the relevance of adrenal hormones (particularly their mineralocorticoid function) to ligament injury has been expanded in the applied kinesiology approach. **(Leaf, 2010; Maffetone, 1999)**

Durlacher **(Durlacher, 1977)** points out the importance of evaluating athletes for this reaction. A rapidly moving individual places strain on the ligaments, appearing to cause immediate weakening of the muscles supporting the articulation just when they are needed. Under these circumstances, weakening appears to be the same as that observed when the articulation is stretched and manual muscle testing is performed immediately afterward. It is possible that the weakened muscle could even be more significant in an athletic endeavor, because muscle demand occurs at the same time the ligament is being stretched.

Clinically it has been observed that patients susceptible to the ligament stretch reaction have exacerbations of symptoms when under considerable stress. Stress is cumulative, and can be classified as emotional, chemical, thermal or physical. The athlete has some -- and probably all -- of these stress factors during competition. Clinical evidence shows that performance is superior, and injury less probable, when all factors known to influence the adrenal gland are functioning normally. This provides an opportunity for the stress of the endeavor to properly enhance performance and not be a possible cause of injury.

Recently Lever (**Lever, 2006**) evaluated 200 asymptomatic patients for the involvement of ligaments in many of the different joints of the foot. The research design consisted of spreading apart the ligament and then manual muscle testing 40 different muscles throughout the body to see how this inhibits or facilitates the remote muscles.

Twenty-one joints and ligaments were tested in these 200 patients and the specific correlations between these joints and ligaments and the muscles they affected were listed. Generally, the calcaneal ligaments were found to affect pelvic and lower limb muscles, while talar ligaments were more involved with neck, upper thoracic and shoulder muscles.

Lever suggests that "Because of the importance of foot proprioception and the foot's relationship to so many body problems from neurological disorganization to gait imbalances, fascial disturbances, and the inhibition of so many muscles when faulted, physical evaluation of patients should include more attention to the feet."

Sprieser (**Sprieser, 2002**) also reports an in-office clinical trial (n=50) where the ligament stretch reaction was present in every case that had confirmatory adrenal stress disorder. Each of the patients in this study showed a drop in systolic blood pressure from lying to sitting or sitting to standing, or a positive Ragland's sign. Nutritional support was needed in all cases to correct the ligament stretch reaction and included adrenal support with choline or adrenal tissue extract, and/or a low dosage of vitamin E from wheat germ oil or octacosanol.

Finally, Hansen (**Hansen, 1999**) reports on a case-series of 5 patients with a medically diagnosed mitral valve prolapse who were also found to demonstrate the ligament stretch reaction. On physical examination, those patients were found to be hypoadrenic. The patients were treated to stabilize the ligament stretch reaction. This included dietary measures such as eliminating stimulants like coffee, tea, cola and refined sugars. AK oral nutrient testing showed a need for adrenal gland nutritional support. Hansen suggests that a mineral imbalance due to depressed adrenal function may cause a systemic weakening of ligaments including those of the heart valves.

In this discussion Selye's observation about the adrenal gland's pervasive influence should be noted. "A general outline of the stress response will not only have to include brain and nerves, pituitary, adrenal, kidney, blood vessels, connective tissue, thyroid, liver, white blood cells and especially muscles, but will also have to indicate the manifold inter-relations between them." (**Selye, 1976**) The relationship of applied kinesiology to the adrenal glands and the endocrine system is one that creates success for the clinician where other manual modalities might fall short. Each of the endocrine organs has been given diagnostic tests, (muscle-organ-gland inter-relationships), therapeutic protocols, nutritional correlations, and treatment monitoring methods. The endocrine glands are of course controlled by the nervous system, and this is why chiropractic and the other manipulative professions have been helpful throughout their history for endocrine-related disorders. (**Masarsky & Masarsky, 2001**) From the diagnostic viewpoint the AK manual muscle test has significance because it makes possible the detection of a "disease" process affecting the ligamentous system in advance of the emergence of symptoms. Whether the muscle inhibitions we find on AK MMT related to the articulations of the lower extremity are primary (as in a postural subluxation) or of secondary reflex origin (as in a ligamentous disturbance due to impaired adrenal gland function), we must recognize that this component in the musculoskeletal system's function is a contributing, exacerbating, and perpetuating influence, that must be specifically diagnosed and given effective treatment regardless of the primary etiology.

First, the muscle to be tested after the ligaments are stretched must be evaluated to determine that its functional quality during a manual muscle test is normal. If the muscle tests weak in the clear, it should be strengthened with the appropriate treatment. Its strength can then be compared after the ligament stretch has been done. If general testing of the body is to be done for the ligament stretch reaction, another general indicator muscle can be chosen.

The stretching procedure should be designed to limit the stretch to the ligaments of the articulation as much as possible as was done in the Lever study described previously. (**Lever, 2006**) This can be done by tractioning the articulation, or attempting to move it in a direction of which it is usually incapable, such as attempting to laterally bend the knee. When attempting to stretch the ligaments, care must be taken not to go through a range of motion which stretches the articulation at the end of the motion; this stretches the muscles as well as the ligaments. A positive test may therefore be for the AK muscle stretch reaction or the ligament stretch reaction.

Care must also be taken that the weakening observed in a muscle is not the result of challenging the articulation; this could be erroneously interpreted as ligament stretch reaction. When there is a ligament stretch reaction, it will be in any direction the ligaments are stretched. In challenging the articulation, one vector of force will weaken the muscle while an opposing vector will strengthen it.

Correction. The therapeutic effort is directed toward support of the adrenal gland. The appropriate reflexes and muscles described in previous applied kinesiology literature should be evaluated and corrected, if involved. (**Leaf, 2010; Garten, 2004; Frost, 2002; Gerz, 2001; Walther, 2000; Goodheart, 1976**) Nutritional support -- usually in the form of adrenal concentrate -- is frequently of value. (**Wilson, 2002; Goodheart, 1973**) Evaluation of stress and the entire endocrine system is important in treating the functional hypoadrenic.

As well as examining the ankle muscles for function with the various means described in this text, one should also evaluate the adrenal and its function, especially under stress.

Foot and Ankle Motion

The structurally sound foot has all metatarsal heads resting on the ground while maintaining good longitudinal and transverse metatarsal arches with a neutral calcaneus position. (**Langer, 2007; Lee et al., 2003; Cailliet, 1997**) A variety of motion is necessary between the hindfoot and forefoot to meet the demands of adjustment during foot action. The foot's numerous articulations have excellent architecture that meets this demand. It is complex but paradoxically simple when each component is considered individually. The important factor is that each simple component must be functioning properly for ideal foot-ankle action.

Movements within the foot are rotations about axes, with the axis located at the articulation; thus the motions in the foot are ginglymus or hinge-type. This causes predetermined motion of each joint, as opposed to a ball-and-socket joint motion in which the plane of movement is determined by the direction of forces acting upon it. **(Levangie & Norkin, 2001; Root et al., 1966)** The predetermined plane of movement in foot articulations is even applicable in the ball-and-socket joint of the talonavicular articulation. In function this joint is half of two different joint complexes. With the talocalcaneal articulation, it forms the talocalcaneonavicular joint complex. The combination of the talocalcaneonavicular and the calcaneocuboid articulations forms the mid-tarsal joint complex, called Chopart's joint. **(Gray's Anatomy, 2004; Jaffe & Laitman, 1982; Elftman, 1960; Hicks, 1953)** The talonavicular hinge movement is different, depending on which of the joint complexes moves. If action takes place at both complexes, there is yet another axis of hinge movement. The calcaneocuboid articulation is slightly saddle-shaped, giving two axes of motion. **(Gray's Anatomy, 2004; Elftman, 1960)** The axis of its movement also depends on whether one or both joint complexes are moving.

The complex arrangement of ginglymus joints between the talus, calcaneus, navicular, and cuboid bones produces the two important movements of the foot: inversion-eversion and pronation-supination. These two movements allow the foot to accommodate to the shape of the substrate and to lock together or unlock the tarsal bones, either forming a stable platform or creating a resilient lever. **(Levangie & Norkin, 2001; Hertling & Kessler, 1996)**

It is easily observed that when the foot pronates, the leg rotates internally; when it supinates, the leg rotates externally. The interdependent relation of leg rotation and foot pronation can be observed by attempting to extend one's toes during weight bearing without allowing external leg rotation. This relates to Hicks' **(Wallden, 2010; Hicks 1954, 1953)** windlass mechanism of the plantar fascia; the toes can easily be extended with external rotation of the tibiotalar column because the fascia is released. With internal rotation of the column, the arch lowers and the fascia becomes taut; toe extension is limited. Balanced weight bearing on the forefoot and hindfoot must be considered in this activity. The toes will rise fully if the weight is shifted to the hindfoot, because there will be minimal weight on the arches and they will easily rise. If the toes rise easily without external leg rotation or shifting one's weight to the hindfoot, the plantar fascia is not providing optimal support and control of the longitudinal arches. This is probably due to midfoot breakdown with plantar fascia stretching. **(Morley et al., 2010)**

Putting further demand on proper foot motion is the internal and external rotation of the pelvis and leg during gait. The foot does not rotate significantly in relation to the substrate during gait. This motion is taken up in pronation and supination. The relationship of leg rotation with pronation and supination was dramatically demonstrated by Inman **(Inman et al., 1981)** by inserting pins in the tibia and midfoot so that the movement could be visualized.

Lundberg et al. **(Lundberg et al., 1989, part 1-3)** studied this motion by roentgen stereophotogrammetry. The subjects had at least three radiopaque markers put into each of the tibia, talus, calcaneus, navicular, medial cuneiform, and 1st metatarsal bones. They stood on a platform capable of tilting the foot in pronation and supination. X-rays were taken in 10° steps, from 20° of pronation to 20° of supination; the three-dimensional motion was calculated from the resulting x-rays. All bones contribute to pronation and supination. The greatest amount is at the talonavicular articulation, with less rotation between the calcaneus and talus. Rotation in the transverse plane at the talocrural joint did not exceed 2°, showing that no significant varus/valgus instability occurred at that location.

There must be structural integrity between the foot and leg that still adapts effectively to the needs for foot flexibility and rigidity. The joints allowing inversion-eversion and pronation-supination give this ability in a manner similar to a universal joint. The standard mechanical universal joint consists of two forks interconnected by a central piece of crossed arms. Each fork is attached to a shaft or similar device for potential rotation. The purpose of the universal joint is to transmit rotation in exact ratio from one rotating member to the other, but in different axes; in other words, it enables rotation to be transmitted around an angle. When the foot rotates in pronation and supination, motion is imparted to the tibiofibular column, but not in the exact ratio as with a standard universal joint. For every degree of foot supination, there is an average of 0.44° of external rotation of the tibia. This is accomplished by the distal fork of the universal joint changing configuration through its range of motion. This distal fork consists of the calcaneus, the cuboid, and the navicular. Together with a central piece, the talus, it constitutes a three-arm link system that explains the configurational change. **(Levangie & Norkin, 2001; Hertling & Kessler, 1996; Olerud & Rosendahl, 1987)**

Universal joint.

Joint action describing motion from the leg to the foot is sometimes described as an oblique hinge, which is an inadequate description. **(Engsberg & Andrews, 1987)** Rotation of the subtalar and transverse tarsal joints is interdependent. **(Levangie & Norkin, 2001; Wright et al., 1964; Manter, 1941)** Normal movement of all the joints is necessary for full range of motion. From a chiropractic and manipulative point of view, fixation of one joint will disrupt normal activity of the other joints. It is important to have the considerable range of pronation and supination for the foot to adapt to body movements and the substrate without imparting great motion to the leg. What motion is normally imparted to the leg can easily be adapted to by rotation at the hip joint, under normal conditions. The mechanism may have greater demands placed upon it by congenital anomalies. For example, when there is a tibial varus or valgus deformity, the joint complex of the foot must compensate. **(Lee et al., 2003; Rothbart & Estabrook, 1988; Ting et al., 1987)** Another example of this interaction is when ankle range of motion is limited by inhibited musculature; many differing compensation patterns have been shown to occur, such as genu recurvatum, early knee flexion, early heel lift or excessive pronation at the subtalar joint. **(Prior, 1999)**

Since the foot must have flexibility or rigidity under different conditions, its action is complex; this makes clear comprehension of foot function difficult. Sarrafian's **(Levangie & Norkin, 2001; Sarrafian, 1987, 1983)** twisted plate model helps explain the mechanism under different circumstances. Here the forefoot is in the transverse plane of the plate, and the hindfoot in the sagittal plane. The joint architecture, ligaments, and aponeurosis are responsible for elevation of the medial longitudinal arch with certain motions and, conversely, arch depression with contrary motions. Additional twisting of the footplate causes the forefoot portion to further pronate in relation to the midfoot to stay horizontal with the substrate; the hindfoot supinates (varus) and the arch rises. In this process, the footplate shortens. If the footplate is untwisted, the forefoot must supinate to stay parallel with the substrate. The hindfoot pronates, the arch drops, and the foot lengthens.

As noted, with the foot in contact with the substrate, the medial longitudinal arch rises and lowers with external and internal leg rotation, respectively. This is due to the architecture of the joint angles and their ligamentous integrity. With external rotation the navicular and calcaneus supinate, and the distal forefoot pronates in relation to the rest of the foot to maintain its contact with the substrate. This causes elevation of the medial longitudinal arch and shortening of the foot, as indicated by the plate model. Conversely, with internal rotation of the tibiotalar column the navicular and calcaneus bones pronate, and the distal forefoot supinates to maintain contact with the substrate; otherwise, the lateral portion of the forefoot would lift from the substrate. The medial longitudinal arch lowers, and the foot elongates.

For optimal function, the components of the foot must be aligned so the axis of the talus body points to the 2nd/3rd toe interspace. This affects axial loading of the metatarsals. "If the axis of the talus points more medially, the foot is described as metatarsus abductus. If the axis of the talus points more laterally, the foot is described as metatarsus adductus." **(Hamilton & Ziemer, 1983)** Abnormal loading of the metatarsals can cause pain in the proximal or distal forefoot.

As a result of the oblique orientation of the subtalar axis of rotation, torsion of the tibia changes the forces in the foot. If external tibial rotation puts force into the foot, weight is shifted to the lateral four metatarsals. Similarly, internal tibial rotation causes increased weight on the 1st metatarsal. **(Morley et al., 2010; Jones, 1945)**

Normally the rotator muscles of the thigh are balanced when the foot is pointed approximately 10° outwardly. One can evaluate this muscular balance by pointing a foot directly forward and then standing on that leg only. Normally, when standing on the right foot, the body will rotate to the left approximately 10° to the neutral position of the hip rotator muscles.

There is a 2:1 ratio of inversion to eversion; thus if there is a normal 30° of subtalar joint motion, 20° would be inversion and 10° eversion. **(Brody, 1980; Subotnick, 1975)** In impairment rating, 30° of inversion and 20° of eversion are considered normal. **(American Medical Association, 2007)** This range is estimated rather than measured with a goniometer. Cailliet **(Cailliet, 1997)** recommends evaluating subtalar motion with ankle dorsiflexion to lock the mortise.

Plantar flexion and dorsiflexion.

The greatest motion in plantar flexion and dorsiflexion is at the talocrural articulation, with additional motion contributed by the joints of the longitudinal arches. This participation is in varying degrees. The distribution of motion between the joints of the arch in plantar flexion

The flat portion of the rectangular plate represents the forefoot, the twisted portion the arch, and the vertical portion the hindfoot.

If the plate's twist increases, the hindfoot must supinate the forefoot to remain horizontal. The arch rises and the the plate shortens.

If the plate's twist decreases, the hindfoot pronates for the forefoot to remain horizontal. The arch lowers and the plate lengthens.

varies greatly among subjects, with some subjects obtaining 40% of plantar flexion in these joints. In dorsiflexion, participation of the joints of the arch is small in normal subjects. **(Kitaoka et al., 1995; Lundberg et al., 1989)** Under certain circumstances there can be excessive dorsiflexion contributed by the longitudinal arches. This abnormal condition is often part of the extended pronation complex. **(Discussed later)**.

Foot Reflexes and Reactions

There are numerous reactions that result from stimulation to the foot nerve receptors. The reactions discussed here are those presented in the general anatomical and physiological literature. One can wonder what future research will bring. With AK examination, correlations are observed that provide excellent opportunities for basic research. An example is Colum's observation of an apparent interaction between ankle dysfunction and carpal tunnel syndrome, an observation confirmed by Mondelli & Cioni. **(Mondelli & Cioni, 1998)** Colum observed in recurrent carpal tunnel syndrome that an interactive AK treatment to the contralateral ankle eliminated the recurrent nature of the carpal tunnel syndrome. **(Colum, 1983)** This interaction is probably on a ligament interlink basis.

The study of Lever discussed earlier confirms interactions between ligaments and muscles. **(Lever, 2006)** His study showed the relationships between the AK sensorimotor challenge to specific ligaments in the foot and muscular impairments resulting throughout the body.

The response to stimulation of foot nerve receptors is an important function to provide for protection. Stimulation of the nociceptors of the lower limb causes coordinated movements of withdrawal. **(Kugelberg et al., 1960)**

Afferent input from the sole of the foot affects postural awareness significantly. In fact sensory receptor information plays numerous roles in creating motor responses. **(Holm et al., 2002)** Cutaneous reflexes from the foot are important to posture and gait. **(Kavounoudias et al., 2001)** Disturbed sensory supply from cutaneous receptors in the feet alters the strength, timing, and velocity of muscle activation as well as altering gait. Lower limb sensory supply alone provides enough information to maintain upright stance and are critical in perceiving postural sway. In addition, movement discrimination in the ankle is better barefoot when compared to wearing shoes **(Waddington & Adams, 2003)**

There are specific reflex patterns for each skin area stimulated that make up an elaborate mechanism. The appropriate withdrawal movement is obtained by integration of flexion and extension reflexes. There is certainly nothing 'primitive' about such a highly purposeful reflex system. **(Holm et al., 2002; Kugelberg et al., 1960)** The highly selective reaction from stimulation to the ball of the foot evokes plantar flexion of the toes; stimulation a few millimeters away produces dorsiflexion. These responses are based on function and are not anatomical. They are responses of multisegmental reflexes where two skin points may be innervated by the same spinal segment and the same cutaneous nerve, yet the responses are the reverse of each other.

Positive support reaction

One tends to think of posture only in a living, dynamic sense. A dead body also has posture — the position gravity impresses on it with no counteracting forces. Sherrington **(Denny-Brown, 1979)** states, "Active posture largely encompasses the counteraction of those effects which gravitation, etc., produce in the dead body. Active postures may be described as those reactions in which the configuration of the body and of its parts is, in spite of forces tending to distract them, preserved by the activity of contractile tissues, these tissues then functioning statically." The tissues functioning statically refer to the muscle turgor, sometimes called tonus. In addition, one should consider the bones and ligaments of the skeleton. When the bones are properly formed and tied together with ligaments, there is strong compressive structure, as demonstrated by the foot's structural integrity. As we will see later, many of the muscles thought to be constantly active in maintaining balanced posture are inactive, as observed by electromyography. It is when there is postural imbalance that the "postural muscles" are active. In normal, balanced posture the body has an economy of muscle function, and balance is maintained with only short, slight twitches of muscle activity.

When the substrate angle changes under the foot, remote muscles contract to maintain balanced posture. When body sway is induced from above, the muscles contract in a different order to maintain posture. **(Fink et al., 2003; Nashner, 1977)** This indicates fixed patterns of muscle organization, with control coming from the foot and many other equilibrium proprioceptors. Excitability of alpha-motoneurons throughout the body is influenced by the supraspinal motor centers of the CNS, segmental spinal interneurons, rhythmic movement pattern generators, involuntary reflexes and sensory input from proprioceptors in muscle, articular structures, and skin. Once again, the diversity of sensorimotor activity going on in the patients being assessed by the clinician requires a system of evaluation that can encompass these various sources of neural activity quickly and reliably: the manual muscle test and the AK sensorimotor "challenge" and "therapy localization" procedures offer this kind of assessment technology.

The positive support reaction, first observed in decerebrate animals, is a basic reflex of posture. When pressure is applied to the plantar surface of the foot, the limb extends strongly enough to support the animal's body weight, leaving the animal standing in a rigid position. **(Hodgson et al., 1994; Langworthy, 1970)** The pathways involve complex circuits in the interneurons, similar to those responsible for the flexor and crossed extensor reflexes. The reaction was originally called the "magnet reaction" or "extensor thrust reaction"; now it is often referred to as the "positive support reaction." **(Guyton & Hall, 2005)** The term "magnet reaction" was used by early experimenters because when pressure was applied to an animal's foot, it followed the examiner's withdrawing hand as if it were attached to a magnet. The term has no reference to electromagnetic activity within the body. Ruch (**Ruch, 1979**) acknowledges that the reaction is more readily demonstrated in a decerebrate preparation, but it is present in normal animals and in man. In the spinal animal the reflex is so sensitive that it can be elicited by simply touching the sole of the foot.

The positive support reaction does not come solely from the myotatic reflex since there is inadequate time for response to the muscle stretch. (**Vedula et al., 2010; Guyton & Hall, 2005**) It appears to come from a combination of stimulation to muscle, joint, and cutaneous mechanoreceptors in the foot. These combine to form a complex union of stimuli to provide facilitation and inhibition of muscles contributing to posture and gait. When the foot is weight-bearing, the force to the interphalangeal joints and stretching of the interosseous, adductor hallucis, and other muscles stimulate the joint receptors and muscle spindles to provide facilitation of the postural extensor muscles, which is the positive support reaction. On the other hand, if the phalanges and metatarsals are squeezed together or flexed instead of being spread apart, there is inhibition to the extensors and all joints of the extremity flex. (**Gowitzke & Milner, 1980; O'Connell & Gardner, 1972**) This takes place if the foot is functioning normally. With many types of foot dysfunction, the positive support reaction fails to function as noted, and apparent confusion develops in the nervous system. Stimulation to the receptors as stated will be used in the applied kinesiology examination of joint neurologic functions.

O'Connell (**O'Connell, 1971**) describes an experiment that has now been done in many physiology laboratories to demonstrate the positive support reaction. She used a swing arranged so that it could be raised and lowered various distances from the floor. The subject sitting in the seat of the swing was raised and lowered, and the seat was randomly tilted to dump the subject toward the floor. The individual landing on his feet maintained an upright posture. After repetitions of the procedure, the individual was blindfolded and the procedure repeated with enough swinging and random elevating and lowering to disorient the subject. Again, random spilling dumped the individual toward the floor. Correct posture was maintained when he landed on his feet, but with greater difficulty and more slowly than when not blindfolded. Finally, the individual's feet were immersed in iced water for a period of twenty minutes to produce local anesthesia, and the blindfolded procedure was repeated. In this instance, the individual was unable to maintain an erect posture and crumpled to the mat. The conclusion, of course, is that chilling of the feet interfered with proprioceptive communication to the body; consequently, there was no facilitation of the extensor muscles.

The positive support reaction, in addition to facilitating upright posture, provides nerve impulses to control lateral sway. This, along with the equilibrium proprioceptors, keeps sway to a minimum in so-called static stance. (**Guyton & Hall, 2005; Hellebrandt & Braun, 1939**)

Location of pressure on the sole of the foot determines the direction the foot turns. (**Guyton, 2005**) In a simple manner, this is observed in the magnet reaction described above. If the experimenter presses flat on the decerebrate animal's paw, there is straight extension. If pressure is applied on one side of the foot, the foot moves toward the source of stimulation in an effort to return balance to the foot position. (**Guyton, 2005**) The ability to maintain balance is enhanced by these reactions, particularly in the biped. A simplified example is seen when one sways to the left; there is adduction of the left femur and abduction of the right one, with eversion of the left foot and inversion of the right one. (**O'Connell, 1958**) With the sway there is stimulation to the muscle, joint, and cutaneous receptors to return the body from lateral sway to neutral. These impulses are interpreted to cause facilitation of the left foot inverter muscles, such as the tibialis anterior. At the right foot and ankle, there is facilitation of the peroneus group to cause eversion. At the pelvic and thigh levels, there is facilitation of the left gluteus medius and tensor fascia lata to cause left hip abduction. There is also facilitation of the right adductors to cause hip adduction. This muscular action brings the body back to neutral from its lateral sway. (**Gowitzke & Milner, 1980**)

Foot dysfunction has been statistically correlated with hip abductor and adductor muscle weakness, as measured by the Cybex II dynamometer. (**Gleim et al., 1978; Nicholas et al., 1976**) The term functional ankle instability is frequently used for recurrent ankle sprains, and functional ankle instability is associated with arthrogenic muscle weakness that occurs throughout the body, including the peroneal muscles of the ankle, (**Palmieri-Smith et al, 2009; Tropp, 1986**) the tensor fascia lata muscle of the hip -- after the AK shock absorber test was administered to the feet (**Zampagni et al., 2009**) -- and in many other muscles throughout the injured limb.

(With kind permission, ICAK-USA)

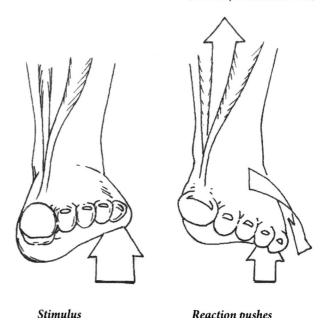

Stimulus   *Reaction pushes against the stimulus.*

The evidence now shows with greater clarity than ever before that inflammation or injury most frequently produces specifically identified inhibited muscles.

Controlled clinical studies have shown that dysfunction and pain specifically in the ankle, **(Nicholas & Marino, 1987)** knee, **(Slemenda et al., 1997; Stokes & Young, 1984)** lumbar spine, **(Hossain & Nokes, 2005; Hodges & Richardson, 1996)** temporomandibular joint, **(Zafar, 2000)** and cervical spine **(Cuthbert et al., 2011; Jull, 2000; Vernon et al., 1992)** will produce *inhibited muscles*. These data indicate that the body's reaction to injury and pain is *not increased* muscular tension and stiffness; rather *muscle inhibition is often more significant*. The use of a clinical tool like the manual muscle test is uniquely designed to detect this important neuromuscular impairment in patients with lower extremity dysfunction.

It appears that almost any type of dysfunction — including subluxations, fixations, extended pronation or intrinsic muscle problems — can cause poor or disorganized remote muscle function. **(Zampagni et al., 2009)** In the first decade of applied kinesiology, Goodheart **(Goodheart, 1973)** reported that Edward Doss, Sr., D.C., of Stuttgart, Arkansas, told him of his frequent observation that a lateral cuboid subluxation correlated with a tensor fascia lata muscle that tested weak. Correction of the subluxation returned normal function to the muscle, as observed by manual muscle testing. Goodheart concurred with the observation, and also found that adjusting the medial transverse arch (medial cuboid) often corrected adductor muscles that tested weak. This logically fits very well with the available information about the magnet or placing reaction. Lateral sway of the body that causes inversion and eversion of the foot to regain balance would stimulate the cutaneous and joint receptors. A medial or lateral subluxation of the transverse tarsal bones would be equivalent to body sway as far as the joint receptors of the cuboid would be able to discern; consequently, impulses would be transmitted to facilitate and inhibit muscles to return balance to the body. Unfortunately, when this neurologic activity is due to a subluxation, it is inappropriate to body needs and creates further imbalance of function.

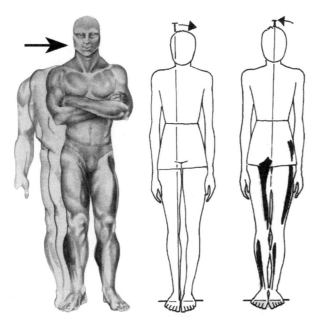

Muscle reaction to bring lateral sway back to balance.

Examination of the positive support reaction

Dysfunction of the positive support reaction can develop as a result of improper stimulation to the muscle, joint, or cutaneous receptors. Typically, but not always, when muscle receptors of the feet are at fault, the dysfunction will be in the extensor muscles. Joint receptors often disturb muscles of ab- and adduction and muscles of gait; cutaneous receptors relate primarily with local muscles and muscles of gait. There are many muscles that must have organized function during gait. **(Landry et al., 2010; Craik & Oatis, 1995; Smidt, 1990)** These include head-turning ones such as the sternocleidomastoid and upper trapezius, shoulder and hip muscles, and muscles of the sacrospinalis. Some dysfunction must be examined for by combination muscle tests, such as in applied kinesiology gait testing. **(Walther, 2000)**

Stimulation of the nerve receptors responsible for the positive support reaction appears to relate with stretching the intrinsic plantar foot muscles. The muscles stretched by weight bearing include the flexor hallucis brevis, abductor hallucis, adductor hallucis, flexor digitorum brevis, lumbricals, and dorsal and plantar interossei. The muscles that are longitudinal to the foot — such as the flexor digitorum brevis and lumbricals — are stretched when the longitudinal arch is flattened; the dorsal interossei and adductor hallucis, especially the transverse head, are stretched when the metatarsal arch flattens and spreads.

In normal weight bearing, when the longitudinal arches and metatarsal arch flatten, there is stimulation to the mechanoreceptors that causes facilitation to the extensor muscles of the body. **(Guyton & Hall, 2005)** The easiest method for evaluating the positive support reaction is with the patient prone, making it easy to evaluate extensor muscles such as the hamstrings, gluteus maximus, deep neck extensors, and upper trapezius. First the muscles should be tested to determine that they are functioning normally. If not, make corrections with the usual applied kinesiology approaches, such as correcting vertebral subluxations and reflexes. When the intrinsic muscles of the foot are stretched to simulate weight bearing, the previously strong postural extensor muscles should remain strong or even test stronger, because these muscles are normally facilitated by stimulation to the receptors of the positive support reaction. Simulation of weight bearing is done by flattening the longitudinal arches and spreading and flattening the metatarsal arch. Immediately after the challenge, one or more of the extensor muscles should be tested. A positive indication of foot involvement is weakening of the extensor muscle(s). Both the gluteus maximus and the hamstring muscles are pelvic extensors, but it is usually best to test the gluteus maximus; the hamstring muscles are both pelvic extensors and knee flexors in their two-joint function.

Another type of challenge for the positive support reaction that is sometimes done in applied kinesiology is squeezing the metatarsal arch together. This frequently shows a weakening of the extensor muscles with positive support dysfunction, but it is not as accurate a challenge as the spreading mechanism. There is a normal action of facilitation of the limb flexors when the phalanges and the metatarsals are squeezed together or flexed instead of being abducted. **(Gowitzke & Milner, 1980)** When there is facilitation of the flexors, there is normally inhibition of the extensors on the basis of reciprocal inhibition. When there is improper stimulation to the positive support reaction mechanoreceptors, the resulting effect of the muscles is unpredictable. Sometimes squeezing the metatarsal arch and phalanges together will cause weakening of the extensor muscles. This may be due to challenging foot subluxations. Forces applied to the normal foot do not cause weakening of the extensor muscles.

When there is weakening of the extensor muscles after challenging the intrinsic foot muscles, treatment is directed to the muscles of the foot. Treatment necessary is usually to the neuromuscular spindle cell, Golgi tendon organ, fascia (fascial release), percussion, trigger point pressure release, and occasionally use of the origin-insertion technique. **(Leaf, 2010; Cuthbert, 2002; Walther, 2000)** The muscle dysfunction is usually secondary to a structural foot problem such as pronation (with or without a short triceps surae), metatarsal subluxations, or other structural foot problems that will be discussed later.

Often the method of flattening the longitudinal arches and spreading the metatarsal arch, then testing the extensor muscles, will provide an adequate examination. In some instances, this simulated weight bearing does not reveal the problem. If body language indicates improper function of the positive support reaction but the prone tests are negative, test the patient in a standing position. Information revealed in standing, sitting, supine and prone positions with manual muscle testing assessments may be combined with other AK assessment measures to reveal distinct patterns of muscular adaptation and/or structural imbalances. When these patterns are assessed and combined with the patient's symptomatology, habits of use, and pain pattern a clearer picture emerges as to what is dysfunctional and how to support the patient.

The standing assessment of the positive support reaction, of course, puts more strain on the longitudinal arches than is usually done in the previously described procedure, and it may cause the extensor muscles to weaken. In addition, the proprioceptors react differently to stimulation, depending on the limb's position at the time of stimulation. Feedback and feed-forward mechanisms regulate motor control by correcting movement after sensory supply has been received. They use closed reflex loops of proprioception from mechanoreceptors and muscles across the joints of the foot. **(Aniss et al., 1992)** Sherrington points this out, stating that the action taken "…is in part determined by the posture already obtaining in the limb at the time of the application of the stimulus." **(Denny-Brown, 1979)** It is usually easier to test the extensor muscles weight bearing, with the patient leaning face forward on an upright hi-lo table, as in this position the positive support reaction is being tested by the patient's positioning.

Pronation

Many terms are often used synonymously to describe foot pronation, such as flatfoot, pes planus, fallen arches, flexible flatfoot, peroneal spastic flatfoot, rigid flatfoot, pes valgo planus, and on and on. **(Shibuya et al., 2010)** What is important is to understand the dynamics of foot function and what causes dysfunction. All too often symptoms in the feet are treated with no regard for the primary cause. Often this is due to the fact that the clinician does not possess a method for measurement or detection of primary causes. Once a foot dysfunction has been identified by virtue of a postural, palpation, orthopedic or other test, it is necessary to define precisely what type of dysfunction exists. The effect of this dysfunction upon other areas and symptom complexes is very important. The effect of this foot dysfunction upon attaching or remote muscle function is offered by the challenge and therapy localization procedures in applied kinesiology. The associated muscle weakness, easily determined by the manual muscle tests as outlined in this text, is then evaluated with the challenge procedure or therapy localization (the AK sensorimotor stimulus). Appropriate angular and pressure stimulation of the articulation produces immediate strengthening of inhibited muscles in remote body areas and symptom complexes due to the foot dysfunction(s).

Orthotics may be prescribed when muscle, joint, or neurologic corrections would more permanently and cost-effectively solve the problem. **(Page et al., 2010; Maffetone, 2003)** On the other hand, treatment by these methods may be applied when there is a congenital anomaly or some other condition requiring orthotics or surgery to correct the primary problem.

Furthermore, one must recognize the influence of the foot on remote body function and of remote body function on the foot. **(Marshall et al., 2009; Mattson, 2008; McVey et al., 2005)** In Lutter's **(Lutter, 1980)** sports medicine clinic, 76% of knee symptoms were related to pronation abnormalities. Structural strain, as in knee problems, is generally recognized; what is often missed is the disturbance to normal neurologic function that may cause dysfunction throughout the body. Researchers have found weakness and changes in muscle activation and velocity in the knee, hip, trunk, shoulders and neck in subjects with ankle instability **(McVey et al., 2005; Beckman & Buchanan, 1995; Bullock-Saxton, 1994)**

These findings point to the importance of manual muscle test examinations beyond the site of injury; the entire muscle system is vulnerable to aberrations in foot function. Applied kinesiology provides methods to find these types of remote dysfunctions and to specifically relate them to the foot for both the doctor and the patient using the sensorimotor challenge and therapy localization procedures.

Treating foot symptoms rather than correcting the cause may adversely affect another part of the body. It is only natural to think that a therapeutic approach that relieves foot symptoms is good for the body; this is not necessarily true. It is not unusual to find improperly fitted orthotics to be the cause of low back, shoulder, or other pain during examination of a new patient. Your credibility may be questioned when the patient says, "My feet have been better since I started wearing orthotics six months ago." Questioning the patient reveals that the back pain slowly developed about two weeks after wearing the new orthotics. Examination may find that the positive support reaction does not function properly when the orthotics are worn, causing the extensor muscles to test weak. Correcting the back problem is accomplished by correcting the foot dysfunction and/or changing the orthotics. (**Sahar et al., 2007; Robbins & Hanna, 1987**)

When working with the feet, it is necessary to be capable of evaluating how remote areas of the body affect them, and also how any therapeutic approach applied to the feet affects the rest of the body. Proper therapeutics and correction enhance function in the rest of the body; improper therapeutics, even though providing relief in the feet, may cause remote disturbances that manifest new symptoms that the patient usually does not associate with foot treatment.

There are many therapeutic approaches to excessive foot pronation; all work to some degree. It is necessary to have a system that evaluates how treatment affects the foot, and also how the rest of the body reacts to the therapeutic approach. As will be seen, not only does applied kinesiology fill this need, it also enables the physician to determine how remote areas of the body may be causing foot dysfunction.

General use of the term "foot pronation" connotes abnormal position or function of the foot to many people. Actually, foot pronation during function is a necessary and important aspect of the gait cycle. Typically, reference to pronation with the patient standing alludes to excessive pronation or "flatfoot." When the term is used in the dynamic sense, it refers to "extended pronation," i.e., the foot comes out of excessive pronation too late in the gait cycle. In this text, when discussing abnormal conditions we tend to use the term "pronation" as a static pronated foot, "extended pronation" in reference to gait, and "excessive pronation" when relating to static and dynamic conditions.

Gait

Volumes have described the individual kinematics, movement patterns and mechanics of gait. (**Cailliet, 1997; Craik & Oatis, 1995; Smidt, 1990; Hoppenfeld, 1976**) The fundamental characteristic of human movement, walking, is a series of prevented catastrophes. Soderberg, Cailliet, Hoppenfeld, Slocum and James, and others (**Soderberg, 1997; Cailliet, 1997; Hoppenfeld, 1976; Slocum & James, 1968**) describe gait as a series of events consisting of the support and swing phases. There are three divisions of the support phase: 1) footstrike, 2) mid-support, and 3) take-off. A common term for the first phase of the support phase is "heelstrike," but there are many patterns of gait, especially in running, when the heel is not and should not be the first part of the foot to strike the ground. (**Abshire, 2010**)

Wallden (**Wallden, 2010**) has described the evolving literature on foot biomechanics indicating that running with a heel strike is not functional after all. In a review of the literature and mechanics on barefoot running and athletic performance, it may be that forefoot and midfoot strike is the natural, functional style of gait at footstrike. Heel striking can lead to many kinds of injuries.

Heel striking during running forces the body to brake slightly thus requiring increased push off forces to maintain velocity. This produces shearing into the lower back and spine, and forces excessive upper body rotation. The ankle becomes unstable when adapting to the substrate. This produces overpronation and oversupination, increased rotational forces into the joints, and an increased vertical bounce with each foot strike. (**Abshire, 2010**) Generally speaking, modern running shoes are traditionally made with lifted heels. The high heel gets in the way of the foot landing parallel to the substrate.

The term "footstrike" is more inclusive for the first part of the support phase. (**Cavanagh, 1982**) Both footstrike and heelstrike will be used in this text; in general the two terms are synonymous primarily for shod feet, indicating the first part of the support or stance phase.

The swing phase of gait can be considered as forward recovery. It, too, is divided into three phases: 1) follow-through, which is immediately after toe-off, 2) forward swing, and 3) foot descent. (**Soderberg, 1997; Slocum & James, 1968**) Gait analysis should take into consideration the lower extremities, the pelvis, spine, head movement on the spine, and shoulder and arm motion. (**Walther, 2000**) Here we will consider only foot function in the stance phase of gait.

At heelstrike in the shod foot, the subtalar joint is supinated and the tibia is in external rotation. Rapid pronation begins as the foot comes into full contact with the substrate, with the subtalar joint moving into a neutral position and the tibia coming out of external rotation. Pronation acts as a shock absorber mechanism to the forces being put into the foot. Authors vary regarding which of the long muscles are active in supporting the medial longitudinal arch. The tibialis anterior and posterior and peroneus longus muscles may all support the medial longitudinal arch during pronation of normal gait, and especially in extended pronation. The joint configurations allow the foot to adapt to the underlying surface. The ligaments, joint constraints, and muscles combine to dissipate stress to the foot and body, similar to a flexible band. The tibia follows the internal rotation of the talus as pronation continues.

At 15-25% of the stance phase, the foot should begin to come out of pronation. At this time it starts to become a rigid

Problems with heel strike in running shoes

- Heel strike forces braking and increased push off forces
- Shearing on the low back & spine
- Excessive upper body rotation
- Ankle becomes unstable with ground
- Overpronation and over-supination
- Increased rotational forces to the joints
- Increased vertical bounce

lever for the toe-off propulsive stage of gait. "The foot that is not re-converted into a rigid lever by 75% weight bearing is defined as having increased pronation." (**Lutter, 1980**) Seventy-five percent weight bearing is at approximately 25% of the stance phase. (**Ramig, 1977**) Up to this point the mid-tarsal joints should be unlocked to allow the foot to adapt to the terrain. At 25% of the gait, the foot should begin supinating to reach neutral at mid-stance (50-65% of stance), and then supinate to become the rigid lever for toe-off. The peroneus longus contributes by everting the foot and flexing the first metatarsal. (**Beardall, 1975**)

Pronation which lingers on beyond 15-25% of the stance phase of gait is abnormal and causes torques in the lower extremities which lead to an overuse syndrome. (**Subotnick, 1991, 1975**) As weight transfers to the metatarsal arch and the toes extend, the arch rises and the foot, as a rigid lever, obtains more support from the plantar aponeurosis, which is tightened by the windlass mechanism described by Hicks. (**Hicks, 1951, 1953, 1954, 1955**) Classically the windlass mechanism operates as the gait cycle moves from mid-stance to toe-off, when the toes move into hyperextension, ideally reaching 65 degrees of extension, and the plantar fascia is thereby drawn tight increasing the arch along the medial aspect of the foot creating a spring like mechanism to push the person forward as they toe-off.

The intrinsic plantar muscles are important in maintaining the arches in the dynamic foot. They must work harder in the foot that pronates excessively. (**Mann & Inman, 1964**) Unfortunately, entrapment of their nerve supply at the tarsal tunnel often accompanies extended pronation, causing atrophy of the intrinsic muscles so they cannot contribute to arch maintenance.

The medial longitudinal arch is important as a shock absorber. During the stance phase of gait it allows pronation from heelstrike to flatfoot. If pronation is extended, difficulties may arise in the midfoot, hindfoot, tibia, and knee from structural stress. (**Morley et al., 2010**)

The knee remaining too long in internal rotation is a good example of structural stress. Maximum knee extension during gait is present when the center of gravity is over the foot, causing strain if the tibia is still in internal rotation. Increased subtalar joint pronation has also been identified as a contributing factor in patellofemoral pain. (**Crossley et al., 2006**) In addition to structural stress, other remote problems may develop from neurologic disorganization due to improper stimulation of the foot mechanoreceptors during extended pronation. Foot pronation during gait, then, is not bad; it is necessary and useful as a shock absorber mechanism. It is when supination and transformation of the foot to a rigid lever are delayed that problems develop.

Although it is important to treat extended pronation, care must be taken not to overcontrol it with orthotics. Pronation in the amount of 4° provides the shock absorption necessary and accommodates the internal rotation of the leg. (**Subotnick, 1975**) Artificially reducing this normal pronation decreases the foot's ability to act as a shock absorber and limits its adaptability to the changing substrate surfaces. Overcorrection of pronation may cause foot or remote problems revealed by symptoms and by applied kinesiology examination.

Many factors are responsible for abnormal foot function. Authors have individually stressed improper bony architecture, poor ligament integrity, or muscle and fascia dysfunction as causes of extended pronation. All these factors come into play, either singly or in combination. If the initial cause is only one component, the other two will probably be secondarily involved if the condition is allowed to continue.

A flatfoot is classified as dynamic or static. The dynamic flatfoot has a normal arch when non-weight bearing and a flat one when weight bearing. This is observable by x-ray. The static flatfoot does not change appreciably from non-weight bearing to weight bearing. The abnormality is present on x-ray, whether weight bearing or not. A normal arch can be formed in the dynamic flatfoot by digitally molding the foot. In the static flatfoot, a normal arch cannot be formed by the examiner's efforts. (**Subotnick, 1991, 1975**)

Etiology

The etiology of flatfoot can be congenital maldevelopment. There are several gross congenital variations that are usually diagnosed and treated in infancy or early childhood. Those conditions — most often treated by orthopedic surgeons — are out of the scope of this text. Full descriptions of them are found elsewhere. (**Galois et al., 2002; Hamilton, 1985; LeNoir, 1982; Tachdjian, 1982**) Here we will deal with the less severe congenital deficiencies that can be managed with conservative treatment, and conditions that develop from misuse and trauma.

Family history provides a good indication of the possibility of congenital flatfeet in children that is not of the severe type. (**Harris, 2010; Cobey, 1958**) Family members with hypermobility and a tendency to an everted weight-bearing position of the entire foot are examples. As a child develops, flatfoot becomes more apparent. As he first begins to walk, the flatfoot becomes easily recognizable.

Congenital anomalies must be put into proper perspective as the etiology for foot dysfunction. They can be overemphasized as the cause of the patient's complaints. Dysfunction, when attributed to a congenital anomaly, may cause the physician to cease efforts to obtain functional correction and rely only on orthotics, shoe correction, and life-style change; much better function could be obtained if dysfunctioning joints, muscles, ligaments, fascia, and skin were examined and corrected. On the other hand, one may fail to observe a congenital anomaly and persist in trying to obtain correction by these methods when an orthotic, shoe correction, or surgery would be more appropriate.

Short triceps surae

The short triceps surae is often referred to as a short Achilles tendon, equinus foot, gastrocnemius equinus, or hypermobile flatfoot with short Achilles tendon. Emphasis is placed on the Achilles tendon when, in reality, it is more often the muscle bellies that are short. Ryerson (**Ryerson, 1948**) points this out in his objection to surgically lengthening the Achilles tendon, stating, "The tendo achillis is not short; the muscle bellies are short, but the tendo achillis is long enough." His treatment approach is directed toward lengthening the muscle. Rarely is it necessary to surgically lengthen the Achilles tendon. **(Greenhagen et al., 2010; Logan, 1995; Lillich & Baxter, 1986)** Of additional concern are the associated complications of any surgical intervention. The risk of any operative procedure is estimated to be as high as 5% with infection and wound breakdown being the predominant concerns. **(Soma et al., 1995)** Pulmonary embolism and sensory deficits due to injury to the sural and other cutaneous nerves can also occur. **(Fierro & Sallis, 1995)**

The triceps surae is composed of the gastrocnemius, the soleus, and their shared tendon. Its length must be adequate to allow at least 10° of dorsiflexion at the ankle. **(Wright et al., 1964)** A short triceps surae was associated with flatfeet in early observations. Royal Whitman, **(Whitman, 1970)** in his classic 1888 presentation on flatfeet, observed "…that there seemed to be an abnormal resistance in the calf muscle…."

Limitation of dorsiflexion can be congenital or acquired. When the triceps surae is congenitally short, the deformity is called talipes equinus. Gastrocnemius equinus is a lack of 10° of dorsiflexion at the ankle with the knee flexed. Gastrocnemius-soleus equinus is a lack of 10° of dorsiflexion with the knee both extended and flexed. The flatfoot associated with short triceps surae is dynamic; that is, the arch appears normal when non-weight bearing and flat when weight bearing.

There are many methods in applied kinesiology that lengthen shortened or hypertonic muscles. Treatment, such as percussion, fascial release or stretch and spray, may be locally directed to the muscle or to the muscle proprioceptors. Remote treatment of many types increases the range of motion throughout the body. The patient should be examined for conditions such as dural tension and modular disorganization, and treated with PRYT, cloacal synchronization, and gait techniques. The initial AK treatment generally increases range of motion; stretching exercises, if needed, provide much better results when the AK procedures are done first.

If there is not 10° of ankle dorsiflexion, compensation must take place elsewhere during the mid-stance phase of gait. **(Craik, 1995; Nicholas & Marino, 1987; Subotnick, 1971)** Stress to the foot develops when the last amount of dorsiflexion needed in walking is obtained at the subtalar and mid-tarsal joints, rather than at the ankle mortise. **(Milgram, 1983)**

As body weight is forced into the midfoot, compensation takes place by pronation of the subtalar and mid-tarsal articulations. In order to compensate for the lack of ankle dorsiflexion, these joints must pronate more than normal, causing the ligaments to stretch and flexible flatfoot to develop. This breakdown of the midfoot is to accommodate the short triceps surae.

When compensation does not take place by midfoot breakdown, it may occur by early heel-off, which happens when the heel leaves the ground before 65-75% of the stance phase is completed. **(Hamilton, 1985; Subotnick, 1971)** The propulsive phase of gait is thus lengthened, which puts more stress on the ball of the foot than it is prepared to accept. As a result, metatarsalgia, intrinsic muscle fatigue, and calluses develop. **(Nicholas & Marino, 1987)** Leaf and Mennell state that metatarsal head pain may be the result of triceps surae shortening. **(Leaf, 2010; Mennell, 1969)**

Major stress is put on the plantar fascia when there is no midfoot breakdown. Myers and Waller **(Myers, 2001; Waller, 1982)** describes the plantar fascia as an extension of the triceps surae-Achilles tendon complex. "The gastrocsoleus muscle complex, Achilles tendon, and plantar fascia should be viewed as a single linkage system." According to this analysis, tension is created in the plantar fascia from the posterior as one rises on his toes or otherwise contracts the calf muscles, and from the windlass mechanism described by Hicks. **(Hicks, 1951, 1953, 1954, 1955)** The strain can be great enough to rupture the plantar fascia. **(Kim et al., 2010)**

The newborn child should have 30° dorsiflexion at the ankle joint with the foot in a neutral position. Subotnick **(Subotnick, 1973)** believes that any child with less than 5° of dorsiflexion is a poor candidate for conservative treatment with neutral control by orthotics, and considers lengthening of the Achilles tendon by surgery. Lengthening when there is more than 5° might have a detrimental effect on the knee joint following the establishment of neutral control. Lengthening of the tendon, which provides for more than 13° to 15° dorsiflexion, appears to be excessive.

According to Subotnick, **(Subotnick, 1973)** when there is definite congenital shortness of the triceps surae a child will not tolerate neutral control by rigid orthotics. The foot's effort to pronate in orthotics will be painful, and the child will refuse to wear the device. If the child does wear the device, there may be adaptation to it by keeping the knee in a more flexed attitude throughout gait. Since the gastrocnemius crosses the knee, this takes some of the strain off the triceps surae, which may allow more ankle dorsiflexion.

During the Second World War, Harris and Beath **(Harris & Beath, 1948)** studied 3,619 military enlistees and described the condition found as "…hypermobile flatfoot with short tendo-Achillis." They attributed the flatfeet in the soldiers to congenital anomaly of the calcaneus and talus, where the calcaneus provides insufficient support to the head of the talus; thus, weight bearing from childhood on causes the talus to be pushed downward and inward, while the forepart of the foot twists upward and outward. Hypermobility develops in the midfoot from the process

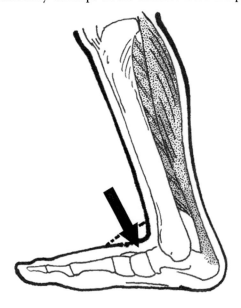

Midfoot breakdown from short triceps surae.

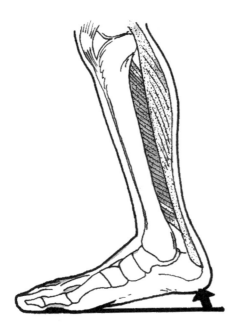

Without extended pronation and midfoot breakdown, there will be early heel-off because of a short triceps surae.

of walking. They state that the short triceps surae is not primary but secondary, and "…probably develops because of the structure of the foot and the laxity of the tarsal joints deprive [the triceps surae] of tension stresses which normally in use would facilitate its elongation."

Flatfoot that results from congenital anomaly of the calcaneus and talus develops early in childhood, but it usually does not cause symptoms until later, possibly in the early teens.

Tarsal coalition

Tarsal coalition is absent or restricted movement between two or more tarsal bones. (**Schenkel et al., 2010; Zaw & Calder, 2010**) It is usually a congenital anomaly with a fibrous (syndesmosis), cartilaginous (synchondrosis), or bony (synostosis) union between the adjoining involved bones, and it may be bilateral in 80% of cases. (**Leonard, 1974**) The most common bridges are between the calcaneus and navicular, and the talus and calcaneus; they are less common between the talus and navicular bones. (**Murray & Jacobson, 1977**) Other conditions, such as trauma, arthritis, or tumor, can limit the motion in the hindfoot.

Tarsal coalition causes a lack of foot mobility, specifically in the subtalar joints. This may result in repeated ankle sprains and strain. Snyder et al. (**Snyder et al., 1981**) found retrospectively that 63% of patients who had ankle sprains had calcaneonavicular coalition. This is a much higher incidence than expected, indicating there may be a predisposition to ankle sprains from this condition.

The ankle joint may be a ball-and-socket type, in which the talus appears on x-ray as a convex dome shape on both the AP and lateral projections. Sixty-five percent of patients with a ball-and-socket ankle joint have associated tarsal coalitions. (**Morgan & Crawford, 1986**)

There is no general consensus regarding the appearance of the foot. There is limited motion that may be great enough to class the foot as rigid flatfoot; (**Zaw & Calder, 2010; Percy & Mann, 1988**) however, as few as 10% of those who have tarsal coalition may have markedly pronated feet. (**Elkus, 1986**) The more subtle cases of coalition may be recognized in an AK practice during the motion palpation phase of examining the tarsal joints.

Tarsal coalition may be asymptomatic or cause symptoms, apparently from the stress that it places on other articulations in the gait cycle. Symptoms resulting from the disorder are aggravated by activities that put additional stress into the feet.

The condition is often recognized in adolescent athletes, probably because of its progression toward ossification of cartilage and the increasing demands on the articulation from athletic endeavors. (**Jack, 1954**) However, it is not necessarily diagnosed at an earlier age in athletes. (**O'Neill & Micheli, 1989**) "Talonavicular coalition ossifies at approximately 2 to 5 years of age, the calcaneonavicular at 8 to 12 years, and the talocalcaneal in early adolescence." (**Percy & Mann, 1988**) The greater general mobility of the foot and ankle in early

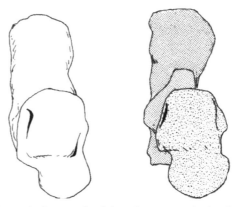

Congenital anomaly of the calcaneus and talus fails to provide balanced suport.

childhood accounts for the lack of symptoms until more rigidity develops and stress is applied in athletics.

Morgan and Crawford (**Morgan & Crawford, 1986**) state, "The majority of patients with tarsal coalition in the general population require only conservative methods of treatment such as shoe modifications (wedges, inserts, counters) or foot orthotics." Many methods of non-operative treatment are aimed at reducing foot stress in tarsal coalition. Many are directed toward relaxing the peroneal muscles, and supporting the foot by use of pes planus shoes, molded inserts, ankle-foot orthoses, physical therapy, and plaster casts. (**Zaw & Calder, 2010; Elkus, 1986**) Most of these adjuncts are not needed with the applied kinesiology examination and treatment methods discussed in this chapter. Some may provide additional help in resistant cases.

The adolescent athlete with tarsal coalition is a more select patient. (**Schenkel et al., 2010**) The bones are ossifying at the time that requires much mobility of the foot and ankle complex for maximum performance. In the absence of favorable conservative treatment, a complete resection of the coalition may be necessary. (**Morgan & Crawford, 1986**) Arthrodesis may be the proper treatment for tarsal coalition when it is associated with advanced degenerative changes or deformity. (**Zaw & Calder, 2010; Peterson, 1989**)

Peroneal spastic flatfoot. Lowy as well as Harris and Beath (**Lowy, 1998; Harris & Beath, 1948**) draw attention to tarsal coalition and its association with peroneal spastic flatfoot. They indicate that the peroneal muscles are not really in spasm, but are secondarily shortened as a result of the foot dysfunction. Jack (**Jack, 1954**) disagrees with this, pointing out the relief obtained when the muscles are anesthetized. It appears that the peroneal muscles go into reflex spasm as a result of the abnormal stresses on the other tarsal articulations. (**Turek, 1984**)

There are three types of peroneal spastic and rigid flatfoot: 1) rigid flatfoot due to talocalcaneal bridge, 2) rigid flatfoot due to calcaneonavicular bar, and 3) other factors that can cause reflex spasm. It should be recognized that when arthritis is the cause for the third type, it would be more accurately designated as "arthritic flatfoot with peroneal spasm."

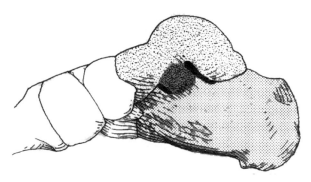

Calcaneonavicular coalition.

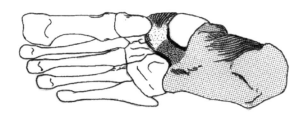

Talocalcaneal coalition.

Peroneal spastic flatfoot occurs much less than other types of flat-footedness. In Harris and Beath's study of 3,619 military enlistees, 2% had the condition. Symptoms are rarely present before twelve to fifteen years of age, when bone growth has been rapid and is more complete. It often develops as a result of another injury leaving residual problems. Continued walking when there is peroneal spasm will continue to break down the arch. If the condition has been present for a prolonged time, there may be actual shortening of the peroneal muscles. (**Lowy, 1998; Jack, 1954**)

Internal tibial torsion or malleolar torsion. Internal leg rotation, whether it is due to tibial or malleolar torsion or even femoral torsion, may be a cause of flatfoot. The foot is adducted into the position of pigeon-toeing. In order to walk with the foot pointed forward, it must excessively pronate. Unless there is severe leg rotation, excessive pronation is not great in this condition. If there is inherent weakness or laxity of the ligaments, or there are other factors that cause excessive pronation, internal leg rotation compounds the problem. In a study by Morley, (**Morley, 1957**) no correlation of genu valgus to foot pronation was found in the age range of one to eleven years.

Prehallux or accessory navicular

The accessory navicular appears between ten and twelve years of age and is the most common accessory bone of the foot; (**Leonard & Fortin, 2010; Turek, 1984**) it is present in 4-10% of the population. (**Logan, 1995; Hamilton, 1985**) (The terms "prehallux" and "accessory navicular" or "accessory scaphoid" are often used interchangeably in the literature.) There are three types of prehallux or accessory navicular bones. (**Sella et al., 1986**) Type 1 is a sesamoid in the tibialis posterior tendon. Approximately 30% of accessory naviculars are this type. Type 2 accessory naviculars are united to the navicular by a cartilaginous synchondrosis measuring 1-3 mm. This is the type that creates the major problem of accessory naviculars. Type 3 is an accessory navicular united to the parent navicular by a bony bridge, producing a cornuate navicular.

The accessory navicular can be responsible for flatfoot or, because of its bulk, give the appearance of flatfoot when the arches are well-formed. (**Leonard & Fortin, 2010; Hamilton, 1985**) In the latter case, x-ray investigation and the diagnostic approach for extended pronation (discussed later) provide the differential diagnosis. If a normally functioning foot with prehallux is treated with navicular pads or orthotics, it will be painful to the patient and probably cause remote neurologic dysfunction, as observed by applied kinesiology testing. Usually x-ray examination will demonstrate the existence of the prehallux/accessory navicular at about ten years of age.

The type 2 accessory navicular creates problems because of biomechanical forces acting on the synchondrosis. This frequently develops as a result of mild increase in recreational activities. (**Sella & Lawson, 1987**) Sella et al. (**Sella et al., 1986**) list three types of force that act simultaneously on the synchondrosis in varying degrees. Tension and shear are produced by action of the tibialis posterior to its tendon insertion at the accessory navicular. When foot pronation is present, it adds compression. This trauma causes a continual breakdown and repair of the synchondrosis, with neither prevailing. Since cartilage has a limited capacity for repair, a painful non-union will result.

The tibialis posterior tendon may be transposed by continual deforming forces of walking. Prehallux develops, and the tendon ceases to be in a position under and medial to the navicular bone; it assumes a position medial and superior to the bone itself. (**Cobey, 1958**) Thus the tibialis posterior action does not move the navicular to increase the anterior strut angle for added medial longitudinal arch strength during the latter portion of the stance phase.

Other factors — such as osteochondritis, breakdown of the midfoot from extended pronation, and stress fractures — can cause pain in this area. A bone scan provides definitive diagnosis for determining the symptomatic or non-symptomatic accessory navicular. (**Sella et al., 1986**)

Hypermobility Syndrome

Sprains that cause tearing of ligaments supporting the arch cause breakdown and, ultimately, prolonged pronation. **(Duarte, 2004)** Systemic conditions, such as Ehlers-Danlos syndrome, cause ligamentous insufficiency. Of course, if there is a systemic condition it should be treated, if possible. Rather than simply accepting the flatfoot as a result of ligament laxity, examine for and correct any muscular or other dysfunction affecting the foot. The foot will probably require support with a suitable orthotic.

Hypermobility syndrome is recognized as a connective tissue variant, though it can relate to specific disease processes like Marfan's and Ehlers-Danlos syndromes. A possible explanation for recurrent joint injuries in hypermobile people may be the proprioceptive impairments that disturb muscle function observed in hypermobilie joints. **(Fatoye et al., 2009; Hall et al., 1995; Mallik et al., 1994)** In the treatment of patients with hypermobility syndrome, many of their most-common musculoskeletal complaints (including recurrent dislocations) are effectively treated with the muscle strengthening procedures used in applied kinesiology practice. Hudson et al **(Hudson et al., 1998)** suggest that physical conditioning and regular exercise are probably protective from the effects of hypermobility syndrome. Colloca and Polkinghorn suggest that chiropractic care may benefit some patients with connective tissue disorders, including Ehlers-Danlos syndrome. **(Colloca & Polkinghorn, 2003)**

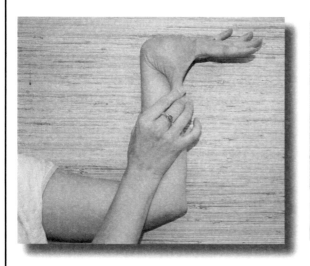

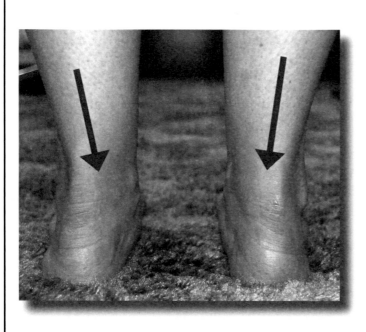
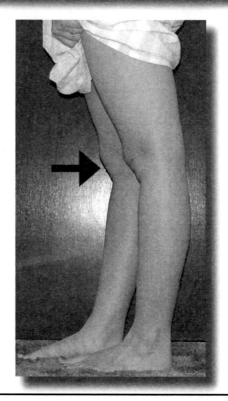

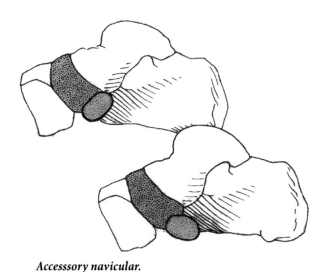

Accesssory navicular.

The accessory navicular is usually asymptomatic, but it may be traumatized to cause pain. Injury usually occurs from repeated small injuries during recreational athletics. Conservative treatment is directed toward obtaining optimal foot and ankle function, as indicated in this and the next chapter; additionally, soft orthotics, rest, and physiotherapy may be of value. If these fail, surgical excision is necessary.

The Kidner surgical procedure (**Prichasuk & Sinphurmsukskul, 1995; Kidner, 1929**) is indicated when there is pain over a prominent accessory navicular bone, or enlarged medial protuberance over the navicular. This procedure removes the accessory navicular and/or re-shapes the navicular, and re-attaches the tibialis posterior tendon. Early in Kidner's use of this procedure, he observed a marked improvement of a flatfoot condition in a patient from whom an accessory navicular had been removed. During the operation the tendon of the tibialis posterior, which originally attached to the accessory navicular, was changed to attach to the navicular as it would under normal circumstances. Kidner attributed improvement in the flatfoot to the re-directed force of the tibialis posterior, giving a greater lifting effect on the arch. It has been observed that the transplanted tendon, even under general anesthetic, will hold the foot in a correct position. The importance of the tibialis posterior muscle for proper arch support has been noted throughout this volume, and dysfunction in this muscle in AK evaluation often produces many foot, ankle, and knee problems. Secondary to posterior tibialis inhibition there often results tightness of the soleus and gastrocnemius muscles, with resulting pain in the Achilles tendon. Without the manual muscle test, the wide variety of nonspecific symptoms resulting from tibialis posterior dysfunction makes the diagnosis of this disorder far more difficult.

Talipes calcaneovalgus

Talipes calcaneovalgus is a congenital, flexible flatfoot deformity with characteristic findings at birth. (**Paton & Choudry, 2009**) This is pronation of the heel combined with supination of the forefoot. It is congenital, and the suggested cause is intrauterine malposture.

Bresnahan as well as Ferciot, (**Bresnahan, 2000; Ferciot, 1972**) studied the balance of infants' feet in this condition and their propensity to develop flatfoot, found muscle imbalance to be the etiology. The combination muscle imbalance is an overactivity of the tibialis anterior and peroneus brevis, and weakness of the peroneus longus and tibialis posterior muscles. Ferciot states that 5% of all newborns have this condition.

The examination for functional muscle imbalance is part of the applied kinesiologist's training and adds much to the examination and treatment of children and young athletes in particular. (**Cuthbert & Rosner, 2010; Cuthbert & Barras, 2009; Blum & Cuthbert, 2009; Karpouzis et al., 2009; Cuthbert, 2008, 2007; Pauli, 2007; Goodheart, 2003; Maykel, 2003; Mathews et al., 1999; Froehle, 1996; Cammisa, 1994; Mathews & Thomas, 1993**) The examination detects inhibited muscles, compensatory movement patterns, temporal aberrations in muscle firing, and muscles recruited in an abnormal sequence. The mixture of weakness and tightness seen in the muscle imbalances involved in foot and ankle disorders alters body alignment and gait and changes the equilibrium points of the joints. There is a growing body of literature suggesting that athletic performance can be enhanced by chiropractic and other manual interventions. (**Hoskins & Pollard, 2010; Costa et al., 2009; Sandell et al., 2008; Shrier et al., 2006; Schwartzbauer et al., 1997; Lauro & Mouch, 1991**)

Talipes calcaneovalgus is indicated by excessive dorsiflexion of the foot at birth. Diagnostically, the foot cannot be brought down from the dorsiflexed position to below a right angle with the pressure of one finger. The dorsiflexed position is maintained by overactivity of the tibialis anterior muscle, which maintains the foot in supination along with the dorsiflexion. This position overstretches and overpowers activity of the peroneus longus muscle. The tibialis posterior muscle is also elongated.

At approximately one month, an active thrust reflex develops. The triceps surae muscles become active, being resisted by the short tibialis anterior muscle. Ferciot (**Ferciot, 1972**) observes that at four to six weeks of age the peroneus brevis muscle begins to be active, along with the strong triceps surae pulling the heel into marked eversion. At three to four months, if the baby is held in a position where the feet press on the floor, the lateral aspect of the foot does not touch the floor. A fixed supination of the forefoot with eversion of the heel develops.

Ferciot states that the condition is aggravated as babies begin to crawl and weight is borne on the medial aspect of the foot, further causing the forefoot bones to develop in a supinated position. He also states that belly sleeping tends to sustain and aggravate this deformity. Walking is usually delayed in these children until twelve to fourteen months of age, and it is marked by poor balance and a toeing-out position.

Subotnick (**Subotnick, 1973**) states that the calcaneovalgus foot responds well to conservative treatment if initiated early and carried out with persistence. Subotnick begins with manipulative reduction of the deformity. The foot is manipulated into plantar flexion and inversion 15 to 20 times, 3 times daily. If manipulations are not entirely successful, he applies corrective casts to hold the feet in the position acquired by manipulative stretching exercises.

Applied kinesiology muscle balancing techniques should be added to the manipulations and should include both muscle lengthening and strengthening procedures, as discussed later and in other AK references. (**Walther, 2000, 1981**)

Subotnick states that 85-90% of these feet can be adequately corrected if treated in infancy. After eight months of age the prognosis is poor for conservative treatment.

Postural influence

Foot pronation is almost always associated with a postural fault. Either can be primary, with the excessive pronation causing the postural fault or vice versa. (**Prior, 1999**) As with all other conditions, optimal treatment is directed toward the primary fault.

A common postural fault is accentuation of the spinal AP curves, with an anteriorly rotated pelvis and hyperlordosis in the lumbar spine. This may result from failure of the foot's positive support reaction, previously discussed. If the gluteus maximus muscles, which support the posterior pelvis, are strong in the clear, one can easily challenge the positive support reaction by stretching the longitudinal arches and spreading the transverse arch; then re-test the gluteus maximus muscles. If they weaken, it is affirmative evidence that the positive support reaction is dysfunctioning. If the gluteus maximus muscles are bilaterally weak in the clear, an upper cervical fixation is indicated which, when corrected, will immediately cause the muscles to test strong. In the latter case, the postural distortion is probably primarily contributing to the extended foot pronation. In addition to stabilizing the posterior pelvis, the gluteus maximus muscles are external hip rotators. Their weakness allows internal hip and leg rotation, taking the forefoot into adduction. The body accommodates this with additional pronation for forefoot abduction. Although there is generally a primary connection between pronation and remote structural distortion, both conditions are usually present. If the primary condition is not corrected, the secondary problem will continue to recur. For example, if the patient has a facet syndrome from an anterior pelvic tilt and hyperlordosis of the lumbar spine, the primary complaint may be low back pain. If extended pronation and dysfunction of the positive support reaction are present, one will be unsuccessful in obtaining lasting improvement of the low back pain.

Lee (**Lee, 1997**) points out that the consequences of gluteus maximus muscle inhibition (in this example due to failure of the positive support reaction from a foot dysfunction) during gait.

> "The consequences to gait can be catastrophic when gluteus maximus is weak. The stride length shortens and the hamstrings are overused to compensate for the loss of hip extensor power. The hamstrings are not ideally situated to provide a force closure mechanism and, in time, the sacroiliac joint can become hypermobile. This is often seen in athletes with repetitive hamstring strains. The hamstrings remain overused and vulnerable to intramuscular tears."

Bullock-Saxton has also shown that the gluteus maximus muscle was significantly delayed in activation in both the injured and uninjured limbs in subjects following ankle sprain. (**Bullock-Saxton, 1994; Bullock-Saxton et al., 1994**)

The common interaction between foot dysfunction and postural distortion makes it difficult to determine the primary problem. Similarly, the relationship between pain and motor control and muscular inhibition is so intimate that pain and muscle strength may represent two dimensions of a common neural event. (**Melzack, 1999**)

Failure of the positive support reaction, with increased AP curves, may ultimately cause an upper cervical fixation, in turn causing greater weakness of the bilateral gluteus maximus muscles. The patient's primary complaint may be suboccipital headaches. When one examines the prone patient, there is no stress on the positive support mechanism in the feet; it may appear that the upper cervical fixation is the primary condition. When it is corrected, the bilateral gluteus maximus muscles test strong, but as soon as the patient walks on the feet, the correction will probably be lost; the entire complex will return to its original status if the feet are not corrected.

Goodheart (**Goodheart, 1967**) first described psoas muscle weakness as a result of foot pronation. This probably results from the excessive internal leg and thigh rotation that excessive pronation causes. This situation is particularly involved during walking and running. The iliopsoas produces its main activity during the swing phase, but there is a secondary peak of activity at mid-stance. (**Basmajian & DeLuca, 1985**) The latter occurs when the foot is coming out of pronation with external leg rotation. With extended pronation, it appears there is excessive stress on external hip-rotating muscles, such as the psoas. The psoas insertion on the lesser trochanter gives minimal leverage for external rotation, and it appears the psoas is working in a vain effort to correct the excessive internal leg and thigh rotation.

As with many conditions in the body, there seems to be reversible interaction between foot pronation and psoas muscle dysfunction. In some cases, psoas muscle dysfunction can cause loss of foot corrections, perpetuating foot pronation. There are applied kinesiology tests that determine and differentiate psoas muscle involvement; these are discussed later in the section on examination of muscle involvement.

Examination

The incidence of flat foot (pes planus) is 20% in adults, the majority of which are flexible. (**Chaitow & DeLany, 2002**) The previous discussions of general examination and body language of the foot provide the basis for foot examination. Here are more specific factors of the pronated foot and extended foot pronation during gait.

The first impression of excessive pronation comes from visual observation, as described in the section on body language of foot dysfunction. What may appear to be flat-footedness needs to be put into proper perspective. An infant's foot has some arch, but it appears flat because of the fat pad. **(Langer, 2007)** Full arch development comes with foot use. A study by Staheli et al **(Staheli, 1986)** showed that the medial longitudinal arch usually becomes evident during the first decade of life. Persistent flatfeet into the teenage years are still within normal range.

Flattening of the arch is not pathognomonic of extended pronation. An individual can have a high arch that is pronated, or the foot can be flat without extended pronation. **(Aronow & Solomone-Aronow, 1986)** To be considered physiologically normal flatfeet, there must be the abundant amount of subcutaneous fat and joint laxity, as opposed to the pathologic, rigid flatfoot that requires treatment. Staheli et al. **(Staheli et al., 1986)** state, "The common practice of 'treating' physiologic flatfeet with shoe modifications, orthotics, or surgery is unnecessary and inappropriate."

Observation of the calcaneal and other bone positions gives indication the foot has moved into pronation. Also, both shod and unshod gait analysis should be done to determine if pronation is extended, and whether walking causes detrimental effects to remote body function.

Static stance. In this discussion, static stance refers to standing still. It is acknowledged that there is no absolute static standing posture, because there are normally small balancing movements.

Observe for three basic factors in the static stance: 1) eversion of the calcaneus, 2) depression and strain of the medial longitudinal arch, and 3) medial bulging of the talonavicular articulation.

Looking at an individual from behind, one observes first for Helbing's sign, **(Dorland's Illustrated Medical Dictionary, 2007)** which is a medial bowing of the Achilles tendon, i.e., the convexity of the curve faces medially. This is most easily observed by visualizing an imaginary line down the center of the leg into the center of the Achilles tendon, continuing into the center of the calcaneus. When Helbing's sign is positive, there will be a break in the line as it deviates laterally into the calcaneus. Confirming Helbing's sign is an appearance of the calcaneus in eversion.

Usually with foot pronation there is depression of the medial longitudinal arch in static stance. As previously noted, a high arch can be pronated in static stance and have extended pronation. If so, there will be tension on the plantar fascia in the weight-bearing position. Normally the plantar fascia is at its greatest tension with metatarsophalangeal extension, which tightens it by the windlass mechanism. In the case of a high arch, the examiner can place his fingers under the medial longitudinal arch and palpate for tension. Have the patient externally rotate his leg to put weight on the lateral longitudinal arch. If there is considerable relaxation of the plantar fascia in this position, pronation is probably present. There will usually be excessive tenderness of the medial longitudinal arch ligaments and of the plantar fascia with digital pressure.

In pronation the talus adducts, moves anteriorly, and plantar flexes in its relationship to the calcaneus. **(Lau & Daniels, 1998; Ramig, 1977)** As the patient is viewed from behind, this creates an appearance of more foot medial to the leg than lateral. The entire appearance from behind is due to Chopart's joint moving medially, which causes a bulge of the talonavicular articulation and a loss of the straight line along the lateral aspect of the foot, with a break at the calcaneocuboid articulation. This causes the forefoot to appear abducted. **(Prior, 1999; Aronow & Solomone-Aronow, 1985)**

Dynamic evaluation. A patient should be evaluated for pronation in a relaxed, normal gait, both with and without shoes. There should be enough steps included in the walk for normal gait to develop. The limited space in most offices prohibits this unless a hallway is used. A treadmill has some value in evaluating for pronation; however, it must be

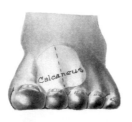

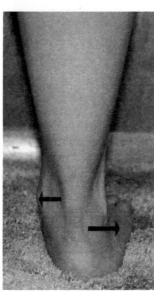

Helbing's sign and calcaneal eversion with Chopart's joint moving medially.

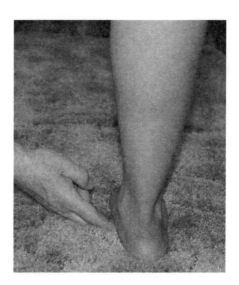

Palpation of medial longitudinal arch while standing

realized that the foot does not strike the moving treadmill in the same way it does solid substrate. After the patient walks on the treadmill for some time, the gait becomes more natural. Evaluating leg and foot motion is similar, whether the patient walks on a treadmill or the substrate.

Movement throughout the body should be symmetrical. Observe the temporal pattern of foot movement during the gait cycle. When the barefooted patient walks, one may observe tendons of the toe extensors rise to aid the tibialis anterior with foot dorsiflexion against a tight triceps surae. (**Cailliet, 1997; Craik & Oatis, 1995; Milgram, 1983, 1964**) This is not always present, but it is a sure sign of extended pronation. If the foot does not yield to a short triceps surae, the individual may walk with a slightly flexed knee. (**Dananberg, 2007; Subotnick, 1975**)

When there is hypermobility of the midfoot with a normal Achilles tendon and triceps surae, heel-off is delayed, causing the arch to maintain flatness for a longer time with excessive medial bulging at the talonavicular area. The flatfoot position may be maintained almost to the point of weight being transferred to the metatarsal heads. As the toes begin to extend, the windlass mechanism tightens the plantar aponeurosis to give the arch some stability, aiding the heel in rising. This is in contrast to the rigid lever the foot should become at 50-65% of the stance phase.

More broadly, with foot pronation the talus dips downward and flattens the medial longitudinal arch due to its tri-planar motion being disturbed. In extended pronation, there is increased force along the medial side of the foot near the time of toe-off. (**Prior, 1999; Mann, 1983**) This causes push-off to appear to be more from the medial side of the hallux. Hallux valgus or a callus on the medial plantar aspect of the hallux confirms chronicity of this condition.

In pronation, heelstrike is usually lateral. As full weight is borne by the foot, it appears to flop over onto the medial side. The lateral strike of the heel appears to follow the rule that for every action there is a reaction. The pronated foot, as it goes into the swing phase, inverts so that heelstrike is on the outer, slightly posterior, portion, with a quick snapping back into pronation as the stance phase begins. The lateral heelstrike is confirmed by excessive shoe wear in that area.

As the foot goes into extended pronation, there will be excessive internal patella rotation at the same time that the foot appears to flop onto the medial arch. Immediately after toe-off, one may see the patella quickly move laterally.

Palpation for pain

When pronation is extended during walking, there are specific areas in which the ligaments are constantly strained if the foot is not completely broken down. Exquisite tenderness of these areas on digital pressure is indicative of excessive pronation. Evaluation for this pain is usually done in combination with evaluating joint motion and challenging the joints.

If there are other signs of excessive pronation and there is no tenderness of the ligaments, they have already been strained and stretched to the point that they no longer maintain the arch. Thus, *painful ligaments are a good prognostic sign* for the possibility of regaining foot integrity without permanent use of orthotics to provide support where there is ligament insufficiency. (**Solomonow, 2009**)

With collapse of the medial longitudinal arch, the head of the talus will drop medially and inferiorly. It will be palpable between the medial malleolus and the navicular tubercle. There will probably be exquisite tenderness in that area. In the individual without foot pronation, the head of the talus is not usually palpable. Digitally press into the plantar area of the talonavicular and cuboideonavicular articulations. If the arch is in strain, these ligaments will be tender.

AK versus Palpation in Foot-Ankle Diagnosis

An interesting comparison of the ideas of Mennell in the examination of the foot with applied kinesiology methods can be made. (**Mennell, 1964**) Mennell stated that it was important to differentiate subtalar problems from those involving the talotibiofibular problems. He differentiated these by stabilizing various components of these complex joint surfaces and introducing movements into them. "If these movements are full, free and painless, there is obviously no pathological condition of this joint…Pain in the performance of this movement indicates [dysfunction] giving rise to pain at the subtalar joint." The applied kinesiology challenge procedure introduces similar forces into the patient's sensorimotor, articular, and ligamentous system of the foot and examines for changes in muscle function. This permits the examiner to bring Panjabi's stability model (**Panjabi, 1992**) into the examination, and does not leave the examiner dependent upon subjective pain and subjective palpation sensations for diagnosis of disturbances in the foot. In the AK approach, if the challenge produces weakness in a strong indicator muscle or strength in a weak indicator muscle, then evidence has been provided that confirms the patient's complaint of pain and muscular dysfunction (local or remote). Additionally, one of the consequences of the dysfunction found (muscle inhibition) can be reassessed after correction for improvements. The reliability of the manual muscle test is superior to those for both motion and static palpation. (**Cuthbert & Goodheart, 2007**)

The muscles are also in strain in their effort to counteract extended pronation. Microavulsions of the tibialis posterior muscle have been described. **(Holmes et al., 1990)** The major area of tenderness is at a muscle's insertion. **(Mense & Simons, 2001)** The tibialis posterior has its major insertion into the tuberosity of the navicular bone, and the tendon spreads out to insert into the plantar surface of the 1st cuneiform, bases of the 2nd, 3rd, and 4th metatarsals, and cuboid bones. It also has a slip that inserts onto the plantar surface of the sustentaculum tali. The tibialis anterior inserts into the medial and plantar surfaces of the 1st cuneiform bone and the base of the 1st metatarsal bone. Both of these muscles are more active when there is excessive pronation **(Gray's Anatomy, 2004; Gray, 1969; Gresczyk, 1965)** in an effort to control it. Usually there will also be tenderness at the origins and the muscle bellies.

With the patient weight-bearing, evaluate pain on digital pressure at the insertions of the tibialis posterior and anterior. Have the patient externally rotate his leg to take his foot out of pronation. Some diminishment of pain on digital pressure is indicative of excessive pronation, because these muscles are active in the standing position when pronation is present; they are not in a normal foot. **(Gray's Anatomy, 2004; Gray, 1969; Gresczyk, 1965)**

In excessive pronation the talus moves medially, anteriorly, and plantar flexes in its relationship to the calcaneus. **(Prior, 1999; Ramig, 1977)** In doing so, the trochlear surface internally rotates, carrying with it the tibia and fibula. This causes a strain at the ankle mortise, manifested as tenderness of its ligaments. Digital pressure applied around the internal and external malleolar ligaments will usually elicit pain when there is excessive pronation.

Joint motion and challenge

Treatment of excessive pronation includes joint manipulation, correction of muscle malfunction, and possibly structural support, physical therapy measures, and rehabilitation. The first order of importance is to correct subluxations, fixations, and muscle dysfunction. The order in which these corrections are made depends on the primary or secondary nature of the dysfunction, which is easy to determine with the applied kinesiology techniques of challenge and therapy localization. **(Walther, 2000)** Often examination techniques of muscle testing, joint motion palpation and challenge, skin challenge, therapy localization, and possibly even other AK techniques — such as respiration assist during evaluation — are used in combination. First the joints and then the muscles are considered, followed by the interaction of various structures and functions.

Evidence of foot subluxations or fixations is usually elicited first by the shock absorber technique, described earlier. Joint challenge is applied to evaluate limited or excessive motion of an articulation, and to determine if movement of the joint stimulates joint receptors to adversely affect muscle function, as determined by manual muscle testing. In many cases, the neurologic and motion aspects can be evaluated at the same time by joint challenge. In other cases, it is necessary to specifically stabilize a bone to determine excessive motion or fixation.

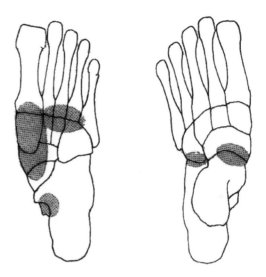

Palpatory pain locations with extended pronation.

Care must be taken to adequately stabilize the navicular bone when evaluating motion between the cuneiform bones and the navicular. **(Gillet & Liekens, 1981; Mennell, 1964)** Failure to do so may lead to evaluating motion at the talonavicular articulation, which is often hypermobile. This may cause one to miss immobility between the navicular bone and cuneiforms. The hypermobility of the talonavicular articulation results from foot hyperpronation when there is lack of dorsiflexion from short triceps surae muscles.

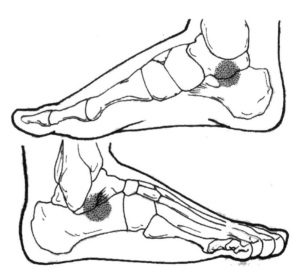

Pain to digital pressure around internal and external malleoli in extended pronation condition.

It is necessary to have motion between all articulations of the foot to allow for flexibility, springiness, and resilience of the foot to meet the substrate. The arch that is fixed, whether it is high or flat, is abnormal. **(Maffetone, 2003; Gillet & Liekens, 1981; Mennell, 1964)**

To examine motion of the subtalar articulation, the body of the talus can be locked between the tibia and fibula. This is accomplished by maximum dorsiflexion if the plantar flexor muscles are not too tight. The anterior wide wedge of the talus body locks between the tibia and fibula as it spreads them with dorsiflexion. **(Cailliet, 1997)** In excessive pronation, especially that associated with a short triceps surae, there is excessive motion in the subtalar articulation. **(Logan, 1995; Harris & Beath, 1948)** This increased range of motion, along with increased motion at Chopart's joint, allows the heel to come into contact with the ground in spite of the short triceps surae.

Chopart's joint should be evaluated for excessive range of motion by holding the hindfoot solidly and moving the midfoot with the other hand contacting around the proximal metatarsals. As mentioned, excessive range of motion in any of these joints is often present with excessive pronation. Prognosis for correcting the foot without the use of support lessens with increased excessive joint motion.

Motion between the metatarsals should be evaluated at both the proximal and distal ends. Although there is no actual joint between the distal metatarsals, there is often restricted movement that must be mobilized. Both ends of the metatarsals can be evaluated for motion with a scissors-type action. At the proximal end, the physician contacts the bases of two adjacent metatarsal bones with his thumbs and forefingers and attempts to move one plantarly and the other dorsally. The heads are contacted at the distal end in a similar manner, with one metatarsal moved into flexion and the other into extension.

With excessive pronation, there is nearly always a subluxation of the talus in the ankle mortise. Applied kinesiology challenge indicates that a lateral-to-medial adjustment is often needed. Because the talus moves medially in pronation, this may seem incorrect. The more common lateral subluxation may be due to the body's effort for correction, or from the excessive supination during the swing phase of gait. In any event, best results are obtained when the talus is adjusted in the direction of challenge that strengthens associated weak muscles. When challenging the talus, contact should be immediately under the malleolus. This is especially important when challenging from lateral to medial because of the limited amount of talus extending below the malleolus.

Muscle evaluation

Evaluating the muscles is done with manual muscle testing and range of motion for the triceps surae. An equinus foot occurs when there is inability to reach 6°-8° of ankle dorsiflexion. **(Charles et al., 2010; Logan, 1995; Subotnick, 1975)** Runners tend to overdevelop the calf muscles compared to the anterior ones. Sometimes they are able to get by for a while, because pre-running stretching procedures increase dorsiflexion to the minimum 10°. There are many other situations responsible for the short triceps surae, e.g., high-heeled shoes, prolonged sitting, general poor conditioning, and many of the remote conditions treated by applied kinesiology.

Myofascial pain in the calf muscles has been documented to cause a biomechanical abnormality of gait, resulting in an excessive knee flexion angle during the stance phase of gait **(Wu et al., 2005)**. Adequate ankle dorsiflexion (>10°) is also necessary in midstance so that the tibia can advance over the foot permitting forward body movement **(Norkin and White, 2003)**. If this ankle ROM is restricted by inhibited or hypertonic musculature, compensation may occur in the form of genu recurvatum, early knee flexion, early heel lift or excessive pronation at the subtalar joint **(Prior, 1999)**. These compensatory mechanisms put undue stresses on many structures superior to them throughout the leg and may lead to foot disorders such as plantar fasciitis, Achilles tendonitis and metatarsalgia **(Hill, 1995)**. Increased subtalar joint pronation has also been identified as a "remote" contributing factor in patellofemoral pain **(Crossley et al., 2006)**.

A short Achilles tendon may be present in children; when present, the range of motion should be increased even if the child has no symptoms. **(Chen & Greisberg, 2009; Wickstrom & Williams, 1970)** As some of the general flexibility usually present in children is lost, symptoms and increased pronation will probably develop. When there is a short triceps surae in children and adolescents, the arch can be as often elevated as depressed, and there is slight pes valgus. Clawing and a tendency toward hammertoes develop, often to the point that there is negligible activity of the toes in weight bearing. **(Kwon et al., 2009; Milgram, 1983)**

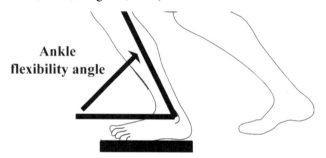

Measurement of loaded dorsiflexion.

When measuring dorsiflexion at the ankle, the measurement should be limited to motion at the ankle mortise. A minimum dorsiflexion of 10° at the ankle mortise is necessary for normal function; **(Charles et al., 2010; Subotnick, 1991, 1975, 1971)** in the infant there should be 30°. **(Subotnick, 1973)** There is usually increased laxity of ligaments in the subtalar and mid-tarsal articulations in the hypermobile flatfoot. In the normal foot, ligaments limit mid-foot dorsiflexion. **(Hamilton, 1985; Hicks, 1951; Manter, 1941)** Ligament breakdown allows increased range of motion that often conceals limited dorsiflexion at the ankle. The examiner should grasp the calcaneus and midfoot to stabilize the calcaneus, navicular, and cuboid bones against the talus before moving the foot into dorsiflexion. **(Harris & Beath, 1948)** Examination of ankle dorsiflexion should be done in subtalar inversion and in neutral. **(Prior, 1999; Milgram, 1983)** Limitation in either test is evidence of a short triceps surae.

When the patient's knee is maintained in extension, the soleus or gastrocnemius or both can limit dorsiflexion. When the knee is flexed, only the soleus can limit dorsiflexion. In the former case, stretching procedures are done with knee extension; in the latter they are done with knee flexion.

Evaluating range of dorsiflexion is a static test that

does not indicate how the muscle will elongate during an eccentric contraction. **(Prior, 1999)** There is no specific test to determine this. Cramping of these or other muscles at night, after physical activity, or during muscle testing is indication for further systemic evaluation, probably of calcium metabolism.

One is often amazed at the limited amount of motion at the ankle. The foot may fail to reach a right angle with the tibia by as much as 25°. In normal feet and ankles, dorsiflexion may reach 20° beyond the right angle. **(Charles et al., 2010; Logan, 1995; Harris & Beath, 1948)**

In impairment rating, 20° of dorsiflexion is considered normal, with a 4% impairment rating of the lower extremity when there is only 10° of dorsiflexion. **(American Medical Association, Guides to the Evaluation of Permanent Impairment, 2007)** It must be put into perspective that impairment measurement is of the goniometer base placed in alignment with the axis of the tibia; the degree of dorsiflexion is based on the goniometer arm placed parallel to the sole of the foot. This includes motion at the subtalar and midfoot joints, which gives a varying amount of additional dorsiflexion. When range of dorsiflexion is used to consider the length of the triceps surae, only motion at the ankle should be measured.

Although it is important to evaluate the range of motion at the ankle for assessing the triceps surae length, it is difficult to obtain an accurate measurement of the degree of foot dorsiflexion. The method recommended by Charles et al. **(Charles et al., 2010)** and Lindsjo et al. **(Lindsjo et al., 1985)** provide reproducible measurements of foot dorsiflexion and plantar flexion. This is necessary for assessment of the progress from stretching procedures, discussed later. Their method measures the range of motion in a weight-bearing position. An individual's foot is put on a 12"-18" (30-46 cm) high stool. He then leans forward to dorsiflex the ankle to the maximum amount while carrying most of the body weight on the side being evaluated. To obtain reproducibility of the measurement, no rotation is allowed in the lower leg; internal rotation increases the range of dorsiflexion by allowing increased pronation. A goniometer is used to measure the angle between the support line of the foot and the long axis of the leg. The amount of dorsiflexion measured by this method will be greater than that obtained when the movement is limited to the ankle, because foot breakdown is included in the former.

Plantar flexion is measured in a similar manner; however, the individual plantar flexes the foot to the maximum amount while maintaining as much body weight on the foot as possible. Again, the measurement is made between the plantar surface of the foot and the long axis of the leg. This method can be combined with x-ray to determine if the mid-arch is breaking down with triceps surae stretching procedures. (X-ray procedures are described next. Triceps surae stretching is described in the foot and ankle rehabilitation section.)

Psoas inihibition & foot pronation

Psoas muscle dysfunction can develop from foot pronation. **(Greenawalt, 1992)** In some cases, the muscle weakness can perpetuate extended pronation by causing loss of foot corrections. When psoas muscle weakness is secondary to excessive pronation, one will often find its neurolymphatic, neurovascular, and other reflexes active. When indicated, treatment to these reflexes will cause the psoas muscle to test normal; however, when the muscle weakness is due to pronation, there will nearly always be a return of muscle dysfunction as soon as the individual stands and walks in his customary manner.

To determine if foot pronation is indeed responsible for psoas dysfunction, one can correct factors influencing the muscle without any foot correction. This may include reflexes, spinal subluxations, the muscle itself, or any other associated factor. Have the patient walk in a figure-eight pattern. If foot pronation is the probable cause for the psoas dysfunction, it will again test weak. This is not pathognomonic; other factors may be involved.

Further evidence that pronation may be responsible for psoas dysfunction can be obtained by re-correcting the muscle, again not correcting the foot. Have the patient stand and walk on the outer borders of his feet to avoid excessive pronation. After walking in this manner, even with turning right and left as in the figure-eight walk, the psoas muscle frequently will not weaken if foot pronation is actually the cause of the muscle dysfunction. This seems to support the hypothesis that psoas dysfunction develops with pronation because of the muscle's inability to control excessive leg and thigh internal rotation.

Sometimes foot corrections are lost as soon as the patient walks or runs. Schmitt **(Schmitt, 1988)** describes a method to determine when psoas muscle dysfunction is perpetuating foot pronation. This can be done during the initial examination of the foot. It should always be done when foot corrections are lost after the patient walks. When a positive shock absorber test or challenge to the foot is positive, evaluate the test again while the patient is therapy localizing to a muscle-organ associated point, such as the kidney alarm point, neurolymphatic reflex, or neurovascular reflex. If this cancels the previously positive test, it indicates that psoas muscle weakness may be a contributing factor to the foot pronation, rather than the psoas dysfunction being secondary to pronation. A similar rationale to that given for psoas muscle weakness secondary to foot pronation is applicable in this situation. If the psoas fails in its role of bringing the leg out of internal rotation, the foot must accommodate by pronation. There will often be no return of foot dysfunction after correcting the psoas muscle. The psoas muscle is the one most often involved in this relation, but other external thigh muscles, such as the piriformis or gluteus maximus, may have the same relationship.

Important muscles stabilizing the arches during gait are the intrinsic muscles of the feet, which receive their nerve supply from branches of the posterior tibial nerve after it has traversed the tarsal tunnel. If there is peripheral nerve entrapment at the tunnel, there may be secondary atrophy of the intrinsic muscles contributing to foot problems. Removing the entrapment and rehabilitating the muscles are necessary if atrophy is present (**The tarsal tunnel syndrome was discussed earlier in *Peripheral Nerve Entrapment of Lower Extremity*).**

At this point in the examination, the tibialis and peroneal muscles should be evaluated with standard manual muscle testing. With knowledge of joint mobility and function of the major muscles of the foot and ankle, one can begin the corrective procedures resulting from excessive pronation.

X-ray Evaluation

Postural lateral x-ray of the foot is taken in a specially constructed cassette holder, with full weight on the foot being examined. This is done by relaxing the contralateral knee (Hanch position). The central ray is parallel to the floor and aimed at the anterior superior corner of the cuboid. **(Yochum & Rowe, 2004; Logan, 1995; Aronow & Solomone-Aronow, 1986)**

A line along the talonavicular and calcaneocuboid articulations should represent a gentle reverse "S" curve, called the cyma line. It is seen on both the lateral and AP projections, and separates the forefoot from the hindfoot. **(Hlavac, 1967)** In a normal foot, the curve is unbroken and undistorted. In the lateral projection, it is shallower in a high arch than in a low one; in a pronated foot, the curve assumes a question-mark shape as a result of the forward slide of the talus in relation to the calcaneus during its triaxial motion.

For an AP projection, the cassette is in the holder parallel with the weight-bearing patient's plantar surface. The central ray is aimed at the base of the 2nd metatarsal, with the tube angled at 10°-15° off vertical, depending on whether the arch height is high, medium, or low. This keeps the ray perpendicular to the metatarsal bone shafts.

Hypertrophy of the two medial metatarsals may be seen as compensation for long-standing extended pronation. This is more evident in the 2nd metatarsal, in which the cortex will be hypertrophied and its diameter increased by about half the normal.

Tarsal coalition. Standard foot x-rays will not usually demonstrate a calcaneonavicular coalition. It is generally observed by the 45° medial oblique projection. The central ray is directed to the center of the everted midfoot. This is a valuable view for studying the tarsus and the tarsometatarsal articulations. **(Yochum & Rowe, 2004; Murray & Jacobson, 1977)** Caution is necessary in interpreting the 45° oblique view for the calcaneonavicular bar. Incorrect positioning may cause bony overlap and simulate a bony coalition. **(Yochum & Rowe, 2004; Elkus, 1986)**

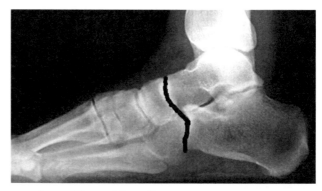

Lateral Normal Cyma line

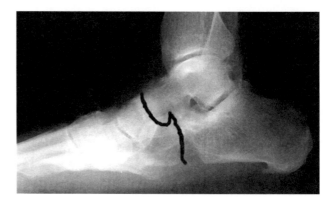

Lateral Broken Cyma line of pes planus

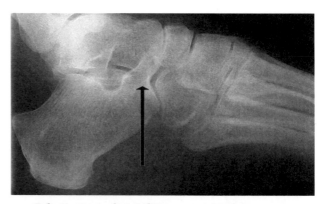

Calcaneonavicular coalition, ant-eater sign

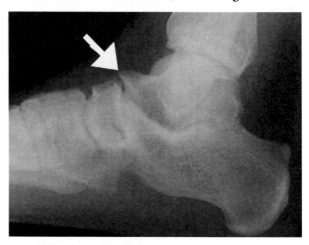

Talocalcaneal coalition

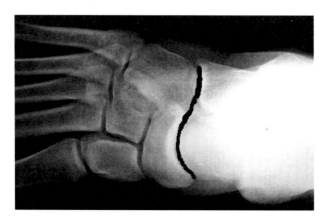

Normal Cyma line

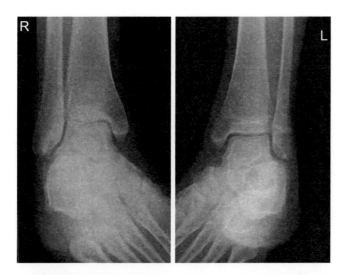
Medial oblique projection of foot

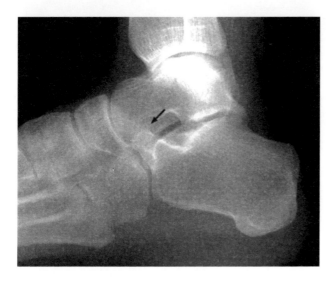
The arrow indicates the location of calcaneonavicular coalition when it is present.

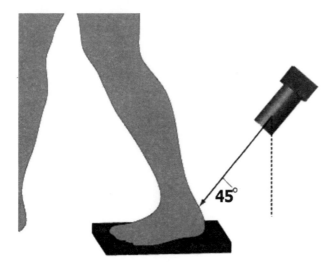
Harris Beath X-ray position.

A view described by Harris and Beath (**Harris & Beath, 1948**) is used to visualize the talocalcaneal coalition. It is an axial view with the central ray angled at 45° and centered between the two malleoli. The properly angled central ray passes between the sustentaculum tali and neck of the talus, (**Murray & Jacobson, 1977**) clearly showing the joint space. When there is a talocalcaneal bridge, the joint is obliterated. This view, or slight variations of it (40°-50 if necessary), will usually reveal a bony coalition of the medial or posterior facet. X-rays of both feet are taken simultaneously to compare symmetry. The talocalcaneal joint angle can be quite variable. When adequate visualization of the joint facet is not obtained, lateral standing views are used to obtain the precise joint angle to determine the exact projection. (**Yochum & Rowe, 2004; Percy & Mann, 1988**)

Subtle changes must be sought in considering fibrous and cartilaginous unions. These changes include close proximity of the two bones, irregular and sclerotic articular surfaces, and hypoplasia of the talus head. (**Conway & Cowell, 1969**)

Plain x-rays are usually the first approach when investigating for tarsal coalition, with computerized tomography reserved for questionable cases. (**Yochum & Rowe, 2004; Elkus, 1986**) The latter can distinguish between bony, cartilaginous, and fibrous fusion. CT is the choice method for talocalcaneal coalition; it is not as useful in other types. (**Logan, 1995; Pineda et al., 1986**)

Tibiotalar motion. Ankle dorsiflexion and plantar flexion range of motion can be measured at the tibiotalar articulation on lateral x-rays. (**Weseley et al., 1969**) The transverse anatomic axis of motion can be established on the same x-ray. This is the amount of articular surface available for dorsiflexion and plantar flexion. Usually dorsiflexion is limited by a tight triceps surae; however, it can be limited by a bony abutment. Arthroplasty may allow 10°-15° of additional dorsiflexion. (**Lillich & Baxter, 1986**) Using this method, one can determine if limited dorsiflexion is due to bony encroachment or caused by soft tissue limitation, such as a short triceps surae.

The accuracy of this method is limited by the inherent problem of measuring movement of a three-dimensional structure with two-dimensional plain x-ray, and by potential placement and position errors of the patient's foot and leg. (**Yochum & Rowe, 2004**) Since the angles being measured are relatively large, the distortion inherent in the x-ray is acceptable and certainly gives more accurate information than any of the goniometer methods of measurement. Placement and position errors can be controlled by carefully following a protocol to enable comparative studies to be made. Placement of the patient refers to putting the patient's foot in a particular relation to the x-ray central ray to provide a base position from which to measure. This enables one to reproduce the starting point of the examination. Positioning refers to moving the patient into maximum dorsiflexion or plantar flexion without disturbing the base position of the foot, thus maintaining a static source from which to measure.

The patient's foot is placed on a cassette holder approximately 15" high. The surface on which the patient's foot rests must be visible on the resulting x-ray, as it will be the base from which measurements will be taken. In relation to a 10" x 12" film, place the foot so the metatarsal heads will be close to the edge of the film, yet clearly visible. Align the foot so that a line from the center of the ankle mortise through the 2nd and 3rd toe interspace parallels the film, with minimal distance between the lateral malleolus and the film. Most of the body weight is put on the target foot by slightly bending the other knee and using that leg to provide stabilization. The central ray projects perpendicular to the center of the medial malleolus.

Positioning of the patient is in neutral, dorsiflexion, and plantar flexion. The neutral projection requires the patient to stand mainly on the foot being examined, with balanced posture. The dorsiflexion view can be done with the knee flexed or extended. In the flexed position, only the soleus is contributing to the triceps surae dorsiflexion limitation; with knee extension the gastrocnemius is also included. For the former view, the patient maximally flexes at his hip and knee and dorsiflexes at the ankle by crouching down. To include both muscles, the patient's stabilizing leg is put on a bench about the same height as the special cassette holder. It is moved forward as necessary with hip and knee flexion to maintain balance. Take care that the patient does not lift his heel from the bench or internally rotate his tibia. The latter is accomplished by increasing pronation, which is to be avoided.

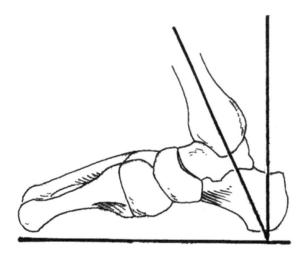

Line drawing dorsiflexion

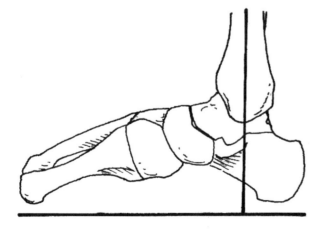

Line drawing extension

The extension view can be taken by having the patient move backward while keeping his foot static; this view is not usually taken.

The potential range of motion is determined from the potential articular surface and the anatomic axis of motion. The anatomic axis of motion is the meeting point of the anterior and posterior margins of the talus' lateral articular surfaces. The anterior and posterior limits of the trochlear articular surface are marked anteriorly at the junction of the neck and trochlea, and posteriorly at the depression adjacent to the posterior talar tubercle. A perpendicular line is drawn through the anatomic axis to measure the potential dorsiflexion and plantar flexion.

Tibiotalar dorsiflexion is determined by the angle of a line from the anterior tip of the tibial articular surface to the anatomic axis with the perpendicular anatomic axis line. In a similar manner, plantar flexion can be measured by projecting a line from the posterior tip of the tibial articular surface to the anatomic axis.

The combined tibiotalar and foot dorsiflexion range of motion is determined by the angle of a line projecting through the center of the tibial shaft and the center of its articular surface with the substrate. This amount is always greater than the tibiotalar motion alone. The difference indicates the amount of midfoot breakdown with extended pronation.

X-ray can be used to determine if the midfoot is breaking down with triceps surae stretching procedures. The foot can be x-rayed in a non-weight-bearing position, weight bearing, and then with full ankle dorsiflexion while weight bearing before starting the stretching program, and after performing it for a specified time.

Joint and Muscle Correction

Ideally, correction is directed toward primary factors. Primary factors can be found by applying numerous examination techniques at once or in succession to determine what eliminates dysfunction. (**Motyka & Yanuck, 1999; Schmitt & Yanuck, 1999**) For example, if the tibialis posterior tests weak and challenge to the talus or some other bone causes it to test strong, the subluxation is primary. Although the tibialis posterior could probably be strengthened by stimulating the neurolymphatic or neurovascular reflexes or with other types of treatment, it very likely would immediately lose its correction with walking or some other structural stress to the foot. If so, the primary cause of the tibialis posterior's weakness appears to be the result of improper stimulation to receptors by the subluxated joint. It might be due to the muscle vainly trying to stabilize the talus or some other bone in subluxation, and weakening because of its inability to do so. It is more likely that the neurologic model is correct. This, along with many other hypotheses, remains to be tested in basic research. Therapeutically the corrective sequence allowing corrections to be maintained after the articulation or muscle is stressed is the appropriate approach.

Muscles that control and support an articulation

should be evaluated to determine if they contribute to the development of the subluxation or fixation. Reversing the preceding scenario, one might find a subluxation of the talus by challenge that causes a previously normal muscle to test weak. An example is a lateral-to-medial challenge of the talus that causes a previously strong tensor fascia lata to weaken. Medial ankle support and integrity of the ankle mortise are largely the responsibility of the tibialis posterior. In addition to the associated ligaments, one can therapy localize to factors of the tibialis posterior such as the neurolymphatic reflex. While continuing the therapy localization, re-challenge the talus in the same manner as before and re-test the tensor fascia lata. If the challenge is no longer positive, the tibialis posterior is probably responsible for the subluxation. The same principle applies if the articular problem is a subluxation or a fixation. The talus in the ankle mortise rarely, if ever, fixates, but other areas can have either. In case of a fixation, the muscle(s) antagonistic to the weak one is probably hypertonic or shortened, which maintains the fixation consistent with Sherrington's Law of Reciprocal Innervation. (**Denny-Brown, 1979**)

When evaluating for articular fixations by motion palpation or applied kinesiology challenge, one of the bones must be adequately stabilized, especially when evaluating the cuneonavicular articulations. These are frequently in fixation, but in combination with a hypermobile talonavicular articulation. (**Chaitow & DeLany, 2002; Hammer, 1999; Gillet & Liekens, 1981**) Stabilize the navicular bone, then attempt motion between the 1st and 2nd cuneiform bones and the navicular bone.

Specific muscular dysfunction is found with several foot subluxations. A medial navicular subluxation often causes adductor muscle group weakness, while a lateral cuboid subluxation causes weakness of the tensor fascia lata and/or gluteus medius. These two subluxations appear to relate with improper nerve receptor stimulation, which correlates with lateral or medial sway of the standing extremity as explained under the magnet reaction. To test for tensor fascia lata or gluteus medius weakness resulting from a lateral cuboid, challenge the cuboid from lateral to medial. The tensor fascia lata and/or gluteus medius will immediately test strong with manual muscles testing if the association exists. Likewise, challenging the navicular bone from medial to lateral strengthens the adductor muscles. With different vectors of challenge, the associated muscle(s) performs to varying degrees during the manual muscle test. There will be one specific vector that provides optimal muscle function; this is the direction in which the bone must be adjusted.

These remarks are designed to help make sense of muscular, postural, orthopedic, and palpation findings and to offer an ideal method for confirmation. The changes that occur on MMT with specific challenges to the painful joints demonstrate for the patient and the doctor what is wrong and whether or not a HVLA manipulation has made a difference.

Once a dysfunction has been identified by virtue of a manual muscle, postural, palpation, orthopedic or other test, it is necessary to define precisely what type of dysfunction exists. The affect of this dysfunction upon attaching or remote muscle function is uniquely demonstrated for the doctor and patient by the challenge and therapy localization procedures in applied kinesiology. The associated muscle weakness, easily determined by the manual muscle test, is then investigated with a challenge procedure which seeks to find the articular change that corrects the muscle inhibition. Appropriate angular and pressure stimulation of the articulation usually produces immediate strengthening of inhibited muscles due to the dysfunction.

An interesting finding in some chronic musculoskeletal pathology is bilateral dysfunction in unilateral injury. Bullock-Saxton (**Bullock-Saxton et al., 1994**) found that subjects with chronic ankle sprain exhibit altered muscle activation patterns on both the injured and the uninjured sides. This supports the view that chronic pain is mediated by the CNS and suggests that clinicians remember to consider areas beyond the pain zone when addressing any chronic joint pain.

Articular adjusting. Subluxations and fixations are common in the foot and ankle. Unfortunately, this is also one of the most overlooked and unsuccessfully adjusted areas. Applied kinesiology challenge determines exactly how the correction should be made. With that information, plus an excellent knowledge of foot anatomy, manipulation should be relatively easy. Standard techniques of adjusting the feet can often be used to apply the optimal challenge vector for a successful correction. One advantage of applied kinesiology challenge is that it determines the individual characteristic of a subluxation. Sometimes it is necessary to modify or develop a different technique for adjusting the foot to properly apply the challenge vector for a particular patient.

It is a characteristic of applied kinesiology methods as used in clinical practice to move seamlessly from the gathering of information toward treatment. As a doctor searches for information through the manual muscle test, the appropriate challenge or therapy localization on the involved subluxation can turn "finding" into "fixing" in a moment. One treatment modality follows another as a rather "custom made" application is created that not only varies from patient to patient but should vary from one session to the next for a particular individual as their dynamic condition changes and improves. (**Motyka & Yanuck, 1999; Schmitt & Yanuck, 1999**)

In adjusting the feet the authors have consulted over a dozen books, DVDs, and some 50 articles on specific manipulative approaches to foot dysfunction. With applied kinesiology, you can modify the methods of manipulation to fit the particular needs of the patient. For instance in some standard lateral adjustments of the talus bone, the line of drive is nothing like that illustrated in standard textbooks for the manipulative maneuver. Applied kinesiology permits the discovery of the precise angle of manipulation determined by the sensorimotor challenge system of diagnosis. This information will change the contact points, the line of drive and the patient positioning significantly. The sensorimotor challenge system of diagnosis in AK allows us to be more specific in terms of the manipulative effort. In a bone as complex and multi-planar as the talus is for instance, challenge may show that only one or two of the articulations of the talus bone are suffering from subluxation or fixation. In this case, a broad general manipulation to the talus will not be helpful and may cause iatrogenic problems. In a case where a small portion of the talus is in a subluxated state with the adjoining bones, ligaments and muscles, a general manipulation that creates great crepitus may be inducing further trauma into the already injured joint. **(Ndetan et al., 2009)**

The most common factor causing corrective effort failure is not obtaining proper relaxation in the foot prior to the manipulative thrust. Only a small force is required for effective correction when proper relaxation has been obtained. Most chiropractors can relate to obtaining proper relaxation when adjusting the cervical spine. Sometimes the doctor can feel the patient signal the proper time by a sensation of "letting go" as relaxation is obtained. At other times it is necessary to distract the patient from the cervical area by having him place his hands on his abdomen, visual synkinesis just prior to the HVLA manipulation, or otherwise drawing attention away from the cervical spine.

Similar relaxation is required in the foot/ankle area prior to an adjustive thrust. Patients tend to hold the foot rigid while contact is being made prior to the adjustment. Sometimes it is adequate to simply say to the patient, "Let me have your foot; just make it loose." Other methods include having the patient move the foot against resistance and then letting it go, or distracting the patient by having him move an upper extremity. The latter method, however, is not as successful as when used to help relax the cervical spine. This is probably because of reciprocal inhibition when the arm and shoulder move toward the abdomen and the neck extensors - especially the upper trapezius - relax. The same principle of reciprocal inhibition can be used to aid manipulation of the foot. Ask the patient to move the foot in the direction opposite that of the manipulative effort just prior to applying the adjustive thrust. An example can be seen when the calcaneus is to be moved inferiorly and anteriorly, as in a typical tarsal tunnel syndrome. Have the patient dorsiflex his foot, then make the adjustive thrust as the patient relaxes to neutral; this causes reciprocal inhibition of the soleus and gastrocnemius.

It is necessary to recognize the importance of a correct vector of adjustment. Challenge provides information for the precise direction in which the correction should be made. Applying the technique to meet that need provides better results and makes adjusting easier. The only disadvantage is that sometimes the standard techniques do not precisely fit the pattern required by the patient's foot. Standard techniques can very often be adapted to fit the needs of the patient's joint. Manipulation of some of the smaller tarsal bones and their articulations is often best accomplished with the activator or Arthrostim instruments. **(Fuhr, 2008; IMPAC, 2012)** Sometimes it is necessary to develop an entirely new technique to meet the conditions presented.

It is not necessary to obtain an articular sound of separation when adjusting an articulation of the foot or ankle. Whether or not articular release is obtained, one should always re-evaluate the articulation with therapy localization and challenge to determine the success of the attempt. Continuing to attempt to obtain articular release when therapy localization and challenge are negative does not aid in the correction; the only result can be joint trauma.

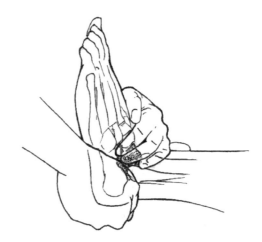

There is limited talus contact area inferior to the malleolus for challenge and adjustment.

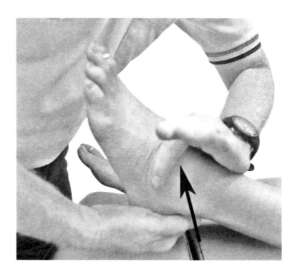

Contact for talus adjustment.

Described here are some standard techniques commonly used. Many manipulative techniques have been described in the literature. **(Bergmann & Peterson, 2010; Leaf, 2010; Greenman, 2003; Chaitow & DeLany, 2002; Hammer, 1999; Goodheart, 1998-1964; Logan, 1995; Hearon, 1981; Gillet & Liekens, 1981; Gertler, 1981, 1978; Stierwalt, 1976; DeJarnette, 1973)**

Lateral or medial talus. Most cases of excessive foot pronation require an adjustment of the talus in the ankle mortise. The direction of correction is usually lateral-to-medial, with a slight superior vector at the end of an extension maneuver. If the talus is subluxated medially, the correction is the reverse of that described here, which is for a left lateral talus.

With the patient supine, the physician cradles the calcaneus in his right fingers and palm. The thenar eminence wraps tightly under the lateral malleolus to contact the talus. It is emphasized that the contact must be as tight under the lateral malleolus as possible because of the limited lateral exposure of the talus. The physician's left hand contacts the dorsum of the foot. The little finger contacts the talus just distal to the tibia. Again, this contact should be as tight against the tibia as possible for the major contact to be on the talus rather than distal to Chopart's joint. When proper patient relaxation has been obtained, the manipulation is performed in two quick maneuvers. First, both hands perform traction on the foot, with the left hand directing its force into the talus to open the ankle mortise. At the end of the traction thrust the physician's right hand moves the talus medially and slightly posteriorly, resulting in slight eversion of the foot.

Rotated cuboid. The cuboid often has a rotational component. It rotates about its longitudinal axis so that the plantar surface of the cuboid rotates laterally and the medial surface drops plantarward. Positive challenge that strengthens a weak associated muscle is a medial vector on the lateral plantar surface of the cuboid or a dorsal vector on the medial plantar surface. When challenging the cuboid, use the base of the 5th metatarsal as a landmark.

Correction of the rotated cuboid often requires adjustment in two steps. It is usually best to correct the lateral component of the plantar surface first. With the patient supine, the physician cradles the left calcaneus in his right palm. The left middle finger contacts the lateral dorsal surface of the cuboid. The index and ring fingers slightly overlap

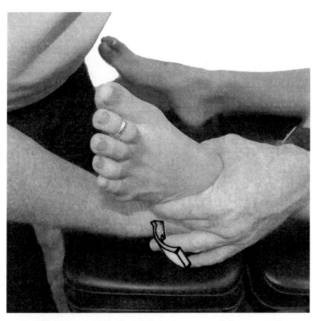

Adjustment of the lateral component of the rotated cuboid subluxation.

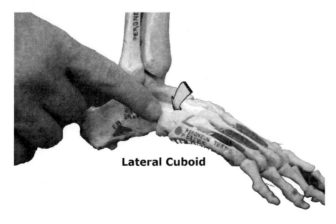

Contact for lateral cuboid correction

the middle finger to provide extra support. The rest of the physician's right hand wraps around the foot with the thenar eminence over the left middle finger, which covers the lateral dorsal aspect of the cuboid. While maintaining a firm contact on the cuboid with a firm wrist and hand, the physician adducts his right elbow to transmit the force through the wrist into the lateral superior aspect of the cuboid.

Sometimes the medial component of the rotated cuboid can best be adjusted by contacting the medial plantar surface of the cuboid with an activator or Arthrostim instrument. **(IMPAC, 2012; Fuhr, 2008)** This may be needed following the adjustment described above when there is still positive challenge on the medial plantar surface of the cuboid bone, and challenge to the lateral portion of the bone is negative. The adjustment is simple; all that is needed is proper alignment of the instrument in the optimal direction of challenge.

Lateral cuboid. The supine patient's left calcaneus is cradled in the physician's right hand. The physician's left thumb contacts over the lateral cuboid, and the thenar eminence of his right hand provides support over the thumb. The rest of the physician's left hand broadly contacts over the dorsal surface of the forefoot for general support.

The patient's leg is raised from the able with approximately 45° internal rotation. In doing so the doctor's arms flex. The corrective thrust is made by rapid contraction of the physician's triceps muscles, similar to a toggle recoil. With slight wrist action the thumb and thenar eminence move the cuboid in the direction of positive challenge.

Inferior cuboid. The prone patient's left knee is flexed, and the physician holds the patient's dorsal foot in his left hand with his fingers wrapping around the plantar surface of the foot to support the metatarsals. The physician's left thumb contacts the inferior cuboid, and the pisiform of his right hand is placed over the thumb. The thrust is a toggle recoil-type in the direction of maximum challenge. The force is directed primarily to the cuboid bone, with support being provided only for the forefoot. Care should be taken to avoid direct torsion into the foot or ankle by twisting the patient's metatarsal bones or hindfoot.

Inferior navicular. The prone patient's left knee is flexed, and the physician cradles the dorsum of the foot in his left hand with his fingers wrapping around the dorsum of the foot. His index finger contacts the talus and his 4th and 5th fingers the cuneiform cuboid, and bases of the metatarsal bones. This leaves minimal contact on the dorsal aspect of the navicular bone by the middle finger. The left thumb contacts the navicular plantar surface, and the pisiform of the right hand is placed over the left thumb. A toggle recoil-type adjustment is made with a rapid triceps contraction. The dorsal contact by the physician's left hand supports and protects the talus and distal aspect of the midfoot.

Superior mid-tarsal bones. The talus, cuneiforms, and/or cuboid may have superior subluxations, i.e., toward the dorsal (anterior) surface of the foot. In some techniques these are adjusted by contacting the dorsum of the supine patient's foot and applying a traction thrust. In most cases it is better to avoid this technique because of the frequent hypermobility of the subtalar articulation and Chopart's joint.

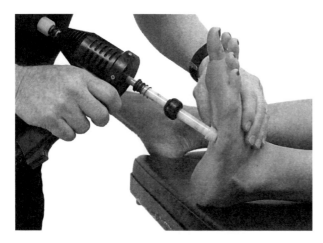

Adjustment of the medial component of the rotated cuboid subluxation.

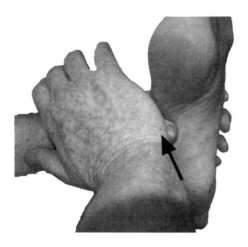

Inferior cuboid adjustment.

An easy correction for superior mid-tarsal subluxations is to have the patient seated, leaning back. And supported by his hands. The hip and knee are flexed, and the foot is flat on the examination table. Contact the superior tarsal bone with a pisiform contact and use a quick toggle recoil-type thrust in the direction of positive challenge.

Another correction very often effective is cradling the patient's foot in any comfortable position and using an activator or Arthrostim instrument (**IMPAC, 2012; Fuhr, 2008**) to direct specific force in the direction of positive challenge.

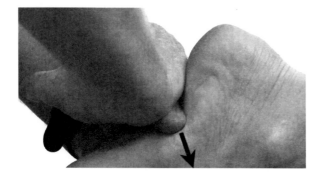

Inferior navicular adjustment.

Inferior cuneiforms. A positive challenge to the 2nd or 3rd cuneiform is usually accompanied by a cuboid, navicular, or 1st cuneiform subluxation. Frequently when these corrections are made, the 2nd and 3rd cuneiform bones will no longer be subluxated. If there is a persistent inferior cuneiform subluxation, it is generally easily corrected by using an Arthrostim or Activator instrument (**IMPAC, 2012; Fuhr, 2008**) on the bone in the direction of positive challenge while the physician cradles the foot in any comfortable position.

Plantar Fascitis and Heel Spurs

Etiology and symptoms

Painful heel is a relatively common condition and the most common cause of heel pain, affecting 10% of the population. (**Crawford, Thomson, 2003**) Plantar fasciitis was first attributed to tuberculosis by Wood in 1812, and since then has been called by many pseudonyms. Plantar fascial insertitis is another, as well as calcaneal enthesopathy, stone bruise, calcaneal periostitis, heel spur syndrome, jogger's heel, subcalcaneal bursitis, subcalcaneal pain syndrome, neuritis and calcaneodynia. (**DeMaio et al., 1993**) As will be seen, plantar fasciitis often presents as a combination of clinical realities rather than one discrete pathophysiological disturbance. For this reason plantar fasciitis should be considered as a syndrome rather than a single condition.

Painful heel is usually caused by plantar fasciitis, and the pain is intensified where the origin of the fascia attaches to the medial calcaneal tubercle.

There may also be diffuse tenderness up the medial or lateral sides of the calcaneus, and is typical of more severe inflammatory processes. Pain is more severe when first rising in the morning. Many patients have a problem putting weight on their feet, and the pain may be severe for the first 50 to 100 steps. Sometimes the pain is so great upon rising that it is impossible to walk; the patient may have to sit on the edge of his bed for 10-15 minutes, slowly putting more weight on his feet with rest periods in between. The pain will eventually decrease to the point that the patient can walk. It continues to gradually decrease with ordinary walking, only to again be increased the next morning.

This pain upon arising after rest may be due to the accumulation of inflammatory byproducts which are compressed into the nerve endings with weight bearing. Another theory is that the muscles and fascia involved in the production of plantar fasciitis are quiescent at night and non-weightbearing, so part of the heel pain occurs when the weight is placed upon the injured muscles and fascia. During periods of immobility, the plantar fascia relaxes and is relieved of its muscular activity and inflammation. The foot basically is re-injured each morning. (**Langer, 2007**)

Generally, plantar fasciitis has an insidious development. Most patients will tolerate the pain in the

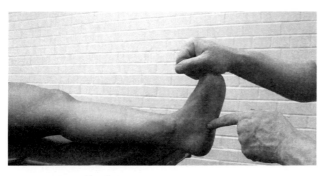

Position of heel spurs

earlier stages, expecting that it will improve. The condition typically intensifies to the point treatment is sought, with approximately 1 million physician visits per year for the condition. (**Riddle & Schappert, 2004**)

Both sedentary and active individuals can develop plantar fasciitis, but it is more common in runners. Lutter (**Lutter, 1997**) reports that 65% of plantar fasciitis sufferers are overweight, with unilateral involvement of the foot most common in 70% of the cases. Approximately 10% of all running athletes have the condition as well, which is about the same rate reporting it in the non-athletic population. (**Bartold, 2004**) In another sports medicine clinic, it is the fourth most common diagnosis. (**James et al., 1978**) The running and jumping in most athletic activities cause the condition to advance more rapidly. A runner's symptoms are further exacerbated by hill running and sprinting. If a runner tries to "run through" his pain during a workout it may improve, only to increase in severity at the end of the workout. To minimize symptoms, a patient tends to maintain his foot in supination from heelstrike to toe-off to avoid tension on the fascia. (**Bartold, 2004; Leach et al., 1986**) This disturbs the gait pattern, the strain of which may cause remote secondary problems.

Plantar fasciitis must be differentiated from generalized trauma to the soft tissue of the calcaneus, plantar bursitis, abductor hallucis myositis, and some types of arthritides. Plantar fasciitis has been associated with differing enthesopathies that occur with connective tissue diseases, particularly rheumatoid arthritis, psoriatic arthritis and ankylosing spondylitis. Men under 40 presenting with bilateral heel pain should be evaluated for Reiter's syndrome and ankylosing spondylitis. The foot is second only to the knee as the site of rheumatoid arthritis symptomatology. When pain is bilateral, the chance is greater that the condition has a systemic background. It must further be differentiated from peripheral nerve entrapment. Of particular interest is entrapment of the medial plantar nerve as it passes under the abductor hallucis muscle. It causes a burning heel pain, aching in the arch, and reduced sensation on the sole of the foot behind the great toe. It is believed to be caused by running with extended pronation, which puts tension on the nerve to stretch it against the tunnel whose superior surface is the navicular bone. It is at this point that maximum tenderness will be found on digital pressure. (**Bartold, 2004; Rask, 1978**)

Painful heel may also be due to tarsal tunnel syndrome. For example, the mixed nerve to the abductor digiti minimi can be entrapped as it rises from the tibial nerve in the tarsal tunnel. Plantar fasciitis and calcaneal spur formation are associated with abductor digiti minimi atrophy on MRI of the foot. (**Chundru et al., 2008**) The nerve courses deep to the plantar aponeurosis near the muscle's origin at the calcaneus. (**Hamilton, 1985**) Many of the foots' vital neurovascular structures are in close proximity to the plantar aponeurosis. Both of these entrapment conditions are usually caused or perpetuated by excessive pronation. (**Cornwall & McPoil, 1999**) (Further discussion of peripheral nerve entrapment in the lower extremity is presented in **Chapter 3**.)

Conditions likely to increase the chance of plantar fasciitis are excessive pronation, cavus feet, leg length inequality, and obesity. (**Cornwall & McPoil, 1999**) When excessive pronation is a factor, it is usually related with a short triceps surae. This causes excessive strain on the plantar fascia, in turn producing chronic inflammation. Running and other athletic activities that stress the foot increase the chance of developing plantar fasciitis, especially in an older athlete. (**Langer, 2007; Bartold, 2004; Hill & Cutting, 1989; Warren & Jones, 1987; Brody, 1987; Krissoff & Ferris, 1979**)

Although plantar fasciitis is common, rupture of the plantar fascia is not often seen. The propensity toward rupture of the plantar fascia may be accelerated by previous treatment of plantar fasciitis with steroid injections. (**Brinks et al., 2010; Leach et al., 1978**) When there is rupture, the patient may report having had a snapping feeling in his foot, along with sharp pain. There is severe localized swelling and acute tenderness. As the swelling diminishes, a palpable defect in the plantar fascia is observed. Conservative treatment is preferable if the plantar fascia ruptures, but occasionally surgical repair is needed. Rupture and partial or complete surgical sectioning of the plantar fascia, may lead to progressive pes planus with associated complications. (**Sharkey et al., 1998**)

Treatment to reduce pressure or shock to the heel is directed toward symptoms rather than the cause of the problem. Two examples of this are use of a heel pad with a depression and shock absorption material. In both cases temporary relief may be obtained, but the basic problem usually continues to get worse. Use of shock absorption material in heel pads or shoe construction may even ultimately increase the problem.

Wallden, Squadrone, Robbins and co-workers (**Wallden, 2010; Squadrone & Gallozzi, 2009; Robbins et al., 1987, 1988, 1989**) present strong evidence that shoes overprotect the foot, isolating it from actively contacting

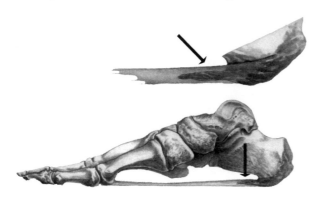

Heel spur

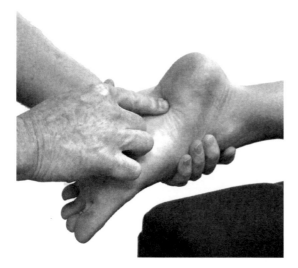

Palpation for signs of heel spurs.

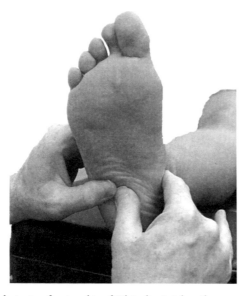

the substrate of natural turf. This diminishes the neurologic stimulation necessary to develop proper arch function. The observation that barefooted runners in international competition have fewer foot injuries stimulated research on the subject. Use of the recently developed shock absorption materials for athletic shoes decreases stimulation to the foot even more.

The need for manufactured shock-absorbing material under the foot can be questioned on the basis of the excellent absorption quality the natural heel pad appears to have. An *in vitro* (**Jorgensen & Bojsen-Moller, 1989**) study found the cadaver heel pad to have significantly greater shock-absorbing quality than 1.0 cm thick ethyl vinyl acetate (EVA) foam or Sorbothane (R) shock absorbers. It seems reasonable that the living heel pad tissue would perform even better. In those who have a painful heel syndrome, there is an increase in the natural heel pad thickness. (**Wearing et al., 2007; Amis et al., 1988**) However, a study by Tsai et al. (**Tsai et al., 2000**) investigated with ultrasound the heel fat pad in patients with plantar fasciitis. They concluded that the heel pad thickness was not altered in the control group compared to subjects with plantar fasciitis.

It may be that other mechanical properties of the older foot and heel pad, for example the relative diminishment of shock absorbency (due to fixations or subluxations of the heel and foot), or to changes in the plantar aponeurosis connective tissue that occurs with aging, may all combine to increase the prevalence of plantar fasciitis with age.

The importance of shock absorption by the heel pad is questioned by Robbins et al., (**Robbins et al., 1989**) who point out that shock absorption of the heel pad diminishes when it "bottoms out" by tissue compression. They conclude "...that the heelpad does not impart significant shock absorption to the body during locomotion, and particularly during running." They propose that the role of the fat pad is protection of the calcaneus by selective tissue deformation. The plantar and palmar surfaces of the feet and hands are similar histologically. They have a pad of adipose tissue with a dense network of fibrous trabeculae providing protection to the bones. At the heel there is less penetration from uneven forces, probably due to tighter tethering of the fibrous trabeculae. The tissue is looser at the metatarsophalangeal articulations for joint movement. Robbins et al. (**Robbins et al., 1989**) suggest that the primary protection from concussive forces to which the foot is subjected comes from plantar sensory feedback. The key factor of the natural heel pad, as opposed to manufactured ones, is that it does not insulate the nerve endings from receiving normal stimulation from the substrate.

In the feet and hands there is a high density of mechanoreceptors and nociceptors, including Meissner's corpuscles that are found only in the digits of higher primates. There is greater sensitivity at the plantar surface of the metatarsophalangeal articulation than at the heel or distal toe. (**Squadrone & Gallozzi, 2009; Robbins et al., 1989**) Muscular reaction varies with stimulation to different areas of the foot. (**Lever, 2006; Nakajima et al., 2006; Kugelberg et al., 1960**) Stimulation to the sensitive metatarsophalangeal area, such as in the later portion of the stance phase, causes reflex contraction of the foot's intrinsic flexor muscles. (**Nakajima et al., 2006; Kugelberg et al., 1960**) They show maximum EMG activity from approximately the time of metatarsal strike until the metatarsals break contact with the substrate. (**Basmajian & De Luca, 1985; Sheffield et al., 1956**) This provides muscular support to the longitudinal arches, and aids the plantar fascia (Hicks' windlass mechanism) (**Hicks, 1954, 1955**) in returning the pronating foot to a solid lever by resupination. There is greater activity in the intrinsic muscles in extended pronation during gait than is apparent in the normal foot. (**Mann & Inman, 1964**) This appears to be the body's adaptation when the plantar fascia is not adequate to resupinate the foot by the windlass mechanism.

Stimulation to the hollow of the medial longitudinal arch causes toe extension. (**Nakajima et al., 2006**) This is an area that receives little or no stimulation when a person is barefooted; in the shod foot it is stimulated, especially when the shoe has arch supports. In the normal foot, arch support may be counterproductive because it interferes with the reflex activity of intrinsic muscle contraction provided by stimulation at the metatarsophalangeal area. (**Squadrone & Gallozzi, 2009**) This occurs when the foot needs the added support of the intrinsic muscles to stabilize the subtalar and mid-tarsal joints, making it a solid lever for push-off. The evidence that sensory receptors and reflexes in the foot and lower leg are functionally linked with the lower back (gluteals, erector spinae muscles, among others) has been well-documented. **(Clair et al., 2009)**

In the normal foot, reflex control of muscles provides the complex adaptation the foot needs to meet the extreme stresses of daily activity. Failure or interference with these mechanisms often results in the painful foot or heel syndrome, and may be an etiological factor in recurrent low-back, shoulder and neck pain syndromes as well. **(Lephart & Fu, 2000)**

Heel spurs have little to do with the pain one experiences; they are simply the result of chronic plantar fasciitis. **(Merck Manual, 2011; Bartold, 2004)** There are no clear studies to show the association of heel spurs and plantar fasciitis. There is no relationship between the size of heel spur and the amount of trouble the patient experiences, **(Lewin, 1959)** nor in his ability to recover. **(Leach et al., 1986)** The heel spur develops in extended pronation because the windlass mechanism increases stress on the plantar aponeurosis. **(Bartold, 2004; Marshall, 1978)** What appears to be stress to the plantar aponeurosis and flexor brevis attachments to the calcaneus is evidenced by a high percentage of positive bone scans at that area. In one study, **(Williams, 1987)** 60% had a positive bone scan; those who did had more severe pain than those with a negative scan did. These authors describe the "saddle sign," which is a concavity in the calcaneus just posterior to a heel spur, if one is present. There is not usually an erosive quality to the saddle sign; rather, it appears to represent a healing of the bone.

Treatment

First it is necessary to differentially diagnose the cause of heel pain. There may be more than one condition present, but frequently the problem will be extended pronation. The function of the plantar fascia during gait is augmented by the dynamic actions of several other extrinsic muscles of the foot. Correcting or controlling extended pronation usually requires strengthening muscles that test weak, adjusting subluxations and fixations, and stretching the triceps surae. Lutter **(Lutter, 1983)** stresses strengthening the tibialis posterior muscle. This muscle is frequently found to be an etiological factor in painful heel syndromes in applied kinesiology practice. Tibialis posterior is particularly important in this regard, with the anatomic location and activity profile of the tibialis posterior muscle suggesting that it helps maintain the medial longitudinal arch during locomotion. The actions of flexor digitorum longus and flexor hallucis longus are also critical to arch stability and may assist the actions of the plantar aponeurosis in the later stages of the stance phase of gait.

The applicability of orthotics is an individual consideration. Orthotics may be needed when there is severe breakdown of the midfoot and stretching of the plantar fascia, at least on a temporary basis. In other cases, they may do more harm than good by stimulating the sensory nervous system in the hollow of the medial longitudinal arch, as previously explained.

Conservative treatment is effective for this condition. A report by Pfeffer **(Bartold, 2004)** to the *American Orthopaedic Foot and Ankle Society*, supports this. In this randomized and blinded clinical trial of 256 patients with isolated heel pain syndrome, 72% improved over the 8 week study period with treatment to the muscles (stretching) alone.

Lutter (1997) reports that 85% of patients with symptomatic plantar fasciitis will respond to conservative management, with surgery indicated for the remaining 15%. However this report concludes that plantar fasciitis is a degenerative, not inflammatory process, which contradicts the bulk of the literature and the pathology and imaging studies. Hambrick, **(Hambrick, 2001)** describing the applied kinesiology management of plantar fasciitis, shows how this kind of assessment for inflammatory as well as degenerative plantar fasciitis provides a more comprehensive method of management of this condition. Most researchers and clinicians agree that athletes and non-athletes with insertional plantar fascial pain can achieve good results without resorting to surgery. Toomey **(Toomey, 2009)** reports that 90% of patients will experience a full recovery with conservative medical management but that this may require 6 to 12 months of treatment and positive encouragement by a physician. Loomey et al **(Loomey et al., 2011)** employed Graston instrument soft-tissue treatment for 10 patients with plantar heel pain with 70% of the patients experiencing clinically meaningful improvements. Treatment was directed to the triceps surae, soleus, plantar fascia, and medial calcaneal tubercle. Areas of fibrous adhesions were detected and treated with the Graston instrument, followed by static stretches and finally ice was applied to the plantar surface of the foot for 15 to 20 minutes.

Relief of pain usually comes within the first week or two of appropriate, multi-modal applied kinesiology treatment. **(Wyatt, 2006; Wynne et al., 2006; McDowall, 2004; Hambrick, 2001)** With chronic conditions it takes time to obtain optimal correction. **(Martin et al., 1998)** Lutter **(Lutter, 1983)** states that plantar fasciitis takes longer to recover from than other running injuries to the foot, with an average length of 7.7 weeks away from running. Cycling or swimming can be substituted during the rehabilitation.

In our experience, treatment by surgery or injection is very rarely indicated. Steroid injections cause marked local osteoporosis **(Brinks et al., 2010; Amix et al., 1988)** and increase the chance of plantar aponeurosis rupture. **(Leach et al., 1978)** The effectiveness of proper conservative treatment is indicated by a report from orthopedic surgeons who state, "We have not found it necessary to operate on a calcaneal spur for the past 24 years and we have not injected the painful bursa over such spurs in 10 years." **(Wickstrom & Williams, 1970)** Although many types of treatment, from casting to drugs, are used for the painful heel, **(Bartold, 2004; Gill, 1987)** the conservative approach described here seems best.

Sometimes it is necessary to provide temporary symptomatic support. When early morning walking is intolerable, it can usually be improved by applying moist heat to the foot's plantar surface. **(Greenawalt, 1989)** If a heel pad is used, the portion under the pain should be relieved with a horseshoe opening rather than a hole. This transfers weight to the surrounding painless tissue and takes pressure off the plantar fascia. (**Stuber & Kristmanson, 2006; Milgram, 1983**)

Sometimes elevation of the heel with heelpads or high-heeled shoes is recommended for painful heel. This increases the load on the posterior foot (**Katoh et al., 1983**) and is not recommended. When this method gives relief, it is due to relief of strain caused by a tight triceps surae. The proper treatment is to correct extended pronation, which will probably require stretching the triceps surae.

As the plantar fascia returns to normal, gradually increase the patient's activities and sports. Shoe evaluation with AK techniques is critically important. Using this method, one sees how often orthotics and certain types of supportive shoes create more problems than they help. Unless the patient's foot function is severely broken down, encourage him to increase foot function by training or doing some daily activities barefooted, preferably on natural turf. The glabrous epithelium is quite capable of hyperkeratinization to resist wear, (**Robbins et al., 1989**) but the activity should begin slowly for the adaptation to take place.

Plantar fasciitis due to a cavus foot is from failure of shock absorption because of the foot's rigidity. (**Badlissi et al., 2005**) The stress is then transferred to the plantar fascia. The therapeutic approach is to correct any fixations, begin stretching procedures to regain foot mobility, and stretch the tight fascia. Often the plantar muscles require application of the applied kinesiology fascial release and percussion techniques. (**Leaf, 2010**)

Conservative treatment for plantar fascia rupture is non-weight-bearing, with crutches, for as long as pain persists. Use local application of ice three times a day for fifteen minutes until acute symptoms subside, which usually takes eight to fourteen days. After pain subsides, use a small felt pad and taping to support the ruptured area, with the patient gradually increasing weight bearing over the next week. (**Sharkey et al., 1998; Leach et al., 1978**)

Forefoot in Extended Pronation

As previously indicated, the forefoot participates in pronation. It may be a factor in creating extended pronation, or stressed as the result of extended pronation. There are several types of forefoot problems, such as functional hallux limitus, hallux rigidus, metatarsalgia, stress fractures, and hallux valgus. In these or other dysfunctions of the forefoot, extended pronation should be considered since it causes or contributes to many of these conditions. Treatment to the localized forefoot problem will be ineffective or only partly successful if extended pronation is not corrected.

Transverse Arch in Extended Pronation

The transverse arch is important in stress dissipation. (**Abshire, 2010; Bartold, 2004**) Extended pronation, metatarsal head predominance, or improper footgear may

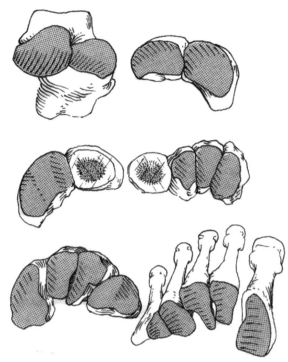

The left illustrations are looking at the distal aspects of the articulations and the right look at the proximal aspect.

result in pain from repeated microtrauma to the area. Both flexibility and strength of the metatarsal arch must be present.

It helps to think of the transverse arch as being most of the length of the foot. (**Kapandji, 2010**) It begins with the calcaneus and talus articulating with the cuboid and navicular bones, and extends to include the three cuneiform and proximal metatarsal bones. In addition to the ligamentous support of the arch at this point, muscle action is important. The tibialis posterior tendon has slips that cross under the foot, with varying attachments to the cuboid, cuneiforms, and metatarsal heads. Kapandji (**Kapandji, 2010**) attributes maintenance of the cuboid-navicular transverse arch to action of the peroneus longus and the tibialis posterior. With medial longitudinal arch breakdown, there typically is weakness of the tibialis posterior. Excessive pronation worsens, and the scene is set for the transverse arch to break down as well.

Continued analysis of the transverse arch distally reveals that the wedge shape of the cuneiform and the cuboid bones provides Roman arch bony support for the transverse arch. At this point there is continued support from the peroneus longus and tibialis posterior muscles, which have intercrossing tendons. These tendons extend to the wedge-shaped base of the metatarsals. Kogler (**Kogler et al., 1996**) observes that although the foot manifests an arch-like appearance, it is not a true arch structurally and cannot maintain its arched shape solely as a result of its own architecture. Rather, the arch of the foot is primarily dependent upon adjacent soft tissues to maintain its arched position. As the transverse arch progresses distally into the metatarsals, it flattens with less support from ligaments and bony architecture. The transverse metatarsal ligament is a reinforcement of the deep plantar aponeurosis. (**Gray's Anatomy, 2004; Viladot, 1982**) It helps keep the heads of the metatarsals together. The interosseous and adductor hallucis muscles hold the metatarsals together, with major contribution from the transverse head of the adductor hallucis.

A basic surgical principle, enunciated recently by the *American College of Foot and Ankle Surgeons'* heel pain committee, is that the dome of all the foot's arches should be maintained. (**Thomas et al., 2010**) This objective extends to conservative care. With loss of the arches, there are usually many subluxations and/or fixations of the articulations. Much can be accomplished in maintaining or rebuilding the arches by correcting fixations and subluxations, and improving muscular function with applied kinesiology techniques and rehabilitation exercises. (**Logan, 1995**)

There have been three theories about the forefoot's weight bearing and these are reviewed by Viladot. (**Viladot, 1982**) The tripod theory relates to primary weight bearing on the calcaneus and the 5th and 1st metatarsals. A second theory is that only the central metatarsals bear weight, based on the high incidence of callus formation under the 2nd, 3rd, and 4th metatarsals. These theories have been abandoned in favor of one that states all metatarsals bear weight. Using a split force plate measurement method, Morton (**Morton, 1935**) substantiated the theory. He also found that the 1st metatarsal carries twice as much weight as the others. He reasoned that since the 1st metatarsal has two sesamoid bones, each making contact, there are actually six contact points, each carrying equal weight. Using more sophisticated measuring devices, Cavanagh et al. (**Cavanagh et al., 1987**) substantiated that normally all metatarsals bear weight, but the maximum weight is born by the 2nd metatarsal head.

The metatarsals carry most of the weight of the forefoot, but balance and function are greatly dependent upon the action of the toes. The metatarsals make firm contact with the substrate because they do not have long, tendinous insertions. The toes, because of the grasping action of the flexor longus and brevis, give dynamic action to weight bearing of the foot. (**Squadrone & Gallozzi, 2009; Warburton, 2001; Viladot, 1973, 1982**)

The rays of the metatarsal arch vary in their flexibility. The 1st and 5th rays are most flexible and, because of the anatomical arrangement of Lisfranc's joint, the middle metatarsals — especially the 2nd and 3rd — are the most rigid. (**Viladot, 1982**) This can readily be observed in motion palpation of the normal foot.

The heads of all metatarsals are normally protected by a fat pad during their weight-bearing function. The 1st metatarsal, because of its role during the thrusting action of toe-off, has additional protection in the form of the sesamoid bones. These are embedded in the tendons of the flexor hallucis brevis muscle. Their ossification occurs between the ages of nine and eleven. The head of the 1st metatarsal is protected as it rolls on the sesamoid bones, while they have solid contact with the substrate as one stands on his toes or at the toe-off position of gait. A graphic picture of the sesamoid bones' role is painted by Viladot. (**Viladot, 1982**) He quotes the pictorial description of Hohmann, who describes the function of the sesamoid pad structure as acting like "a shoe that nature would have created for the support of the first metatarsals."

The lateral four metatarsals rarely have sesamoid bones, but their heads, like the 1st, have fat-pad protection. The fat pads are located between fibrous tissue attached to the proximal phalanges and the flexor digitorum longus tendon, to which they are attached by fibrous slips. The fat pads for the 1st and 5th metatarsals are generally separate from the pads for the 2nd-4th metatarsals, which may join to form one single pad.

With flexion of the proximal phalanx, as in toe-off, the fat pad is pulled forward by its fibrous attachment to the proximal phalanx. In normal function, this is advantageous because it keeps the fat pad under the metatarsal head as the toes flex during the stance phase of gait, thus providing protection. As described earlier, a clawing deformity of the toes develops with tarsal tunnel syndrome or any other condition that causes weakening of the short flexors while the long flexors maintain strength. In this case, the proximal phalanx hyperextension keeps the fat pad forward, causing loss of protection between the metatarsal head and the substrate. (**Hlavac & Schoenhaus, 1970**)

Types of Metatarsalgia

Pain under the forefoot is not necessarily metatarsalgia. Tarsal tunnel syndrome, Morton's neuroma, and Buerger's disease can cause pain under the forefoot, but they are distinct clinical entities in themselves and must be treated as such for a successful outcome. Sometimes symptomatic padding under the metatarsal bones will relieve the forefoot pain, but the condition will probably worsen until the proper treatment is given. (**Cailliet, 1997**)

In a study by Scranton (**Scranton, 1981, 1980**) of 98 cases of pain under the forefoot, 23 different diagnoses were made. He classified these as primary, secondary, and from other than pressure in the forefoot.

Primary metatarsalgia is a condition in which pain under the forefoot is secondary to static imbalance in the weight-bearing distribution across the metatarsophalangeal joint(s).
- Static disorders
- Iatrogenic (post surgical)
- Secondary to hallux valgus
- Hallux rigidus
- D.J. "Morton's foot"
- Congenital
- Long 1st ray
- Freiberg's disease

Secondary metatarsalgia manifests increased pressure under the metatarsal heads, secondary calluses, and pain. Although there is obviously abnormal pressure under the metatarsal head, both the metatarsophalangeal problem and the primary problem must be addressed.
- Rheumatoid arthritis
- Sesamoiditis
- Post-traumatic
- Neurogenic
- Stress fractures
- Gout
- Short ipsilateral leg

There is pain under the forefoot, but not from direct pressure in the region. Primary treatment is directed to the originating cause of the forefoot pain.

- Morton's neuroma
- Plantar fasciitis
- Causalgia
- Tarsal tunnel syndrome
- Tumors
- Intermittent claudication
- Buerger's disease
- Plantar verruca (wart)

Metatarsalgia should not be used as a diagnostic term because it only describes the symptomatic picture, not the cause of the condition. One must determine the cause of the problem before proper therapeutics can be applied. Often the cause is found in several types of dysfunction. As is applicable in nearly all body function, all possible relating factors must be evaluated or only partial results may be obtained.

Metatarsalgia is very common, with a much higher incidence seen in women. (**Latinovic et al., 2006**) In a study of over 1,000 feet with metatarsalgia, 88.5% of those involved were women and 11.5% were men. (**Viladot, 1982**) Many indict women's shoes as a major cause of the condition. The reasonableness of this accusation will be clearly seen with the investigation of the mechanics of a frequent cause of metatarsalgia.

As the cause of metatarsalgia is corrected, there may be other benefits to the patient. As with all foot dysfunction, nerve receptors can be improperly stimulated to cause remote dysfunction in the body. The foot dysfunction may change the individual's gait, such as greatly reducing the time and force of weight bearing in the stance phase. (**Aniss et al., 1992; Bard et al., 1992; Grundy et al., 1975**)

Pronation. Although extended pronation is not mentioned in the three categories listed previously, it is a common predisposing factor. (**Cailliet, 1997**) Metatarsalgia usually results from breakdown of the midfoot. There may seem to be limited evidence of a short triceps surae. (**Hamilton, 1955**) This is usually due to the dorsiflexion range of motion examination taking into account the movement of the midfoot when only ankle dorsiflexion should be measured. Adequate ankle dorsiflexion is of particular importance in the athlete. (**Whitting et al., 2010; Lillich & Baxter, 1986**)

In correcting extended pronation, one must take into account bony subluxations and fixations and examine the function of the peroneus longus, tibialis posterior, and the intrinsic muscles of the foot. Of course, in correcting extended pronation, potential proximal problems at the pelvis, hip, knee, and ankle must be evaluated and corrected, as well as modular and gait dysfunction.

Loss of transverse arch. Goldthwait (**Goldthwait, 1894**) considers loss of the transverse arch to be one cause of metatarsalgia. An imprint of the normal foot reveals what he calls the re-entering angle. This is the sharp indentation of the imprint posterior to the ball of the foot, which joins with the imprint made by the lateral aspect of the foot. When there is loss of the transverse arch with a normal lateral longitudinal arch, there is a bulge rather than the sharp re-entering angle. Quite often loss of the transverse arch is in combination with loss of the medial longitudinal arch, and no re-entering angle is observable.

The re-entering angle can be observed on imprints of the foot made on paper by various methods like "The Wet Test", or by observing the foot imprint while a patient stands on elevated glass, or by observing imprints made in impression material, such as that supplied by Foot Levelers, Inc., to fabricate Spinal Pelvic Stabilizers®. (**Foot Levelers, Inc.**)

The wet test involves imprinting a wet footprint on a piece of paper to determine what kind of arch you have. (**Abshire, 2010**)

Fill a pan with water and submerge just the bottom of the foot. Quickly remove the foot and step onto a brown grocery sack, newspaper or dark-colored construction paper. Push the foot straight down and quickly remove it. Look at the shape left by the wet foot to identify the heel, midfoot, and forefoot and see if it fits one of the images.

You might see the other extreme on the paper with your wet footprint: a fairly smallish space for your forefoot that tapers back sharply to the lateral (outside) of the foot, so much that you might see the heel print as a separate imprint, just a round ball that is nearly or completely disconnected from the rest of the footprint. This means you have a high to very high arch and could be an under-pronator or supinator, but you could also be a fairly neutral runner if you have strong, developed feet.

If you see something that's somewhere in between the two prints just described, such as a wide forefoot print that tapers down gradually to the midfoot/arch region (but still leaves a wet spot that's an inch or wider) and then widens back to a round heel imprint, you likely have a medium arch. It's called a normal arch because it typically leads to normal pronation.

Footwear One of the most common causes of

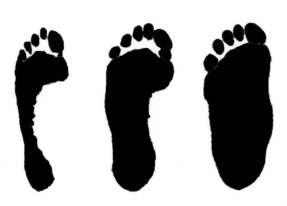

Re-entering angle

Metatarsalgia from high-heeled shoes.

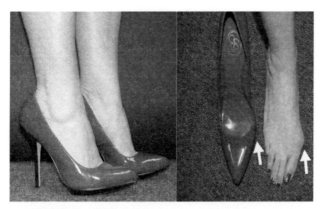
High heels and foot shape

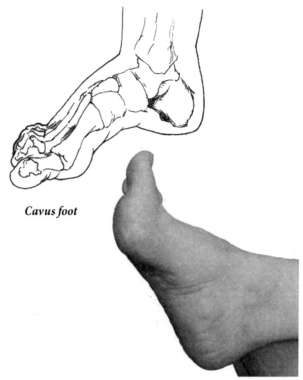
Cavus foot

Cavus foot patient

metatarsalgia is women's high-heeled shoes. Whenever there is an increase in weight bearing at the metatarsal heads, there is the likelihood of metatarsalgia. (**Ko et al., 2009**) Three times the weight is carried on the metatarsal heads when high heels are worn as opposed to flat heels. (**Hong et al., 2005; Viladot, 1982**)

Pain at the metatarsal heads can be caused by tight shoes. (**Chaitow & DeLany, 2002; Mennell, 1969**) Women's shoes are further indicted here because of the pointed toes that fashion so often dictates. The foot is forced down into the wedge shape of the toe-box, cramping the metatarsal heads together. The pain is caused from chronic irritation of the bursa between the metatarsal heads.

Metatarsalgia may be due to improper weight bearing on the distal aspects of the metatarsal heads, from jamming the metatarsal heads together, or from both. Differential diagnosis can be obtained by palpating the plantar and distal surfaces of the metatarsal heads, taking care not to apply pressure between the metatarsal heads. Compare this with squeezing the metatarsal heads together. If pain is elicited on the plantar and/or distal aspects of the heads, the problem is from weight bearing. If no pain occurs there but does when the metatarsal heads are squeezed together, the problem is bursitis due to a tight toe-box. Pain on squeezing the metatarsal heads together is unique to bursitis. A common misconception is that pain from Morton's neuritis can be elicited by squeezing the metatarsal heads together. (**Hoppenfeld, 1982**)

Cavus foot. In the cavus foot, the anterior pillars of the longitudinal arches are too steep; consequently, there is an overload of weight bearing at the metatarsal heads. Typically, there is hyperextension of the metacarpophalangeal articulations and flexion of one or more interphalangeal articulations; however, all cases of pes cavus do not have this deformity. In this condition, there is poor tolerance to distance running. (**Fields et al., 2010; James et al., 1978**) In the severe cavus foot, the first stage of the stance phase is on the toes and metatarsal heads rather than on the heel. (**Fields et al., 2010**)

The cavus foot may be associated with severe neurologic disease, creating a need for a complete neurologic work-up. (**Hsu & Imbus, 1982**) It usually develops as a result of weakness of the triceps surae in neurologic conditions. The calcaneus moves into a cavus position because the intrinsic muscles are unopposed. (**Houtz & Walsh, 1959**) Shortening of the plantar fascia accentuating the cavus foot follows. (**Cailliet, 1997**) In the pathological cavus foot, surgery may be required for the individual to have a functional foot.

A milder form of cavus foot can develop from non-weight bearing over a prolonged period, such as in a protracted illness from which one finally recovers and can resume weight bearing. Mild pes cavus can be treated conservatively, and sometimes more progressive cases must be treated because they are poor surgical candidates.

Conservative treatment of the cavus foot requires regaining mobility of the articulations. Fixations and subluxations must be corrected. Foot and toe stretching exercises may be needed to stretch the ligaments and aponeurosis. Stretching the triceps surae and peroneal muscles is usually necessary. Vitamin B, large shoes, barefoot exercises, massage, diathermy, and hydrotherapy are all applicable therapeutic approaches. (**Maffetone, 2003; Viladot, 1982**)

Imbalance of weight distribution. The balance between the length of the metatarsals and toes may be a factor in metatarsalgia. As weight bearing moves through the forefoot with heel rise, it must move in a smooth arc around the metatarsal heads. (**Elftman, 1969**) Each ray must perform its function in weight bearing and ambulation. Failure to do so places additional strain on the other rays.

First ray insufficiency syndrome (**Viladot, 1982**) occurs when the 1st ray fails to carry its share of forefoot weight. This can be caused by a short 1st metatarsal, commonly called Morton's foot after Dudley Morton. (**Morton, 1935**) The condition may be confused with Morton's neuritis, which was first described by Thomas Morton. (**Morton, 1876**) The two conditions are totally independent, although each can cause metatarsalgia. When Dudley Morton's name is used to describe 1st ray insufficiency syndrome, it is called Morton's syndrome. When Thomas Morton's name

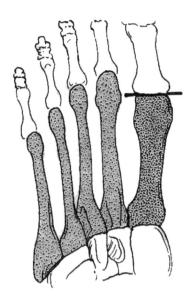

Morton's foot - short 1st metatarsal

Sesamoid and fat pad of hallux

is used for the nerve condition, which is usually between the 3rd and 4th toes, it is called Morton's neuritis, Morton's neuralgia, Morton's neuroma, or Morton's metatarsalgia, as well as several other names. In this text, the preferred term is plantar interdigital neuralgia. (**Plantar interdigital neuralgia is discussed in Chapter 3**).

In Morton's (**Morton, 1935**) description of the 1st ray insufficiency syndrome, the 1st ray was considered to carry twice the weight of the other individual rays. It has been demonstrated by more recent measuring devices that the 2nd ray carries more weight in both the Morton and non-Morton foot, but more in the former. (**Rodgers & Cavanagh, 1989**)

Metatarsalgia may be secondary to untreated hallux valgus (**Wang et al., 2009**) or the result of surgical treatment for the condition. Many types of operations for hallux valgus shorten the 1st metatarsal, creating an insufficiency syndrome like that of Dudley Morton's. Also described by Morton is a varus position of the 1st metatarsal of more than 15°, and posterior displacement of the sesamoids, both of which cause reduced weight bearing of the 1st ray.

The peroneus longus muscle, by its insertion on the plantar base of the 1st metatarsal, provides the strength of 1st ray flexion during the last portion of the stance phase. (**Beardall, 1975; Bartold, 2004; Lusskin, 1982**) Weakness of the peroneus longus leads to increased pressure on the 2nd and 3rd metatarsal heads, which may cause metatarsalgia. This weakness may also be responsible for inferior subluxation of the 2nd and 3rd metatarsal heads.

Normal adjustment of pressure on the forefoot is controlled in part by differential action of the toe flexors. (**Elftman, 1969**) Thus, when there is muscular imbalance of the intrinsic muscles, the dynamics of pressure on the substrate change; this may cause functional overload to certain portions of the foot. Accurate testing of the intrinsic and long toe flexors is limited. What often appears to be dysfunction of the intrinsic muscles can be observed by positive plantar therapy localization that does not associate with a subluxation. Treatment by neuromuscular spindle cell, Golgi tendon, or origin and insertion techniques often eliminates the positive therapy localization and appears to improve muscle function.

Sesamoiditis. The sesamoid bones of the hallux are covered with hyaline cartilage, providing an articular surface. (**Gray's Anatomy, 2004**) They increase the metatarsophalangeal flexion power by increasing the moment arm between flexion and extension to maintain a functional arc of motion. They also serve as shock absorbers by dispersing impact forces on the metatarsal head. The bones are concave in contact with the metatarsal head, and convex on the plantar surface. Between this convex surface and the skin is an adipose cushion of varying thickness. The mobility of the skin under the sesamoids helps reduce shear stress of the tissue itself and on the sesamoids during acceleration and deceleration. Between the plantar medial capsule of the 1st metatarsophalangeal joint and the medial sesamoid bone there is an inconsistent bursa. (**Richardson, 1987**)

Heavy activity such as running, jumping, and dancing on the balls of the feet can cause damage to the sesamoid bones. The resulting painful inflammation is called sesamoiditis. It heals slowly and has a tendency to recur. Sesamoiditis is typically found in the recreational long-distance runner, less frequently in other sports, and not usually in the sedentary person. The unusual factor is that generally there is no specific instance of trauma the athlete can describe to account for the beginning of his troubles. The complaint is that of increasing, poorly localized pain over several weeks. The condition tends to advance, and chondromalacia or roughening may result. Osteoarthritis may occur in these joints as a result of recurrent trauma. (**Cohen, 2009**)

Gentle palpation over the sesamoid area will generally localize the pain effectively. Swelling is present only in the more advanced chronic cases. Differential diagnosis is made by x-ray, and possibly a bone scan. Possible conditions are bursitis, sesamoiditis, osteochondritis, chondromalacia, degenerative arthritis, and fracture. Bursitis is indicated by swelling and redness. Other non-localized conditions, such as one of the rheumatoid variants, should be considered. A conservative trial is justified, except in displaced fractures. (**Cohen, 2009; Richardson, 1987**)

Treatment involves shifting the weight elsewhere by means of pads until the sesamoid heals. Padding, made from adhesive sponge rubber, is placed under the shaft of the 1st metatarsal for primary support to relieve the pressure on the sesamoid bones. The pad is extended under the shafts and heads of the 2nd and 3rd metatarsals to further reduce weight carrying of the hallux. (**Cohen, 2009; Bartold, 2004**) This type of J-shaped pad can be incorporated into an orthotic, and can even give relief in the case of sesamoid fracture. The support is not effective

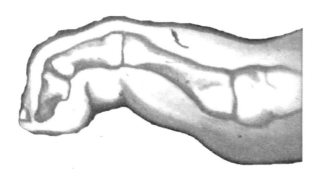

Typical appearance of metatarsophalangeal hyperextension and interphalangeal flexion. Note that only the tips of the toes are visible.

in activities that require toe walking, such as dancing and skating. **(Axe & Ray, 1988)**

Non-surgical treatment is best. **(Cohen, 2009; Viladot, 1973)** Excising sesamoid bones is not as popular as it once was because of the unpredictable success rate. It may be needed in fracture cases, **(Axe & Ray, 1988)** such as acute fracture, stress fracture, or spontaneous osteonecrosis. **(Guebert & Thompson, 1987)**

Metatarsalgia differential diagnosis. Metatarsalgia from improper weight distribution or function of the structures must be differentiated from interdigital nerve entrapment, functional hallux limitus, hallux rigidus, hallux valgus, Morton's foot, 1st metatarsal jam or turf toe, **(Maffetone, 1989)** intermetatarsophalangeal bursitis, various types of arthritis, and direct trauma to the area. These are some of the more common factors. The lists at the beginning of this section cover most of the conditions that can cause pain under the forefoot, but even the lists are not all-inclusive.

Consider factors peculiar to the patient, such as stress fractures in runners and Freiberg's disease in adolescents, which may require x-ray examination. A special view to determine standardized quantitative measurement of distal metatarsal head distance from substrate with weight-bearing has been described, **(Dreeben et al., 1987)** but its greatest value is in research.

Diagnosis of local conditions is made by careful palpation of the plantar and distal aspects of the metatarsal heads for pain, and the metatarsals for position, flexibility, and joint play. Anterior displacement of the fat pad(s), as previously described, is a common cause of metatarsalgia. The metatarsal heads are tender to palpation. There is puffiness just anterior to the metatarsal heads. Because metatarsophalangeal hyperextension is usually the cause of the fat pad anterior displacement, one may see only the tips of the toes from the plantar view, with no visualization of the shaft of the phalanges.

Light palpation along the plantar surface of the metatarsal heads will reveal the inferiorly displaced bone. This is confirmed by either pressing the head inferiorly to weaken a strong associated muscle, or pressing it superiorly to strengthen a weak associated muscle.

These findings are correlated with the evaluation of the entire foot and remote influences on the foot, such as leg, knee, thigh, hip, and pelvic position and function. Applied

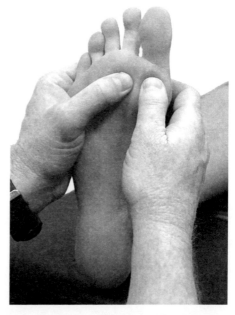

Palpate inferior to the metatarsal heads for bony displacement and pain.

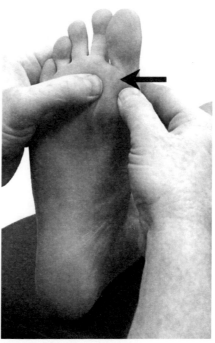

Palpate the distal aspect of the metatarsal head for pain, which indicates fat pad displacement.

kinesiology articular challenge and therapy localization, as well as the function of muscles, provide additional evaluation of the functional status of the total complex. In addition to evaluating the five factors of the IVF for the cause of muscle weakness, consider the possibility of peripheral nerve entrapment. Examples are: entrapment of the common peroneal nerve may be responsible for peroneus longus muscle weakness; tarsal tunnel entrapment may cause intrinsic muscle weakness. Both of these conditions are often associated with foot dysfunction. In the final analysis, all examination findings must correlate to arrive at the proper diagnosis for the cause of metatarsalgia.

Treatment

Since metatarsalgia is often only part of a total dysfunctioning complex, most of the conservative treatment is discussed elsewhere (pronation, proper shoes, and rehabilitation). The most severe cases of metatarsalgia are usually found in runners and others who similarly stress their feet. Conservative treatment will usually be effective even in these, with surgery only occasionally needed. **(Abshire, 2010; Maffetone, 2003; Lillich & Baxter, 1986)** Presented here are manipulative procedures for subluxations and fixations.

Sometimes, when challenging the transverse metatarsal arch to improve muscle function, one finds that re-forming the entire arch from the cuboid-navicular area all the way to the distal metatarsals best strengthens the associated muscles. In this case, there are probably many subluxations that must be corrected within the transverse arch and in the entire foot. One must progressively find and correct the subluxations and fixations to obtain maximum improvement of the associated weak muscles.

Positive challenge can be observed as weak associated muscles strengthen, or strong muscles weaken when a specific vector of force is applied to the articulation. It is best to find a muscle that is weak as a result of the subluxation and strengthen it with challenge. This determines the exact vector of force needed to correct the subluxation. When a previously strong muscle weakens, the opposite vector of force is required to make the correction. The problem with using this approach is that one cannot as specifically determine the precisely opposite vector of a positive challenge.

There are many descriptions of metatarsal manipulation and adjustment in the literature. **(Bergmann & Peterson, 2010; Leaf, 2010; Greenman, 2003; Chaitow & DeLany, 2002; Walther, 2000; Hearon, 1994; Viladot, 1982; Gertler, 1978, 1981; Gillet & Liekens, 1981; Maitland, 1977; Stierwalt, 1976; DeJarnette, 1973; Mennell, 1969, 1964; Strachan, 1954; Laylock, 1953)** Most are effective; each individual must find the procedures easiest for him to perform.

Because of the AK concept of the five factors of the IVF influencing each muscle, the determining factor regarding which treatment method to employ is reduced to the ones which practitioner has mastered and feels most confident. One technique may work as well as another so long as it is designed for the condition being addressed, muscular imbalance, and the principles of the treatment modality used are borne in mind.

Two types of manipulative procedures are necessary. One is to mobilize articulations that are fixed; the other is to correct subluxations. Motion should be available at each articulation. **(Petty & Moore, 1998; Gillet & Liekens, 1981)** Joint fixation is often found when challenging the various articulations for subluxations when the joint does not yield to the challenge. This is further evaluated by holding one bone and attempting to move its articulating partner. This type of palpation is described by Mennell **(Mennell, 1969)** Gillet, **(Gillet & Liekens, 1981)** Chaitow & DeLany, **(Chaitow & DeLany, 2002)** and many others.

A fixation can be corrected by direct manipulation of the articulation, with the precise vector determined by challenge. Experienced clinicians will realize that many times the vector of misalignment is not the one commonly described in textbooks; the applied kinesiology challenge method will reveal the biomechanical individuality of patients with clarity. Another effective approach is to mobilize the joint by holding one bone and maneuvering the adjacent one in all directions.

The golf ball exercise, described under rehabilitation, is very effective in aiding foot mobilization. Restricted movement of the bones is often caused by tight shoes, which should be eliminated or the condition will recur.

One of the most common positive metatarsal challenges is a dorsal base with the head plantarly positioned. There may or may not be a subluxation of the metatarsophalangeal articulation. The two metatarsals most commonly subluxated are the 2nd and 4th. There will be a positive subluxation of one or both ends of the metatarsal. They will show a positive challenge independently; if both are present, they will show a positive simultaneous challenge.

Adjustment of the metatarsal whose proximal end has moved dorsally and distal end plantarly is accomplished in one maneuver. The physician's thumb is placed longitudinally over the plantar surface of the dropped metatarsal head. Depending upon the size of the patient's foot, the physician's 3rd or 4th finger is wrapped over the dorsal surface of the metatarsal base. The other thumb is placed crosswise over the plantar-contacting thumb, and

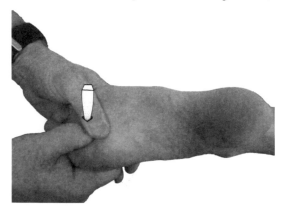

1st contact step - thumb contact on metatarsal head.

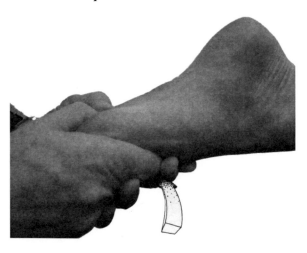

2nd contact step - finger contact on metatarsal base.

the fingers are wrapped over the finger contacting the dorsal base of the metatarsal. Solid, specific contact is maintained on the head, and specific contact is obtained on the base by emphasizing pressure on the contact finger directly over the base with the support finger. The patient's foot is slightly plantar-flexed so that when traction is applied, a separation is obtained at the base of the metatarsal. Continued traction moves the base plantarly, and there is a quick follow-through of the thumb to move the head of the metatarsal dorsally. This is a quick, continuous maneuver with three successive actions: 1) traction for separation of the base from the cuneiform(s), 2) plantar movement of the base, and 3) dorsal movement of the head. There will often be an audible release, but it is not necessary.

Support to the metatarsal bones is often done with padding, or it is built into an orthotic. Specialized x-ray has shown the effective lifting of the metatarsal heads by a pad placed just proximal to the heads. **(Dreeben et al., 1987)** Sometimes the pad is inadequately placed, thus creating new problems. For example, a pad to shift weight away from the sesamoid bones in sesamoiditis may provide adequate support under the 2nd and 3rd metatarsal bones and abruptly stop, causing the 4th metatarsal to drop inferiorly and resulting in a new subluxation. After placing new padding or obtaining a new orthotic, have the patient walk; then use applied kinesiology testing procedures to determine the support's effectiveness. No new muscle weakness should develop.

Freiberg's Infraction and Bone Fragmentation

Pain in the foot may be due to fragmentation of bone, generally of the metatarsals. The condition is usually found in the juvenile during the rapid growth period; in this case, it is considered an osteochondrosis called Freiberg's disease. It is a dorsal trabecular stress injury of the metatarsal head. Overuse and repetitive stress are initiating factors. **(Thordarson, 1996; Scartozzi et al., 1989; Trott, 1983)** Although frequently found in athletes, it is also common in those who do considerable marching, such as in drum and bugle corps and in the military. **(Trott, 1983)** Freiberg's disease occurs in the 2nd and 3rd metatarsals; it has rarely been found in the 4th. Young et al. **(Young et al., 1987)** describe a mechanism whereby the condition in adults may be due to a shearing compression-type recurrent injury at the interface between mineralized and non-mineralized articular cartilage (tidewater mark), rather than a true avascular necrosis.

Smillie **(Smillie, 1967)** described five stages of the disease, beginning with Stage I in which a fissure fracture develops in the ischemic epiphysis, to the final Stage V where there is total flattening, deformity, and arthritis. He found that the condition developed in structurally weak feet with short, varus, or hypermobile 1st metatarsals. **(Smillie, 1957)**

In the early stage, the only physical sign of Freiberg's disease may be pain on weight bearing and tenderness over the head of the metatarsal or over the metatarsophalangeal joint on digital pressure. **(Thordarson, 1996; Katcherian, 1994; Helal & Gibb, 1987)** Symptoms may subside, perhaps for many years, only to return with athletic activities, excessive walking, or with a girl wearing her first pair of high-heeled shoes. **(Maresca et al., 1996; Turek, 1984)**

Plain x-rays may show no changes. A "hot spot" will be revealed over the affected head with a radionucleotide bone scan. A stress fracture of the metatarsal head is often mistaken for Freiberg's disease. **(Lutter, 1983)**

The condition may be present in a higher number of persons than those with symptoms in their feet. In routine x-rays of children attending Pottenger's **(Pottenger, 1945)** clinic, fragmentation of foot bones was found in a high percentage. Pottenger's procedure was to routinely take skull and foot x-rays to determine bone development in all the children he examined. This report is of the first 400 children studied between the ages of six weeks and eleven years. Most of the children were in the clinic because of health problems, with the following conditions diagnosed: asthma, 190; chronic bronchitis, 59; eczema, 45; allergic rhinitis, 11; gastric allergy, 9; rheumatic fever, 15; tuberculosis, 3; developmental problems, 39; severe malnutrition, 7; and "normal" children, 22. The study was divided into two groups. There were 221 subjects from six weeks to six years old, of which 182 had active fragmentation. Of the 179 in the second group, from six to eleven years, 103 had active bone fragmentation in their feet. The fragmentation was found in many areas of the foot, including the cuboid, navicular, cuneiforms, and metatarsals.

The interesting aspect of this study is that during the period of marked foot bone fragmentation, the children showed little or no outward evidence of extensive bone damage. There was no pain, limping, or other evidence of foot disturbance. Pottenger **(Pottenger, 1945)** attributed the bone fragmentation to repeated physiologic stress, as in early infant walking, and to "…severe metabolic upset such as an acute illness, or, of even greater importance, the recurrent insult of allergic manifestations."

Pottenger's study seems to indicate the need for general health care, including proper nutrition and correction of childhood illness. According to Trott, **(Trott, 1983)** proper foot support and cessation of the causative activity help resolve the situation in many instances. Surgery is usually required when the metatarsal head becomes involved, but it should not be removed. **(Katcherian, 1994; Drez, 1982)**

The usefulness of the applied kinesiology approach is that it focuses on factors which may be amenable to change and which make up the constellation of dysfunctions affecting the individual's total health picture. In applied kinesiology manual hands on diagnostic and treatment methods are used to assess and treat neuromusculoskeletal imbalances, and exercise and life style changes are given to help rehabilitate the biomechanical, biochemical, and psychosocial factors in a patient's total health picture. Simultaneously and integratively for the dysfunctions found, nutritional modifications are employed to address biochemical imbalances. Then psychological approaches are used to deal with psychosocial influences. Each of these modalities are employed in the care of most patients to address the totality of dysfunction that are found in most

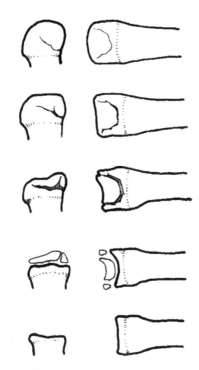

Five stages of Frieberg's infraction.

patients, as was reported by Pottenger. It is necessary to address which ever of these (or additional) influences on musculoskeletal pain that can be identified in order to remove as many etiological and perpetuating factors as possible.

Pottenger, like Goodheart in the development of applied kinesiology methods, found that no tissue exists in isolation but it acts upon and is interwoven with many other structures. The body is inter-related from top to bottom, side to side and back to front, by the integrated muscular and nervous systems. When we work on a local area in the foot, we need to maintain a constant awareness that we are potentially influencing the whole body. Manual muscle testing evaluation can demonstrate this integrated functional relationship readily for both the doctor and the patient.

In the earlier stages, a surgical procedure developed by Smillie (**Smillie, 1967**) is effective in correcting the condition; for later use, a prosthesis of a small bi-stemmed universal joint spacer made of silicone elastomer has been used satisfactorily. (**Helal & Gibb, 1987**)

Although not as frequently, other areas of the foot can have bone fragmentation. There may be avascular necrosis of the talar head; (**Early, 2004; Thordarson, 1996; Katcherian, 1994; Schmidt & Romash, 1988**) this, along with osteochondritis dissecans of the midfoot, is difficult to diagnose. The condition is not typically demonstrable on plane x-rays. It is characterized by complaints of pain exacerbated by activity, followed by aching discomfort at rest that fails to respond to the usual applied kinesiology treatment. There may be loss of motion and swelling. A definitive diagnosis is made by radionucleotide bone scan. If the bone scan is normal, no further tests are needed. If the scan is positive, tomograms or a CT scan should be ordered. (**Early, 2004; Lehman & Gregg, 1986**) Treatment is usually conservative in the skeletally immature patient; surgery is performed in the skeletally mature.

Stress Fractures

Stress fractures tend to develop in untrained individuals who begin running or some other athletic programs too rapidly. Stress fractures can develop without athletic activity or increased training. McBryde (**McBryde, 1975**) reports that 95% of all stress fractures occurred in the lower extremity for athletes. New shoes, a change of running surface, or a new sport can cause stress fractures, especially of the metatarsals, upper third of the tibia and the fibula. (**Yochum & Rowe, 2004; Aspegren et al., 1989**) Unusual stress fractures can result from structural faults, such as hallux valgus causing stress fracture of the 1st toe proximal phalanx. (**Yokoe & Mannoji, 1986**) Activities that put unusual stress on the structure may result in stress fractures. The most common pathologic bone problem in dancers is the stress fracture. (**Goulart et al., 2008; Sammarco, 1986**)

Military marching and other training procedures are examples of stress fractures that develop with rapidly increasing activity. Two-hundred ninety-five Israeli infantry recruits were evaluated for stress fractures that developed during basic training. (**Milgrom, 1985**) Thirty-one percent sustained stress fractures during basic training. Seventy-two percent of those who sustained stress fractures were followed for a minimum of one year to determine their continued status. Those who initially sustained stress fractures had more subjective complaints of bone pain during subsequent training. They were high risks for recurrent stress fractures, being over six times more susceptible than those who had not had stress fractures. This study reveals a markedly different recovery course in the individuals. The commonly recommended rest/recovery schedule must be adapted to the individual. Evaluation of recurrent stress fractures revealed that it was rare to have a stress fracture in a previously affected anatomical site.

Stress fractures are more common in the forefoot than in the mid- or hindfoot. A common area is in the center of the 2nd or 3rd metatarsal shaft, (**Lutter, 1983**) because the highest bending strain in shod distance running is in the 2nd and 3rd metatarsals. (**Gross & Bunch, 1989**) The frequency of occurrence is 2, 3, 4, 1, 5. (**McBryde, 1975**) Patients with low longitudinal arches are more likely to develop stress fractures in the forefoot. (**Simkin et al., 1989**)

The body adapts to stress when training is done properly. Evidence of the body preventing stress fractures is seen in hypertrophy of the 2nd metatarsal that develops in runners. (**Madjarevic et al., 2009**) This is the body's innate method of preventing the problem.

Generalized pain and unexplained doughy pitting edema over dorsal metatarsals indicate the possibility of a stress fracture. (**Merck Manual, 2011; Jahss, 1982**) X-ray may be negative for up to six weeks after initial onset of the condition; (**Yochum & Rowe, 2004; Jahn, 1985**) radionucleotide bone scan provides the definitive diagnosis.

Stress fracture of the 5th metatarsal is less common than in the other four. It may occur with no history of acute injury. (**De Lee et al., 1983**) When it does occur, the usual treatment is intramedullary screw fixation. (**Delee et al., 1983**) Three cases that refused surgery and were treated by casting are described. (**Acker & Drez, 1986**)

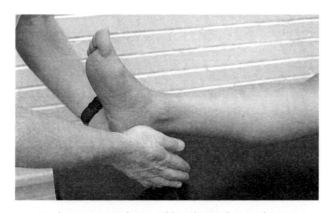
Heel squeeze test for possible calcaneal stress fracture.

They returned early to their regular athletic participation without recurrent symptoms or refracture.

It has been found that ballet dancers, female distance runners and gymnasts with irregular menstrual periods have calcium and other nutrient deficiencies in the diet that contribute to stress fractures due to loss of bone density. **(Lloyd & Triantafyllou, 1986)** Training in running shoes that are older than 6 months is also a risk factor for stress fractures. **(Gardner et al., 1988)** Once again it is important to note that in this text and its companion volume, substantial attention is given to musculoskeletal stress resulting from postural, emotional, chemical, and other influences. As will become clear in these discussions, there is a constant merging and mixing of such fundamental influences on the particular AK findings each patient presents. In making sense of the patient's problems, it is clinically essential to differentiate between these interacting environmental factors. The model which is used in applied kinesiology classifies negative influences into three categories:
- Structural
- Biochemical or nutritional
- Mental or psychosocial

Each of these influences can be assessed specifically (using the AK sensorimotor challenge and therapy localization procedures) in relationship to the muscular imbalances found using the manual muscle test.

Differential diagnosis of calcaneal stress fracture must be made from Achilles tendinitis and retrocalcaneal bursitis. More difficult to differentially diagnose is calcaneal stress fracture from plantar fasciitis. Pain from digital compression applied to the medial and lateral aspects of the calcaneus and positive x-ray findings at three weeks confirm calcaneal stress fracture. Treatment is not critical, and symptoms usually permit resumption of a full running schedule after three weeks. **(McBryde, 1975)**

Functional Hallux Limitus

Functional hallux limitus (FHL) was first described by Dananberg, a friend and patient of Goodheart. **(Goodheart, 1996; Dananberg, 1986)** FHL is the inability of the hallux to properly extend at the metatarsophalangeal articulation at the proper stage during the stance phase of the gait cycle, despite normal extension of this joint when non-weight bearing. Dananberg describes a new method of viewing gait efficiency, and presents a new entity not visible to even the most trained observer of gait. In evaluating disturbances in gait and their possible relationship to disturbances in the low back, Dananberg notes: **(Dananberg, 1997)**

> "When viewing X-rays of the patient, it is well known that a single view of the body is not acceptable. Generally, three views provide a far more accurate picture of a three-dimensional being. Viewing a patient walk is no different. Simply watching a subject walk back and forth in a hallway loses the entire sagittal plane view. Although most offices are not equipped for gait analysis, the use of the treadmill can be helpful in providing the multiple viewpoints necessary for accurate determination of cause and effect."

It is to Dananberg's credit that a podiatric blockage in the first toe is no longer seen only as a blockage of the joint or a podiatric problem alone, but as a neuromusculoskeletal dysfunction affecting the entire ambulant organism. Even a slight alteration in the mechanics of the foot can induce effects throughout the musculoskeletal system; the extreme sensitivity of the muscle spindles is responsible for this. **(Cramer & Darby, 2005; Mense & Simons, 2001)** The most common approach for diagnosing gait imbalances (for practical purposes) is based primarily on the visual observation of standing and gaiting posture. While this approach may sometimes be useful, it does mean that treatment based on this finding may be somewhat non-specific. Additionally, aside from subjective symptomatology, progress is difficult to gauge with such subjective approaches. Dananberg cautions that the visual diagnosis of muscular and gait problems is difficult. The different elements within the chain of events occurring with any particular movement that a patient undertakes before the examiner can occur within a fraction of a second, and too rapidly to be accessed separately in the absence of laboratory tools. Therefore, what is actually observed by the examiner is the grand total of how rapidly and smoothly a person can change between two activities – inaccurate, but for many clinicians, good enough. The AK manual muscle testing approach for dysfunction in the strength and movement of the great toe, and testing the muscles that move the great toe during the stance position of gait, as well the influence of functional hallux limitus upon remote muscle function throughout the body, is a great help in diagnosing this subtle but critical disorder.

Functional hallux limitus had gone unrecognized because there is complete range of motion at the metatarsophalangeal articulation when non-weight-bearing, as it is most often examined. The term "functional" differentiates this condition from one in which there is limited range of motion of the hallux at all times, such as in hallux rigidus.

It is obvious there is still some hallux extension in gait even with FHL, as noted by the upper shoe crease in this area. The failure of metatarsophalangeal joint extension is the timing. The restriction can be objectively observed with the Electrodynogram™ by the Langer Biomechanics Group, **(Langer Biomechanics Group)** which records foot forces with ground contact. Six sensors are applied to the following areas: the medial heel; lateral heel; 1st, 2nd, and 5th metatarsal heads; and the interphalangeal articulation of the hallux. A seventh sensor can be placed at the will of the investigator. The Electrodynogram™ makes a computer recording of the temporal and force patterns of foot contact. The first metatarsophalangeal restriction varies in length and may be less than 100 milliseconds in duration and *invisible to the naked eye*. (**Dananberg, 1986**) Symptoms caused by FHL can be almost anywhere in the body, but they are rarely in the foot. Dananberg shows that many cases of acute or chronic low back pain are related to gait abnormalities that result from FHL. (**Dananberg, 2007**) To understand this it is necessary to consider the mechanics and dynamics of gait.

The gait mechanism is an excellent example of muscle action being conserved by the body when possible. It might seem that forward propulsion is primarily powered by muscle contraction, such as by the hip extensors of the stance leg; this is not the case. For example, the gluteus maximus is inactive except at the beginning and end of the stance phase. (**Dananberg, 1995; Basmajian & DeLuca, 1985**) Forward movement power is from the kinetic energy of the swing leg. This pull phase is sufficient to move the center of the body up and over the ipsilateral leg. (**Dananberg, 2007; Mann et al., 1986**) Gluteus maximus contraction at the end of the stance phase may be to control the swing phase energy.

The gait cycle begins with a slight backward and then forward body movement as the swing limb flexes at the hip and knee. As the swing limb moves forward, movement through gait is powered by the kinetic energy of the swing leg. The forward motion of the limb moves the body's center of gravity forward, advancing it over the stance limb. The pendular kinetic energy of the stance leg is called the "pull force." The entire body weight pivots over the metatarsophalangeal articulation, which is fixed with the ground. The kinetic energy of the swing limb continues to advance the body's center of gravity past the ground contact. The stance leg produces rearward thrust on the ground, attempting to push it backward. Since the ground can't be moved, the pressure on the stance limb provides propulsion for forward motion without muscle action, except for stabilization. The greatest power input to the ground force for forward motion is at the point of weight bearing of the metatarsophalangeal pivot point. It is at this phase of stance limb motion that a momentary halt of hallux extension can disrupt the gait pattern. (**Dananberg, 1993**) There is maximum power to move the body forward at this point; limitation of hallux extension interrupts the proper sequential dissipation of energy, which must take alternate paths not in keeping with normal function. The relationship between FHL and more proximal postural, gait, and muscle problems have been discussed by Lewit, Chaitow & DeLany, and Liebenson. (**Liebenson, 2007; Chaitow & DeLany, 2002; Lewit, 1999**)

Chaitow & DeLany (**Chaitow & DeLany, 2002**) note:

> "This condition limits the rocker phase since 1st MTP joint dorsiflexion promotes plantar flexion. If plantarflexion fails to occur, there will be early knee joint flexion prior to the extension of that leg. The result of early knee flexion prevents the hip flexors from gaining mechanical advantage, thereby reducing the efficiency of the motion of

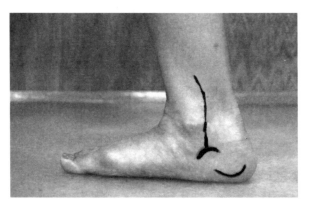

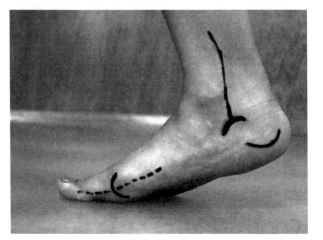

Upon heel strike, the round underside of the calcaneus serves as the pivot for gait movement. Once the flat foot is achieved, forward motion occurs at the ankle joint with dorsiflexion. Upon heel lift, ankle motion reverses to plantarflexion as the metararsophalangeal joint provides the balance for the forward plane motion.

the swing limb. A further effect is that gluteals and quadratus lumborum on the contralateral side become overactive in order to pull the limb into its swing action. Overactivity of quadratus lumborum and/or the gluteals may encourage overactivity of piriformis. 'The reduced hip extension converts the stance limb into a dead weight for swing, which is exacerbated by hip flexor activity…resulting in ipsilateral rotation of the spine, stressing the intervertebral discs.'" **(Prior, 1999)**

The 1st metatarsophalangeal joint combines ginglymus and arthrodial type movement; that is, it has both hinge and glide functions. The hinge motion is approximately 15-20° of the total motion and the glide makes up approximately 50°, providing the total of 65-70° extension range of motion. With this amount of action, the heel can adequately rise to advance the leg while maintaining digital ground contact. The mechanisms that stabilize the foot come into play, and the plantar aponeurosis is shortened to raise and resupinate the arch and externally rotate the lower leg. This keeps the lower leg synchronous with the external thigh and pelvic rotation brought about by the swing limb pull force, and the energy is sequentially dissipated within the foot and leg. When full extension of the 1st metatarsophalangeal joint is limited and/or the temporal pattern is disturbed, the forces are directed elsewhere; this ultimately causes remote dysfunction and symptoms that are often not recognized as being caused by foot dysfunction.

The paradox of FHL is that the first MTP joint's sagittal plane pivotal motion is locked during all or portions of the single support phase of the gait cycle. This is true even though there is full range of motion of the MTP joint during the non-weight bearing examination. Compensatory movements for pain or dysfunction eventually become ingrained in the motor cortex, essentially reprogramming normal movement patterns. This aberrant gaiting, repeated at least 2,500 times per day (or 1 million strides per year), occurs over and over for the patient within the time span of approximately 600-750 milliseconds. **(Dananberg, 2007)** Pain will manifest most notably in those articulations moved by the inhibited muscles such as the medial and lateral knee, greater trochanter, sacroiliac, lumbosacral, lumbodorsal, cervicodorsal and upper cervical joints. The theory of the spinal engine **(Gracovetsky, 1989)** may be implicated in the FHL syndrome and lead to premature exhaustion of the spinal engine during gait. The alterations in the spinal engine with dysfunctional feet will presumably initiate pain, and pain forces the CNS to unload the overstressed articulations, i.e. produce inhibited muscles throughout the body during gait.

Foot Stabilization

Dananberg **(Dananberg, 1993)** describes three distinct mechanisms that permit the foot to support the applied stress during gait: (1) calcaneocuboid locking secondary to aponeurosis tightening, (2) the locking wedge and truss effect, and (3) the windlass effect. They rely little, if at all, on muscle function and are referred to as "autosupport."

Calcaneocuboid Locking.

Dananberg cites Bojsen-Møller's observation that the transition from the anthropoid to human foot to provide a rigid lever for propulsion and protection of the tissues from the extreme repetitive forces of walking and running. **(Bojsen-Møller, 1979)**

Relaxed, the ball of the foot is a soft and pliable pad. The plantar skin can be moved from side to side as well as proximally and distally. Rigidity of the tissue and bones comes with toe dorsiflexion. This is accomplished by the tissue arrangement divided into three transverse areas, each with a different mechanical function: (1) a series of transverse bands proximal to the metatarsal head in which the deep fibers of the plantar aponeurosis form 10 sagittal septa, eventually connecting to the proximal phalanges, (2) inferior to the head where vertical fibers form the joint capsules and the sides of the fibrous flexor sheaths to form a cushion below each metatarsal head with fat bodies, and (3) a distal area where the superficial fibers of the plantar aponeurosis insert into the skin. Metatarsophalangeal extension tenses the three areas, anchoring the skin firmly to the skeleton so that forces to the skin during push-off and breaking are transferred to the skeleton. **(Kapandji, 2010; Bojsen-Møller, 1979; Bojsen-Moller & Flagstad, 1976)**

The tissues of the ball of the foot are a complex network that provides plantar fascia tension to the calcaneocuboid articulation that becomes close-packed by pronation of the forefoot in relation to the hindfoot. It is congruency between the joint surfaces obtained in this position that provides strength. The calcaneus overhangs the cuboid dorsally, which stops the movement. The peroneus longus is a key to pronating the forefoot for high gear push-off and locking of the calcaneocuboid articulation. It assists in internal rotation of the crus, forcing the foot to use the transverse axes. **(Dananberg, 1993)**

Truss and Locking Wedge Effect.

In engineering a truss is a structure usually formed by a triangle or series of triangles. This is a stable arrangement because a triangle cannot be distorted by stress. (*Encyclopedia Britannica*, **2011**) The triangles of the foot are made by the bones of the foot in combination with the plantar aponeurosis. The wedge effect is demonstrated by the stones in a Roman arch which, like some of the bones in the foot wedge, combine with compressive force to provide foot strength.

Windlass Effect.

The British research physician Hicks **(Hicks, 1951, 1953, 1954, 1955)** studied the dynamic function of the foot longitudinal arches by dissection, by x-rays of movement at sequential steps of the dissection, and by x-rays of the living foot. He concluded that the normal foot arch is maintained by the plantar aponeurosis, with no contribution from the intrinsic or extrinsic muscles. The operative words here are "normal foot." There must be normal activity of the intrinsic and extrinsic muscles during running and walking. The mechanism that Hicks describes elevates the arch by tightening the aponeurosis with extension of the toes. The influence of toe extension on the plantar aponeurosis results from the attachment of the aponeurosis to the proximal phalanges. As the toe extends it pulls with it the plantar pad, which is an extension of the aponeurosis. The plantar pad, sliding around the metatarsal head, pulls on the plantar aponeurosis and tightens it as if it were a cable arrangement being pulled around the drum of a windlass. Maximum toe extension shortens the effective length of the aponeurosis by approximately 1 cm. This mechanism exists in all five toes, but it is most marked in the hallux. When the hallux is amputated, its role in the windlass is lost; the forefoot weight bearing shifts laterally throughout gait, and the height of the medial longitudinal arch diminishes. **(Mann et al., 1988)**

The mechanism Hicks describes can be observed in the normally functioning subject. With the person standing in a neutral position, the arch will rise and the tibia will externally rotate. Failure of this action may be due to abnormal alignment of the first ray. **(Rose, 1982)** As described later, there is failure of the hallux to extend properly in this test when FHL is present.

Sequence of Motion. Upon heel strike the subtalar articulation rotates into pronation for shock absorption and to accommodate internal leg rotation. As the heel unloads, weight is transferred through the foot with supination to accommodate external limb rotation. The plantar aponeurosis tightens, causing the calcaneus and cuboid to align in such a manner as to "lock" the foot and prevent arch collapse. The weight continues to move toward the toes and transfers to the second and first rays. Dananberg **(Dananberg, 1994)** calls the body weight flow an automatic natural arch support, noting the importance of its efficiency because forces can be two to three times body weight. As forward motion continues, the weight of the body moves over the metatarsophalangeal articulation. This extends and tenses the aponeurosis to bring the calcaneus closer to the toes, supporting the medial longitudinal arch by the windlass action. The windlass action also rotates the posterior part of the foot externally (supinates) to synchronize with the tibia external rotation. **(Hicks, 1954)** Strands of the aponeurosis attach to the skin of the ball of the foot, which tightens the tissue as the metatarsal head rotates within it to prevent tissue trauma.

As the body moves over the stance limb all of the forward movement is centered on the ball of the foot, with the most important weight bearing being carried on the firmly planted 1st metatarsophalangeal articulation. The metatarsals and midfoot become more vertical, and the body weight is carried more by compression of a column than by the arch.

The Effects of Functional Hallux Limitus

Failure of hallux extension when the body is being pulled over the 1st metatarsophalangeal articulation during gait causes the kinetic energy to be dissipated in compensations that usually stress the weakest links. In the foot and ankle there are five alterations in movement that may be combined or individually present to adapt to functional hallux limitus: (1) altered heel lift, (2) vertical toe-off, (3) inverted step, (4) abducted toe-off, and (5) adducted toe-off. **(Dananberg, 1993)**

Heel lift occurs as the body weight moves forward over the metatarsophalangeal articulation. If there is failure of hallux extension the adjacent midfoot joints become obliged to move, giving the appearance of excessive pronation. Over a prolonged period the bones will remodel according to Wolff's law, **(Wolff, 1986)** causing greater instability and extended pronation. Since the average person takes 5,000 steps per day, or 2,500 steps per foot, this subtle imbalance is repeated thousands of times per day. When the heel cannot lift because of hallux extension failure, the next form of compensation — vertical toe-off — develops. Direct lifting of the toe from the substrate eliminates the thrusting forward of normal forward motion often seen as the slow, shuffling gait of the elderly.

The last three variations due to functional hallux limitus are avoidance compensations. Normally weight is transferred through the foot from the heel along the lateral longitudinal arch, across the metatarsals to the hallux. In the presence of hallux limitus there is prolonged weight bearing along the lateral longitudinal arch, with failure to move over to the hallux because of its inability to properly extend. The aponeurosis is never tightened by Hicks' mechanism. With heel lift there is a rapid stretching of the aponeurosis that has not been properly tightened. This results in trauma to the aponeurosis attachment at the calcaneus that may cause inflammation and ultimately a heel spur with the consequent pain. Typically there will be excessive lateral wear on the forefoot of the shoe. In this case functional pathologies within the foot and elsewhere are not due to the ineffectiveness of the windlass mechanism, but the functional inability of the great toe to extend, producing another reason for plantar fasciitis and heel spurs to develop.

It must be emphasized that pain may be located almost anywhere in the body, and there may be no foot complaint. **(Dananberg, 2007)** Remote pain develops as a result of compensation that takes place to dissipate the kinetic force that is not properly transmitted through the foot and leg for forward propulsion. When the forces of gait are dissipated abnormally in the body, there is recurring remote strain with each step, creating thousands of microtraumas at the area each day. The body responds to the trauma by inflammatory reaction, swelling, and stiffness. The resulting chronic inflammation is the hallmark of degenerative joint disease of old age, especially when it is combined with systemic problems that cause failure of tissue regeneration. **(Hurley & Newman, 1993; Dananberg et al., 1990)**

The following table lists the normal motion that occurs during the second half of single support during gait and the compensations when functional hallux limitus is present. **(Dananberg, 1994)**

Symptoms

Functional hallux limitus has been overlooked for several reasons. Symptoms are rarely in the foot until well advanced; they are then attributed to the extended pronation, not to its original cause. In many cases gait may appear normal on visual observation, yet objective recording by the Electrodynogram™ clearly shows the disturbance. Finally, the hallux range of motion in the non-weight-bearing foot examination is completely normal.

Symptom possibilities are as follows: lower leg, knee, thigh, sciatic-type pain, lower back, neck, TMJ, **(Dananberg, 2007; 1994; 1988)** and chronic headache among others. Many other corrections may not remain intact as a result of this remote imbalance. Treatment to the areas of pain and dysfunction may help the symptoms; however, if the primary cause of gait dysfunction is not corrected, symptomatic relief is not long-lasting or may manifest in some other area of the body…a consistent

Joint	Normal Motion	Compensatory Motion
Mid-tarsal joint arch	Supination	Pronation
Ankle	Plantar flexion	Dorsiflexion
Knee	Extension	Flexion
Hip	Extension	Flexion
Lumbar spine	Lordosis	Lumbar flexion
Cervical spine	Lordosis	Cervical flexion

Compensatory motion that may take place during the second half of single support due to FHL.

finding in applied kinesiology, reiterated throughout the applied kinesiology literature. The ability to specifically test the muscles in a kinetic chain commonly involved in FHL allows the AK practitioner to explore the many ramifications of this condition with greater ease and specificity than before.

Visualizing the weakness or imbalance in the hamstrings, or psoas, or contralateral shoulder or cervical flexor muscles for example is quite challenging. **(Goodheart, 1996; Dananberg, 1986)** Visualizing whether the movement pattern anomalies resulting from the FHL are due to inhibition or over-facilitation of these muscles' antagonists is impossible without the MMT. Furthermore is there an associated joint subluxation or fixation specifically relating to these muscular imbalances related to the FHL in the body? The AK method of challenge and/or therapy localization can detect this related factor as well.

By broadening the examination process to include the entire kinetic chain resulting from FHL, the AK practitioner has a tool that permits the discovery of the many consequences of FHL that a local focus on the 1st metatarsophalangeal disorder would limit.

Examination

In the non-weight-bearing foot there should be full hallux range of motion. The condition is found with the patient weight bearing. With the patient standing in a neutral position, i.e., with the feet in the normal position of gait, the hallux is passively lifted into extension by the physician. Normally the medial arch will rise and the tibia will externally rotate. **(Kitaoka et al., 1995)** An easy way to test for this is to have the patient stand on a platform, such as a 1"-thick board, with the ball of the foot close to the edge so the toes hang over. The patient stabilizes himself by holding on to the physician's shoulder or something else and stands only on the foot being examined. The physician attempts to lift the hallux into extension; failure of passive extension when weight bearing is a positive test.

Weight bearing can be simulated with the patient supine. With one thumb push up on the ball of the foot directly under the 1st metatarsal head until some resistance is felt. With the other thumb, attempt to extend the hallux while maintaining the constant upward pressure on the 1st metatarsal head. Resistance of the hallux to move into extension indicates the standing test will probably be positive.

The extensor hallucis longus and brevis muscles will test weak in the presence of flexor hallux limitus. This weakness causes secondary contraction of the flexor hallucis longus and brevis muscles. The contracted muscle is secondary to the weak antagonist muscle, a familiar finding in applied kinesiology. Applied kinesiology treatment for FHL is directed to returning the weak extensor hallucis muscle(s) to normal.

Podiatry

There are several modes of treatment by podiatrists that specialize in gait analysis and the foot's relation to remote body function. Treatment may consist of orthotics, various pads and strapping, manipulation, and shoe modification. **(Dananberg et al., 1996)** Dananberg describes the Kinetic Wedge® orthotics **(Langer Biomechanics Group)** for correction of FHL. The Kinetic Wedge® is also built into athletic shoes. **(Dananberg, 1988)**

Dananberg and Guiliano **(Dananberg & Guiliano, 1999)** have demonstrated that 84% of patients with chronic lower back pain, who have been prescribed the orthotics described above, significantly improved. Another study of podiatric treatment of functional hallux limitus resulted in 46% of the patients reporting they were 75-100% better and 35% reporting 50-75% better, making a total of 77% to be at least 50% better. **(Dananberg et al., 1990)** The most interesting aspect of this study is that none of the patients complained of foot pain or discomfort. The primary symptoms were all remote from the feet.

Dananberg **(Dananberg, 2001)** also shows the effect of podiatric treatment of FHL upon hip extension range of motion. He retrospectively reviewed hip joint range of motion on 20 subjects walking before and after intervention with custom foot orthotics. He demonstrated a 50% increase in hip joint extension by the conclusion of single support phase during gaiting as a result of the functional changes brought about by the orthotics.

Applied Kinesiology and FHL

Examination and treatment of functional hallux limitus in applied kinesiology is directed to finding the cause and making lasting correction of extensor hallucis muscle weakness. Disturbances in muscle function due to subluxation or fixations of the foot or ankle, myofascial trigger points, low back or pelvic joint dysfunctions producing muscle imbalances that alter the gait cycle are not sufficiently treated with the use of orthotics alone. Supporting the 1st MTP joint when it is not required (because its cause is elsewhere) is also treating the symptom not the cause. A major factor in all applied kinesiology corrections is that the weakness in the extensor hallucis muscle does not return after walking. Test the extensor hallucis with the ankle in dorsiflexion and the hallux in extension. Apply the testing force to flex the hallux only at the metatarsophalangeal articulation. The weakness may be uni- or bilateral. The extensor digitorum muscles are also tested and may or may not be weak. The weak muscles are treated with the five factors of the IVF involved with the weakness. All foot subluxations should be corrected before examining for FHL. Correcting foot subluxations and the muscles will not usually correct FHL; it is a separate entity. Any or most of the following may need attention for lasting correction.

Liver meridian. The liver meridian begins at the lateral nail point of the hallux. LV 3 is where the 1st and 2nd proximal metatarsal bones join, just lateral to the extensor hallucis tendon. Mann (**Mann et al., 1992**) places a high degree of importance on the LV 3 area. He discusses it as an area rather than an acupuncture point. Goodheart (**Goodheart, 1998**) noted that activity at LV 3 is often present in FHL. Often a weak extensor hallucis temporarily strengthens with therapy localization to the liver alarm point (LV 14); when it does, stimulation of LV 3 strengthens the muscle. Stimulation can be done by any standard method of acu-point stimulation, e.g., teishein, fingertip tapping, laser, needle, electrical, and others. When experience indicates that FHL is present but the extensor hallucis is not weak, it will often show a subclinical weakness with therapy localization to the liver alarm point. When the extensor hallucis is strengthened by stimulation of LV 3, the improved function will often be lost after walking if other corrections are not made.

Origin and insertion technique. Therapy localize the origin and insertion of the extensor hallucis; if it strengthens, apply the hard digital pressure used in that technique. Most often the involvement will be at the origin along the middle one-half of the medial aspect of the fibula.

Repeated muscle activity – patient induced. If the extensor muscles weaken after the patient actively extends the toes ten times, origin and insertion technique is needed along with nutritional support with water for additional hydration and wheat germ oil three times a day.

Muscle stretch reaction. If the extensor muscles weaken following stretching, apply fascial release, trigger point, or myofascial gelosis technique as indicated by examination. The most common need is for fascial release. Vitamin B_{12} may be necessary for lasting correction.

Maximum muscle contraction. If the extensor muscles weaken following maximum contraction, apply strain/counterstrain technique. Chaitow has presented Goodheart's techniques for strain/counterstrain approvingly in a number of his books. (**Chaitow, 2008, 2005, 2002, 1988**) The tender point will usually be at a more proximal area of the muscle. A collagen source of glycine may be needed for lasting correction.

Rib pump technique. Determine the need for rib pump technique by therapy localizing in the 4th and 5th anterior rib area. Strain/counterstrain technique often needs to be applied in this condition to ribs 4 and 5, both anteriorly and posteriorly, for adequate rib pump activity.

Deep peroneal nerve. The deep peroneal nerve may be entrapped by distal tibia and fibula spreading, much like nerve entrapment at the carpal tunnel by the spreading of the distal radius and ulna. Test for extensor hallucis strengthening by having the patient approximate the proximal one-third of the tibia and fibula by holding them together. The approximation could also be done at the distal ends, but most patients cannot reach that far down. The distal tibia is adjusted toward the fibula with the patient side-lying and the physician using a high velocity thrust.

Support. Most often the patient will lose the corrections immediately upon walking. If the corrections are lost with walking, support is added to give the corrections an opportunity to stabilize.

The correction at the tibia and fibula is supported by taping with porous adhesive elastic tape, such as Elastikon by Johnson and Johnson. Use two layers, taking care not to apply too much tension to the elastic and impede circulation. The patient should wear the support over the distal tibia and fibula for one week. He can bathe or shower without taking the tape off because it dries rapidly and will not loosen. Patients are not usually allergic to this type of tape. If there is a problem with allergy, prevent the adhesive dermatitis by placing the first circular wrap with the sticky side out and then apply two more wraps over the first one in the usual manner.

Finally the podiatric approach involves a triangular adhesive felt or foam pad placed under the 2nd-5th metatarsal heads. The felt pad provides a more solid support than the foam pad, but the foam is tolerated better than felt regarding comfort and usually provides adequate support. The pads are available in 1/8" or 1/4" thickness. They are 2" wide at the point under the metatarsal heads. The lateral border of the pad extends along the 5th metatarsal as the pad narrows to about 1/2" width at the end pointing toward the calcaneus. Most patients can easily accommodate the pad in their regular shoes. The metatarsal foam pad survives bathing or showering better than the

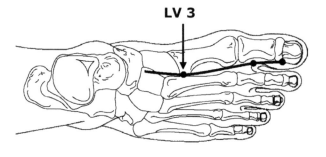

Liver 3   (With kind permission, ICAK-USA)

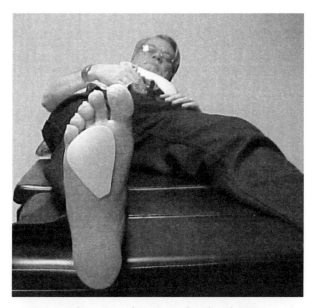
Correct placement of metatarsal pad with FHL

felt one. Neither lasts as well as the adhesive tape wrap. It is advisable to give the patient several pads to take home with instructions for application. Caution the patient to not walk without the pad(s). If it is removed for bathing it should be replaced before walking. The pad(s) can be left in place when showering and then replaced before dressing. The triangular pads are available from podiatric suppliers.

The immediate prescription of an orthotics or metatarsal pad should not be the first line of therapy for FHL. The benefits of these devices is well documented, however the need for other therapies for causative factors in the FHL syndrome may be neglected.

Hallux Rigidus

Hallux rigidus is the most common degenerative arthritis of the foot affecting most commonly people between 30 to 60 years of age, but it is a common disorder in young athletes as well. It is a premature arthritis of the first metatarsophalangeal (MTP) articulation, and the disability resulting from hallux rigidus is greater than that seen with hallux valgus. (**Zammit et al., 2009**)

Hallux rigidus is commonly preceded by some form of physical trauma to the first MTP joint; the patient may or may not remember the incident. Manral (**Manral, 2004**) lists the causative factors: physical trauma is primary, followed by systemic conditions affecting the first MTP. Secondary etiological factors include pronation of the foot, improper shoes (with a narrow toe box), interposed sesamoid bones, age related arthritis, disorders of the proximal parts of the lower limb, gait abnormality due to any cause, obesity, and occupational ergonomic factors. Anatomically it involves the periosteum, joint capsule, synovium and subchondral bone. Occasionally the MTP joint of the smaller toes is affected as well; in that case it is called Freiberg's disease. (**Turek, 1984**) Besides the visual inspection of the first MTP joint (arthritic inflammation and swelling are not always present in FHL), an important diagnostic factor distinguishing hallux rigidus from FHL is that there is decreased range of motion of the first MTP joint weight-bearing *as well as* non-weight-bearing. In FHL, the hallux can still extend when passively examined or when the patient is non-weight-bearing, except during a very brief portion of stance phase of gait that may be only 100 milliseconds in duration. (**Dananberg, 2007**) In hallux rigidus, X-ray shows joint degeneration, dorsal osteophyte formation, and MTP joint narrowing. (**Zammit et al., 2009**) Passive extension of the great toe is also painful with swelling and tenderness present, and the toe held in slight flexion.

Because the great toe of the patient with hallux rigidus has limited dorsiflexion, push-off during gait can be painful. Cheilectomy (the chiseling away of bony irregularities on the lips of a joint cavity) is commonly performed in which not only the dorsal spur but also the dorsal third of the metatarsal head is removed. (**Mulier et al., 1999**) Mulier et al claim that this provides long-term pain relief in most patients. If this fails, arthrodesis (surgical fusion of the joint) is suggested. Surgeons suggest that mobilization of the toe should be initiated soon after surgery, emphasizing the functional importance of movement therapy in the treatment of disorders like this one. The widespread negative influences of FHL (see section above) also suggest that the musculoskeletal impairments resulting from hallux rigidus may also benefit from the functional improvements delivered for the first MTP, via conservative treatment or other means. (**Dananberg, 2007; Manral, 2004**) Conservative treatment is worth a clinical trial in the early stages of this disorder. (**Smith et al., 2000**) Smith et al. reviewed 24 cases with hallux rigidus, and showed that in 75% cases, the patients would "still choose not to have surgery" if they had to make the decision again.

Hallux Valgus

Hallux valgus is the lateral deviation of the great toe with secondary medial deviation of the 1st metatarsal. The condition modifies foot function so that the weight load of the forefoot shifts toward the lateral side of the foot. The 2nd and other metatarsals begin to take more of the load, and metatarsalgia develops. (**Nix et al., 2010; Mann, 1983**) The bursa located on the medial aspect of the first metatarsal head may become inflamed (usually due to rubbing on the shoe), resulting in the formation of a bunion.

Nix et al (**Nix et al., 2010**) conducted a meta-analysis of the literature involving a total of 78 papers reporting results of 76 surveys (total 496,957 participants). Prevalence estimates for hallux valgus were 23% in adults aged 18-65 years, and 35.7% in elderly people aged over 65 years. Prevalence increased with age and was higher in females (30%) compared to males (13%).

The terms "hallux valgus" and "bunion" are often used interchangeably in the literature. Here, reference to hallux valgus means the disrelation or subluxation of the articulation(s). Characteristic features are the bony overgrowth on the medial distal head of the 1st metatarsal, and the lateral deviation of the great toe at the metatarsophalangeal articulation (rarely the interphalangeal articulation). General agreement is that hallux valgus is present when there is 8-10° of lateral

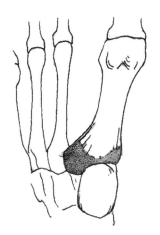
Enlarged 1st metatarsal base.

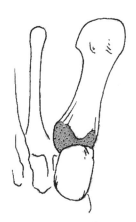
Angled 1st metarsal base.

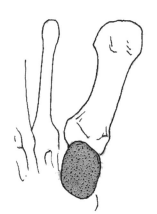
Angled 1st cuneiform.

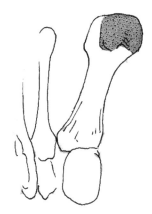
Enlarged 1st metatarsal head.

deviation at the metatarsophalangeal articulation. (**Chhaya et al., 2008; Lidge, 1976**)

The term "bunion" is not as specific. It probably is derived from the irregular French word "bunny," which means swelling. (**Webster's Third New International Dictionary of the English Language, 2002**) Langer, Maffetone, and Kelikian (**Langer, 2007; Maffetone, 2003; Kelikian, 1982**) refer to hallux valgus as a complex. The lateral deviation of the great toe is only occasionally an isolated entity. Often there are also deformities of the lesser toes, and the metatarsal bones are distorted from their normal positions.

Etiology. Throughout the literature on hallux valgus there is strong opinion that the condition's etiology relates to (1) congenital anomaly or function of the bones, (2) muscle anomaly or improper function, and (3) improper footwear. Although not mentioned as often, Nix and Mann (**Nix et al., 2010; Mann, 1983**) emphasize extended foot pronation as a cause of hallux valgus; this puts an increased force on the medial side of the foot near the time of toe-off. Goodheart, Bandy, Leaf, and Lee et al (**Leaf, 2006; Lee et al., 2003; Bandy, 1976; Goodheart, 1998-1964**) recognized these etiologies as applicable to applied kinesiology's therapeutic approach. Conditions such as rheumatoid arthritis must also be taken into consideration as an etiology of hallux valgus. Any condition that causes progressive loss of integrity of the capsular structure supporting the metatarsophalangeal joints, thus providing less lateral support by the lesser toes, must be systemically treated. There is debate as to whether the lateral deviation of the great toe or the medial deflection of the 1st metatarsal is the primary deformity in the hallux valgus complex. (**Mann, 1982**)

Although many authors present convincing evidence that their favorite etiology is the primary cause of hallux valgus, it is obvious that each one — and perhaps a combination — is primary in individual cases. It becomes necessary, then, to analyze each case to determine if surgery, conservative treatment, or palliative care is the optimal approach.

Congenital bone formation. X-ray examination can determine if there are congenital malformations that predispose an individual to hallux valgus. Malformation between the 1st cuneiform and 1st metatarsal can cause the metatarsal to angle medially, causing a secondary hallux valgus. There may be wedging of the 1st cuneiform, or obliquity of the base of the metatarsal so that the latter articulates in a varus position. (**Hammer, 1999; Turek, 1984**) There may also be a large lateral facet or exostosis at the base of the 1st metatarsal that holds it away from the 2nd, causing medial angulation. (**Mann, 1983**)

Footwear. The distortion of feet caused by restriction is classically evident in the no longer practiced tradition of binding Chinese women's feet. (**Chen, 1992**) Any footwear that places lateral pressure on the hallux can be a potential cause of hallux valgus. Two classifications are particularly indicted as causing hallux valgus. Children's shoes are often poorly fitted or not changed often enough during rapid growth periods. Hicks (**Hicks, 1965**) states, "We realize that the onset of hallux valgus and other foot deformities is slow and insidious. Mistakes made in shoe selection when young become painfully apparent only in middle age." Regarding extensive research on the hallux angle in children he states, "Our results show that distortion of the big toe starts at the age of 7 years."

The other major problem is the high-heeled, pointed-toe shoes worn by women. Not only do the pointed toes cause angulation of the hallux, the high heels cause gravity to force the foot into the narrow toe portion of the shoe. Unfortunately, shoes are often selected for style rather than function.

There is conflicting evidence indicating that hallux valgus is caused by wearing shoes. Supporting the shoe-wearing etiology is a 33% incidence of hallux valgus in a shoe-wearing Chinese population versus 1.9% in those who do not wear shoes. In addition, the deformity is nine times more prevalent in females than in males. (**Sim-Fook & Hodgson, 1958**)

Another study was made of most of the population of a small island centered in the south Atlantic Ocean. (**Shine, 1965**) It revealed significant difference between the shoe-wearing and unshod populations. In this study of 3,515 people, hallux valgus of 15 degrees or greater was present in fewer than 2% of those barefoot, and in 16% of the men and 48% of the women who had worn shoes for more than 60 years. There was no significant correlation with age, social class, occupation, or exercise habits. The higher incidence of hallux valgus among women appears to be female-related rather than shoe-related, since both sexes in this population wear similar shoes.

Some consider the higher incidence in women a result of the shoes worn by females, (**Hong et al., 2005; Mann, 1983**) while others believe it is due to females having weaker ligamentous structure. Sports podiatrists have observed that women runners complain of a widening of the foot and a "knot" forming. Those who work with these athletes indicate that it is due to weaker ligaments in females. (**Nguyen et al., 2010; Hlavac, 1977**)

When the 1st metatarsal abducts in spread foot, there is propensity for secondary hallux valgus to develop. In a high-heeled shoe, the metatarsophalangeal articulation is extended throughout gait. The sesamoid bones are thus anteriorly displaced on the 1st metatarsal head at heelstrike. The ridge between the sesamoid bones ordinarily stabilizes them in their distal and proximal movements, but the height of the ridge as it progresses distally is less, and it fails to stabilize the bones adequately. This loss of bony guidance causes resistance to the 1st metatarsal abduction to depend entirely on the ligament of the medial sesamoid bone. At the moment of and immediately after heelstrike the tibialis anterior eccentrically contracts to prevent the forefoot from slapping on the ground. (**Winter & Scott, 1991**) This action is exerted on the 1st metatarsal, which is inadequately stabilized by the usual mechanism. The tibialis anterior in this way contributes to 1st metatarsal abduction in a spread foot. (**Haines, 1947**)

Although evidence points toward improper shoes as a contributing cause of hallux valgus, it certainly is not the only one. Kaplan (**Kaplan, 1955**) considers that shoes are not a major cause of hallux valgus. The condition was observed in World War I and II military personnel who came from a farm background in which shoes were infrequently worn. Some central African tribes and Australian aborigines who never wore shoes developed hallux valgus. This brings us to the primary role that muscles play in the condition.

Muscle role. Two considerations must be made regarding the role muscles play in hallux valgus. First is the anatomical arrangement of the muscles, a factor with which conservative treatment cannot cope. Second is the function of the muscles, as readily determined by applied kinesiology examination methods.

There is considerable congenital variation of the foot muscles. (**Bejjani & Jahss, 1986**) Extensive study has been done by dissection and electromyography of the muscles' action, both in normal and congenital variations. Possibly the reason for the variations is the bipedal stance unique to man. Thomson (**Thomson, 1960**) considers the abductor and adductor hallucis muscles as two of the most important primitive muscles not necessary to the human foot. Since the major motion of the great toe in the human is flexion and extension, he believes that "…these two muscles lose their functions or their insertions had drifted to aid in the normal plane of movement." Of specific interest to us is the action of the abductor hallucis.

Lovejoy (**Lovejoy, 2007**) offers a list of distinct lower limb morphological characteristics relating to the structural and functional anatomical adaptations in the lower leg and feet. These adaptations reflect bipedal gait in early hominids (including *Australopithecus*):

- High anterolateral lip of the lateral femoral condyle providing patellar retention with the fully extended knee
- Transverse distal tibial plafond
- Wedged talar dome
- Longitudinal and transverse pedal arches
- Permanently adducted first ray

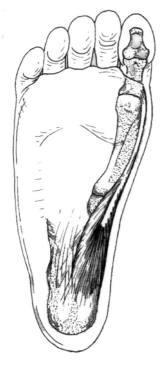

Abductor hallucis.

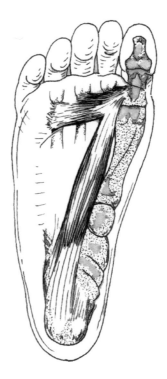

Adductor hallucis.

- Vertical groove for the tibialis anterior
- Distinct dorsiflexion profile in metatarsophalangeal joints for all five rays
- Markedly expanded calcaneal tuber for energy dissipation at heel strike

Thomson (**Thomson, 1960**) discusses an unpublished study by Basmajian and Kerr that is mentioned only briefly in Basmajian and Deluca's text. (**Basmajian & DeLuca, 1985**) Twenty-two adult feet were dissected to investigate the insertion of the abductor hallucis muscle. "In only 5% of the specimens did the tendon lie on the medial border of the foot, and insert into the medial side of the base of the proximal phalanx as an obvious abductor." (**Thomson, 1960**) In 19% of the specimens, the abductor hallucis joined with the medial head of the flexor hallucis brevis, inserting into the base of the medial sesamoid to be an obvious flexor. Between these two extremes, in 19% of the cases the abductor tendon took a plantar position, passing over the medial portion of the sesamoid without attaching to it, finally attaching to the base of the proximal phalanx. In 25% of the cases, the lateral slip of tendon inserted into the medial sesamoid before its insertion on the phalanx. In another 25%, a common slip of the insertion was to the medial head of the short flexor into the sesamoid. In only 19% of the cases was the abductor hallucis anatomically located as a true abductor. In 83%, the adductor hallucis acted to flex the great toe at the metatarsophalangeal articulation. Another study (**Agawany & Meguid, 2010**) investigates the insertion of the abductor hallucis muscle and also shows the variations in the insertion of this muscle.

This makes the role of the abductor hallucis muscle in the prevention of hallux valgus questionable from what one would think after studying the standard anatomy texts, which clearly show the insertion into the medial 1st phalanx base.

Iida and Basmajian (**Iida & Basmajian, 1974**) studied the balance between the adductor and abductor muscles in "idiopathic" hallux valgus and normal feet with electromyography. In hallux valgus, the adductor hallucis muscle becomes markedly weak with time, but the abductor hallucis muscle loses its abduction function completely and works only as a flexor. Because of mechanical misalignment, the flexor hallucis brevis tendon stretches so that the muscle loses some of its function as a flexor. The stresses of weight bearing with time cause the muscle imbalance to become worse. In this study, they could not conclude whether hallux valgus is caused from muscular imbalance, or the muscular imbalance develops secondarily to the structural imbalance. In any event, one must not count on the abductor hallucis' participation in the correction of hallux valgus until one determines that it is actually an abductor.

The length and function of both heads of the adductor hallucis must be evaluated. Their shortness or hypertonicity can contribute to the deformity. It appears important to provide a therapeutic approach to return optimal balance to the intrinsic muscles.

Another congenital variation that may produce hallux valgus was demonstrated by Kaplan. (**Kaplan, 1955**) By dissection he found a consistent expansion of the tibialis posterior tendon into the flexor hallucis brevis and the oblique head of the adductor hallucis in those with hallux valgus. The same expansion of the tendon was not found in normal subjects. Pulling on the tendon caused an increase in the hallux valgus deformity. There was no influence on normal subjects' toes when the tendon was pulled.

Kaplan also demonstrated an increase in the hallux valgus deformity by stimulating with faradic current the tibialis posterior in living subjects. On the other hand, there was no visible movement of the big toe in normal subjects with the stimulation.

If spread foot and hallux valgus begin to develop, muscle action can be responsible for increasing the problem even when there was originally proper muscle attachment to the bones. (**Arinci et al., 2003**) Action of the flexor hallucis longus increases the valgus position of the hallux and the varus position of the 1st metatarsal when hallux valgus is present. (**Snidjers & Philippens, 1986**) For this reason, lateral pressures on the hallux during childhood — from shoes or socks that may start the process that will be continued by the muscle activity — should be avoided. This supports the finding of Hicks (**Hicks, 1965**) "… that if a woman can reach the age of 20 years with a big toe angle of less than 10, she is unlikely to develop bunions in old age."

An unusual cause of hallux valgus is, ironically, surgery to correct hallux varus, due to the change in insertion of the abductor hallucis. (**Thomson, 1960**)

Muscles may play a further role in hallux valgus by developing histologic abnormalities. This appears to develop from "…chronic ischemia caused by the elevated pressure occurring within the foot during gait." (**Hoffmeyer, 1988**) Fifty-three of 57 patients with hallux valgus had histologically abnormal biopsies of muscle taken at the time of surgery for correction of the deformity. There were myogenic and neurogenic alterations, as well as ultrastructural changes. Gait evaluation revealed that this group, in comparison with a control group, had 70% abnormal rollover pattern; in the control group, only 20% failed to have the usual pattern (heelstrike, 5th, and then 1st metatarsal head contact, ending with great toe push-off). Finally, the muscle activity, as indicated by surface EMG, was abnormal in the patient group when compared with the volunteer group. The difference in distribution of the electrical activity patterns was statistically significant ($p<0.0005$).

Histological abnormalities are common for injured muscles. Proprioceptive acuity depends upon intact mechanoreceptors and their peripheral to central pathways. (**Bard et al., 1992**) However in injuries the damage is to the proprioceptive apparatus and mechanoreceptors themselves. Later this may be accompanied by adaptive central re-organization throughout the motor cortex.

Pronation. Eversion of the foot at toe-off in extended pronation is an important contributor to hallux valgus. The examination and therapeutic approach for extended pronation, if present, should be routine in all cases of hallux valgus.

Examination and treatment. Many factors must be taken into account before deciding to attempt conservative management and treatment of hallux valgus. Limiting effective conservative care may be an advanced stage of degenerative development, congenital anomalies such as wedging of the 1st cuneiform, functional hallux limitus, hallux rigidus, and malformation of the base of the 1st metatarsal. **(Chhaya et al., 2008; Maffetone, 2003; Cholmeley, 1958)** Occasionally conservative management is necessitated because the patient is a poor candidate for surgery, such as when there is a circulatory deficit. **(Veves et al., 2002; Kelikian, 1982)** If circulation is a problem, as in diabetes or patients on blood-thinning medications, the case becomes much more critical. One must carefully guard against pressure necrosis and infection. **(Veves et al., 2002; Aronow & Solomone-Aronow, 1986)** Further complicating the case is an insensitive foot. Finally, before deciding on conservative care, factors must be found that contribute to or cause the hallux valgus.

Although hallux valgus may appear to be a localized problem, total body function must be analyzed. Disturbance in gait function, which may relate with pelvic and spinal organization, knee problems, and especially excessive pronation, must be evaluated and corrected. The examination and treatment of these conditions are dealt with elsewhere in these texts.

Local examination of the foot for hallux valgus should begin with x-ray to evaluate for congenital anomalies and general bone condition. Even if the x-ray indicates referral for surgery, intrinsic muscle evaluation and correction are advisable.

Determine abduction capability of the abductor hallucis by having the patient attempt to abduct the hallux. While the patient attempts the action, the physician should palpate the abductor hallucis muscle to determine if it is contracting. If there is simply flexion of the hallux with the muscle's contraction, one cannot count on correcting hallux valgus by improving the muscle's function.

Even in normal individuals, this is a difficult muscle to selectively contract because wearing shoes restricts the activity. In those with hallux valgus, the abductor hallucis is typically weak making isolated contraction even more difficult. The patient should learn this activity by abducting the hallux against his finger, spreading his toes, and sliding the forefoot toward the other foot while keeping the heel stationary. In addition, the hallux can be abducted passively to stretch the adductor hallucis. **(Chaitow & DeLany, 2002; Ramamurti, 1979)** With persistence, an individual will be able to make a small abduction movement if the tendon is inserted to perform the activity.

The distal tendons of the flexor hallucis brevis contain the sesamoid bones and insert into the medial and lateral sides of the proximal phalanx of the hallux. This muscle is often found weak on manual muscle testing. **(Leaf, 2003; Bandy, 1976)** To help align the sesamoids and supply some stability to the hallux, both the abductor hallucis and the flexor hallucis brevis should be strengthened with applied kinesiology techniques.

The intrinsic muscles of the foot often respond to origin and insertion technique. Treatment to the Golgi tendon organ or neuromuscular spindle cell may be necessary. The neurolymphatic reflexes are located bilaterally inferior to the symphysis pubis at the height of the obturator foramen. Posteriorly they are between the posterior superior iliac spine and L5 spinous process. The neurovascular reflexes are located bilaterally on the frontal bone eminences.

The adductor hallucis may be hypertonic. Both the

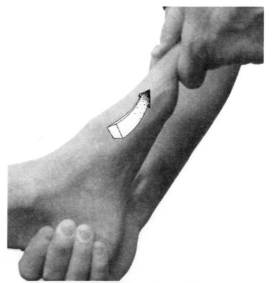

First metatarsal and 1st cuneiform adjustment.

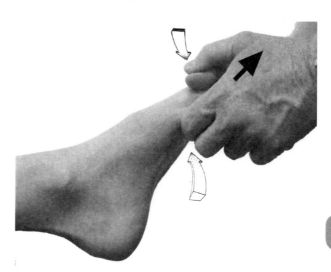

Separation between the 1st and 2nd toes and contact on the dorsal and plantar surface of the 1st toe proximal phalanx.

oblique and transverse heads contribute to adduction of the hallux. Fascial release technique can be applied to help ensure normal function of these muscles. Manipulation of the 1st metatarsal and hallux, when needed, should be done after other foot corrections have been obtained.

Severe arthritis or other joint disease contraindicates HVLA manipulation. **(Bergmann & Peterson, 2010; Chaitow & DeLany, 2002)** Mennell **(Mennell, 1969)** states that it is all right to manipulate a hallux valgus with early osteoarthritic change, but there will probably be failure if there is any dorsal lipping.

First metatarsal and 1st cuneiform adjustment.

The adjustment may be needed to correct a subluxation or fixation of the articulation. Applied kinesiology challenge will determine the necessary direction of correction. To adjust the articulation, the physician holds the supine patient's left calcaneus in his right hand, with the heel of his hand applying firm pressure around the distal calcaneus and cuboid for stabilization. His left hand grasps the 1st metatarsal, with the fingers wrapping underneath and the thumb contacting the bone's dorsal base. Traction is applied with the thumb and fingers moving the metatarsal in the direction of optimal challenge. The base may need to be moved in any direction, but it usually requires a lateral thrust. Often the combination is separation of the base from the 1st cuneiform and lateral movement of the head. It is sometimes difficult to get a good contact on the metatarsal head because of the soreness of the individual's bunion. A towel for padding aids in avoiding pain, and also helps keep the doctor's hand from slipping on the extremity.

First metatarsophalangeal adjustment.

The metatarsophalangeal articulation of the hallux should be challenged and adjusted accordingly. Techniques typically emphasize reduction of the valgus position and traction. **(Hearon, 1994; Schafer, 1982; Bandy; 1976)** Traction of the phalanx is emphasized in the three techniques described by Gillet, **(Gillet & Liekens, 1981)** Bandy, **(Bandy, 1976)** and Schultz. **(Schultz, 1979)** One of these three techniques should meet the needs of the challenge.

Traction is the main factor in Gillet's technique. **(Gillet & Liekens, 1981)** It is applied to the phalanx to separate the articulation. It is necessary to obtain a good grip on the hallux, and the traction must be rapid and strong. There is rarely any pain. If the grasping fingers tend to slip, a better purchase can be obtained by wrapping the toe in a paper towel or facial tissue.

In Bandy's adjustment, **(Bandy, 1976)** the physician stands to the left of the supine patient for correction of the left hallux. Grasp the toe with the left hand and apply pressure in the direction of optimal challenge. The head of the 1st metatarsal is contacted with either the thumb or the softer web between the thumb and index finger. If the head of the 1st metatarsal is exquisitely tender, contact slightly proximal to it. The physician stabilizes the ankle and foot with his right thigh. With a quick traction of the toe, the 1st metatarsal is thrust laterally in a quick one-two motion.

Schultz **(Schultz, 1979)** recommends that care be exercised in the use of his adjustment for the hallux because the digit does not allow as much flexion as the other digits. Description is for the left foot. With the patient supine, the physician puts his right thumb between the 1st and 2nd toes so that the web between the patient's toes and that between the doctor's thumb and index finger approximate each other. The index finger of the doctor's left hand contacts under the hallux. Added support is accomplished by the right finger supporting the left finger, and the left thumb supporting the right. Apply axial traction to the toe to separate the metatarsophalangeal articulation. Without releasing the traction, quickly — by wrist action — bring the fingers up and the thumbs down. The doctor's thumb between the hallux and the 2nd toe brings the toe medially, which is usually the direction challenge indicates for correction. There should be minimal pain to the patient. Pain indicates that too much downward pressure is being applied with the thumbs. Remember, there is less flexion at this articulation.

In addition to the manipulation described above, it is often valuable to mobilize the hallux in all directions. For general stretching and mobilization of the foot, the patient will often relax better when lying supine, with his hip and knee flexed and his foot flat on the table. **(Maitland, 1977)**

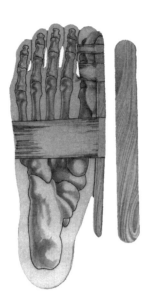 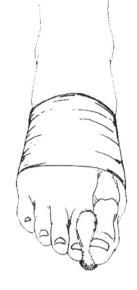

Nighttime hallux valgus support. *Daytime hallux valgus support.*

Support.

Following correction of the muscles and joints, support is added until maximum correction has been obtained. Ramamurti **(Ramamurti, 1979)** recommends that a night abduction splint be worn when passive abduction is limited to an angle short of the longitudinal axis of the foot.

A night splint can be simply made from a tongue depressor taped to the toe and foot; it is especially valuable in juvenile hallux valgus. **(Cailliet, 1997)** The tongue blade is applied to the medial foot, with cotton padding as needed. Daytime taping can be applied with spacers, which are available from suppliers or can be made by building pads out of 2" x 2" sponges or cotton. **(Krissoff & Ferris, 1979)** Acute pain from a bunion is due to acute bursitis, which should be appropriately treated. **(Mennell, 1969)** When bursitis is present, there is an inflammation of the lubricating membrane.

There are four classic signs of inflammation: pain, heat, redness, and swelling. The basic reason inflammatory processes develop is injury to the tissue. When the bursa is inflamed it cannot perform its primary lubricating function properly. This is particularly important because the bursa around the bunion is a point of high stress and wear.

Maffetone **(Maffetone, 2003)** has described the "biochemical foot factors" which should be considered and tested for using applied kinesiology methods in order to control inflammation in patients with foot pain and disorders. Schmitt and McCord, in their applied kinesiology approach called *Quintessential Applications: A(K) Clinical Protocol* **(Schmitt & McCord, 2010)** has provided excellent methods for evaluating and conservatively treating excessive inflammation.

In applied kinesiology the key factor when bursitis occurs is not removal of the symptoms of pain and limited function of the area; rather, it is finding the exact cause of the bursitis and eliminating it. Unfortunately, symptomatic treatment for bursitis in the form of painkillers and anti-inflammatory drugs is very common.

Bursitis very often parallels the different forms of

arthritis and, indeed, is caused by some of the same factors that cause arthritis. Protein and calcium metabolism are very important in the development of certain types of arthritis, as they are in the development of bursitis. Uric acid metabolism is important in gouty arthritis; it can also be the cause of bursitis.

A common cause of bursitis is excessive structural strain. The bursa becomes inflamed as a result of excessive wear in an area that already has excessive use, misuse or abuse. This excessive stress often occurs in the feet because of imbalanced muscular pull. If muscular balance around the forefoot is not present it must be regained or, regardless of the treatment to the bursa itself, the condition will remain and probably flare up again after the medication has worn off.

Bursitis may develop because the membranes comprising the bursa have an inadequate nutritional level; thus tissue is not as strong as it should be. It is very important for the bursal membranes to be healthy because, by the very nature of the membranes, they are subjected to significant stress. A lowered health level results in breakdown during periods of wear. Sometimes an individual will work extra hard in the garden, at sports, or at some other physical activity, and develop bursitis. The physical activity gets blamed for the bursitis; actually, the membranes — in a lowered state of health — were just waiting for extra stress to begin manifesting symptoms. If, upon examination, an inadequate protein level or some other factor causing lowered tissue health is found, it may be necessary to change an individual's diet, add nutritional supplements, or improve the digestive system so the body can correctly use the food ingested. It is very important to remove the cause of bursitis, because long-term bursitis can ultimately result in permanent damage.

Surgery. In 1973, Viladot (**Viladot, 1973**) reported that there were more than 100 operations proposed for the treatment of hallux valgus, with the majority concerned more with foot esthetics foot than with function. There are cases where it is appropriate to refer the patient for surgery; however, one must recognize that surgery is not always successful. Mann (**Mann, 1983**) states, "I believe that only on rare occasions should a hallux valgus deformity be corrected solely for cosmetic reasons. If the patient has a painless hallux valgus deformity, I believe the patient should keep the deformity because even after successful surgery there may be discomfort about the [metatarsophalangeal] joint. I do not believe that the treating orthopedist should ever be in a hurry to correct the hallux valgus deformity because of concern that the deformity will progress. Once the deformity becomes symptomatic, that is the time to contemplate surgery."

A recent Cochrane Systematic Review (**Ferrari et al., 2009**) showed that the numbers of participants who remained dissatisfied at follow up after surgical treatment for hallux valgus, even when the hallux valgus angle and pain had improved, were consistently high (25-33%).

Shoes

Improperly fitted shoes are a major cause of foot dysfunction; perhaps shoes of any kind limit optimal foot development. (**Abshire, 2010; Squadrone & Gallozzi, 2009; Maffetone, 2003; Ramamurti, 1979; Hoppenfeld, 1976**) Abshire, Squadrone & Gallozzi, Warburton, Robbins and others (**Abshire, 2010; Squadrone & Gallozzi, 2009; Warburton, 2001; Robbins et al., 1987, 1988, 1989**) have investigated why barefoot runners in international competition have fewer running injuries than those who have trained in "scientifically" designed running shoes. There is a strong case indicating that foot integrity is developed when adapting to the ground by natural foot contact. In cultures that do not wear shoes, less than 10% of the population will seek medical care for foot pain. (**Langer, 2007**) After studying barefoot populations in China and India, Schulman concluded that shoes were "the cause of most of the ailments of the human foot." Health care access is more limited in these countries, but is has been well-established that shoe-wearing populations are prone to chronic foot problems. In fact, according to the American Podiatric Medical Association, 85% of the U.S. population will seek medical care for foot pain at some point in their lives.

As has been indicated, afferent input from the sole of the foot affects postural awareness significantly. Cutaneous reflexes from the foot are important to posture and gait. (**Kavounoudias et al., 2001**) A shoe dampens the stimulation to foot nerve receptors; it smoothes out the substrate and restricts motion. As far back as 1932, Herzmark (**Herzmark, 1947**) found that providing a mat with uneven surfaces in a playpen helped develop strong arches and good foot function in young children. In 1947, Janse et al. (**Janse et al., 1947**) encouraged barefoot activity for foot rehabilitation in the chiropractic setting. Kerrigan et al (**Kerrigan et al., 2009**) concluded that there was far more impact to the ankle, knee, and hip joints in runners wearing traditional running shoes with elevated heels compared to those running barefoot.

Langer, (**Langer, 2007**) a podiatrist and clinical advisor to the American Running Association, notes that "Walking barefoot allows the feet to function naturally, forcing the muscles to expand and contract, and the joints to bend and stretch to absorb step impact. This, in turn, promotes muscle strength and healthy joint alignment. Conversely, wearing shoes can inhibit the natural function of the feet by providing artificial cushioning and support, constricting the feet into unnatural positions and limiting air flow around the skin and nails." Similar to the way that gloves interfere with the sense of touch in the hands, cushioning in shoes interferes with the sense of touch in the soles.

In the past decade, Masai Barefoot Technology shoes have become a hot item in the marketplace for athletic shoe wear. At the same time Nike, the largest producer of sports shoes, recognized that in order to improve running performance and to decrease running injuries, athletes should frequently train barefoot. (Nike was the originator of the market for "running shoes" with annual revenues of more than $10 billion). (**McDougall, 2009**)

Ramamurti (**Ramamurti, 1979**) calls attention to the effect shoes have on foot development.

> "During barefoot walking, the intrinsic toe flexors invariably participate during push-off. Try to walk barefooted without toe gripping, particularly through sand or over gravelly surfaces. Progress through sand will be tedious and, over a gravelly

surface, painful. Full employment of toe gripping facilitates both tasks. During shoe walking, on the other hand, these provocations to toe gripping are blocked. Hence, the use of shoes will, for most people, decrease the participation by the intrinsic muscles of the foot. However, it will not decrease participation by the long dorsiflexors of the ankles and toes as recovery and heel strike demand their participation, whether in or out of shoes. *As the intrinsic muscles weaken*, the strain on the plantar ligaments and aponeurosis increases, and the tendency of the long extensors and flexors to flex the (interphalangeal) joints while extending the (metatarsophalangeal) joints will be unopposed by the lumbricals and interossei. If, in addition, the heel of the shoe be elevated, the tendency toward talar slip will increase, and if the toe of the shoe be pointed, the adduction forces on the great toe will exceed abduction forces much of the time. Thus, as normal function is prevented, and abnormal postures are forced, painful distortions in the anatomy will begin to emerge."

Maffetone emphasizes this point: (**Maffetone, 2003**)

"An extreme example of this is wearing high-heeled shoes, especially those with very small pointed heels and a small toe box. In this case, all the body's weight is directed into the ground through a very small area – through the heel and the front of the foot. When we wear a flat shoe or are barefoot, the distribution of weight is over a larger area, although even a flat shoe can interfere with our weight bearing."

While society's desire for fashionable shoes drives the footwear industry toward profit, the shoe-wearer's desire for comfort, practicality and diversity of foot use also creates strong countervailing needs. The advent of the modern shoe industry and their new promotion of Barefoot Technologies is a sign that the shoe industry is becoming aware of the science of barefoot walking. An elegant and thorough review of the Barefoot Technologies available to consumers – and the scientific literature regarding life in shoes versus barefoot – has been given by Wallden in the superb *Journal of Bodywork and Movement Therapies*. (**Wallden, 2010**)

At this time there is no need to further discuss the advantages of barefoot function. The constraints of our society require wearing shoes for both protection and comfort, and most of the patients who have foot dysfunctions that present to clinicians for evaluation are wearing shoes during their daily life. However it should be remembered that a growing consensus among runners, fitness experts, podiatrists and researchers suggests that "barefoot therapy" is a critical part of foot rehabilitation. Restoring optimal muscle function in the feet is better achieved barefoot than with many other therapies. Spending ones free time at home, recreationally, and while driving barefoot is an excellent idea. (**Abshire, 2010; Maffetone, 2003**)

There are three prevalent problems with shoes: (1) they are too small for the needs of the feet, (2) they are not properly or soundly constructed, and (3) they are too often selected for style and fashion, rather than the functional features of protection and comfort. (**Abshire, 2010; Maffetone, 2003**)

A Chinese fortune cookie reads, "To forget your troubles, wear tight shoes." Notwithstanding this, Lewin (**Lewin, 1959**) colorfully points out, "There is no shoe too small for an ambitious female's foot." Furthermore, he says, "Fit your feet and not the other person's eye." Unfortunately, it is often difficult to have the proper shoes because the marketability is on style rather than function. (**Abshire, 2010; Robbins & Waked, 1997; Hicks, 1965**) Shoe companies have funded studies hoping to reveal the secrets of shoe comfort, without conclusive results. (**Miller et al., 2000**)

Chantelau & Gede (**Chantelau & Gede, 2002**) noted in their study of foot measurements in subjects 65 and older that two-thirds were wearing shoes that were too narrow. Frey et al. (**Frey et al., 1993**) found in their survey of women with shoe-related pain that most women experienced foot-related problems in their twenties; 88% wore shoes that were too small as measured by an orthopedic surgeon; 79% had not had their feet measured in the last 5 years when buying shoes; 59% wore uncomfortable shoes daily; 76% had foot deformities, with bunions and hammertoes the most common finding; and 80% of women reported significant foot pain while wearing shoes!

Robbins et al (**Robbins et al., 1997**) showed that over-cushioned shoes do not protect from step impact and can even increase impact when compared to firmer footwear. "Expensive athletic shoes are deceptively advertised to safeguard well through 'cushioning impact' yet account for 123% greater injury frequency than the cheapest ones." Maffetone notes "Shoes that are too cushioned can give the nervous system the improper perception that the impact is much less than it actually is, which can result in the inadequate or improper response by the foot (and rest of the body) to the actual impact. In the course of a walk, run, or all day moving about in your shoes, this could add up to be a significant cause of injury." (**Maffetone, 2003**) Nike introduced in 1979 (the company already owned 50% of the U.S. market) its first shoes with Air-Sole cushioning technology (sturdy, sealed pouches of pressurized gas set in the sole that compressed under impact and then sprang back). This new shoe-cushioning technology was an instant success and pushed Nike further into the lead among running shoe corporations and led other brands to develop softer foams, gel packets, and other gimmicks to keep up with Nike's fast-growing promotional and advertising advantage. (**Abshire, 2010**)

Maffetone and Stewart (**Maffetone, 2003; Stewart, 1945**) both provide interesting histories about shoe development. They began as simple sandals of various thicknesses and rigidity to protect the plantar surface from injury by the substrate. The upper portion of the foot needed protection from the cold in more adverse climates. This was originally supplied by loose animal fur and skins. It was not until much later that form-fitted and confining shoes were developed.

Shoes or boots with a true heel did not appear until the 15th century. It is said that Timur, in the battle of Seistan, developed a contracted heel cord as a result of injuries received in battle. After the injury he was known as Timur the Lame, or Tamerlane. By putting a heel under that foot, he was able to walk without pain; to prevent a limp, he put a heel under the other side also. The uniqueness of the heels would be hidden if everyone in his court wore them. This became a historical method for covering royal defects. According to Stewart, it is probable "...that the heel is essentially a prosthesis for shortened heel cords, and that men and especially women are still paying unconscious tribute to the vanity of a wounded Tamerlane...." **(Stewart, 1945)**

The metal shank was introduced in the shoe to maintain the heel in position and give the foot support, which had been removed by the introduction of the heel. The use of a metal shank is not recorded prior to 1875. In 1875 Charles Goodyear, Jr., developed a new machine that made shoes using a new material called rubber, a substance invented by his father...and the modern shoe industry was born.

In his historical account of shoe development, Stewart **(Stewart, 1945)** made observations on the effect of shoes on our current status: "One should remember that the heel is essentially a prosthesis for a deformity. Too many writers assume that heels — any heels — are an essential part of a shoe; *few if any approach the heel analytically from the standpoint of physiology*. What is the effect of the heel on normal foot physiology?" In providing an analytical assessment of the heel, he states, "The toes as they are dorsiflexed tend to spread apart, so that the higher the heel the broader the toes should be, but in modern fashions the higher the heel the more pointed is the toe. Thus the strain is doubled.

"Thus it appears that heels from the physiological standpoint are of primary rather than tertiary importance, being the most destructive factor in foot physiology."

Muscle activity changes with varying heel height. For example, there is increased activity in the lateral gastrocnemius in women with 2-1/2" high heels; this muscle imbalance causes other muscles to lose significant function. **(Butler, 1991; Basmajian & Bentzon, 1954)** Negative heels (earth shoes) decrease activity in the erector spinae and soleus, while the activity increases in the tibialis anterior, rectus femoris, and biceps femoris muscles. **(Soderberg & Staves, 1977)** There have been periodic efforts to

Heel Helper ™

popularize the earth shoe to help low back conditions. In the 1930s there was an attempt to popularize the heel-less shoe in England. **(Mennell, 1939)** Some have varied heel height to regulate the lumbar index. None of these attempts have become popular. Maykel, **(Maykel, 1984)** using applied kinesiology testing, found that there is a point of heel elevation in which general muscle weakness develops. He calls this the "calcaneal tolerance factor." Further testing by Maykel and Manning **(Maykel & Manning, 1987)** found a range of 1-1/2" to 3-1/4" in which the heel can be raised before general weakness is demonstrated on manual muscle testing.

The current concept of a shoe varies from molded plastic slippers to the conventional shoe, consisting of its various parts. Whether a person can function adequately in molded slippers and sandals depends on the individual's

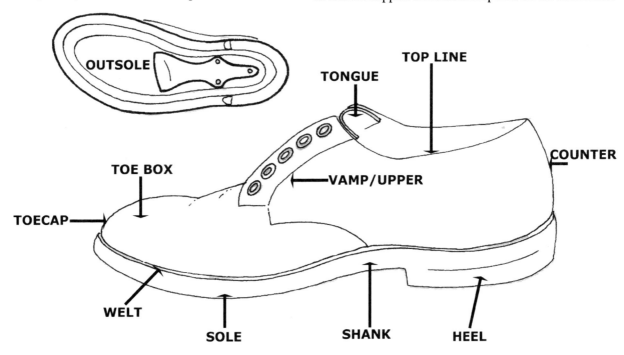

feet. Obviously this type of footwear is inadequate if foot support or orthotics is needed.

The shape of the conventional shoe is determined by its last, which is the form upon which the shoe is built. There are lasts for neutral, adducted, markedly adducted, and abducted feet, as well as for those with bunions. **(Enke et al., 2009; Richards et al., 2009; Milgram, 1964)**

The base of the shoe is the sole; giving it strength is the shank, which is necessary to provide strength from the ball of the foot to the heel. The shank is usually made of steel or some other solid substance. It is anchored at the heel and extends forward under the longitudinal arches. The shoe should bend at the metatarsophalangeal articulations; **(Langer, 2007; Turek, 1984)** this should not be restricted by the shank extending too far forward. In the absence of a strong shank, there is inadequate support under the longitudinal arches. Shank strength can easily be tested by pressing anterior to the heel to determine if the shoe yields. It is common to lose foot corrections as soon as a patient walks in this type of shoe; it can even be the original cause of the foot dysfunction. A weak shank is especially a problem if longitudinal arch support is provided by an orthotic. When the shank yields to the force of the supporting material of the flexible orthotic, there is no support.

A wedge shoe does not have the "bridge" from the ball of the foot to the heel. Support is provided by the shoe's continuous contact with the ground throughout its length. This provides flexibility without yielding to any supporting pressure.

The shoe portion surrounding the heel is the counter. It should be solidly constructed in all shoes, and is especially important when there is extended pronation. Some counters are made of soft, flexible leather; these do not provide proper heel stabilization. Counters made of solid leather, but having weak attachment to the sole, provide poor support and may become misaligned with use. **(Langer, 2007; McMahan, 1983)** Even new shoes may be improperly lasted so the counter is misaligned with the shoe. In either case, the counter will tend to put the calcaneus into inversion or eversion. **(Langer, 2007; Vixie, 1982)** When a person has good foot function the solidity is not as important, but in all cases the counter must be properly aligned.

The front part of the shoe is the vamp; it covers the forefoot and a portion of the midfoot. The anterior portion containing the toes and anterior metatarsals is called the toe box. The tongue extends from the vamp to cover some of the midfoot. The portion of the shoe that generally connects the counter and vamp is the quarter; it extends over the tongue and has eyelets for lacing. This is called a blucher shoe. It provides good support and is generally preferred over slip-on shoes.

Athletic shoes are designed for specific activities. **(McDougall, 2009; Maffetone, 2003)** Court shoes, including those for tennis, basketball, and racquetball, are designed to cushion against jumping, as well as medial and lateral stresses. Running shoes are developed with forward motion as the primary purpose. Heel and arch cushions are the main features of these shoes. In running shoes, pronation is of primary concern. Firm midsoles are considered important to limit extended pronation. **(Langer, 2007; McMahan, 1983)** Increasing lateral heel flare increases initial pronation. **(Abshire, 2010; Nigg & Morlock, 1987)** Since running styles, speed, and distance differ, it should be obvious that one shoe design will not be optimal for all. **(McDougall, 2009; Langer, 2007)**

The most important aspect in selecting shoes is that they aid natural function, not restrict it. Improved performance can usually be determined by a stopwatch or in an athlete's increased general ability. Applied kinesiology examination helps determine when shoes are creating foot dysfunction, or causing loss of structural corrections that were made by adjusting the articulations or treating muscles. What we have seen in patients who have already seen numerous clinicians previously for their foot problems (podiatrists, chiropractors, massage therapists and others), is that many foot subluxations and muscle inhibitions are still present throughout the foot and ankle. In order to help reform the architecture of the foot in this kind of recurrent foot pain patient, often times as many as 6 or 7 subluxations in the foot will be found using the applied kinesiology sensorimotor challenge, and after correction of these persisting factors will improve muscle function in the foot and ankle as well as throughout the body. **(Leaf, 2006; Lee et al., 2003; Goodheart, 1998-1964)**

The athlete can be educated regarding how his shoes affect his performance by applied kinesiology testing. Unfortunately, selection of athletic shoes may not be for optimal function and safety; psychological and financial rewards may be the ruling factors. **(Chaitow & DeLany, 2002; Ellfeldt, 1983)**

Sometimes shoe construction is based on erroneous assumptions. For example, the competitive cycling shoe typically has a rigid sole, with the midsole made of wood or plastic. The purpose has been to distribute the pedal force over a larger area of the foot to prevent ischemia and consequent paresthesia that develop in competitive cycling. In steady speed cycling, it has been demonstrated that the distribution of force is more even in the forefoot with running shoes than in the cycling shoe, thus negating the supposed advantage of the solid shoe. **(Sanderson & Cavanagh, 1987)**

Proper shoe fit. Regardless of how well a shoe is designed and made, all is lost if it does not fit properly. Shoes cause a high percentage of foot deformities, the most common of which is hallux valgus, discussed previously. In general, foot deformities increase when foot mobility is lost and decrease with increased foot mobility. There is greater mobility in those not wearing shoes than in those who do. **(Abshire, 2010; Wallden, 2010; Squadrone & Gallozzi, 2009; Sim-Fook & Hodgson, 1958)** Additionally, there is more foot dysfunction — such as subluxations, fixations, and muscle weakness — when there is loss of mobility. Loss of mobility is primarily due to wearing shoes that are too small. This is a common finding; rarely do people wear shoes that are too large. **(Chaitow & DeLany, 2002)**

A shoe store should have an adequate stock in all shoe styles so that a person can progressively evaluate larger shoes. Sometimes an improper fit occurs because the store doesn't have the right size. Both feet should be measured while standing. Asymmetry is common, but there is no significant correlation between limb dominance and foot breadth. **(Rothbart, 2006; Didia & Nyenwe, 1988)** The shoe fit should be for the larger foot. **(Mennell, 1939)**

Becoming friendly with the owner and salespeople of a quality shoe store is a valuable service to your patients.

This referral service is two-way. You will learn the types of shoes available, and the sales people will know the type of conditions you successfully treat for referring those who have foot problems. For some patients, it may be necessary to find a shoe store that still makes custom shoes that specifically fit a specific patient's anomalous foot. An Internet search for a local custom shoemaker may be the best method for offering custom shoes with precise measurements and service for these patients.

Langer (**Langer, 2007**) offers a shoe-store clerk's perspective:

"I worked in shoe stores while a podiatry student and have seen first hand how difficult it can be to convince some women that they are wearing shoes that are too small for their feet. In my experience, many women have become so accustomed to tight, constricting shoes, that when they wear properly fitting shoes, they are convinced that the shoes feel 'sloppy'. Both as a student and as a practicing podiatrist, numerous women have insisted to me that their shoe width was double-A or triple-A when they clearly needed footwear two or three sizes wider. As a shoe clerk, I had to give them what they wanted, and they almost always wanted the shoes that were too small. As a podiatrist, I can now advise both women and men that their foot pain will only get worse if they insist on wearing poorly fitting shoes."

A shoe must fit the foot properly at the heel, across the metatarsals, and lengthwise. A study by Schwartz et al. (**Schwartz et al., 1935**) shows the need for a snug counter and solid last if extended foot pronation is to be controlled. The snug fit is around the medial and lateral aspects of the calcaneus, not just at the posterior of the heel. A loose or flexible counter allows the calcaneus to slip, especially when an individual makes sharp turns while walking. The importance of good fit in this portion of the shoe is exemplified by the case of phlebitis perpetuated by foot dysfunction, discussed in the **Dynamic Examinaton Section of this Chapter.**

Sometimes one cannot obtain proper fit of the counter around the calcaneus and still have the other aspects of the shoe fit correctly. This is especially true in those who have a wide forefoot and narrow hindfoot. Lack of snugness posteriorly can be improved by applying orthopedic felt to the tongue to force the foot back into the counter. (**Mennell, 1939**) Padding around the calcaneus to make a snug counter fit is usually not adequate.

Many people wear shoes that are too short. This may result from children's feet growing rapidly or adults thinking their feet never change in size, and that all shoes of the same size will fit their feet properly. Wearing shoes that are too small may be the result of vanity, even though probably no one would notice the change if the proper size were worn. Generally, shoes without laces — such as loafers and pumps — are fitted too short to avoid heel slippage. Patients may be accustomed to a poor fit and say that the shoe is comfortable. Frey et al found that the average woman wears shoes that are 1-1 ½ inch too narrow and a half size to a full size too short. (**Frey et al., 1993**) They may be unaware of the problems the shoe is causing because symptoms such as neck pain or headaches are remote from the foot.

Proper shoe fit in the forefoot is different from that of the heel counter, where a snug fit is needed. In the forefoot, adequate length and width are needed for proper foot function.

Shoes should always be fitted for the weight-bearing and functioning foot. This requires consideration of several foot positions. The greatest length of the shoe is required during the stance phase near toe-off; as the shoe flexes the counter stops posterior foot movement, causing the toes to slip forward on the surface of the sole. Mennell (**Mennell, 1939**) considers that 2/3″ over the standing foot length is necessary for this slipping action of the foot in the shoe.

An approximation of proper length can be obtained by standing in stocking feet on a length of cardboard 3/4″ wide. Mark the cardboard straight down from the heel and mark the end of the toe. Cut the cardboard to this length, and put it inside the shoe to be evaluated. When the end of the cardboard touches the front end of the shoe, there should be approximately 1/2″ to 3/4″ spaces between the end of the cardboard and the back of the shoe based on a round toe box. The distance from the ends of the toes to the end of a pointed shoe must be considered on an individual basis. Since children's feet grow sporadically and in very rapid spurts, this measurement should be done on them approximately every three or four weeks. (**Greenawalt, 1985**)

Leaf (**Leaf, 2006**) demonstrates how changes in foot size occurred in 180 people after applied kinesiology procedures were employed to the foot and ankle. First, the patient's foot was placed lightly on paper and the foot outlined. Second, with their foot bearing their body weight another tracing was drawn with another colored pencil. An increase of more than ¼ inch indicated the loss of intrinsic foot support. Only 15% of the participants in this study had a difference of less than ¼ inch weight bearing compared to non-weight bearing. Applied kinesiology testing and treatment procedures were applied to the muscles, joints, and skin in the ankles and feet of all patients. Skin imbalances were treated using Kinesio tape. (**Kinesio Tape**) Proprioceptive neuromuscular facilitation was applied to the involved ankle and foot muscles. Spinal subluxations from L4 to the sacrum were treated if present. The patient was then instructed to walk for 30 steps. A new piece of paper and tracing measurement of the foot was taken and compared to the original. In all the cases with more than ¼ inch difference non-weight bearing compared to weight bearing, after the applied kinesiology protocols were utilized, the second tracing showed markedly less difference in foot size than the original tracing.

Maffetone (**Maffetone 2003, 1989**) has observed a common condition in athletes that he calls the "short-shoe syndrome" or "first metatarsal jam". This results in a 1st metatarsal subluxation that develops insidiously. Because of the slow adaptation the foot is forced to make to the restriction, the athlete does not usually complain of pain at that location; he has other foot complaints. It is important to recognize the 1st metatarsal's problem and its cause, since change in 1st ray function may be the basic cause of gait change resulting in other foot problems or remote dysfunction.

Wearing shoes that are too short tends to jam the metatarsal back into the 1st cuneiform, subluxating the articulation. Indicating the problem may be 1st metatarsal swelling, blisters or calluses of the toe, or discoloration of

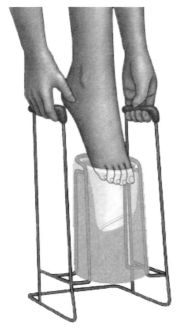
Sock-donner

the nail bed (the "black toenail"). Although the patient does not usually complain of pain in the 1st ray, it will generally be there when the joint is palpated and moved. The subluxation is confirmed by challenge. The vector of correction is usually the 1st metatarsal base in a plantar direction. It may have a more medial or distal vector. Since the cause of the problem is improper shoe fit, the patient's shoes must be evaluated as indicated later, and the patient educated regarding the cause of the problem. Maffetone recommends evaluating an athlete's training shoes, especially for the 1st toe position. It may be necessary to look at the shoe's insole, or to palpate the inside of the shoe to find the wear pattern from the 1st toe. This condition must be differentially diagnosed from functional hallux limitus as discussed earlier.

Before leaving the discussion on shoe length, we should consider short socks or stockings that can be just as problematic as too short a shoe. This foot restriction can cause bony deviation, fixations, subluxations, and ingrown toenails. Socks are an important part of footwear as they protect the skin from friction, keep the feet dry and thereby minimize bacteria and fungal problems. Many older patients do not wear socks because they are too difficult to get on. A device called a sock donner can make this task much easier. This tube-like apparatus stretches the sock and helps pull it up the leg with less strain on the hands or the low back.

The widest portion of the foot at the metatarsophalangeal articulations should match the widest portion of the shoe. This is called a ball-fit; correct fit at this point is crucial. (**Greenawalt, 1989; McMahan, 1983**) The 1st metatarsophalangeal articulation should rest at the widest part of the shoe. If it passes that point and is closer to the toe of the shoe, the shoe is too short. Seldom will you find a shoe too long. (**Greenawalt, 1985**) In addition, the shape of the shoe should match the forefoot. Some people need specially lasted shoes to match forefoot deviation. (**Enke et al., 2009; Wickstrom & Williams, 1970**)

The toe box should be large enough that toe movement is not restricted within the shoe. Width and height are important. The circumference around the metatarsal heads can increase up to 3/4" from non-weight-bearing to standing. The vamp of the shoe should not taper excessively toward the toes or toward the sides of the shoe, decreasing the forefoot space. (**Mennell, 1939**)

Observe whether the foot causes the vamp to spread over both edges of the sole; if it does, the shoe was too narrow for the foot when purchased. If the vamp rides over the sole on the lateral side, a Dutchman's wedge can help restore even sole wear. (**Greenawalt, 1985**) It is a smoothly cut, thin-layered wedge built into the shoe under the lateral forefoot.

It is common to wear shoes that are too small. Over an 18-month period in Maffetone's clinic, (**Maffetone, 1989**) 52% of the new athletic patients were found to be training in shoes too small, confirming a similar finding for patients 65 and older cited earlier. (**Chantelau & Gede, 2002**) Many people automatically buy the same shoe size, even though shoes vary from brand to brand regardless of the same indicated size. They even vary from the same manufacturer. A recent Internet search for "walking shoes" found 864 different models! In addition, people's feet change.

When buying shoes, it may be of value to measure the feet two or even three times during the day. Feet typically are larger at the end of the day, but there should not be more than a half-size difference. A larger difference requires additional examination to determine the cause of the edema fluctuation. Women's feet in pregnancy do not significantly increase in length or width, but they do significantly increase in volume that lasts for up to eight weeks postpartum. (**Wetz et al., 2006; Alvarez et al., 1987**) This indicates no change in ligamentous laxity; the complaint of tight shoes is due to edema.

Try on several pairs of shoes and spend enough time to make a thorough evaluation. You may not find the best shoe for you in the first store you visit. Most shoe stores carry only a few of the 864 different models available. Walk on a hard surface rather than on the soft carpet often found in shoe stores. Maffetone (**Maffetone, 2010, 1989**) recommends shoe fitting by "trying on the size you normally wear. Even if that feels fine, try on a half size larger. If that one feels the same (or even better), try on another half size larger. Continue trying on larger half sizes until you find the shoes that are obviously too large. Then go back to the previous half-size and usually that's the best one." Use comfort as the main criteria. Similarly, try various widths. The width is checked easily by placing the thumb and forefinger across the widest part of the forefoot and squeezing. If the foot is stretching the upper tightly, then the shoe is too narrow. "Don't let anyone say that you will have to 'break them in' before they feel good." Shoes should not be purchased unless they feel comfortable after walking in them sufficiently at the store.

Foot Support

The patient's foot and its supporting muscles should be thoroughly examined and treated before foot support is considered. With a comprehensive understanding of the foot and any remote dysfunction, the decision regarding support can be made. Support is needed when corrections do not hold, or when immediate relief of pain is necessary. In either case, the need for support means that correction has not been made or is not permanent. The best approach is to obtain a lasting correction and a properly functioning foot without the need of any support. Obviously, this is not always possible; knowledge of padding, strapping, and orthotics is of value for temporary use or on a permanent basis. As greater knowledge of how to correct and rehabilitate the foot is gained, less and less support is needed.

As previously discussed in this chapter, foot dysfunction can cause the return of cranial faults, shoulder problems, spinal subluxations, and a myriad of other conditions as soon as the patient walks. The fact that remote conditions return with walking does not necessarily mean that the feet are the problem. Other factors such as gait and pelvic motion must be considered. If foot subluxations or other foot dysfunction returns with walking, it is a good indication that the problem is still in the feet.

A method to help determine if the feet are the cause of a condition returning immediately after walking is to provide tape support to the foot, or to have the patient walk on the outside edges of his feet to eliminate extended pronation. If the remote condition does not return under these conditions, further foot correction or support to the feet is probably needed.

Foot supports are prescribed to improve function or relieve symptoms. This may be done to transfer weight from painful areas, separate structures such as in Morton's neuroma, rehabilitate functional hallux limitus, reduce strain on muscles, give support under hammertoes, or support under forefoot varus. The use of pads and supports for any of these examples, as well as for many other reasons, may be necessary, but function is often overlooked. The literature dealing with treatment by padding tends to emphasize the painful foot rather than pain and function. **(Ball & Afheldt, 2002)**

Examination with applied kinesiology methods helps address the question of function. This discussion on pads and taping will deal primarily with the painful foot and providing temporary support. Possibly more important is the portion of the discussion on evaluating the effect of supports that the physician has applied to the foot and ankle. This two-phase discussion then deals with applying the physician's knowledge to the condition, and evaluating how the patient's body reacts to the analysis and the supportive corrections the physician makes.

Material for padding and taping. Foot support may be needed only temporarily, for a moderate time, or on a permanent basis. **(Refshauge et al., 2009)** A temporary support, provided early in treatment, gives the ability to evaluate the need for and proper placement of the material. A few supplies are necessary to be able to provide quick office support.

There are four basic approaches to support: 1) evaluating and correcting, if needed, the patient's shoe selection, as previously discussed; 2) directly applying support to the patient; 3) putting a pad or an orthotic in the shoe; and 4) modifying or changing the patient's shoes.

When support is applied directly to the patient, as in taping or by adhesive pads, the skin should be properly prepared. Any hair in the area should be removed. Small electric clippers used by barbers to trim sideburns and mustaches trim very close to the skin and do not irritate it as shaving does. Compound Tincture of Benzoin applied to the skin increases adhesion of the material to be applied and reduces the possibility of allergic reaction.

Adhesive-backed 1/4" sponge rubber from 8" stock rolls is an easy material to apply directly to the patient. The paper or plastic backing that protects the adhesive surface is left in place until the material is cut to the correct size, edges beveled, and ready for placement. Greater thickness can be obtained by layering the material. When this is necessary, the edges should be smoothly cut into thin layers so there are no lumps in the pad. With weight bearing, the layered pad quickly unites into a single mass. When the pad is properly placed, as indicated by testing described later, it is covered with adhesive material such as moleskin or tape. Moleskin is a smooth, adhesive-backed material that has many uses in padding and supporting the feet.

Adhesive support can be protected during bathing or showering by covering the foot with a plastic bag held in place with a rubber band around the ankle. After bathing, the exposed portion of the foot around the support can be washed separately.

Removal of adhesive pads and tape should be done from distal to proximal to work with the grain of the skin. This is particularly important over calluses. Adhesive solvent is available to make removal easy for both technician and patient. It is usually packaged in a can with a spout for control of application. As the adhesive material is removed, apply a few drops of solvent between the adhesive surface and skin. Continue to remove the support while adding more drops as necessary. After the support is removed, any residual adhesive on the patient can be removed with the solvent applied to a gauze sponge. Wash the skin with soap and water, and dry.

Medial longitudinal arch.

The usual reason for supporting the medial longitudinal arch is to control extended pronation. This can be accomplished with rigid or pliable orthotics. **(Kogler et al., 1996)** One study showed how pronation control was achieved using rigid orthotics in a group classed as pronators. **(Bates et al., 1979)** They had been successfully treated with orthotic appliances. Their function was evaluated with high-speed photography while running barefoot, in a regular shoe, and in a regular shoe plus an orthotic device. The period of pronation and the amount of maximum pronation were significantly reduced using the foot orthotic device. These results were compared with normal or asymptomatic runners. During running, the normal group had greater ankle dorsiflexion and less corresponding values for maximum pronation. This may indicate that a short triceps surae was contributing to the excessive pronation in the orthotic group.

Christensen **(Christensen, 1983)** used high-speed photography to evaluate the effect of Spinal Pelvic Stabilizers®, **(Foot Levelers, Inc.)** a flexible orthotic used for controlling extended pronation. He demonstrated that "…maximum pronation and percentage of support time were significantly decreased in the SPS condition when compared to the shoe condition."

There is a tendency to support the longitudinal arches

in the cavus foot. In most cases this should not be done because it takes away ligament stretching (**Wenger, 1989; Turek, 1984**) and often gives poor relief of symptoms. (**Dugan & D'Ambrosia, 1986**) It is justified only when applied kinesiology testing indicates need for support. If support is given, begin a stretching program and gradually reduce the support elevation as improved foot position is obtained.

When support to the medial longitudinal arch is needed, care must be taken not to err toward overcorrecting the arch. The individual with a short triceps surae should not have support that eliminates pronation, which is necessary to give flexibility to the foot that must adapt to the limited dorsiflexion. (**Wyndow et al., 2010; Warburton, 2001**) First the triceps surae must be lengthened with techniques indicated by AK examination, e.g. pincer palpation followed by percussion, neuromuscular spindle cell, Golgi tendon organ, fascial release, trigger point pressure release, occasionally the use of the origin-insertion technique, Graston Instrument, PRYT, dural tension, and others. (**Leaf, 2010; Cuthbert, 2002; Walther, 2000**) If there is still limited dorsiflexion, the stretching program described later should be taught to the patient.

Rubber pads of various types and sizes can be stocked; they provide good temporary support to the medial longitudinal arch while foot corrections are being obtained. The patient may be evaluated for need of a more permanent type of support, such as orthotics or shoe modification, at a later time. Temporary support for the medial longitudinal arch is put in the shoe under the navicular bone. These are called navicular or scaphoid pads; in the shoe industry, they are called "cookies." They are available in different heights, lengths, and widths. The cookie is temporarily placed in the patient's shoe, and the patient is tested to determine if the support is correct. If the cookie tends to slip while the patient walks, it can be held in place by a small piece of double stick tape. When the correct size cookie and its proper location are determined, solidly fasten it to the shoe with additional double-stick tape. It often helps to cover the cookie with moleskin, which extends about 1/2" to 1" over the edges of the cookie. Rather than gluing the cookie in place it is better to use double-stick tape and moleskin, because any residual adhesive can usually be easily removed with adhesive solvent when the cookie is no longer desired. The cookie does not tend to move in the shoe as much as metatarsal pads.

Versatility of AK diagnosis

Because of the versatility of manual muscle testing it is possible to evaluate patients in many different positions and states (including walking, sitting, supine, prone and while moving any number of parts of the nervous system) while the manual muscle test is being conducted. The AK manual muscle test system for testing patients while moving or in positions of physical stress adds to the ability of the doctor to diagnose the muscular dysfunctions that occur with everyday activities. In the typical AK clinical encounter patients are frequently put into positions of pain during the MMT examination. For instance, many patients suffer their foot, ankle or leg pain primarily when standing. The MMT of the muscles of the foot and ankle are frequently tested after walking in these cases. This is thought to also evaluate dysfunctions in the feet that may be producing weaknesses that are not present in the supine position of normal examination. The muscle weakness connected to these patients' foot and leg pain may only be found after walking or in the standing position. (**Zampagni et al., 2009; Nicholas & Marino, 1987**)

An example of applied kinesiology is offered by Gangemi (**Gangemi, 2011**) who presents an approach for treatment of tibialis posterior muscle dysfunction. Adding extra stress to the tibialis posterior muscle with a specific challenge before the manual muscle test, and then supporting any inhibition found with a simple taping procedure, will assist in the recovery of tibialis posterior dysfunction.

> "The tibialis posterior muscle is often a muscle that is harder to keep corrected (facilitated) after treatment.... Provided the physician has thoroughly evaluated and corrected the cause of the tibialis posterior problem, whether structural, chemical/nutritional, or emotional, and has verified that any and all footwear or supporting devices are not hindering its function, the patient is ready to provide a stress test to the tibialis posterior.
>
> "The physician will ask the patient to stand and put the majority of weight on the affected tibialis posterior leg. Next, the patient should be instructed to shift the majority of weight to the forefoot, lifting the heel slightly off the ground as if trying to push an object through the floor with the ball of the foot. Finally, the patient will be instructed to torque the lower leg by rotating the foot, ankle, and lower leg back and forth three to four times. This movement can be likened to doing the opposite motion of twisting a cork out of wine bottle, by pushing rather than pulling. After this procedure is done, the patient should immediately lay supine on the treatment table and the physician should re-test the tibialis posterior muscle. Re-inhibition to the muscle after the stress test proves that either the interosseous membrane and/or the deep fascia is unable to handle the overwhelming stress of all their functions and allow the muscle to heal. Therefore, a simple taping method is used to help the support the tibialis posterior and the surrounding tissues so the patient may recover much faster.
>
> "Once the tibialis posterior is stressed and the muscle once again shows inhibition, the patient is then asked to assist the physician by placing the hands on either side of the lower leg and compressing (squeezing) them together, with a moderate amount of force. The patient begins at the top of the lower leg at the head of the fibula and the physician tests the tibialis posterior muscle. If the muscle becomes facilitated, the physician notes this and stops. If the muscle does not facilitate, then the patient is instructed to move the hands down about one-inch and the physician retests the muscle. Each time the patient moves distally down the leg, the physician retests the tibialis posterior. There will be one place somewhere

between the head of the fibula and the medial and lateral malleoli, where the muscle will facilitate (become strong). In the example here (**photo 1**), it is approximately half-way down the lower leg.

"Once the exact area of maximum facilitation is found, it is supported with athletic tape. The taping method used here is to simply wrap the tape one to two times around the lower leg, compressing the fibula and tibia together (**photo 2**). The tape should be wrapped tightly enough to provide the needed support but not so tightly that it causes any skin irritation or vascular

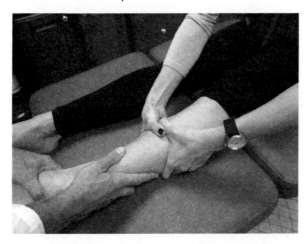

photo 1

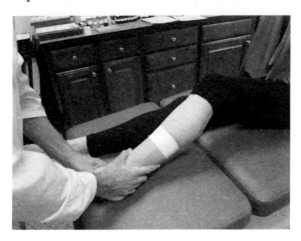

photo 2

compromise.

"The patient should then re-perform the same torque test that previously weakened. This will ensure the tape is properly supporting the tibialis posterior, deep fascia, and interosseous membrane so the patient will not "walk back into the problem." If the tape does not provide support, it is either too loose or on the wrong location of the leg.

"The tape ideally should stay on the rest of the day until the patient is off his or her feet and into bed for the night. It can then be removed. If there is some reoccurrence of pain or return of the problem the next morning, the tape can be reapplied and worn as long as it continues to help. Sometimes two to three days of tape support are needed. In some cases, there is a different spot which needs to be taped. This can only be verified by a return visit to the office to see exactly what area now needs tape application. The author has seen this in cases where the patient has placed significant torque stress on the lower leg…

"Additionally, fascial release technique can often be beneficial to certain tibialis posterior problems. This can be verified by stretching the tibialis posterior and testing for a reoccurrence of the inhibition. (**Walther, 2000**) Fascial release treatment should then be performed and can also be done on the area where patient compression strengthens prior to applying the tape, as previously described." (**Gangemi, 2011**)

Metatarsal support. Metatarsal support is commonly used and offers considerable relief for many patients, particularly for the functional hallux limitus (FHL) described earlier. When used correctly, it can help hold corrections obtained by adjusting subluxations or fixations. Improperly used, it can cause subluxations and pain.

The stock metatarsal pad is pear-shaped, supporting the medial metatarsals. It may be a separate pad or incorporated into off-the-shelf "arch supports." Patients often self-prescribe this type of support and purchase it at a drugstore. It provides support best for the 3rd metatarsal, and supplies some support to the 2nd and 4th metatarsals. When the 1st metatarsal is involved, the stock-type pad provides no support. If specific support is required for the 4th, 2nd, or 1st metatarsal(s), it must be prescribed or custom-made by the physician as a temporary support. It may be best to use the temporary approach over several weeks to determine need for a more permanent appliance. A trial period of support provides information about the width, height, and exact placement prior to having custom supports made.

Metatarsal pads are used to help maintain metatarsal subluxation correction and to control pain. When painful walking is the problem, subluxation of one or more metatarsals — or perhaps a metatarsal fixation — is almost always present.

The thickness of padding can be estimated by determining the amount of motion between the metatarsals. Using scissors-type motion palpation, determine the flexion-extension motion between each two metatarsals. More padding is needed and tolerated under metatarsals that have considerable motion, and less under those that are restricted. (**Nordsiden et al., 2010; Jahss, 1982**)

The first consideration is to pad for recurrent subluxations. The subluxated metatarsal head can usually be easily palpated as displaced plantarly. Palpation may reveal exquisite tenderness, which will help delineate the offending metatarsals. Padding is put under the distal metatarsal shaft and does not extend under the metatarsal head. Outline the area indicated by palpation and tenderness with a marking pencil. The pad is formed with the adhesive backing in place and smoothly cut into thin layers to match the outlined area. Peal off the backing and apply directly to the prepared skin, using the markings as a guide. Cover the pad with moleskin or non-compressive adhesive tape (Dermicel). If you choose to have the tape encircle the foot, it should be applied loosely to avoid impeding circulation with normal forefoot spreading during weight bearing. Covering only the plantar and lateral aspects of the foot is usually adequate.

Donner **(Donner, 1985)** recommends putting the pad in the patient's shoe. To accurately locate the position for a metatarsal pad, put a small dot of water-soluble paint on the offending metatarsal head(s). Have the patient plantar flex his toes while putting his foot into his shoe. This may be difficult, but it protects the paint until the patient's foot is properly placed. The patient now stands comfortably with his toes in neutral position. Remove the shoe and glue a single or multiple metatarsal pads just posterior to the dot to contact the distal metatarsal shaft. A hot glue gun or other fast-drying glue should be used.

Occasionally it is necessary to take pressure off most of the metatarsal heads. This is done with a J-shaped pad made to support the distal shafts and transfer weight to the 1st metatarsal, which is the strongest and most capable of providing support to the weight of the body. **(Kang et al., 2006; Milgram, 1983, 1964)** The long portion of the J is applied transversely across the foot, with the short portion extending to just behind the metatarsal heads that need support. Several variations of this type of right-angle pad can be used to selectively supply support where needed.

Many types of health problems originate from a mechanism in the body that has an integrated function about which the majority of all doctors is unaware. The reasons the function of the foot is not better recognized as a cause of health problems is the difficulty in determining when function is improper. Applied kinesiology has taken a major step forward in the understanding of health problems which relate to the functional gait and foot mechanism; AK provides an effective method of evaluating and treating this system.

Orthotics

Generally orthotics are needed only in chronic conditions that are uncorrectable, or sometimes for a short term while correction is being accomplished. **(Bartold, 2004)** Usually orthotics are not needed or desirable for young children. It has been shown that in flexible flatfoot, there is no statistically significant improvement in three years of wearing corrective shoes or inserts in young children. Wenger et al **(Wenger et al., 1987)** suggest that in most cases orthotics do not change osseous relationships and are ineffective. They suggest that arch supports may make the patient's symptoms worse if the shortened triceps surae is not corrected.

Off-the-shelf orthotics for adults rarely provides proper support. The height, length, and width of the medial longitudinal arch vary so much between people that the chance of obtaining the proper support is very slim. If the patient also needs metatarsal support, there is increased chance the support will be incorrect and may even do harm.

When support is needed on a long-term basis, it is usually best to provide the patient with custom orthotics. **(Prior, 1999)** Wise use of this kind of foot support can turn a patient's poor response into a successful outcome. Although foot support can be the difference between success and failure, its application should always be based on physiologic need. Improper application may be more than unnecessary; harmful effects may result.

An example of the improper use of orthotics is a case of low back pain in a 7-month pregnant dentist treated by SCC. This patient was an active, healthy 30-year-old who continued her running exercise during her pregnancy, as she had during a previous one. Her back pain developed over a 2-month period prior to examination. At the time of examination, the back pain was severe enough that she could stand only for approximately a half hour before having to lie down. She had to quit work, whereas in her previous pregnancy she worked until delivery. Initial examination was relatively unremarkable. The only dysfunction found was a 2nd lumbar subluxation and a pelvic Category I, which were adjusted; she was re-scheduled for follow-up in three days. At that time she said she had been completely free of pain throughout the rest of the day of the initial examination; the pain returned while she was cooking breakfast the next morning. Again I found a 2nd lumbar subluxation. This time I spent more time examining and correcting the intrinsic muscles of the spine at that location. The vertebra was again adjusted, and she was scheduled for follow-up in two days. She came in and again remarked that she'd had no more pain the remainder of the treatment day, but it came back while cooking breakfast the following morning. I questioned her about her activities from the time of rising until cooking breakfast when the pain developed. She said she did her daily run shortly after getting up. She experienced no pain during the run, came home and showered; the pain developed while cooking breakfast. I told her it would be necessary for me to test her gait pattern because the subluxation might be recurring during her run. She said she felt there was no problem with her running; in fact, at a recent running clinic her gait had been analyzed and declared to be nearly perfect, except for a little "flick" of her foot just after toe-off. She further stated this had been corrected with orthotics, which she now wore while running. I was astonished to hear that she used orthotics while running, because during the initial routine foot examination always done on every new patient, I found no problem whatsoever with her feet. Her feet were unusually pretty, showing no stress, calluses, or other body language indicating foot dysfunction. This time I examined her using the applied kinesiology gait testing technique and found no problem. Again I corrected the 2nd lumbar subluxation and told her to come in the next day. She was not to run before the office visit, and I wanted her to bring her running shoes and orthotics. When she came in, she had cooked breakfast and spent half the morning with no back pain developing. Upon examination there was no subluxation in the lumbar spine. I had her put on her running shoes with the orthotics and go across the street to run in the park. When she returned, examination disclosed the presence of the 2nd lumbar subluxation. Upon questioning I learned that her backache had started slowly, approximately one week after the running clinic. I again corrected the 2nd lumbar subluxation and she ran in her running shoes without the orthotics; no subluxation developed. I had her discontinue the orthotic use, and she needed no more adjustments. She is occasionally re-evaluated for other health issues; she has had no more back pain and continues her running program successfully.

The orthotics prescribed in the above case were rigid. In a popular article entitled *The Great Orthotics Debate* Gill (**Gill, 1985**) highlights the controversy and emphasizes the cost, which ranged from $200 to $300. He summarizes by writing, "If you decide to invest in a pair of orthotics, don't expect any miracles. They're not cure-alls, merely aids."

Langer (**Langer, 2007**) concurs: "An insole or orthotics is only as supportive as the shoe in which it is placed. Shoes that do not have room for an insole or orthotics are not capable of helping a painful foot, and unsupportive shoes will not magically become supportive by adding an insole or orthotics."

Our experience indicates that rigid orthotics create problems more often than the flexible orthotic. It appears there are three reasons for this. 1) They are made to conform to what the physician's education indicates the foot balance should be. 2) They are unforgiving, i.e., they do not work with the foot but rather attempt to control it. 3) They are made from a cast of the non-weight-bearing foot.

The non-weight-bearing plaster cast does not represent the foot in its functional position. Prior to forming the cast, the physician puts the patient's foot in what appears to him the correct position. The orthotic is fabricated from the information the cast provides. Rigid orthotics may provide hindfoot or forefoot posting. A post is a wedge in the orthotic to control valgus or varus distortions. Michaud (**Michaud, 1997, 1987**) points out that rigid orthotics designed to post for rearfoot or forefoot deformities must be cast properly, otherwise iatrogenic problems might result. With flexible orthotics, there is less chance of iatrogenic problems because they flex with the foot. Lillich and Baxter (**Lillich & Baxter, 1986**) characterize the ideal orthotic as lightweight, dynamic, and durable. Langer and Subotnick (**Langer, 2007; Subotnick, 1991, 1975**) use rigid, semi-rigid, and flexible orthotics, but they recommend the flexible type for certain types of competition.

Spinal Pelvic Stabilizers® (**Foot Levelers, Inc.**) are flexible orthotics widely prescribed in the chiropractic profession. We have found these very satisfactory when orthotic supports are needed. They provide the advantage of flexibility and most often provide proper support for the patient, confirmed by applied kinesiology testing. The orthotic is fabricated from measurements obtained from a weight-bearing cast. As in measuring for shoe fit, the weight-bearing factor is important. People's feet change in different ways from non-weight-bearing to weight bearing. The measurements are best made in the manner in which the feet function, and feet function as weight-bearing structures. Another apparent advantage of these orthotics is the lateral arch support, which is rarely present in rigid supports.

Most patients who need orthotics have limited motion in the forefoot. This is usually from wearing too-tight shoes. The orthotics will start to mobilize the forefoot, which often causes discomfort. Evaluate the forefoot motion when a patient is molded for orthotics; if it is limited, teach him to do the golf ball exercise, described later. This will help avoid most of the discomfort when he starts wearing the orthotics.

Another cause of discomfort when first wearing new orthotics is limited range of motion in areas remote from the foot. (**Dananberg, 2007; Prior, 1999; Aronow, Solomone-Aronow, 1986**) The body works as an integrated whole. Emphasis has been made on how foot dysfunction can adversely affect remote areas on a neurologic basis. Likewise, symptoms may develop in remote areas of the body that have limited range of motion and cannot accept increased activity created by improved foot function. Examining for the muscle stretch reaction and the need for strain/counterstrain technique will usually demonstrate the type of treatment needed. In addition, the patient may need to do some generalized stretching techniques. (**Looney et al., 2011; Page et al., 2010; Dananberg, 2007; Prior, 1999**) Athletes often overdevelop certain areas relating to their sport. This should give clues indicating where to examine for limited range of motion or antagonist muscle weakness.

Custom-made flexible orthotics, such as Spinal Pelvic Stabilizers®, may occasionally need to be modified. This can be accomplished by adding folded pieces of adhesive tape to the underside of the orthotic. For example, if there is a recurring inferior distal 4th metatarsal subluxation, fold three or four layers of tape into a pad about 1/2" wide by 1" long; with another piece of tape, attach it to the underside of the orthotic. Have the patient use the orthotic for a while, then challenge to determine if the subluxation returns. The temporary pad can be readjusted until the proper amount and placement are found. It is usually best to have the patient use the orthotic for several weeks before sending it in for modification, because the feet will change as you continue to treat the dysfunction. Often it will not be necessary to send the orthotic in for modification, because the tape pad adequately provides the necessary support. If modification is necessary, contact Foot Levelers, Inc., with a description of the procedure you have done and send the orthotic to them for the modification. (**Foot Levelers, Inc.**)

When it is necessary to prescribe an orthotic, it does not mean the patient is destined to wear them permanently. (**Gangemi, 2011; Maffetone, 2003; Michaud, 1997**) Greenawalt (**Greenawalt, 1987**) describes a study in which weight-bearing x-rays were used to evaluate femur head height, sacrovertebral angle, and lumbosacral disc angle before and after wearing Spinal Pelvic Stabilizers® without receiving any other treatment. The postural x-rays were taken with the subject wearing no shoes, and each parameter measured improved. The balancing of femur head height may have been a combination of improvement in the arch and/or a decrease in pelvic rotation.

As previously discussed, there is no muscle activity supporting the normal arch in a static stance position. It is in the abnormal foot with excessive pronation that muscle activity develops. Franettovich et al. and Suzuki (**Franettovich et al., 2008; Suzuki, 1956**) demonstrate electromyographic activity in the peroneus longus in conditions of flatfoot. When support is put under the arch, the electrical activity of the peroneus longus ceases. Suzuki observed, "In the patient whom symptoms of flat foot have been alleviated by applying an arch-supporting instrument for a long period, action currents from the (peroneus longus) were not observed while standing without use of the instrument." Possibly the improvement in function developed from the body repairing the ligaments and plantar fascia as they were relieved from stress while wearing the orthotic. If foot corrections are made and an effective rehabilitation program is followed, one can often improve foot function enough that support is no longer needed.

A very important factor in evaluating the effectiveness of orthotics is that the patient's shoes be adequate. This has been previously discussed, but it needs to be highlighted here because this is the reason that properly made flexible orthotics are often ineffective. The counter should be solid and fit relatively tightly around the calcaneus. The shoe should be wide enough in the forefoot, with a large toe box, and be a ball fit. Finally — and very important when using longitudinal arch support — the shoe should have a solid shank. When the shank collapses, support provided by the orthotic is lost.

An ancient Indian book of fables and wisdom tales called the *Pancha Tantra* says that for a foot in a shoe, the whole world seems paved with leather.

Foot Rehabilitation

Triceps surae stretch.

The triceps surae should allow 10° of dorsiflexion at the ankle. Dorsiflexion measurement should always be done passively. When the movement is patient-assisted, it is always greater. **(Bohannon et al., 1989)** One must also limit motion to ankle joint movement only. The examiner should grasp the calcaneus and midfoot to stabilize the calcaneus, navicular, and cuboid bones against the talus before moving the foot into dorsiflexion. It is in this action that there should be a minimum of 10° dorsiflexion. Remember that in excessive pronation, there is increased laxity of the subtalar and mid-tarsal articulations. If the entire foot is brought into dorsiflexion by applying pressure at the metatarsals, it will appear that there is much greater dorsiflexion than is being allowed at the ankle mortise. When motion measurement is limited to the ankle, one is often amazed at how little there is. The foot may fail to reach a right angle with the tibia by as much as 25°. In normal feet and ankles, dorsiflexion may reach 20° beyond the right angle.

When the triceps surae is short, a surgical procedure to lengthen the Achilles tendon is sometimes done. At one time this was the most common procedure performed by orthopedic surgeons. **(White, 1943)** Ryerson **(Ryerson, 1948)** points out, "The tendo-Achilles is not short. The muscle bellies are short." He recommends stretching the muscle to lengthen it.

If there is less than 10° dorsiflexion at the ankle mortise, the triceps surae must be lengthened. **(Logan, 1995)** First obtain as much increase as possible in range of motion by using applied kinesiology techniques directed to the muscular and remote neurologic factors. On a local basis percussion, stretch and spray, trigger point pressure release and fascial release, and treatment to the muscle and cutaneous receptors will often increase the muscle's length. Folic acid helps relieve the pain associated with the tender points in strain-counterstrain involvements, and vitamin B-12 relieves the pain of Travell's myofascial trigger points. There are several techniques in applied kinesiology — such as PRYT, cloacal synchronization, and dural tension relief — that increase range of motion throughout the body. **(Leaf, 2010; Walther, 2000)**

It may be necessary to apply stretching procedures to the triceps surae. The major problem with stretching the triceps surae is that the foot must be used as a lever to apply the stretch. This tends to compound the problem of stretching the ligaments of the midfoot and breaking it down further.

The muscles can best be stretched by a contraction, relaxation, and stretch technique, which has been shown more effective than a ballistic stretch. **(Simons, 2002; Wallin, 1985)** In a ballistic stretch the muscle is rapidly stretched, activating the intrafusal muscle spindles to reflexly cause a protective muscle contraction. The proper stretching procedure is done with the patient facing a wall with his feet approximately three feet from it; his legs are in internal rotation so the feet are in a "pigeon-toed" position, **(Logan, 1995; Travell & Simons, 1992)** and weight is directed to the lateral borders of the feet. This helps lock the midfoot to reduce stretch on its ligaments. The foot is further locked by contracting the intrinsic plantar muscles. **(Janse, 1947)** The patient should be taught to isolate these muscles as much as possible in order to flex the metatarsophalangeal articulations and not the interphalangeal articulations. The latter are flexed with the long flexors, which usually do not need strengthening, especially when claw toes are present. The hands are placed on the wall and the patient leans toward it, stretching the gastrocnemius and soleus almost to their limit. In this position, an isometric contraction of the triceps surae is held for seven seconds. The muscles are then relaxed for two to five seconds, followed by moving the hips forward, keeping the knees extended, and flexing the arms to place maximum stretch on the gastrocnemius and soleus for seven to eight seconds. If the major stretch is desired at the soleus, the knees are allowed to bend slightly. The cycle of contract-stretch is repeated five times on a daily basis. After thirty days of stretching, one can expect approximately 10° increase in dorsiflexion. When the desired range of motion is obtained, it can be maintained by a stretch series once a week. **(Wallin et al., 1985)** It is important to emphasize that MTrPs in the soleus muscle specifically relate to a restriction of dorsiflexion of the ankle **(Grieve et al., 2011; Travell and Simons, 1992)**. Additionally, metatarsal head pain may be the result of a shortened triceps surae. **(Mennell, 1969)**

Toft et al. **(Toft et al., 1989)** evaluated the contract-stretch method of triceps surae stretching by measuring passive tension on the muscles. This may be an improved method of evaluating for increased length of the triceps surae over measuring range of motion, since it appears to be more objective. It is reproducible and unaffected by the subjects' desire to demonstrate progress from their flexibility program. Range of motion can also be influenced by a raised muscle pain threshold. The method of contract-stretch evaluation began with eight seconds of maximum contraction of the plantar flexors, followed by two seconds of relaxation, and — finally — eight seconds of slow stretch with the knee flexed. Straight knee stretching was not evaluated. The stretching procedure had a short-term effect lasting for at least 90 minutes after stretching, and a long-term effect lasting for three weeks after twice daily stretching. This method of stretching muscle is effective, regardless of whether the person has considerable motion restriction or good flexibility. Stretching may remove all the symptoms caused by the shortness. Logan notes that it is not uncommon for symptoms as far away as in the mid-thoracic spine to improve after stretching of the triceps

Triceps Surae Stretch

Your doctor has determined that it is necessary for you to lengthen the calf muscles in your lower leg. The purpose for the stretching exercise is to increase flexibility in your ankles. Increased flexibility will permit better function of your feet. You will receive the greatest benefits if you follow these directions exactly as given.

Position:
1. Remove your shoes and socks.
2. Stand 2 ½ to 3 feet from a wall.
3. Place your hands on the wall.
4. Point your toes inward (pigeon-toe).
5. Place your weight on the outer edges of your feet.

Step 1
1. Contract the muscles on the bottom of your feet by curling your toes as if you were trying to pick up the carpet with your toes.

Step 2
1. Contract your calf muscles by pushing down on the floor with your forefoot as if you were going to stand up on your toes, but do not let your heels come off the floor. This is an isometric contraction of your calf muscles, which means "muscle contraction without movement."
2. Hold this contraction for eight seconds.
3. Relax for a second, but continue leaning against the wall to be ready for the next step.

Step 3
1. Lean forward toward the wall, keeping your back and pelvis straight until you feel significant stretching of your calf muscles and behind your knees.
2. Hold this stretch for eight seconds.
3. Push back from the wall and rest for five seconds.
4. Do not bounce to increase the stretching action. Research has shown the procedure described will give the best and quickest lengthening of your muscles.
5. Repeat steps 1-3 five times.
   Repeat the series ___ times daily.

The average person will gain 10° of increased ankle movement (dorsiflexion) in one month by doing these procedures, if done properly. Once adequate range of motion has been gained, you can maintain it with one series of stretching exercises per week.

In the clinic we offer patients the instructions above to specifically achieve the triceps surae lengthening.

surae. (**Logan, 1995**)

The passive tension method of evaluating the contract-stretch procedure provides information about the muscles throughout the joint's range of motion, i.e., the amount of tension on the muscle halfway through the motion can be compared. Since tension was reduced throughout range of motion, it can be seen that the muscle was affected at the lengths used during walking and running.

Travell & Simons and Janse et al. (**Travell & Simons, 1992; Janse, 1947**) describe a seated method for stretching the triceps surae and exercising the intrinsic plantar muscles. This is done by active muscle contraction to obtain maximum ankle dorsiflexion and toe flexion. This position is held with maximum contraction for about ten seconds and then relaxed. It is repeated 10-20 times with each foot. Each contraction can have a slightly varying amount of forefoot adduction or abduction. This method enables one to exercise while doing some other activity, such as working at a desk. Additionally, it exercises the dorsiflexor muscles, which are underdeveloped in many runners.

The soleus can be stretched by having the patient stand with his feet slightly toed-in and close together. The patient then assumes a squat position without raising his heels from the floor. As the buttocks approximate the calf of the legs, a strong pull in the Achilles tendon should be felt. Hold the squatting position 10-15 seconds and repeat six times. (**Greenawalt, 1988**) This procedure does not stretch the gastrocnemius because it is shortened with knee flexion.

Foot rehabilitation with yoga

The classic yoga asana downward facing dog achieves this kind of triceps surae lengthening with elegance and safety. A growing body of western scientific evidence indicates that yoga practice is associated with improvements in overall physical fitness, (**Cowen & Adams, 2005**) along with increases in muscular strength, endurance, and flexibility. (**Tran et al., 2001**) – each of these outcomes are consistent with the AK approach.

This is probably the most widely taught forward bending posture in hatha yoga. In its ideal form the downward dog asana takes on the shape of an upside-down V, with only the hands and feet touching the floor. The hips are flexed sharply, the ankles are flexed 45 degrees, and the lumbar lordosis is kept intact. The pose enables the gastrocnemius and soleus muscles, as well as the ankle joints, to increase their flexibility, counteracting lifelong habits for functioning within limited lengths and ranges of motion.

Besides the downward facing dog, the Standing Lunge is one of the best practices for correcting an incapacity for flexion at the ankle.

The experience of lengthening the entire posterior myofascial channels of the body is impressively felt in the forward bend. Entering and exiting the forward bend from the hips permits the experience of lengthening the entire posterior body that envelops you from head to toe. Each inhalation releases the stretch slightly, and each exhalation lowers the head further down increasing the myofascial lengthening.

Logan (**Logan, 1995**) reports on another effective way

Downward Facing Dog
Anusara yoga instructor MG Ballantyne
(with kind permission)

to stretch the triceps surae is by using a slant board.

"At home the patient may stand on the board 10 to 15 times per day. Each time the patient should spend only 1 to 2 minutes, attempting to stand up as straight as possible. This allows a gradual elongation of the muscles. Usually, 10 to 14 days of home stretching on the slant board will restore full range of motion in most patients."

When the stretching program results in poor dorsiflexion increase, one should suspect formation of osteophytes that can develop on the anterior aspect of the talar neck. A similar lesion can develop on the tibialis anterior. (Hontas et al., 1986) These lesions are more apt to develop in athletes whose major role is running and jumping. Along with the decreased dorsiflexion, there is usually

Standing Lunge

Forward bend

tenderness to digital pressure at the anterior ankle. Lateral x-ray of the ankle provides the definitive diagnosis. The joint itself is usually free of degenerative changes. "This condition can easily be treated by surgical excision of the offending spur. There is usually not any problem with returning to previous activity levels." (**Hontas et al., 1986**)

Thigh and spinal muscles. When one is rehabilitating from foot injury or dysfunction, the muscles of the extremities and spine must be considered. Disuse, abuse, misuse, postural change, or trauma may cause muscle shortening when there is distal dysfunction. (**Nicholas & Marino, 1987**) Altered feedback from sensory receptors in the feet and ankle alters gait, strength, and patterns of muscle activation throughout the body above. (**Nurse & Nigg, 2001; Freeman & Wyke, 1967**) Patients with supinated or pronated feet exhibit poor postural control compared to people with neutral feet. (**Tsai et al., 2006**) Any shortening or contraction – usually producing the tightness-weakness pattern on manual muscle testing described by Janda (**Janda, 1993**) -- must be eliminated during rehabilitation. When returning to activity, tight muscles may improperly transfer the load onto the weight-bearing foot; this can lead to sprains, strains, or stress fractures, especially in athletes. Tight quadriceps may lead to increased patellofemoral joint compression and resist free patella excursion distally. (**Sprieser, 2003; Duffy, 1999; Zatkin, 1990**) Tight iliotibial bands may cause patellar and/or trochanteric pain; they can lead to iliotibial band tendinitis, especially in distance runners. (**Leaf, 2003**) Chinn, Nicholas, Schafer and co-workers (**Chinn & Hertel, 2010; Nicholas & Marino, 1987; Schafer, 1986**) routinely use a number of tests for flexibility: the Thomas test for hip flexors, Ober test for iliotibial band, Ely test for quadriceps, and the sit-and-reach for lumbodorsal fascia and hamstrings.

Plantar muscle exercise. The plantar muscles are often atrophied in foot dysfunction, especially when tarsal tunnel syndrome causes poor nerve control of the muscles. Loss of strength in the flexor hallucis brevis muscle may lead to an unstable first ray. Consequently, the movement of the foot into pronation and the effectiveness of the windless mechanism of Hicks (**Hicks, 1951**) can be compromised if the first ray becomes unstable and lifts during load bearing. This is particularly evident in cases of functional hallux limitus as discussed previously.

Explain to the patient the difference between the intrinsic and extrinsic flexor muscles, how flexion of the base of his toes is done by the muscles that need exercising, and how flexion of the distal joints is done by the muscles that do not need exercising.

After correcting the tarsal tunnel syndrome (**discussed in Chapter 3**) and other foot conditions, teach the patient to contract the intrinsic foot muscles by flexing at the metatarsophalangeal articulations with minimal or no flexing of the interphalangeal articulations. For most patients this will be difficult at first, but with a little practice it can be easily accomplished.

There are two exercises to strengthen the plantar muscle when the patient can easily control them. Picking up marbles from the floor by flexing the toes will exercise the plantar muscles and help develop mobility of the forefoot. Have the patient pick up a marble from a pile, rotate the foot to its limit, and put the marble down. Turn the foot back as far as it will go in the direction of the original pile, and pick up another marble to put in the second pile. With practice the patient will be able to pick up a marble with any toe, gaining great dexterity in his feet. Make certain the patient avoids using only the distal interphalangeal joints of his toes; rather, he should use the entire toe, with emphasis on bending at the metatarsophalangeal articulation.

Another exercise that requires muscle control to

flex at the metatarsophalangeal joint is towel gathering. **(Gangemi, 2011; Crawford & Thomson, 2003; Maffetone, 2003)** Put a hand towel on a hard surface floor and have your seated patient put his foot on one end of the towel. The towel is gathered beneath his foot with toe action. Again, it is necessary for the toe action to be primarily at the metatarsophalangeal articulations. Encourage the patient to relax his calf muscles and concentrate on contracting the muscles in the bottom of his foot.

Toe-rising exercises can also strengthen the intrinsic muscles, but they have the disadvantage in some cases of also exercising the plantar-flexing muscles. This may be a problem when the triceps surae needs stretching, as in some runners who have overdeveloped the calf muscles in comparison to the dorsiflexor muscles. **(Logan, 1995)** Toe rising can be done by simply putting the forefoot on a raised area, such as a stair step, and raising up on the toes. The patient drops his heel down as far as possible, and again rises up on his toes. This exercise is contraindicated if there is anterior displacement of the metatarsal fat pad, and in some other cases of metatarsalgia.

Obtaining foot flexibility with a golf ball. Normally the foot is very flexible and has no painful areas. The foot that has been functioning abnormally for a prolonged period will have many tight and painful areas. The best way to regain motion in these areas is to work the foot on a golf ball. The extremely painful areas are the ones needing the most work.

Have your patient roll her foot back and forth over a golf ball, then from side to side. The uncomfortable areas give indication of where to concentrate the pressure. As the patient's foot loosens and gains mobility, have her apply more pressure on the golf ball until she reaches the point where she can stand and put considerable pressure on the ball. If the ball tends to be lost from under the foot and scoots across the floor, it can be put inside a Mason jar ring for containment.

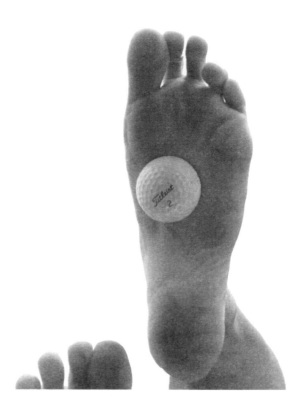

References

- Abshire D. Natural Running: the simple path to stronger, healthier running. VELO Press: Boulder; 2010.
- Acker JH, Drez D, Jr. Nonoperative treatment of stress fractures of the proximal shaft of the fifth metatarsal (Jones' fracture). Foot & Ankle. 1986;7(3):152-5.
- Activator Methods, Inc., P.O. Box 80317, 3714 E. Indian School Rd., Phoenix, AZ 85060—0317.
- Agawany AE, Meguid EA. Mode of insertion of the abductor hallucis muscle in human feet and its arterial supply. Folia Morphol (Warsz). 2010;69(1):54-61.
- Alvarez R, et al. The effect of pregnancy on women's feet. Foot & Ankle. 1987;7(5).
- Amis J, et al. Painful heel syndrome: Radiographic and treatment assessment. Foot & Ankle. 1988;9(2):91-5.
- American Medical Association, Guides to the Evaluation of Permanent Impairment, 6th Ed. American Medical Association Press; 2007.
- American Podiatric Medical Association. [http://www.apma.org/.]
- Aniss AM, Gandevia SC, Burke D. Reflex responses in active muscles elicited by stimulation of low-threshold afferents from the human foot. J Neurophysiol. 1992;67(5):1375-84.
- Arinci Incel N, Genç H, Erdem HR, Yorgancioglu ZR. Muscle imbalance in hallux valgus: an electromyographic study. Am J Phys Med Rehabil. 2003;82(5):345-9.
- Arnold BL, Linens SW, de la Motte SJ, Ross SE. Concentric evertor strength differences and functional ankle instability: a meta-analysis. J Athl Train. 2009;44(6):653-62.
- Aronow R, Solomone—Aronow B. Backache relief and postural control factors from the foot up, Part 1. Chiro Econ. 1985;28(3).
- Aronow R, Solomone—Aronow B. The foot—fascial approach to low back pain care. Am Chiro. 1986.
- Aronow R, Solomone—Aronow B. Backache relief & postural control factors from the foot up, Part 4. Chiro Econ. 1986;28(6).
- Aspegren D, Cox JM, Benak DR. Detection of stress fractures in athletes and nonathletes. J Manip Physiol Ther. 1989;12(4):298-303.
- Axe MJ, Ray RL. Orthotic treatment of sesamoid pain. Am J Sports Med. 1988;16(4):411-6.
- Badlissi F, Dunn JE, Link CL, Keysor JJ, McKinlay JB, Felson DT. Foot musculoskeletal disorders, pain, and foot-related functional limitation in older persons. J Am Geriatr Soc. 2005;53(6):1029-33.
- Bahler A. The biomechanics of the foot. Am Chiro. 1986.
- Ball KA, Afheldt MJ. Evolution of foot orthotics--part 1: coherent theory or coherent practice? J Manipulative Physiol Ther. 2002;25(2):116-24.
- Bandy JV. Bunions. Collected Papers International College of Applied Kinesiology: Detroit; 1976.
- Bard C, Paillard J, Lajoie Y, Fleury M, Teasdale N, Forget R, Lamarre Y. Role of afferent information in the timing of motor commands: a comparative study with a deafferented patient. Neuropsychologia. 1992;30(2):201-6.
- Bartold SJ. The plantar fascia as a source of pain – biomechanics, presentation and treatment. J Bodyw Mov Ther. 2004;8:214-226.
- Basmajian JV. Weight—bearing by ligaments and muscles. Can J Surg. 1961;4:166-70.
- Basmajian JV, Bentzon JW. An electromyographic study of certain muscles of the leg and foot in the standing position. Surg Gynecol Obstet. 1954;98(6):662-6.
- Basmajian JV, De Luca CJ. Muscles Alive — Their Functions Revealed by Electromyography, 5th Ed. Williams & Wilkins: Baltimore; 1985.
- Basmajian JV, Stecko G. The role of muscles in arch support of the foot — an electromyographic study. J Bone Joint Surg. 1963;45:1184-90.
- Bates BT, Osternig LR, Mason B, James LS. Foot orthotic devices to modify selected aspects of lower extremity mechanics. Am J Sports Med. 1979;7(6):338-42.
- Bauman JH, Brand PW. Measurement of pressure between foot and shoe. Lancet. 1963;1(7282):629-32.
- BBC News. Sierra Leone amputees turn to art. 11 June, 2002. http://news.bbc.co.uk/2/hi/africa/2038949.stm.
- Beardall AG. The foot. Collected Papers International College of Applied Kinesiology: Gaylord, MI; 1975.
- Beckman SM, Buchanan TS. Ankle inversion injury and hypermobility: effect on hip and ankle muscle electromyography onset latency. Arch Phys Med Rehabil. 1995;76(12):1138-43.
- Bejjani FJ, Jahss MH. Translators. "Le Double's study of muscle variations of the human body, Part II: Muscle variations of the foot. Foot & Ankle. 1986;6(4):157-76.
- Bergmann TF, Peterson DH. Chiropractic Technique: Principles and Procedures, 3rd Ed. Mosby: St. Louis; 2010.
- Berkson D. The Foot Book — Healing the Body through Foot Reflexology. Funk & Wagnalls: New York; 1977.
- Berthoz A. The Brain's Sense of Movement. Harvard University Press: Cambridge, MA; 2000.
- Blechschmidt E. The Ontogenetic Basis of Human Anataomy: A Biodynamic Approach to Development from Conception to Birth. North Atlantic Books: Berkeley, CA;2004.
- Blum C, Cuthbert S. Developmental Delay Syndromes and Chiropractic: A Case Report. J Ped Matern Fam Health; 2009:3.
- Bohannon RW, Tiberio D, Zito M. Selected measures of ankle dorsiflexion range of motion: Differences and intercorrelations. Foot & Ankle. 1989;10(2):99-103.
- Bojsen-Møller F. Calcaneocuboid joint and stability of the longitudinal arch of the foot at high and low gear push off. J Anat. 1979;129(1):165-76.
- Bojsen-Møller F, Flagstad KE. Plantar aponeurosis and internal architecture of the ball of the foot. J Anat. 1976;121(3):599-611.
- Bordelon RL. Foot first — evolution of man. Foot & Ankle. 1987;8(3):125-6.
- Botte RR. An interpretation of the pronation syndrome and foot types of patients with low back pain. J Am Podiatry Assoc. 1981;71(5):243-53.
- Brantingham JW, Globe G, Pollard H, Hicks M, Korporaal C, Hoskins W. Manipulative therapy for lower extremity conditions: expansion of literature review. J Manipulative Physiol Ther. 2009;32(1):53-71.
- Bresnahan P. Flatfoot deformity pathogenesis. A trilogy. Clin Podiatr Med Surg. 2000;17(3):505-12.
- Brinks A, Koes BW, Volkers AC, Verhaar JA, Bierma-Zeinstra SM. Adverse effects of extra-articular corticosteroid injections: a systematic review. BMC Musculoskelet Disord. 2010;11:206.
- Brody DM. Running injuries. Clin Symp. 1980;32(4):1-36.

- Brody DM. Running injuries — prevention and management. Clin Symp. 1987;39(3):1-36.
- Brooks VB: The Neural Basis of Motor Control. New York: Oxford University Press; 1986.
- Brukner P, Khan K. Clinical Sports Medicine. 3rd Ed. McGraw-Hill: Roseville, NSW, Australia; 2006.
- Bullock-Saxton JE. Local sensation changes and altered hip muscle function following severe ankle sprain. Phys Ther. 1994;74(1):17-28.
- Bullock-Saxton JE, Janda V, Bullock MI. The influence of ankle sprain injury on muscle activation during hip extension. Int J Sports Med. 1994;15(6):330-4.
- Burke D, McKeon B, Skuse NF. The muscle spindle, muscle tone and proprioceptive reflexes in normal man. In: Proprioception, Posture and Emotion. Ed. Garlick D. Univ New South Wales: Kensington, NSW; 1982.
- Butler D. Mobilisation of the nervous system. Churchill Livingstone: Edinburgh; 1991.
- Cailliet R. Low Back Pain Syndrome, 5th Ed. (Philadelphia: F.A. Davis Co, 1995).
- Cailliet R. Foot and Ankle Pain, 3rd Ed. (Philadelphia: F.A. Davis Co, 1997).
- Cammisa KM. Educational Kinesiology with learning disabled children: an efficacy study. Percept Mot Skills. 1994;78(1):105-6.
- Campbell KM, Biggs NL, Blanton PL, Lehr RP. Electromyographic investigation of the relative activity among four components of the triceps surae. Am J Phys Med. 1973;52(1):30-41.
- Carter M. Helping Yourself with Foot Reflexology. Parker Pub Co, Inc: W. Nyack, NY; 1969.
- Cavanagh PR. The shoe—ground interface in running. In: Symposium on The Foot and Leg in Running Sports. Ed. Mack RP. The CV Mosby Co.: St. Louis; 1982.
- Cavanagh PR, Lafortune MA. Ground reaction forces in distance running. J Biomechan. 1980;13(5):397-406.
- Cavanagh PR, Rodgers MM, Liboshi A. Pressure distribution under symptom—free feet during barefoot standing. Foot & Ankle. 1987;7(5):262-76.
- Ceffa G, Chio C, Gandini G. Importance of thermography and plantar supports and spine. In: Chiropractic Interprofessional Research. Ed. Mazzarelli JP. (Edizioni Minerva Medica: Torino, Italy; 1982.
- Chaitow L, DeLany JW. Clinical Application of Neuromuscular Techniques, Volume 1, 2nd Edition: The Upper Body. Elsevier: Edinburgh; 2008.
- Chaitow L, et al. Naturopathic Physical Medicine. Elsevier: Edinburgh; 2008.
- Chaitow L, et al. Cranial Manipulation: Theory and Practice, 2nd Edition: Osseous and Soft Tissue Approaches. Elsevier: Edinburgh; 2005.
- Chaitow L, DeLany JW. Clinical Application of Neuromuscular Techniques: Volume 2, The Lower Body. Elsevier: London; 2002.
- Chaitow L. Soft-Tissue Manipulation. Healing Arts Press: Rochester, VT; 1988.
- Chantelau E, Gede A. Foot dimensions of elderly people with and without diabetes mellitus - a data basis for shoe design. Gerontology. 2002;48(4):241-4.
- Charles J, Scutter SD, Buckley J. Static ankle joint equinus: toward a standard definition and diagnosis. J Am Podiatr Med Assoc. 2010;100(3):195-203.
- Chen L, Greisberg J. Achilles lengthening procedures. Foot Ankle Clin. 2009;14(4):627-37:627-37.
- Chen BX. Treatment of hallux valgus in China. Chin Med J (Engl). 1992;105(4):334-9.
- Chhaya SA, Brawner M, Hobbs P, Chhaya N, Garcia G, Loredo R. Understanding hallux valgus deformity: what the surgeon wants to know from the conventional radiograph. Curr Probl Diagn Radiol. 2008;37(3):127-37.
- Chinn L, Hertel J. Rehabilitation of ankle and foot injuries in athletes. Clin Sports Med. 2010;29(1):157-67, table of contents.
- Cholmeley JA. Hallux valgus in adolescents. Proc R Soc Med. 1958;51(11):903-6.
- Christensen K. Effectiveness of Spinal Pelvic Stabilizers to modify pronation. Success Express. 1983;6(2).
- Chundru U, Liebeskind A, Seidelmann F, Fogel J, Franklin P, Beltran J. Plantar fasciitis and calcaneal spur formation are associated with abductor digiti minimi atrophy on MRI of the foot. Skeletal Radiol. 2008;37(6):505-10.
- Clair JM, Okuma Y, Misiaszek JE, Collins DF. Reflex pathways connect receptors in the human lower leg to the erector spinae muscles of the lower back. Exp Brain Res. 2009;196(2):217-27.
- Clement DB,. Taunton JE. A guide to the prevention of running injuries. Aust Fam Physician. 1981;10(3):156-61, 163-4.
- Cobey MC. The care and correction of the congenital flatfoot. Southern Med J. 1958;5(5):586-590.
- Cohen BE. Hallux sesamoid disorders. Foot Ankle Clin. 2009;14(1):91-104.
- Collis WJMF, Jayson MIV. Measurement of pedal pressures — an illustration of a method. Ann Rheumat Disord. 1972;31:215-7.
- Colloca CJ, Polkinghorn BS. Chiropractic management of Ehlers-Danlos syndrome: a report of two cases. J Manipulative Physiol Ther. 2003;26(7):448-59.
- Colum EL. Carpal tunnel & ankle weakness. Collected Papers International College of Applied Kinesiology: Detroit, MI; 1983.
- Conway JJ, Cowell HR. Tarsal coalition: Clinical significance and roentgenographic demonstration. Radiology. 1969;92(4):799-811.
- Corneal JM, Dick R. An attempt to quantify muscle testing using meridian therapy/acupuncture techniques. Collected Papers International College of Applied Kinesiology. 1987;Winter:59-78.
- Cornwall MW, McPoil TG. Plantar fasciitis: etiology and treatment. J Orthop Sports Phys Ther. 1999;29(12):756-60.
- Coskun N, Yuksel M, Cevener M, Arican RY, Ozdemir H, Bircan O, Sindel T, Ilgi S, Sindel M. Incidence of accessory ossicles and sesamoid bones in the feet: a radiographic study of the Turkish subjects. Surg Radiol Anat. 2009;31(1):19-24.
- Costa SM, Chibana YE, Giavarotti L, Compagnoni DS, Shiono AH, Satie J, Bracher ES. Effect of spinal manipulative therapy with stretching compared with stretching alone on full-swing performance of golf players: a randomized pilot trial. J Chiropr Med. 2009;8(4):165-70.
- Costa LA, de Araujo JE. The immediate effects of local and adjacent acupuncture on the tibialis anterior muscle: a human study. Chin Med. 2008;3(1):17.
- Cowen VS, Adams TB. Physical and perceptual benefits of yoga asana practice: results of a pilot study. Journal of Bodywork and Movement Therapies. 2005;I9:211e219.
- Craik RL, Oatis CA. Gait Analysis: Theory and Application. CV Mosby: St. Louis; 1995.
- Cramer GB, Darby SA. Basic and Clinical Anatomy of the Spine, Spinal Cord, and ANS, 2nd Ed. Mosby: St. Louis; 2005.
- Crawford F, Thomson C. Interventions for treating plantar heel pain. Cochrane Database Syst Rev. 2003;(3):CD000416.
- Crossley K, Cook J, Cowan S, McConnell J. Anterior knee pain. In: Clinical Sports Medicine, 3rd Eds. Brukner P, Khan K. McGraw-Hill: Roseville, NSW, Australia; 2006.
- Csapo R, Maganaris CN, Seynnes OR, Narici MV. On muscle, tendon and high heels. J Exp Biol. 2010;213(Pt 15):2582-8.

- Cuccia AM. Interrelationships between dental occlusion and plantar arch. J Bodyw Mov Ther. 2011;15(2):242-50.
- Cuthbert S, Rosner AL, McDowall D. Association of manual muscle tests and mechanical neck pain: Results from a prospective pilot study. J Bodyw Mov Ther. 2011;15(2):192-200.
- Cuthbert S, Rosner A. Applied kinesiology methods for a 10-year-old child with headaches, neck pain, asthma, and reading disabilities. J Chiro Med. 2010; 9(3):138-145.
- Cuthbert S, Rosner A. Applied Kinesiology Management of Candidiasis and Chronic Ear Infections: A Case History. J. Pediatric, Maternal & Family Health; 2010.
- Cuthbert SC, Barras M. Developmental delay syndromes: psychometric testing before and after chiropractic treatment of 157 children. J Manipulative Physiol Ther. 2009;32(8):660-9.
- Cuthbert S. A Multi-Modal Chiropractic Treatment Approach for Asthma: a 10-Patient Retrospective Case Series. Chiropr J Aust 2008;38:17-27.
- Cuthbert SC, Goodheart GJ Jr. On the reliability and validity of manual muscle testing: a literature review. Chiropr Osteopat. 2007;15:4.
- Cuthbert S. Applied Kinesiology: An Effective Complementary Treatment for Children with Down Syndrome. Townsend Letter. 2007;288:94-107.
- Cuthbert S. Applied Kinesiology and the Myofascia. Int J AK and Kinesio Med, 2002;13-14.
- Dale RA. The Systems, Holograms and Theory of Micro-Acupuncture. Am J Acupunct, 1999;27(3-4):207-242.
- Dale RA. The demystification of Chinese pulse diagnosis: An overview of the validations, holograms, and systematics for learning the principles and techniques. Am J Acupunct. 1993;21(1):63-80.
- Damm JA, Waugaman CH, Eds. The Practical and Technical Encyclopedia. Wm. H. Wise & Co, Inc: New York; 1948.
- Dananberg H. Lower back pain as a gait-related repetitive motion injury. In: Vleeming A et al (Eds.) Movement, Stability and Low Back Pain. New York, NY: Churchill Livingstone;2007:253-264.
- Dananberg HJ. Gait style and its relevance in the management of chronic lower back pain. In: Vleeming A, et al. (Eds.) Proceedings of the 4th Interdisciplinary World Congress of Low Back and Pelvic Pain. ECO: Rotterdam. 2001;225-230.
- Dananberg HJ, Guiliano M. Chronic low-back pain and its response to custom-made foot orthoses. Journal of the American Podiatric Medical Association. 1999; 89(3): 109-117.
- Dananberg H. Lower back pain as a gait-related repetitive motion injury. In: Vleeming et al., Eds. Movement, stability and low back pain. Churchill Livingstone: New York; 1997.
- Dananberg HJ, Phillips AJ, Blaakman HE. A rational approach to the nonsurgical treatment of hallux limitus. In: Advances in Podiatric Medicine and Surgery, Vol 2. Mosby-Year Book, Inc; 1996.
- Dananberg HJ. Lower extremity mechanics and their effect on lumbosacral function. Spine: State of the Art Reviews. 1995;9(2).
- Dananberg HJ. Foot malfunction and its relationship to craniomandibular disorders. In: New Concepts in Craniomandibular and Chronic Pain Management. Ed. Gelb H. Mosby-Wolfe: London; 1994.
- Dananberg HJ. Gait style as an etiology to chronic postural pain. Part I. Functional hallux limitus. J Am Podiatr Med Assoc. 1993;83(8).
- Dananberg HJ, Lawton M,. DiNapoli DR. Hallux limitus and non-specific gait related bodily trauma. In: Reconstructive Surgery of the Foot and Leg, Update. Ed. DiNapoli DR. Podiatry Institute Publishing Co: Tucker, GA; 1990.
- Dananberg HJ. The Kinetic Wedge. (Letter to the Editor). J Am Podiatr Med Assoc. 1988;78(2):98-9.
- Dananberg HJ. Functional hallux limitus and its relationship to gait efficiency. J Amer Podiatric Med Assoc. 1986;76(11):648-52.
- Davey JR, Rorabeck CH, Fowler PJ. The tibialis posterior muscle compartment. An unrecognized cause of exertional compartment syndrome. Am J Sports Med. 1984;12(5):391-7.
- Degirmen N, Ozerdogan N, Sayiner D, Kosgeroglu N, Ayranci U. Effectiveness of foot and hand massage in postcesarean pain control in a group of Turkish pregnant women. Appl Nurs Res. 2010;23(3):153-8.
- DeJarnette MB. Extremity Technique, 3rd ed. Privately published: Nebraska City, NE; 1973.
- Delahunt E, Monaghan K, Caulfield B. Altered neuromuscular control and ankle joint kinematics during walking in subjects with functional instability of the ankle joint. Am J Sports Med. 2006;34(12):1970-1976.
- Delee JC, Evans JP, Julian J. Stress fracture of the fifth metatarsal. Am J Sports Med. 1983;11(5):349-53.
- DeMaio M, Paine R, Mangine RE, Drez D Jr. Plantar fasciitis. Orthopedics. 1993;16(10):1153-63.
- Denny-Brown D. Ed. Selected Writings of Sir Charles Sherrington. Oxford Univ Press: New York; 1979.
- De Wit B, De Clercq D, Aerts P. Biomechanical analysis of the stance phase during barefoot and shod running. J Biomech. 2000;33(3):269-78.
- Didia BC, Nyenwe EA. Foot breadth in children — its relationship to limb dominance and age. Foot & Ankle. 1988;8(4):198-202.
- Didia BC, Omu ET, Obuoforibo AA. The use of Footprint Contact Index II for classification of flat feet in a Nigerian population. Foot & Ankle. 1987;7(5):285-9.
- Donatelli RA. The Biomechanics of the Foot and Ankle. 2nd Ed. FA Davis Company: Philadelphia; 1996.
- Donatelli RA, Wooden MJ. Biomechanical orthotics. In: Donatelli RA editors. The Biomechanics of the Foot and Ankle. 2nd ed.. FA Davis Company: Philadelphia; 1996.
- Donner LF. The metatarsal heads. Collected Papers International College of Applied Kinesiology: Santa Monica, CA; 1985.
- Dorland's Illustrated Medical Dictionary, 31st Ed. Saunders; 2007.
- Dreeben S, Thomas PB, Noble PC, Tuilos HS. A new method for radiography of weight—bearing metatarsal heads. Clin Orthop. 1987;224:260-7.
- Drez D, Jr. Forefoot problems in runners. In: Symposium on The Foot and Leg in Running Sports. Ed. Mack RP. The C.V. Mosby Co: St. Louis; 1982.
- Duarte M. J Am Chiropr Assoc. Lax ligaments: a source of common foot problems.2004;41(1):48-49
- Duffy C. Applied kinesiology management of chronic Osgood-Schlatter disease: a case history. Collected Papers International College of Applied Kinesiology, 1998-1999;1:171-172.
- Dugan RC, D'Ambrosia RD The effect of orthotics in the treatment of selected running injuries. Foot & Ankle. 1986;6(6).
- Durlacher JV. Injury Susceptibility in Hypoadrenia. Collected Papers of the International College of Applied Kinesiology: Detroit, MI; 1977.
- Dvorak J, Dvorak V. Manual Medicine: Diagnostics. Thieme Medical Publishers Inc: New York; 1990.
- Early JS. Management of fractures of the talus: body and head regions. Foot Ankle Clin. 2004;9(4):709-22.
- Edgerton VR, Wolf SL, Levendowski DJ, Roy RR. Theoretical basis for patterning EMG amplitudes to assess muscle dysfunction. Med Sci Sports Exerc. 1996;28(6):744-51.
- Elftman H. The transverse tarsal joint and its control. Clin Orthop. 1960;16:41-6.

- Elftman H. Dynamic structure of the human foot. Artif Limbs. 1969;13(1):49-58.
- Elkus RA. Tarsal coalition in the young athlete. Am J Sports Med. 1986;14(6)477-80.
- Ellfeldt HJ. Problems of the foot and ankle in professional football and basketball players. In: Symposium on The Foot and Ankle. Eds. Kiene RH, Johnson KA. The C.V. Mosby Co: St. Louis; 1983.
- Encyclopedia Brittanica. http://www.britannica.com/
- Engsberg JR, Andrews JG. Kinematic analysis of the talocalcaneal/talocrural joint during running support. Med Sci Sports Exer. 1987;19(3):275-84.
- Enke RC, Laskowski ER, Thomsen KM. Running shoe selection criteria among adolescent cross-country runners. PM R. 2009;1(9):816-9.
- Espinosa N, Brodsky JW, Maceira E. Metatarsalgia. J Am Acad Orthop Surg. 2010;18(8):474-85.
- Evans RC. Illustrated Orthopedic Physical Assessment, 3rd Ed. Mosby: St. Louis; 2008.
- Fatoye F, Palmer S, Macmillan F, Rowe P, van der Linden M. Proprioception and muscle torque deficits in children with hypermobility syndrome. Rheumatology (Oxford). 2009;48(2):152-7.
- Ferciot CF. The etiology of developmental flatfoot. Clin Orthop. 1972;85:7-10.
- Ferguson A. Cranial osteopathy: a new perspective. J Acad Appl Osteopa. 1991;1(4):12-16.
- Fernández-de-las-Peñas C, Carratalá-Tejada M, Luna-Oliva L, Miangolarra-Page JC. The Immediate Effect of Hamstring Muscle Stretching in Subjects' Trigger Points in the Masseter Muscle. Journal of Musculoskeletal Pain. 2006;14:27-35.
- Ferrari J, Higgins JP, Prior TD. Interventions for treating hallux valgus (abductovalgus) and bunions. Cochrane Database Syst Rev. 2009;(2):CD000964.
- Ferrari R. Responsiveness of the short-form 36 and oswestry disability questionnaire in chronic nonspecific low back and lower limb pain treated with customized foot orthotics. Journal of Manipualtive and Physiological Therapeutics 2007; 30(6): 456-458.
- Fields KB, Sykes JC, Walker KM, Jackson JC. Prevention of running injuries. Curr Sports Med Rep. 2010;9(3):176-82.
- Fierro NL, Sallis RE. Achilles tendon rupture. Is casting enough? Postgrad Med. 1995;98(3):145-152.
- Fink M, Wähling K, Stiesch-Scholz M, Tschernitschek H. The functional relationship between the craniomandibular system, cervical spine, and the sacroiliac joint: a preliminary investigation. Cranio. 2003;21(3):202-8.
- Foot Levelers, Inc. http://www.footlevelers.com/. 518 Pocahontas Ave. N.E., P.O. Box 12611, Roanoke, VA 24027—2611.
- Franettovich M, Chapman A, Vicenzino B. Tape that increases medial longitudinal arch height also reduces leg muscle activity: a preliminary study. Med Sci Sports Exerc. 2008;40(4):593-600.
- Freeman MAR. Instability of the foot after injuries to the lateral ligaments of the ankle. J Bone Joint Surg. [Br] 1965;47-(B): 669-77.
- Freeman M, Wyke B. Articular reflexes at the ankle joint: an electromyographic study of normal and abnormal influences of ankle-joint mechanoreceptors upon reflex activity in the leg muscles. J Brit Surg 1967;54:990-1001.
- Frey C, Thompson F, Smith J, Sanders M, Horstman H. American Orthopaedic Foot and Ankle Society women's shoe survey. Foot Ankle. 1993;14(2):78-81.
- Froehle RM. Ear infection: a retrospective study examining improvement from chiropractic care and analyzing for influencing factors. J Manipulative Physiol Ther. 1996;19(3):169-77.
- Frost R. Applied Kinesiology: A training manual and reference book of basic principals and practices. North Atlantic Books: Berkeley, CA; 2002.
- Fuhr A. The Activator Method. Mosby; 2nd Ed.; 2008.
- Galois L, Mainard D, Delagoutte JP. Polydactyly of the foot. Literature review and case presentations. Acta Orthop Belg. 2002;68(4):376-80.
- Gangemi SC. Tibialis posterior torque and stability taping. Collected Papers International College of Applied Kinesiology. 2010-2011.
- Gangemi SC. Expanded gait assessment and evaluation and validation of minimalist footwear. Collected Papers International College of Applied Kinesiology. 2010-2011.
- Gardner LI Jr, Dziados JE, Jones BH, et al. Prevention of lower extremity stress fractures: a controlled trial of a shock absorbent insole. Am J Public Health. 1988;78(12):1563-7.
- Gardner-Morse MG, Stokes IA. The effects of abdominal muscle coactivation on lumbar spine stability.Spine. 1998;23(1):86-91; discussion 91-2.
- Garten H. Lehrbuch Applied Kinesiology Muskelfunktion-Dysfunktion-Therapie. Urban & Fischer: Munich; 2004.
- Garten H. Acupuncture in applied kinesiology: a review. Int J AK and Kinesio Med. 2002;14.
- Gertler L. Illustrated Manual of Extravertebral Technic, 2nd Ed. Privately published: Oakland, CA; 1978.
- Gertler L. Supplement: Lower Extremity Technic. Privately published: Oakland, CA; 1981.
- Gerz W. Lehrbuch der Applied Kinesiology (AK) in der naturheilkundlichen Praxis. AKSE-Verlag: Munchen; 2001.
- Geyer H, Herr H. A muscle-reflex model that encodes principles of legged mechanics produces human walking dynamics and muscle activities. IEEE Trans Neural Syst Rehabil Eng. 2010;18(3):263-73.
- Gill EK. The great orthotics debate. Runner's World; 1985.
- Gill LH. Conservative treatment for painful heel syndrome. Proceedings of 3rd Annual Summer Meeting of Amer Orthop Foot/Ankle Society, Santa Fe, NM, 7/17—19/87. Synopsis of report. In: Foot & Ankle. 1987;8(2).
- Gillet H.. Liekens M. Belgian Chiropractic Research Notes. Motion Palpation Inst: Huntington Beach, CA; 1981.
- Gleim GW, Nicholas JA, Webb JN. Isokinetic evaluation following leg injuries. Phys Sportsmed. 1978;6(8).
- Goldthwait JE. The anterior transverse arch of the foot: Its obliteration as a cause of metatarsalgia. Boston Med Surg J. 1894;CXXXI(10).
- Gondin J, Guette M, Maffiuletti NA, Martin A. Neural activation of the triceps surae is impaired following 2 weeks of immobilization. Eur J Appl Physiol. 2004;93(3):359-65.
- Gong YL, Zhang YB, Han C, Jiang YY, Li Y, Chen SC, Liu ZY. Clinical observation on therapeutic effect of the pressing plantar reflex area with wooden needle for treatment of patients with insomnia. *Zhongguo Zhen Jiu.* 2009;29(11):935-7.
- Goodheart GJ, Jr. The psoas muscle and the foot pronation problem. Chiro Econ. 1967;10(2).
- Goodheart GJ, Jr. Applied Kinesiology 1973 Research Manual, 9th Ed. Privately published: Detroit, MI; 1973.
- Goodheart GJ, Jr. Applied Kinesiology Research Manuals. Detroit, MI: Privately published yearly; 1998-1964.
- Goodheart GJ, Jr. Functional Hallux Limitus and AK Gait Patterns: Seminar Notes. ICAK USA: Shawnee Mission, KS; 1996.
- Goodheart GJ, Jr. Applied Kinesiology 1997-1998 Workshop Procedural Manual. Privately published: Grosse Pointe Woods, MI; 1998.
- Goodheart GJ, Jr. AK classic case management: enuresis. Int J AK and Kinesio Med, 2003;16:22-23.

- Goto K, Ishii N, Kizuka T, Takamatsu K. The impact of metabolic stress on hormonal responses and muscular adaptations. Med Sci Sports Exerc. 2005;37(6):955-63.
- Goulart M, O'Malley MJ, Hodgkins CW, Charlton TP. Foot and ankle fractures in dancers. Clin Sports Med. 2008;27(2):295-304.
- Gould N, et al. Development of the child's arch. Foot & Ankle. 1989;9(5):241-5.
- Gould SJ. The Structure of Evolutionary Theory. Harvard University Press: MA; 2000.
- Gowitzke BA, Milner M. Understanding the Scientific Bases of Human Movement, 2nd Ed. Williams & Wilkins: Baltimore; 1980.
- Gracovetsky S. The Spinal Engine. Springer-Verlag: Berlin;1989.
- Grath G. Widening of the ankle mortise: A clinical and experimental study. ACTA Chir Scand. 1960;Supp 263.
- Graven-Nielsen T, Svensson P, Arendt-Nielsen L. Effects of experimental muscle pain on muscle activity and co-ordination during static and dynamic motor function. Electroencephalogr Clin Neurophysiol. 1997;105(2):156-64.
- Gray ER. The role of leg muscles in variations of the arches in normal and flat feet. Phys Ther. 1969;49(10).
- Gray's Anatomy: The Anatomical Basis of Clinical Practice. Churchill Livingstone: Edinburgh; 2004.
- Greenawalt MH. The role of the weightbearing foot in the kinetic chain. Todays Chiropr. 1992;21(1):52-54.
- Greenawalt MH. Important ... Routine examination of the feet. Success Express. 1980;3(3).
- Greenawalt MH. Bone shock and serial distortion. Success Express. 1982;5(2).
- Greenawalt MH. Let's talk about shoes. Success Express. 1985;9(2).
- Greenawalt MH. The foot, gait and chiropractic. Success Express. 1987;11(1).
- Greenawalt MH. Feet and the dynamic science of chiropractic. Chiro Econ. 1988;30(6).
- Greenawalt MH. Common foot conditions and their orthotic application. Am Chiro. 1989;February.
- Greenawalt MH. If the shoe fits...wear it! Am Chiro. 1989;June.
- Greenhagen RM, Johnson AR, Peterson MC, Rogers LC, Bevilacqua NJ. Gastrocnemius recession as an alternative to tendoAchillis lengthening for relief of forefoot pressure in a patient with peripheral neuropathy: a case report and description of a technical modification. J Foot Ankle Surg. 2010;49(2):159.e9-13.
- Greenman PE. Principles of Manual Medicine. Williams & Wilkins; Baltimore; 2003.
- Gresczyk E. Electromyographic study of the effect of leg muscles on the arches of the normal and flat foot. Thesis, Univ. VT;1965.
- Grieve R, Clark J, Pearson E, Bullock S, Boyer C, Jarrett A. The immediate effect of soleus trigger point pressure release on restricted ankle joint dorsiflexion: A pilot randomised controlled trial. J Bodyw Mov Ther. 2011;15(1):42-9.
- Gross TS, Bunch RP. A mechanical model of metatarsal stress fracture during distance running. Am J Sports Med. 1989;17(5):669-74.
- Grundy M, et al. An investigation of the centres of pressure under the foot while walking. J Bone Joint Surg. 1975;57(1):98-103.
- Guebert GM, Thompson JR. Sesamoid stress fracture of the foot. Chiro Sports Med. 1987;1(1).
- Guyton AC, Hall JE. Textbook of Medical Physiology, 11th ed. W.B. Saunders Co: Philadelphia; 2005.
- Haines RW. The mechanism of the metatarsals and spread foot. Chiropodist. 1947;2(8):197-209.
- Haines T. Walking with Dinosaurs. DK Adult: London; 2000.
- Hall MG, Ferrell WR, Sturrock RD, Hamblen DL, Baxendale RH. The effect of the hypermobility syndrome on knee joint proprioception. Br J Rheumatol. 1995;34(2):121-5.
- Hambrick T. Plantar Fasciitis. J Bodyw Mov Ther. 2001;1:49-55.
- Hamilton AR. The short tendo calcaneus. J Bone Joint Surg. 1955;37B.
- Hamilton JJ, Ziemer LK. Functional anatomy of the human ankle and foot. In: Symposium on The Foot and Ankle, Eds. Kiene RH, Johnson KA. The C.V. Mosby Co: St. Louis; 1983.
- Hamilton WG. Surgical anatomy of the foot and ankle. Clin Symp. 1985;37(3):2-32.
- Hammer WI. Functional Soft Tissue Examination and Treatment by Manual Methods: New Perspectives, 2nd Ed. Aspen Publishers, Inc.: Gaithersburg, MD; 1999.
- Hansen SJ. Collected Papers International College of Applied Kinesiology, 1998-1999;1:63-65.
- Harris EJ. The natural history and pathophysiology of flexible flatfoot. Clin Podiatr Med Surg. 2010;27(1):1-23.
- Harris RI, Beath T. Hypermobile flat—foot with short tendo Achillis. J Bone Joint Surg. 1948;30A(1):116-40.
- Harris RI, Beath T. Etiology of peroneal spastic flat foot. J Bone Joint Surg. 1948;30B(4):624-34.
- Harrison N. Postural foot imbalance and back pain. Chiro Econ. 1964;7(3).
- Headley BJ: Muscle inhibition. Physical Therapy Forum. 1993;1:24-26.
- Hearon KG. Advanced principles of lower extremity adjusting. Hearon Seminars, Inc.: Boise, ID; 1994.
- Hearon KG. What You Should Know About Extremity Adjusting. K.G. Hearon: Sequim, WA; 1981.
- Helal B, Gibb P. Freiberg's disease: A suggested pattern of management. Foot & Ankle. 1987;8(2):94-102.
- Hellebrandt FA, Braun GL. The influence of sex and age on the postural sway of man. Am J Phys Anthropol. 1939;24(3).
- Hennig EM, Cavanagh PR. Pressure distribution under the impacting human foot. In: Biomechanics X—A. Ed Jonsson B. Human Kinetics Pub, Inc: Champaign, IL; 1987.
- Herman R, Bragin S. Function of the gastrocnemius and soleus muscles. Phys Ther. 1967;47(2).
- Hertling D, Kessler RM. Management of common musculoskeletal disorders, 3rd Ed. Lippincott Williams & Wilkins; 1996.
- Herzmark MH. Floor pad for foot—exercising. J Bone Joint Surg. 1947;29(4):1098.
- Hicks JF. The health of children's feet. Chiropodist. 1965;August.
- Hicks JH. The function of the plantar aponeurosis. J Anat. 1951;85(1):25-30.
- Hicks JH. The mechanics of the foot, I. The joints. J Anat. 1953;87(4):345-57.
- Hicks JH. The mechanics of the foot II. The plantar aponeurosis and the arch. J Anat. 1954;88(1):25-30.
- Hicks JH. The foot as a support. Acta Anat. 1955;25(1):34-45.
- Hill RS. Ankle equinus: prevalence and linkage to common foot pathology. Journal of American Podiatric Medical Association. 1995;85(6):295–300.
- Hill JJ, Jr., Cutting PJ. Heel pain and body weight. Foot & Ankle. 1989;9(5):254-6.
- Hlavac HF. Differences in x—ray finding with varied positioning of the foot. J Am Podiatry Assoc. 1967;57(10).
- Hlavac HF. The Foot Book — Advice for Athletes. World Publications: Mtn. View, CA; 1977.

- Hlavac H, Schoenhaus H. The plantar fat pad and some related problems. J Am Podiatry Assoc. 1970;60(4):151-5.
- Hodges PW, Richardson CA. Inefficient muscular stabilization of the lumbar spine associated with low back pain. Spine 1996;21(22):2640-2650.
- Hodgson JA, Roy RR, de Leon R, Dobkin B, Edgerton VR. Can the mammalian lumbar spinal cord learn a motor task? Med Sci Sports Exerc. 1994;26(12):1491-7.
- Hoffmeyer P, et al. Muscle in hallux valgus. Clin Orthop. 1988;232:112-8.
- Holm S, Indahl A, Solomonow M. Sensorimotor control of the spine. J Electromyogr Kinesiol. 2002;12(3):219-34.
- Holmes GB, Cracchiolo A, Goldner JL, et al. Current practices in the management of posterior tibial tendon rupture. Contemp Orthop. 1990;20:70-108.
- Hong WH, Lee YH, Chen HC, Pei YC, Wu CY. Influence of heel height and shoe insert on comfort perception and biomechanical performance of young female adults during walking. Foot Ankle Int. 2005;26(12):1042-8.
- Hontas MJ, Jaddad RJ, Schlesinger LC. Conditions of the talus in the runner. Am J Sports Med. 1986;14(6):486-90.
- Hoppenfeld S. Physical examination of the foot by complaint. In: Disorders of the Foot, Vol 1. Ed. Jahss MH. W.B. Saunders Co: Philadelphia; 1982.
- Hoppenfeld S. Physical examination of the spine and extremities. Prentice-Hall, Inc.: New York; 1976:133-141.
- Hoskins W, Pollard H. The effect of a sports chiropractic manual therapy intervention on the prevention of back pain, hamstring and lower limb injuries in semi-elite Australian Rules footballers: a randomized controlled trial. BMC Musculoskelet Disord. 2010;11:64.
- Hossain M, Nokes LDM. A model of dynamic sacro–iliac joint instability from malrecruitment of gluteus maximus and biceps femoris muscles resulting in low back pain. Medical Hypotheses. 2005;65(2):278-281.
- Houtz SJ, Fischer FJ. Function of leg muscles acting on foot as modified by body movements. J Applied Physiol. 1961;16:597-605.
- Houtz SJ, Walsh F. Electromyographic analysis of function of muscles acting on the ankle during weight—bearing with special reference to triceps surae. J Bone Joint Surg. 1959;41A:1469-81.
- Hsu JD, Imbus CE. Pes cavus. In: Disorders of the Foot, Vol 1. Ed. Jahss MH. W.B. Saunders Co: Philadelphia 1982.
- Hudson N, Fitzcharles MA, Cohen M, Starr MR, Esdaile JM. The association of soft-tissue rheumatism and hypermobility. Br J Rheumatol. 1998;37(4):382-6.
- Hughes J, Kriss S, Klenerman L. A clinician's view of foot pressure: A comparison of three different methods of measurement. Foot & Ankle. 1987;7(5):277-84.
- Hurley MV, Newham DJ. The influence of arthrogenous muscle inhibition on quadriceps rehabilitation of patients with early, unilateral osteoarthritic knees. Br J Rheumatol 1993;32:127-131.
- Iida M, Basmajian JV. Electromyography of hallux valgus. Clin Orthop. 1974;101:220-4.
- IMPAC. Salem, OR. http://www.impacinc.net/client.
- Inman V, Ralston HJ, Todd F. Human Walking. Williams & Wilkins: Baltimore; 1981.
- Inman V. Joints of the ankle. Williams & Wilkins: Baltimore; 1976.
- International College of Applied Kinesiology. www.icak.com.
- Jack EA. Bone anomalies of the tarsus in relation to 'peroneal spastic flat foot. J Bone Joint Surg. 1954;36B(4):530-42.
- Jaffe WL, Laitman JT. The evolution and anatomy of the human foot. In: Disorders of the Foot, Vol 1. Ed. Jahss Mh. W.B. Saunders Co.: Philadelphia; 1982.
- Jahn WT, Sr. Ed. Stress fractures. Ortho Brief;1985.
- Jahss MH. Examination. In: Disorders of the Foot, Vol 1. Ed. Jahss MH. W.B. Saunders Co.: Philadelphia; 1982.
- James SL, Bates BT, Osternig LR. Injuries to runners. Am J Sports Med, Vol 6. 1978(2):40-50.
- Janda V. Muscle strength in relation to muscle length, pain and muscle imbalance, Chapter 6. In Muscle Strength, Harms-Ringdahl K, Ed. New York: Churchill Livingstone; 1993.
- Janse J. Principles and Practice of Chiropractic. Ed. Hildebrandt RW. National Coll of Chiro: Lombard, IL; 1976.
- Janse J, Houser RH, Wells BF. Chiropractic Principles and Technic. National Coll of Chiro: Chicago;1947.
- Jaskoviak PA. A look at the sprained ankle. Success Express. 1983;6(1).
- Johanson D. *Lucy: The Beginnings of Mankind*. Simon & Schuster;1981:309.
- Jones RL. The human foot. An experimental study of its mechanics, and the role of its muscles and ligaments in the support of the arch. Am J Anat. 1941;68(1).
- Jones RL. The functional significance of the declination of the axis of the subtalar joint. Anat Rec. 1945;93:151-9.
- Jorgensen U, Bojsen-Moller F. Shock absorbency of factors in the shoe/heel interaction – with special focus on role of the heel pad. Foot & Ankle. 1989;9(6):294-9.
- Jull GA. Deep cervical flexor muscle dysfunction in whiplash. J Musculoskel. Pain. 2000;8:143-154.
- Kalivas J. Treatment of skin disorders of the feet and nails. In: Symposium on The Foot and Ankle. Eds. Kiene RH, Johnson KA. The C.V. Mosby Co.: St. Louis; 1983.
- Kang JH, Chen MD, Chen SC, Hsi WL. Correlations between subjective treatment responses and plantar pressure parameters of metatarsal pad treatment in metatarsalgia patients: a prospective study. BMC Musculoskelet Disord. 2006;7:95.
- Kapandji IA. The Physiology of the Joints, Vol. 2 – The Lower Limb, 6th Ed. Churchill Livingstone: New York; 2010.
- Kaplan EB. The tibialis posterior muscle in relation to hallux valgus. Bull Hosp Joint Dis. 1955;16(1):88-93.
- Kaplan M, Kaplan T. Flat foot. Radiology. 1935;25.
- Karpouzis F, Pollard H, Bonello R. A randomised controlled trial of the Neuro Emotional Technique (NET) for childhood Attention Deficit Hyperactivity Disorder (ADHD): a protocol. Trials. 2009;10(1):6.
- Katcherian DA. Treatment of Freiberg's disease. Orthop Clin North Am. 1994;25(1):69-81.
- Katoh Y, et al. Objective evaluation of painful heel syndrome by gait analysis. In: Biomechanics VIII-A, Ed. Matsui H, Kobayashi K. Human Kinetics Pub, Inc: Champaign, IL; 1983.
- Kavounoudias A, Roll R, Roll JP. Foot sole and ankle muscle inputs contribute jointly to human erect posture regulation. J Physiol. 2001;532 (Pt 3):869-78.
- Kayano J, et al. Dynamic changes of the medial arch of the foot during walking. In: Biomechanics X-A. Ed. Jonsson B. Human Kinetics Pub, Inc: Champaign, IL; 1987.
- Keating JC, Jr. From the Ground Up: Monte Greenawalt and the Early Growth of the Foot Levelers' Tradition. Chiropractic History. 2002;22(1):79-91.
- Kelikian H. The hallux. In: Disorders of the Foot, Vol 1. Ed. Jahss MH. W.B. Saunders Co.: Philadelphia; 1982.
- Kerrigan DC, Franz JR, Keenan GS, Dicharry J, Della Croce U, Wilder RP. The effect of running shoes on lower extremity joint torques. PM R. 2009;1(12):1058-63.

- Kharrazian D. The role of the transverse ligament in suprascapular nerve entrapments. Collected Papers International College of Applied Kinesiology, 2000-2001;1:81-83.
- Kidner FC. The prehallux (accessory scaphoid) in its relation to flat-foot. J Bone Joint Surg. 1929;11.
- Kim C, Cashdollar MR, Mendicino RW, Catanzariti AR, Fuge L. Incidence of plantar fascia ruptures following corticosteroid injection. Foot Ankle Spec. 2010;3(6):335-7.
- Kinesio tape. http://www.kinesiotaping.com/.
- Kitaoka HB, Lundberg A, Luo ZP, An KN. Kinematics of the normal arch of the foot and ankle under physiologic loading. Foot Ankle Int. 1995;16(8):492-9.
- Kleiger B. The mechanism of ankle injuries. J Bone Joint Surg. 1956;38A(1):59-70.
- Ko PH, Hsiao TY, Kang JH, Wang TG, Shau YW, Wang CL. Relationship between plantar pressure and soft tissue strain under metatarsal heads with different heel heights. Foot Ankle Int. 2009;30(11):1111-6.
- Kogler GF, Solomonidis SE, Paul JP. Biomechanics of longitudinal arch support mechanisms in foot orthoses and their effect on plantar aponeurosis strain. Clin Biomech (Bristol, Avon). 1996;11(5):243-252.
- Korr IM. Proprioceptors, and somatic dysfunction. JAOA. 1975;74(7):638-50.
- Korr IM. The spinal cord as organizer of disease processes: Some preliminary perspectives. JAOA. 1976;76(1):35-45.
- Krissoff WB, FerrisWD. Runners' injuries. Physician Sportsmed. 1979;7(12).
- Kugelberg E, Eklund K, Grimby L. An electromyographic study of the nociceptive reflexes of the lower limb. Mechanism of plantar responses. Brain. 1960;83:394-410.
- Kunz B, Kunz K. The Complete Guide to Foot Reflexology, 3rd Ed. Rrp Press: Albuquerque, NM; 2005.
- Kwon OY, Tuttle LJ, Johnson JE, Mueller MJ. Muscle imbalance and reduced ankle joint motion in people with hammer toe deformity. Clin Biomech (Bristol, Avon). 2009;24(8):670-5.
- Lake N. The arches of the feet. Lancet. 1937;2.
- Landry SC, Nigg BM, Tecante KE. Standing in an unstable shoe increases postural sway and muscle activity of selected smaller extrinsic foot muscles. Gait Posture. 2010;32(2):215-9.
- Langer P. Great Feet for Life: Footcare and Footwear for Healthy Aging. Fairview Press: Minneapolis, MN; 2007.
- Langer Biomechanics Group. The Electrodynogram SystemTM: Deer Park, NY; 11729.
- Langworthy OR. The Sensory Control of Posture and Movement – A Review of the Studies of Derek Denny-Brown. Williams & Wilkins: Baltimore; 1970.
- Lapidus PW. Misconception about the 'springiness' of the longitudinal arch of the foot. Arch Surg. 1943;46.
- Larson D. Physical balancing: Acupuncture and Applied Kinesiology. Am J Acupunct. 1985;13(2):159-162.
- Latinovic R, Gulliford MC, Hughes RA. Incidence of common compressive neuropathies in primary care. J Neurol Neurosurg Psychiatry. 2006;77(2):263-5.
- Lau JT, Daniels TR. Effects of tarsal tunnel release and stabilization procedures on tibial nerve tension in a surgically created pes planus foot. Foot Ankle Int. 1998;19(11):770-7.
- Lauro A, Mouch B. Chiropractic effects on athletic ability. J Chiropr Res Clin Inv. 1991;6(4) :84-87.
- Lawrance DJ. Pes planus: A review of etiology, diagnosis and chiropractic management. J Manip Physiol Ther. 1983;6(4):185-8.
- Laycock BE, Ed. Manual of Joint Manipulation.Still College: Des Moines, IA; 1953.
- LeNoir JL. Inverted talipes and rotational deformities of the lower extremities. In: Disorders of the Foot, Vol 1. Ed.. Jahss MH. W.B. Saunders Co: Philadelphia; 1982.
- Leach R, Jones R, SilvaT. Rupture of the plantar fascia in athletes. J Bone and Joint Surg. 1978;60A(4):537-9.
- Leach RE, Seavey MS, Salter DK. Results of surgery in athletes with plantar fasciitis. Foot & Ankle. 1986;7(3)156-61.
- Leaf D. Applied Kinesiology Flowchart Manual, 4th Ed. Privately Published: David W. Leaf: Plymouth, MA; 2010.
- Leaf D. Effectiveness of applied kinesiology procedures on foot size. Collected Papers International College of Applied Kinesiology, 2005-2006;1:99-100.
- Leaf D. Comments on utilizing some of the concepts of Janda. Collected Papers International College of Applied Kinesiology: Shawnee Mission, KS; 2003;1:25-26.
- Leahy PM. Active Release Techniques: Logical Soft Tissue Treatment. Chapter 17. In: Functional Soft Tissue Examination and Treatment by Manual Methods, 2nd Ed. Ed. Hammer WI. Gaithersburg, MD: Aspen Publishers; 1999:551.
- Lederman E. Neuromuscular rehabilitation in manual and physical therapies: principles to practice. Churchill Livingstone: Edinburgh; 2010.
- Lee SW, Lee JW, Park FI. The Change of the Asymmetry of Resting Calcaneal Stance Position by Applied Kinesiology. Collected Papers International College of Applied Kinesiology, 2002-2003;1:27-32.
- Lee D. Treatment of pelvic instability. In: Vleeming A, Mooney V, Dorman T, et al. (Eds.) Movement, stability and low back pain. Churchill Livingstone: Edinburgh; 1997.
- Lehman RC, Gregg JR. Osteochondritis dissecans of the midfoot. Foot & Ankle. 1986;7(3):177-82.
- Leonard ZC, Fortin PT. Adolescent accessory navicular. Foot Ankle Clin. 2010;15(2):337-47.
- Leonard MA. The inheritance of tarsal coalition and its relationship to spastic flat foot. J Bone Joint Surg Br. 1974;56B(3):520-6.
- Lephart SM, Fu FH. Proprioception and Neuromuscular Control in Joint Stability. Human Kinetics: Champaign, IL; 2000.
- Levangie C, Norkin P. Joint structure and function: a comprehensive analysis, 3rd Ed. FA Davis: Philadelphia, PA; 2001.
- Lever JP. Collected Papers International College of Applied Kinesiology, 2006-2007;1:79-90.
- Lewin P. The Foot and Ankle. Lea & Febiger: Philadelphia; 1959.
- Lewit K. Manipulative therapy in rehabilitation of the locomotor system, 3rd Ed.. Butterworth-Heinemann: Oxford; 1999.
- Lidge RT. Hallux valgus – surgical correction by three-in-one technique. In: Foot Science. Ed. Bateman JE. W. B. Saunders Co: Philadelphia; 1976.
- Liebenson C. Ed: Rehabilitation of the Spine: A Practitioner's Manual, 2nd ed. Lippincott, Williams & Wilkins: Philadelphia; 2007.
- Lieberman DE, et al. Foot strike patterns and collision forces in habitually barefoot versus shod runners. Nature. 2010;463:531-535.
- Lillich JS, Baxter DE. Common forefoot problems in runners. Foot & Ankle. 1986;7(3):145-51.
- Lindsjo U, Danckwardt-Lilliestrom G, Sahlstedt B. Measurement of the motion range in the loaded ankle. Clin Orthop. 1985;199:68-71.
- Lloyd T, Triantafyllou SJ, Baker ER, Houts PS, Whiteside JA, Kalenak A, Stumpf PG. Women athletes with menstrual irregularity have increased musculoskeletal injuries. Med Sci Sports Exerc. 1986;18(4):374-9.

- Logan AL. The Foot and Ankle: Clinical Applications. (A.L. Logan Series in Chiropractic Technique). Gaithersburg, MD: Aspen Publishers, Inc; 1995.
- Looney B, Srokose T, Fernández-de-las-Peñas C, Cleland JA. Graston instrument soft tissue mobilization and home stretching for the management of plantar heel pain: a case series. J Manipulative Physiol Ther. 2011;34(2):138-42.
- Lord M. Foot pressure measurement: A review of methodology. J. Biomed Eng. 1981;3(2):91-9.
- Lovejoy CO. Evolution of the human lumbopelvic region and its relationship to some clinical deficits of the spine and pelvis. In: Movement, Stability & Lumbopelvic Pain: Integration of research and therapy. Eds.: Vleeming A, Mooney V MD, Stoeckart R. Churchill Livingstone: Edinburgh; 2007:141-158.
- Lovejoy CO. Evolution of Human Walking, Scientific American. 1988;259:118-125.
- Lowy LJ. Pediatric peroneal spastic flatfoot in the absence of coalition. A suggested protocol. J Am Podiatr Med Assoc. 1998;88(4):181-91.
- Luger EJ, Nissan M, Karpf A, Steinberg EL, Dekel S. Patterns of weight distribution under the metatarsal heads. J Bone Joint Surg Br. 1999;81(2):199-202.
- Lundberg A, et al. Kinematics of the ankle/foot complex: Plantarflexion and dorsiflexion. Foot and Ankle. 1989;9(4):194-200.
- Lundberg A, et al. Kinematics of the ankle/foot complex – Part 2: Pronation and supination. Foot and Ankle. 1989;9(5):248-53.
- Lundberg A, et al. Kinematics of the ankle/foot complex – Part 3: Influence of leg rotation. Foot & Ankle. 1989;9(6):304-9.
- Lusskin R. Peripheral neuropathies affecting the foot: Traumatic, ischemic, and compressive disorders. In: Disorders of the Foot, Vol 2. Ed. Jahss MH. W. B. Saunders Co: Philadelphia; 1982.
- Lutter LD. Pronation biomechanics in runners. Contemp Orthop. 1980;2(8).
- Lutter LD. Forefoot injuries. In: Symposium on The Foot and Ankle. Eds. Kiene RH, Johnson KA. The C.V. Mosby Co: St. Louis; 1983.
- Lutter LD. Problems in the foot and ankle in runners. In: Symposium on The Foot and Ankle. Eds. Kiene RH, Johnson KA. The C.V. Mosby Co: St. Louis; 1983.
- Lutter LD. Plantar fasciitis: A guide to appropriate diagnosis and treatment. The Medical Journal of Allina. 1997;6(2). http://www.allina.com.
- MacConaill M, Basmajian J. Muscles and Movement: A Basis for Human Kinesiology. Williams & Wilkins: Baltimore; 1969.
- Madjarevic M, Kolundzic R, Trkulja V, Mirkovic M, Pecina M. Biomechanical analysis of functional adaptation of metatarsal bones in statically deformed feet. Int Orthop. 2009;33(1):157-63.
- Maffetone P. The Big Book of Endurance Training and Racing. Skyhorse Publishing: New York, NY; 2010.
- Maffetone P. Fix Your Feet. The Lyons Press: Guilford, CT; 2003.
- Maffetone P. Complementary Sports Medicine: Balancing traditional and nontraditional treatments. Human Kinetics: Champaign, IL; 1999.
- Maffetone P. First metatarsal subluxation: The short shoe syndrome. Chiro Econ. 1989;32(1).
- Maffetone P. Everyone is an Athlete. David Barmore Pub: New York; 1989.
- Maitland GD. Peripheral Manipulation, 2nd Ed. Butterworths: Boston; 1977.
- Mallik AK, Ferrell WR, McDonald AG, Sturrock RD. Impaired proprioceptive acuity at the proximal interphalangeal joint in patients with the hypermobility syndrome. Br J Rheumatol. 1994;33(7):631-7.
- Mann F. Reinventing Acupuncture – A New Concept of Ancient Medicine. Butterworth-Heinemann Ltd.: Oxford; 1992.
- Mann RA. Biomechanics of running. In: Symposium on The Foot and Leg in Running Sports. Ed. Mack RP. The C.V. Mosby Co: St. Louis; 1982.
- Mann RA. Biomechanics. In: Disorders of the Foot, Vol 1. Ed. Jahss MH. W.B. Saunders Co: Philadelphia; 1982.
- Mann RA. Hallux valgus. In: Symposium on The Foot and Ankle. Ed. Kiene RH, Johnson KA. The C.V. Mosby Co: St. Louis; 1983.
- Mann RA, Inman VT. Phasic activity of intrinsic muscles of the foot. J Bone Joint Surg. 1964;46:469-81.
- Mann RA, Poppen NK, O'Konski M. Amputation of the great toe – a clinical and biomechanical study. Clin Orthop. 1988(226):192-205.
- Mann RA, Moran GT, Dougherty SE. Comparative electromyography of the lower extremity in jogging, running, and sprinting. Am J Sports Med. 1986;14(6):501-10.
- Manral DB. Hallux rigidus: A case report of successful chiropractic management and review of the literature. J Chiropr Med. 2004;3(1):6-11.
- Manter JT. Movements of the subtalar and transverse tarsal joints. Anat Rec. 1941;80(4).
- Manter J. Distribution of compression forces in joints of the human foot. Anat Rec. 1946;96(3):313-21.
- Maresca G, Adriani E, Falez F, Mariani PP. Arthroscopic treatment of bilateral Freiberg's infraction. Arthroscopy. 1996;12(1):103-8.
- Marshall PW, McKee AD, Murphy BA. Impaired trunk and ankle stability in subjects with functional ankle instability. Med Sci Sports Exerc. 2009;41(8):1549-57.
- Marshall RN. Foot mechanics and joggers' injuries. NZ Med J. 1978;88(621):288-90.
- Martin RL, Irrgang JJ, Conti SF. Outcome study of subjects with insertional plantar fasciitis. Foot Ankle Int. 1998;19(12):803-11.
- Masarsky C S, Masarsky MT. Somatovisceral Aspects of Chiropractic: An Evidence-Based Approach. Churchill Livingstone: Philadelphia; 2001.
- Mathews MO, Thomas E, Court L. Applied Kinesiology Helping Children with Learning Disabilities. Int J AK and Kinesio Med. 1999;4.
- Mathews MO, Thomas E. A pilot study on the value of applied kinesiology in helping children with learning difficulties. Br Osteopathic J. 1993;XII.
- Mattson RM. Resolution of chronic back, leg and ankle pain following chiropractic intervention and the use of orthotics. Journal of Vertebral Subluxation Research. 2008;March 20:1-4.
- Mauch M, Mickle KJ, Munro BJ, Dowling AM, Grau S, Steele JR. Do the feet of German and Australian children differ in structure? Implications for children's shoe design. Ergonomics. 2008;51(4):527-39.
- Maykel W. The effect of calcaneal elevation on peripheral muscle strength. Collected Papers of the International College of Applied Kinesiology; 1984.
- Maykel W, Manning JB, Jr. Heal helper update. Collected Papers of the International College of Applied Kinesiology; 1987.
- Maykel W. Pediatric case history: cost effective treatment of block naso-lacrimal canal utilizing applied kinesiology tenets. Int J AK and Kinesio Med. 2003;16:34.
- McBryde AM, Jr. Stress fractures in athletes. J Sports Med. 1975;3(5):212-7.
- McCormack RG, Leith JM. Ankle fractures in diabetics. Complications of surgical management. J Bone Joint Surg Br. 1998;80(4):689-92.
- McDougall C. Born to Run: A Hidden Tribe, Superathletes, and the Greatest Race the World Has Never Seen. Knopf: New York; 2009.
- McDowall D. Fix foot problems without orthotics. Int J AK and Kinesio Med. 2004;18.

- McMahan JO. Proper footwear for play and the fitting of painful, deformed feet. In: Symposium on The Foot and Ankle. Eds. Kiene RH, Johnson KA. The C.V. Mosby Co: St. Louis; 1983.
- McVey ED, Palmieri RM, Docherty CL, Zinder SM, Ingersoll CD. Arthrogenic muscle inhibition in the leg muscles of subjects exhibiting functional ankle instability. Foot Ankle Int. 2005;26(12):1055-61.
- Melzack R. Pain--an overview. Acta Anaesthesiol Scand. 1999;43(9):880-4.
- Mennell J. Foot-gear. Proc R Soc Med. 1939;33(2):105-10.
- Mennell J. Joint Pain. Little, Brown & Co.; 1964.
- Mennell J. Foot Pain. Little, Brown & Co: Boston; 1969.
- Mense S, Simons DG. Muscle Pain: Understanding Its Nature, Diagnosis, and Treatment. Lippincott Williams & Wilkins: Philadelphia; 2001:131-157.
- Merck Manuals. www.merck.com; 2011.
- Meyer PF, Oddsson LI, De Luca CJ. The role of plantar cutaneous sensation in unperturbed stance. Exp Brain Res. 2004;156(4):505-12.
- Michael RH, Holder LE. The soleus syndrome: A cause of medial tibial stress (shin splints). Am J Sports Med. 1985;13(2):87-94.
- Michaud T. Foot Orthoses and Other Forms of Conservative Foot Care. Williams & Wilkins: Baltimore;1997.
- Michaud T. Pedal biomechanics as related to orthotic therapies: An overview – Part II. Chiro Econ. 1987;29(4).
- Milgram JE. Office measures for relief of the painful foot. J Bone Joint Surg. 1964;46:1095-116.
- Milgram JE. Design and use of pads and strappings for office relief of the painful foot. In: Symposium on The Foot and Ankle. Eds. Kiene RH, Johnson KA. The C.V. Mosby Co: St. Louis; 1983.
- Milgrom C, et al. The long-term followup of soldiers with stress fractures. Am J Sports Med. 1985;13(6):398-400.
- Miller JE, Nigg BM, Liu W, Stefanyshyn DJ, Nurse MA. Influence of foot, leg and shoe characteristics on subjective comfort. Foot Ankle Int. 2000;21(9):759-67.
- Milne H. The Heart of Listening: A Visionary Approach to Craniosacral Work. North Atlantic Press: Berkeley, CA; 1995.
- Moncayo R, Moncayo H. Evaluation of Applied Kinesiology meridian techniques by means of surface electromyography (sEMG): demonstration of the regulatory influence of antique acupuncture points. Chin Med. 2009;4:9.
- Mondelli M, Cioni R. Electrophysiological evidence of a relationship between idiopathic carpal and tarsal tunnel syndromes. Neurophysiol Clin. 1998;28(5):391-7.
- Morgan RC, Jr., Crawford AH. Surgical management of tarsal coalition in adolescent athletes. Foot & Ankle. 1986;7(3):183-193.
- Morley JB, Decker LM, Dierks T, Blanke D, French JA, Stergiou N. Effects of varying amounts of pronation on the mediolateral ground reaction forces during barefoot versus shod running. J Appl Biomech. 2010;26(2):205-14.
- Morley AJM. Knock-knee in children. Br Med J. 1957;2.
- Morton DJ. The Human Foot. Columbia Univ Press: New York; 1935.
- Morton TG. A peculiar and painful affection of the fourth metatarso-phalangeal articulation. Am J Med Sci. 1876;71.
- Motyka T, Yanuck SF. Expanding the Neurological Examination Using Functional Neurologic Assessment, Part I: Methodological Considerations. International Journal of Neuroscience. 1999;97:61-76.
- Mulier T, Steenwerckx A, Thienpont E, Sioen W, Hoore KD, Peeraer L, Dereymaeker G. Results after cheilectomy in athletes with hallux rigidus. Foot Ankle Int. 1999;20(4):232-7.
- Munro RR. Electromyography of the masseter and anterior temporalis muscles in subjects with potential temporomandibular joint dysfunction. Aust Dent J. 1972;17(3):209-18.
- Murray RO. Jacobson HG. The radiology of Skeletal Disorders, Vol I, 2nd Ed. Churchill Livingstone: New York; 1977.
- Murray RO, Jacobson HG. The Radiology of Skeletal Disorders, Vol III, 2nd Ed. Churchill Livingstone: New York; 1977.
- Myers TW. Anatomy Trains: Myofascial Meridians for Manual and Movement Therapists. Churchill-Livingstone; 2001.
- Nakajima T, Sakamoto M, Tazoe T, Endoh T, Komiyama T. Location specificity of plantar cutaneous reflexes involving lower limb muscles in humans. Exp Brain Res. 2006;175(3):514-25.
- Napier J. Hands. Princeton University Press: Princeton, NJ; 1993.
- Nashner LM. Fixed patterns of rapid postural responses among leg muscles during stance. Exp Brain Res. 1977;30(1):13-24.
- Ndetan HT, Rupert RL, Bae S, Singh KP. Prevalence of musculoskeletal injuries sustained by students while attending a chiropractic college. J Manipulative Physiol Ther. 2009;32(2):140-8.
- Nguyen US, Hillstrom HJ, Li W, Dufour AB, Kiel DP, Procter-Gray E, Gagnon MM, Hannan MT. Factors associated with hallux valgus in a population-based study of older women and men: the MOBILIZE Boston Study. Osteoarthritis Cartilage. 2010;18(1):41-6.
- Nicholas JA, Marino M. The relationship of injuries of the leg, foot, and ankle to proximal thigh strength in athletes. Foot & Ankle. 1987;7(4):218-28.
- Nicholas JA, Strizak AM, Veras G. A study of thigh muscle weakness in different pathological states of the lower extremity. Am J Sports Med. 1976;4(6):241-8.
- Nigg BM, Morlock M. The influence of lateral heel flare of running shoes on pronation and impact forces. Med Sci Sports Exer. 1987;19(3):294-302.
- Nix S, Smith M, Vicenzino B. Prevalence of hallux valgus in the general population: a systematic review and meta-analysis. J Foot Ankle Res. 2010;3:21.
- Nordsiden L, Van Lunen BL, Walker ML, Cortes N, Pasquale M, Onate JA. The effect of 3 foot pads on plantar pressure of pes planus foot type. J Sport Rehabil. 2010;19(1):71-85.
- Norkin C, White D. Measurement of Joint Motion – A Guide to Goniometry. 3rd Ed.. F.A Davis Company: Philadelphia; 2003.
- Nurse MA, Nigg BM. The effect of changes in foot sensation on plantar pressure and muscle activity. Clin Biomech (Bristol, Avon). 2001;16(9):719-27.
- O'Brien M. The anatomy of the Achilles tendon. Foot Ankle Clin. 2005;10(2):225-38.
- O'Connell AL. Electromyographic study of certain leg muscles during movements of the free foot and during standing. Am J Phys Med. 1958;37(6):289-301.
- O'Connor J, Bensky D. Acupuncture: A Comprehensive Text. Eastland Press: Seattle, WA; 1981.
- O'Connell AL. Effect of sensory deprivation on postural reflexes. Electromyography. 1971;11(5)519-27.
- O'Connell AL, Gardner EB. Understanding the Scientific Basis of Human Movement. Williams & Wilkins: Baltimore; 1972.
- O'Neill DB, Micheli LJ. Tarsal coalition – a followup of adolescent athletes. Am J Sports Med. 1989;17(4):544-9.
- Olerud C, Rosendahl Y. Torsion-transmitting properties of the hind foot. Clin Orthop. 1987(214):285-94.
- Page P, Frank CC, Lardner R. Assessment and treatment of muscle imbalance: The Janda Approach. Human Kinetics: Champaign, IL; 2010.
- Page LE. The role of the fascae in the maintenance of structural integrity. Acad Appl Osteo Yr Bk. 1952.

- Palmieri-Smith RM, Hopkins JT, Brown TN. Peroneal activation deficits in persons with functional ankle instability. Am J Sports Med. 2009;37(5):982-8.
- Panjabi MM. The stabilizing system of the spine. Part II. Neutral zone and instability hypothesis. J Spinal Disord. 1992;5(4):390-6; discussion 397.
- Paton RW, Choudry Q. Neonatal foot deformities and their relationship to developmental dysplasia of the hip: an 11-year prospective, longitudinal observational study. J Bone Joint Surg Br. 2009;91(5):655-8.
- Pauli Y. The Effects of Chiropractic Care on Individuals Suffering from Learning Disabilities and Dyslexia: A Review of the Literature. J Vertebral Subluxation Res. 2007;1:1-12.
- Percy EC, Mann DL. Tarsal coalition: A review of the literature and presentation of 13 cases. Foot & Ankle. 1988;9(1):40-4.
- Peterson HA. Dowel bone graft technique for triple arthrodesis in talocalcaneal coalition – report of a case with 12-year follow-up. Foot and Ankle. 1989;9(4):201-3.
- Petty N, Moore A. Neuromusculoskeletal examination and assessment. Churchill-Livingstone: Edinburgh; 1998.
- Pick M. Cranial Sutures. Eastland Press: Seattle; 1999.
- Pineda C, Resnick D, Greenway G. Diagnosis of tarsal coalition with computed tomography. Clin Orthop. 1986;208:282-8.
- Platzer W. Color atlas/text of human anatomy. Georg Thieme: Stuttgart; 1992.
- Pollard JP, LeQuesne LP, Tappin JW. Forces under the foot. J Biomed Eng. 1983;5.
- Pottenger FM, Jr. Studies of fragmentation of the tarsal and metatarsal bones resulting from recurrent metabolic insults. Transact Am Ther Soc. 1945;45.
- Prichasuk S, Sinphurmsukskul O. Kidner procedure for symptomatic accessory navicular and its relation to pes planus. Foot Ankle Int. 1995;16(8):500-3.
- Prior TD. Biomechanical foot function: a podiatric perspective: part 2. J Bodyw Mov Ther. 1999;3(3):169-184.
- Raffelock D. An investigation of applied kinesiology's manual mucle testing by three dimensional computerized force-plate analysis. Collected Papers International College of Applied Kinesiology, Winter. 1987:213-230.
- Ramamurti CP. Orthopaedics in Primary Care. Williams & Wilkins: Baltimore; 1979.
- Ramig D, et al. The foot and sports medicine – biomechanical foot faults as related to chondromalacia patellae. Bull Sports Med Section, APTA. 1977;3(2).
- Rao UB, Joseph B. The influence of footwear on the prevalence of flat foot. A survey of 2300 children. J Bone Joint Surg Br. 1992;74(4):525-7.
- Rasch PJ, Burke RK. Kinesiology and Applied Anatomy, 6th Ed. Lea & Febiger: Philadelphia; 1978.
- Rask MR. Medial plantar neurapraxia (jogger's foot). Clin Orthop. 1978;134:193-5.
- Refshauge KM, Raymond J, Kilbreath SL, Pengel L, Heijnen I. The effect of ankle taping on detection of inversion-eversion movements in participants with recurrent ankle sprain. Am J Sports Med. 2009;37(2):371-5.
- Richards CE, Magin PJ, Callister R. Is your prescription of distance running shoes evidence-based? Br J Sports Med. 2009;43(3):159-62.
- Richardson EG. Injuries to the hallucal sesamoids in the athlete. Foot & Ankle. 1987;7(4):229-44.
- Richie DH, Jr. Functional instability of the ankle and the role of neuromuscular control: a comprehensive review. J Foot Ankle Surg. 2001;40(4):240-251.
- Riddle DL, Schappert SM. Volume of ambulatory care visits and patterns of care for patients diagnosed with plantar fasciitis: a national study of medical doctors. Foot Ankle Int. 2004;25(5):303-10.
- Robbins SE, Hanna AM. Running-related injury prevention through barefoot adaptations. Med Sci Sports Exerc. 1987;19(2):148-56.
- Robbins SE, Gouw GJ, Hanna AM. Running-related injury prevention through innate impact-moderating behavior. Med Sci Sports Exer. 1989;21(2):130-9.
- Robbins SE, Hanna AM, Gouw GJ. Overload protection: Avoidance response to heavy plantar surface loading. Med Sci Sports Exer. 1988;20(1):85-92.
- Robbins SE, Hanna A, Jones LA. Sensory attenuation induced by modern athletic footwear. J. Test Eval. 1988;16.
- Robbins S, Waked E. Hazard of deceptive advertising of athletic footwear. Br J Sports Med. 1997;31(4):299-303.
- Robbins S, Waked E, Allard P, McClaran J, Krouglicof N. Foot position awareness in younger and older men: the influence of footwear sole properties. J Am Geriatr Soc. 1997;45(1):61-6.
- Roberts AC, McClure RD, Weiner RI, Brooks GA. Overtraining affects male reproductive status. Fertil Steril. 1993;60(4):686-92.
- Rodgers MM, Cavanagh PR. Pressure distribution in Morton's foot structure. Med Sci Sports Exer. 1989;21(1):23-8.
- Rogers MM, Cavanagh PR, Sanders LJ. Plantar pressure distribution of diabetic feet. In: Biomechanics X-A. Ed. Jonsson B. Human Kinetics Pub, Inc: Champaign, IL; 1987.
- Rolf IP. Rolfing: The Integration of Human Structures. Dennis-Landman Publishers: Santa Monica; 1977.
- Root ML, et al. Axis of motion of the subtalar joint – an anatomical study. J Am Podiatry Assoc. 1966;56(4).
- Rose GK. Pes Planus. In: Disorders of the Foot, Vol 1. Ed. Jahss MH. W.B. Saunders Co: Philadelphia; 1982.
- Rothbart BA. Vertical facial dimensions linked to abnormal foot motion. J Am Podiatr Med Assoc. 2008;98(3):189-96.
- Rothbart BA, Estabrook L. Excessive pronation: a major biomechanical determinant in the development of chondromalacia and pelvic lists. J Manipulative Physiol Ther. 1988;11(5):373-9.
- Rothbart BA. Relationship of functional leg-length discrepancy to abnormal pronation. J Am Podiatr Med Assoc. 2006;96(6):499-504; discussion:505-7.
- Ruch TC. Brain stem control of posture and orientation in space. In: Physiology and Biophysics, I -- The brain and Neural Function, 20th Ed. Eds. Ruch T, Patton HD. W.B. Saunders Co: Philadelphia; 1979.
- Running USA. www.Runningusa.org.; 2008.
- Ryerson EW. Comment in discussion of "Hypermobile flat-foot with short tendo Achillis" by Harris RI, Beath T. J Bone Joint Surg. 1948;30A(1).
- Sahar T, Cohen MJ, Ne'eman V, Kandel L, Odebiyi DO, Lev I, Brezis M, Lahad A. Insoles for prevention and treatment of back pain. Cochrane Database Syst Rev. 2007;(4):CD005275.
- Sammarco GJ. Foot and ankle injuries in sports. Am J Sports Med. 1986;14(6).
- Sammarco JG. Dancing injuries. Foot & Ankle. 1986;6(6).
- Sandell J, Palmgren PJ, Björndahl L. Effect of chiropractic treatment on hip extension ability and running velocity among young male running athletes. J Chiropr Med. 2008;7(2):39-47.
- Sanderson DJ, Cavanagh PR. An investigation of the in-shoe pressure distribution during cycling in conventional cycling shoes or running shoes. In: Biomechanics X-B. Ed. Jonsson B. Human Kinetics Pub, Inc: Champaign, IL; 1987.
- Sarrafian SK. Functional characteristics of the foot and plantar aponeurosis under tibiotalar loading. Foot & Ankle. 1987;8(1):4-18.

- Sarrafian SK. Anatomy of the Foot and Ankle, 2nd Ed. JB Lippincott: Philadelphia; 1983.
- Sartoris DJ, Resnick D. Pictorial review: Cross-sectional imaging of the foot and ankle. Foot & Ankle. 1987;8(2):59-80.
- Scartozzi G, Schram A, Janigian J. Freiberg's infraction of the second metatarsal head with formation of multiple loose bodies. J Foot Surg. 1989;28(3):195-9.
- Schafer RC. Chiropractic management of sports and recreational injuries, 2nd Ed. Williams & Wilkins: Baltimore; 1986.
- Schenkel D, deGraauw J, deGraauw C. Talocalcaneal coalition in a 15 year old female basketball player. J Can Chiropr Assoc. 2010;54(4):222–228.
- Schmitt WH, McCord KM. Quintessential Applications: A(K) Clinical Protocol (QA). www.quintessentialapplications.com; 2010.
- Schmitt W, Yanuck S. Expanding the Neurological Examination Using Functional Neurologic Assessment Part II: Neurologic Basis of Applied Kinesiology. International Journal of Neuroscience. 1999;97:77-108.
- Schmitt WH. The Ligament Stretch-Adrenal Stress Syndrome: Summary of Clinical Investigation. Proceedings of Summer Meeting, International College of Applied Kinesiology. Detroit, MI; 1977.
- Schwartzbauer J, Kolber J, Schwartzbauer M, Hart J, Zhang J. Athletic performance and physiological measures in baseball players following upper cervical chiropractic care: a pilot study. J Vertebral Subluxation Res. 1997;1(4):33-39.
- Schmidhofer R. Knee and allergy. Medical Journal for Applied Kinesiology. 1997;1:47-50.
- Schmidt DM, Romash MM. Atraumatic avascular necrosis of the head of the talus: A case report. Foot & Ankle. 1988;8(4).
- Schmitt WH, Jr. One common cause of recurrent foot subluxations. Chiro Econ. 1977;19(6).
- Schultz AL. Athletic and Industrial Injuries of the Foot and Ankle. Argus Printers: Stickney, SD; 1979.
- Schwartz RP, Heath AL, Misiek W. The influence of the shoe on gait as recorded by eletrobasograph and slow-motion moving pictures. J Bone Joint Surg. 1935;17.
- Scranton PE, Jr. The Metatarsalgia: Diagnosis and treatment. J Bone Joint Surg. 1980;62(5):723-32.
- Scranton PE, Jr. Metatarsalgia: A clinical review of diagnosis and management. Foot & Ankle. 1981;1(4):229-34.
- Sella EJ, Lawson JP. Biomechanics of the accessory navicular synchondrosis. Foot & Ankle. 1987;8(3):156-63.
- Sella EJ, Lawson JP, Ogden JA. The accessory navicular synchondrosis. Clin Orthop. 1986;209:280-5.
- Selye H. The Stress of Life, Revised Edition. McGraw-Hill: New York; 1976.
- Sgarlato TE. Transplantation of the flexor digitorum longus muscle tendon in hammertoes. J Am Podiatry Assoc. 1970;60(10):383-8.
- Sgarlato TE, Sokoloff TH, Mosher M. Anomalous insertion of extensor hallucis longus tendon – a case report. J Am Podiatry Assoc. 1969;59(5):192-3.
- Sharkey NA, Ferris L, Donahue SW. Biomechanical consequences of plantar fascial release or rupture during gait: part I--disruptions in longitudinal arch conformation. Foot Ankle Int. 1998;19(12):812-20.
- Sheffield FJ, Gersten JW, Mastellone AF. Electromyographic study of the muscles of the foot in normal walking. Amer J Phys Med. 1956;35(4):223-36.
- Shibuya N, Jupiter DC, Ciliberti LJ, VanBuren V, La Fontaine J. Characteristics of adult flatfoot in the United States. J Foot Ankle Surg. 2010;49(4):363-8.
- Shine IB. Incidence of hallus valgus in a partially shoe-wearing community. Br Med J. 1965;1(5451):1648-50.
- Shrier I, Macdonald D, Uchacz G. A pilot study on the effects of pre-event manipulation on jump height and running velocity. Br J Sports Med. 2006;40(11):947-9.
- Silverman JJ. Asymmetrical pronation. ACA J Chiro. 1986;23(11).
- Sim-Fook L, Hodgson AR. A comparison of foot forms among the non-shoe and shoe-wearing Chinese population. J Bone Joint Surg Am. 1958;40-A(5):1058-62.
- Simkin A, et al. Combined effect of foot arch structure and an orthotics devise on stress fractures. Foot & Ankle. 1989;10(1):25-9.
- Simons DG. Understanding effective treatments of myofascial trigger points. J Bodyw Mov Ther. 2002;6(2):81-88.
- Simons DG, Travell J, Simons L. Myofascial pain and dysfunction: The trigger point manual, Vol. 1: Upper half of the body, 2nd Ed. Williams & Wilkins: Baltimore; 1999.
- Simons DG. Referred phenomena of myofascial trigger points. In: New Trends in Referred Pain and Hyperalgesia. Vecchiet L, Albe-Fessard D, Lindlom U. Elsevier: Amsterdam; 1993.
- Slemenda C, Brandt KD, Heilman DK, Mazzuca S, Braunstein EM, Katz BP, Wolinsky FD. Quadriceps weakness and osteoarthritis of the knee. Ann Intern Med. 1997;127(2):97-104.
- Slocum DB, James SL. Biomechanics of running. JAMA. 1968;205(11):721-8.
- Smidt GL. Gait in Rehabilitation. Churchill Livingstone: New York; 1990.
- Smillie IS. Freiberg's infraction (Kohler's second disease). J Bone Joint Surg. 1957 ;39B.
- Smille IS. Treatment of Freiberg's infraction. Proc Royal Soc Med. 1967;60.
- Smith JW. The activity of leg and foot muscles in standing. J Anat. 1951;85.
- Smith JW. Muscular control of the arches of the foot in standing: An electromyographic assessment. J Anat. 1954;88(2):152-63.
- Smith RW, Katchis SD, Ayson LC. Outcomes in hallux rigidus patients treated nonoperatively: a long-term follow-up study. Foot Ankle Int. 2000;21(11):906-13.
- Snijders CJ, Philippens MMGM. Biomechanics of hallux valgus and spread foot. Foot & Ankle. 1986;7(1):26-39.
- Snyder RB, Lipscomb AB, Johnston RK. The relationship of tarsal coalitions to ankle sprains in athletes. Am J Sports Med. 1981;9(5):313-7.
- Soderberg GL. Kinesiology: Application to Pathological Motion, 2nd Ed. Williams & Wilkins: Baltimore; 1997.
- Soderberg GL, Staves SA. The effect of negative heel footwear upon postural muscle activity. Electromyogr Clin Neurophysiol. 1977;17(3-4).
- Solomonow M. Ligaments: a source of musculoskeletal disorders. J Bodyw Mov Ther. 2009;13(2):136-54.
- Solomonow M, Zhou B, Harris M, Lu Y, Baratta RV. The ligamento-muscular stabilizing system of the spine. Spine. 1998;23(23):2552-62.
- Solomonow M, Baratta RV, Zhou BH, Shoki H, Bose W, Beck C, D'Ambrosia R. The synergistic action of the ACL and thigh muscles in maintaining joint stability. Am J Sports Med. 1987;15:20-213.
- Soma CA, Mandelbaum BR. Repair of acute Achilles tendon ruptures. Orthop Clin North Am. 1995;26(2):239-247.
- Soma staff. Diagnosing diabetics' feet. Soma. 1988;2(4).
- Sprieser P. A new epidemic of knee injuries: anterior cruciate ligament in women athletes. Collected Papers International College of Applied Kinesiology, 2002-2003;1:45-49.
- Sprieser P. The association of repeat muscle activation patient induced (R.M.A.P.I.) to hypoadrenia. Collected Papers International College of Applied Kinesiology, 2001-2002;1:181-182.

- Squadrone R, Gallozzi C. Biomechanical and physiological comparison of barefoot and two shod conditions in experienced barefoot runners. J Sports Med Phys Fitness. 2009;49(1):6-13.
- Staal A, van Gijn J, Spaans F. Mononeuropathies: Examination, Diagnosis and Treatment. WB Saunders: London; 1999.
- Staheli LT, Chew DE, Corbett M. The spontaneous resolution of childhood flatfeet: Implications regarding the need for treatment. Foot and Ankle. 1986;6(6).
- Steindler A. Kinesiology of the Human Body Under Normal and Pathological Conditions. Charles C. Thomas Pub: Springfield, IL; 1955.
- Stewart SF. Physiology of the unshod and shod foot with an evolutionary history of footgear. Am J Surg. 1945;58.
- Stierwalt DD. Extremity Adjusting. Privately published: Davenport, IA; 1976.
- Stokes M, Young A. Investigations of quadriceps inhibition: implications for clinical practice. Physiotherapy. 1984;70:425-428.
- Stormont DM, <prreu NF. Am LM. Cass JR. Stability of the loaded ankle. Am J Sports Med. 1985;13(5):295-300.
- Strachan WF. Appendicular technic. In: Principles of Osteopathic Technic. Ed. Fryette HH. Acad Applied Osteopathy: Carmel, CA; 1954.
- Stuber K, Kristmanson K. Conservative therapy for plantar fasciitis: a narrative review of randomized controlled trials. J Can Chiropr Assoc. 2006;50(2):118-33.
- Subotnick SI. Equinus deformity as it affects the forefoot. J Am Podiatry Assoc. 1971;61(11):423-7.
- Subotnick SI. The flexible flatfoot – diagnosis conservative and surgical treatment. Arch Podiatric Med Foot Surg. 1973;1(1).
- Subotnick S. Sports & Exercise Injuries: Conventional, homeopathic & alternative treatments. North Atlantic Books: Berkeley, CA; 1991.
- Subotnick S. Orthotic foot control and the overuse syndrome. Physician Sportsmed. 1975;3(1).
- Subotnick S. Podiatric Sports Medicine. Futura Publishing Co: Mt. Kisco, NY; 1975.
- Suzuki R. Function of the leg and foot muscles from the viewpoint of the electromyogram. J Japanese Orthop Surg Soc. 1956;30.
- Tachdjian MO. Congenital deformities. In: Disorders of the Foot, Vol 1. Ed. Jahss MH. W.B. Saunders Co: Philadelphia; 1982.
- Tecco S, Salini V, Calvisi V, Colucci C, Orso CA, Festa F, D'Attilio M. Effects of anterior cruciate ligament (ACL) injury on postural control and muscle activity of head, neck and trunk muscles. J Oral Rehabil. 2006;33(8):576-87.
- Tecco S, Polimeni A, Saccucci M, Festa F. Postural loads during walking after an imbalance of occlusion created with unilateral cotton rolls. BMC Research Notes. 2010;3:141.
- Thomas JL, Christensen JC, Kravitz SR, Mendicino RW, Schuberth JM, Vanore JV, Weil LS Sr, Zlotoff HJ, Bouché R, Baker J; American College of Foot and Ankle Surgeons heel pain committee. The diagnosis and treatment of heel pain: a clinical practice guideline-revision 2010. J Foot Ankle Surg. 2010;49(3 Suppl):S1-19.
- Thomson SA. Hallus varus and metatarsus varus. Clinic Orthop. 1960.
- Thordarson DB. Detecting and treating common fractures of the foot and ankle part 2: the midfoot and forefoot. Phys Sportsmed. 1996;24(10):58-64.
- Thordarson DB, Schmotzer H, Chon J, Peters J. Dynamic support of the human longitudinal arch. A biomechanical evaluation. Clin Orthop Relat Res. 1995;(316):165-72.
- Ting AJ, et al. The role of subtalar motion and ankle contact pressure changes from angular deformities of the tibia. Foot & Ankle. 1987;7(5):290-9.
- Tobias PV. Man the tottering biped: The evolution of his erect posture. In: Proprioception, Posture and Emotion. Ed. Garlick D. Univ New South Wales: Kensington, NSW; 1982.
- Toft E, et al. Passive tension of the ankle before and after stretching. Am J Sports Med. 1989;17(4):489-94.
- Toomey EP. Plantar heel pain. Foot Ankle Clin. 2009;14(2):229-45.
- Tourné Y, Besse JL, Mabit C; Sofcot. Chronic ankle instability. Which tests to assess the lesions? Which therapeutic options? Orthop Traumatol Surg Res. 2010;96(4):433-46.
- Tran MD, Holly RG, Lashbrook J, Amsterdam EA. Effects of hatha yoga practice on the health-related aspects of physical fitness. Preventive Cardiology. 2001;4(4):165e170.
- Travell JG, Simons DG. Myofascial Pain and Dysfunction: The Trigger Point Manual: The Lower Extremities. Williams & Wilkins: Baltimore; 1992.
- Trott AW. Foot and ankle problems in children and adolescents: Sports aspects. In: Symposium on The Foot and Ankle. Eds. Kiene RH, Johnson KA. CV Mosby Co: St. Louis; 1983.
- Tropp H. Pronator muscle weakness in functional instability of the ankle joint. Int J Sports Med. 1986;7(5):291-4.
- Tsai LC, Yu B, Mercer VS, Gross MT. Comparison of different structural foot types for measures of standing postural control. J Orthop Sports Phys Ther. 2006;36(12):942-53.
- Tsai WC, Chiu MF, Wang CL, Tang FT, Wong MK. Ultrasound evaluation of plantar fasciitis. Scand J Rheumatol. 2000;29(4):255-9.
- Turek SL. Orthopaedics – Principles and Their Application, Vol 1, 4th Ed. J.B. Lippincott Co.: Philadelphia; 1984.
- Turek SL. Orthopaedics – Principles and Their Application, Vol 2, 4th Ed. J.B. Lippincott Co.: Philadelphia; 1984.
- Vedula S, Kearney RE, Wagner R, Stapley PJ. Decoupling of stretch reflex and background muscle activity during anticipatory postural adjustments in humans. Exp Brain Res. 2010;205(2):205-13.
- Vernon HT, Aker P, Aramenko M, Battershill D, Alepin A, Penner T. Evaluation of neck muscle strength with a modified sphygmomanometer dynamometer: reliability and validity. J Manipulative Physiol Ther. 1992;15(6):343-9.
- Veves A, Giurini JM, LoGerfo FW. The Diabetic Foot: Medical and Surgical Management. Humana Press: Totowa, NJ; 2002.
- Viladot A. Metatarsalgia due to biomechanical alterations of the forefoot. Orthop Clin North Am. 1973;4(1):165-78.
- Viladot A. The metatarsals. In: Disorders of the Foot, Vol 1. Ed. Jahss MH. W.B. Saunders Co.: Philadelphia; 1982.
- Vixie DE. Orthotics and shoe corrections. In: Symposium on The Foot and Leg in Running Sports. Ed. Mack RP. The C.V. Mosby Co: St. Louis; 1982.
- Vleeming A, Mooney V, Dorman T, et al. (Eds.) Movement, stability and low back pain. Churchill Livingstone: Edinburgh; 1997.
- Vukasinović Z, Zivković Z, Vucetić C. Flat feet in children. Srp Arh Celok Lek. 2009;137(5-6):320-2.
- Waddington G, Adams R. Football boot insoles and sensitivity to extent of ankle inversion movement. Br J Sports Med. 2003 Apr;37(2):170-4; discussion 175.
- Wagner FW, Jr. The insensitive foot. In: Symposium on The Foot and Ankle. Ed. Kiene RH, Johnson KA. The C.V. Mosby Co: St. Louis; 1983.
- Wallden M. Chains, trains and contractile fields. J Bodyw Mov Ther. 2010;14(4):403-10.
- Wallden M. Shifting paradigms. J Bodyw Mov Ther. 2010;14(2):185-94.
- Waller JF, Jr. Biomechanics and rehabilitation of the gastroc-soleus complex. In: Symposium on The Foot and Leg in Running Sports. Ed. Mack RP. The C.V. Mosby Co: St. Louis; 1982.

- Wallin D, et al. Improvement of muscle flexibility – a comparison between two techniques. Am J Sports Med. 1985;13(4):263-8.
- Walther DS. Applied Kinesiology, Volume I – Basic Procedures and Muscle Testing. Systems DC: Pueblo, CO; 1981.
- Walther DS. Applied Kinesiology, Volume II – Head, Neck and Jaw Pain in Dysfunction-The Stomatognathic System. Systems DC: Pueblo, CO; 1983.
- Walther DS. Locating the Cause of Neurologic Disorganization. Selected Papers of the International College of Applied Kinesiology. ICAK USA: Shawnee Mission, KS; 1983.
- Walther DS. Applied Kinesiology: Synopsis. 2nd Ed. International College of Applied Kinesiology: Shawnee Mission, KS; 2000.
- Wang X, Jiang JY, Ma X, Huang JZ, Gu XJ. Management of the second and third metatarsal in moderate and severe hallux valgus. Orthopedics. 2009;32(12):892-6.
- Wanivenhaus A, Pretterklieber M. First tarsometatarsal joint: Anatomical biomechanical study. Foot & Ankle. 1989;9(4):153-7.
- Warburton M. Barefoot running. Sportsscience. 2001;5(3).sportsci.org.
- Warren BL, Jones CJ. Predicting plantar fascitis in runners. Med Sci Sports Exer. 1987;19(1):71-3.
- Wearing SC, Smeathers JE, Sullivan PM, Yates B, Urry SR, Dubois P. Plantar fasciitis: are pain and fascial thickness associated with arch shape and loading? Phys Ther. 2007;87(8):1002-8.
- Webster's Third New International Dictionary of the English Language (1976).
- Weinert CR, Jr., et al. Human fibular dynamics. In: Foot Science. Ed. Bateman JE. W.B. Saunders Co.: Philadelphia, 1976.
- Wenger DR, Mauldin D, Speck G, Morgan D, Lieber RL. Corrective shoes and inserts as treatment for flexible flatfoot in infants and children. J Bone Joint Surg Am. 1989;71(6):800-10.
- Weseley MS, Koval R, Kleiger B. Roentgen measurement of ankle flexion-extension motion. Clin Orthop. 1969;65:165-74.
- Wetz HH, Hentschel J, Drerup B, Kiesel L, Osada N, Veltmann U. Changes in shape and size of the foot during pregnancy. Orthopade. 2006;35(11):1124, 1126-30.
- White JW. Torsion of the Achilles tendon: Its surgical significance. Arch Surg. 1943;47.
- Whitman R. Observations of forty-five cases of flat-foot with particular reference to etiology and treatment. Clin Orthop. 1970;70.
- Whitting JW, Steele JR, McGhee DE, Munro BJ. Dorsiflexion Capacity Affects Achilles Tendon Loading During Drop Landings. Med Sci Sports Exerc. 2011;43(4):706-13.
- Wickstron J, Williams RA. Shoe corrections and orthopaedic foot supports. Clin Orthop. 1970;70.
- Wilk B, Nau S, Valero B. Physical therapy management of running injuries using an evidence-based functional approach. American Medical Athletic Association Journal. 2009;Jan.:36-38.
- Williams PL, et al. Imaging study of the painful heel syndrome. Foot & Ankle. 1987;7(6):345-9.
- Wilson JL. Adrenal Fatigue: The 21st Century Stress Syndrome. Smart Publications; 2002.
- Winter DA, Scott SH. Technique for interpretation of electromyography for concentric and eccentric contractions in gait. J Electromyogr Kinesiol. 1991;1(4):263-9.
- Wolff J. The Law of Bone Remodelling. Springer-Verlag: New York; 1986.
- Wright DG, Desai SM, Henderson WH. Action of the subtalar and ankle-joint complex during the stance phase of walking. J Bone Joint Surg. 1964;46A(2).
- Wright PB,. Brady LP. An anatomic evaluation of whiplash injuries. Clin Orthop. 1958;11:120-31.
- Wu S, Hong C, You J, Chen C, Wang L, Su F. Therapeutic effect on the change of gait performance in chronic calf myofascial pain syndrome: a time series case study. Journal of Musculoskeletal Pain. 2006;13(3):33–43.
- Wyatt LH. Conservative chiropractic management of recalcitrant foot pain after fasciotomy: a retrospective case review. J Manipulative Physiol Ther. 2006;29(5):398-402.
- Wyndow N, Cowan SM, Wrigley TV, Crossley KM. Neuromotor control of the lower limb in Achilles tendinopathy: implications for foot orthotic therapy. Sports Med. 2010;40(9):715-27.
- Wynne MW, Burns JM, Eland DC, Conatser RR, Howell JN. Effect of Counterstrain on Stretch Reflexes, Hoffmann Reflexes, and Clinical Outcomes in Subjects With Plantar Fasciitis. JAOA. 2006;106(9):547-556.
- Yavuz M, Husni E, Botek G, Davis BL. Plantar shear stress distribution in patients with rheumatoid arthritis: relevance to foot pain. J Am Podiatr Med Assoc. 2010;100(4):265-9.
- Yeung MS, Chan KM, So CH, Yuan WY. An epidemiological survey on ankle sprain. Br J Sports Med. 1994;28(2):112-6.
- Yochum TR, Rowe LJ. Essentials of Skeletal Radiology, 3rd Ed. Lippincott Williams & Wilkins: Baltimore; 2004.
- Yokoe K, Mannoji T. Stress fracture of the proximal phalanx of the great toe – a report of three cases. Am J Sports Med. 1986;14(3):240-2.
- Young MC, Fornasier VL, Cameron HU. Osteochondral disruption of the second metatarsal: A variant of Freiberg's. Foot & Ankle. 1987;8(2)103-9.
- Zafar H. Integrated jaw and neck function in man. Studies of mandibular and head-neck movements during jaw opening-closing tasks. Swed Dent J Suppl. 2000;(143):1-41.
- Zammit GV, Menz HB, Munteanu SE. Structural factors associated with hallux limitus/rigidus: a systematic review of case control studies. J Orthop Sports Phys Ther. 2009;39(10):733-42.
- Zampagni ML, Corazza I, Molgora AP, Marcacci M. Can ankle imbalance be a risk factor for tensor fascia lata muscle weakness? J Electromyogr Kinesiol. 2009;19(4):651-9.
- Zatkin A. An observation of a knee. Collected Papers International College of Applied Kinesiology, 1989-1990;1:83-86.
- Zaw H, Calder JD. Tarsal coalitions. Foot Ankle Clin. 2010;15(2):349-64.
- Zwipp H, Dahlen C, Amlang M, Rammelt S. Injuries of the tibialis posterior tendon: diagnosis and therapy. Orthopade. 2000;29(3):251-9.

"WATCH where the patient holds himself during a postural analysis."
- George J. Goodheart, Jr.

"…the results indicate, no difference between relief of pain and reduction of swelling was demonstrated. We cannot recommend routine treatment with ibuprofen for acute ankle joint injuries."
- Fredberg et al., 1989

"If you don't have a leg to stand on, you can't put your foot down."

~ Robert Altman

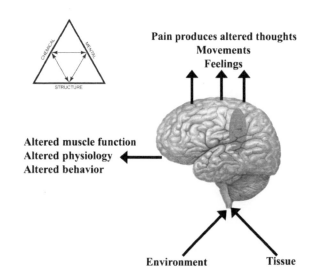

CHAPTER FIVE

Leg and Ankle

This chapter deals with the leg's anatomy and conditions that develop in the leg and ankle. It excludes ankle anatomy and function, since those are covered in the previous chapter on the foot. In anatomical and functional discussions, the leg is typically considered that portion of the lower extremity from below the knee to the ankle.

The tibia is the second longest bone in the body and with its companion, the fibula, articulates superiorly and inferiorly; in addition, there is attachment along the shafts by the fibrous interosseous membrane. Normal motion between the bones is necessary for proper function of the ankle mortise and action between the bones during gait. The distal end of the tibia is rotated laterally (tibial torsion) compared to its proximal end by about 30 degrees, with this torsion accentuated in Africans. **(Eckhoff et al., 1994)**

The superior tibiofibular articulation is a plane joint with a synovial membrane and fibrous capsule. The tibial articular facet is on the posterolateral aspect of the rim of the tibial condyle, facing obliquely posteriorly, inferiorly, and laterally. The fibular facet is on the upper surface of the head of the fibula, meeting the tibial facet anteriorly, superiorly, and medially. A portion of the biceps femoris tendon inserts into the styloid process of the fibula. The movement of the fibular head is powerfully influenced by the biceps femoris muscle. **(Greenman, 2003; Walther, 1981)**

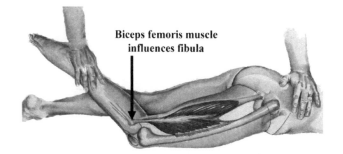

(With kind permission, ICAK-USA)

The inferior tibiofibular articulation is a syndesmosis, having no articular cartilage. This permits the talus to distribute loads over the entire foot, and is entirely covered by articular surfaces and ligamentous insertions and has no muscular attachments, giving it the characteristic of a 'relay station'. The body weight is effectively distributed through the talus in the manner. The two bones have no contact with each other, giving flexibility between the internal and external malleoli that is necessary for them to follow the tapered head of the talus.

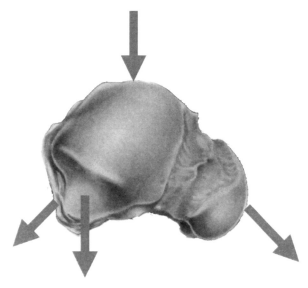

The distribution of body weight through the talus (after Kapandji, 2010)

Ankle mortise function is discussed in the previous chapter. Briefly, ankle dorsiflexion separates the lateral and medial malleoli to accept the wider anterior portion of the talus body. During plantar flexion the distance between the malleoli narrows. **(Dubbeldam et al., 2010)** The contraction of the tibialis posterior, having its bipennate origin from both the tibia and fibula, causes an approximation of these bones, thus tightening the ankle mortise when the tibialis posterior contracts during plantar flexion. **(Gray's Anatomy, 2007)**

The bones and muscles of the leg are surrounded by a strong fascial sheath called the crural fascia. There are fibrous septa that separate the muscles of the leg into compartments. The anterior compartment contains the tibialis anterior, extensor digitorum longus, extensor hallucis longus, and peroneus tertius muscles. The anterior tibial artery and vein and the deep peroneal nerve are also contained in this compartment. The lateral compartment contains the peroneus longus and brevis muscles; the superficial posterior compartment contains the gastrocnemius and soleus, and the almost vestigial plantaris muscle. The deep posterior compartment contains the flexor digitorum longus, flexor hallucis longus, and popliteus muscles. It also contains the posterior tibial and peroneal arteries and veins and the tibial nerve. A fifth compartment has been described **(Gray's Anatomy, 2007)** that contains only the tibialis posterior muscle. The muscle compartments and their significance are described later.

Ankle

Ankle Joint Strain

The general use of the terms "sprain" and "strain" denotes a sprain as a partial or complete rupture of ligament fibers, **(Dorland's Illustrated Medical Dictionary, 2007)** and a strain as an overstretching or overexertion of some part of the musculature. **(Chaitow & DeLany, 2002; Cailliet, 1997)** In this discussion, "joint strain" refers to mechanical stress between a joint's components caused from remote dysfunctioning forces. As a result of joint strain, muscle strain may develop.

Movement throughout the lower extremity must be properly integrated to allow unstrained action of the joints. Imbalanced function in the lower extremity can result from conditions such as tibial torsion, muscle imbalance, or foot dysfunction. Goodheart and Greenman both observe that recurrent ankle sprains are frequently due to weakness of the peroneal and anterior tibial muscles. **(Greenman, 2003; Goodheart, 1980)**

If increased torsion is transmitted into the ankle or knee from a proximal or distal source, the ankle and/or knee receive greater strain during weight bearing, especially when one walks and runs.

Tibial torsion or hip rotation may be responsible for excessive internal leg rotation. Leg rotation puts torsional forces into the ankle mortise from proximal to distal. On the other hand, extended foot pronation puts torsional forces into the ankle mortise from distal to proximal. The torsion in the ankle puts strain on the ligaments, typically causing tenderness around the malleolus on digital pressure. The tibialis posterior is also strained in its role of

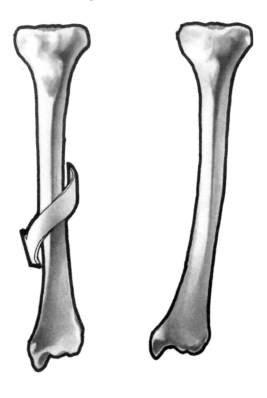

Tibial torsion

maintaining joint congruity by adjusting the width of the ankle mortise. Strain is indicated by tenderness of the ankle ligaments and of the tibialis posterior's belly as well as the muscle's origin and insertion on digital pressure. Under these circumstances the tibialis posterior may test weak, having been stressed and failing in its usual function. Frequently when the muscle is corrected with applied kinesiology techniques, it will not hold the correction when the patient walks or runs. Effective treatment to the remote cause of ankle joint strain may be needed before the tibialis posterior will hold its correction.

When the tibialis posterior muscle continues to function poorly, the overall problem may compound. The tibialis posterior is important in maintaining the medial longitudinal arch when a person walks and runs. **(Neville et al., 2010)** If the arch breaks down, internal leg rotation during the stance phase of walking increases. All causes of increased internal rotation during gait must be determined. Extended pronation, discussed in the previous chapter, is a common dynamic cause of increased internal leg rotation, as are weak external or hypertonic internal hip rotators. **(Gangemi, 2011)** The applied kinesiology method offered by Gangemi for insuring the stability of tibialis posterior muscle should be reviewed. **(See chapter 4)**

Normally the hip rotator muscles are balanced when each foot points laterally, approximately 10° from the sagittal plane. One can observe this muscular balance by pointing the feet directly forward and then standing on one leg only. The body will rotate away from the weight-bearing leg approximately 10° to the neutral position of the hip rotators. **(Jones, 1945)**

When foot pronation is present, its influence on rotation of the lower extremity can be observed by measuring rotation while a person stands on one leg, as indicated above, and then repeating the procedure when padding has been placed under the medial longitudinal arch to properly orient the calcaneus, as indicated by Helbing's sign. If forefoot varus is indicated, it will also be necessary to place a wedge under the forefoot that tapers from medial to lateral. **(Rose, 1962)**

The ankle joint is liberally supplied with mechanoreceptors capable of producing powerful reflex responses from the leg muscles. **(Dananberg, 2007; Goodheart, 1967)** Proprioceptors (mechanoreceptors) are found in the skin, muscles, tendons, ligaments and joints. Afferent fibers from mechanoreceptors converge segmentally on the dorsal horn of the spinal cord. The afferent fibers tend to diverge in an ascending and descending manner, over several segments, synapsing with different neuronal pools and spinal interneurons. This sharing of afference by motor centers has also been demonstrated in the cortex. **(Ginanneschi et al., 2005)** For this reason many synergistic muscle groups share common afferent inputs. **(Luscher & Clamann, 1992; Eccles et al., 1957)** This means that muscle spindle afference from one group of muscles supply the motor neurons in which they are embedded, as well as other synergistic muscles, consistent with Hilton's Law that the nerve supplying a joint supplies also the muscles which move the joint and the skin covering the articular insertion of those muscles.

In this way strain in the ankle from proximal or distal torsion may cause some of the local or remote aberrant muscle function observed in applied kinesiology testing. Subluxations of the talus and proximal or distal tibiofibular articulations may also result. Elimination or reduction of torsion causing ankle strain is usually accomplished by correcting extended foot pronation, pelvic and hip conditions, dural tension, and other modular distortions of the body. **(Leaf, 2010; Walther, 2000)**

Ankle Sprains

The acute trauma of ankle sprains may or may not be treated by an applied kinesiologist, depending on his or her scope of practice. Many prefer to limit practice to treating functional conditions. In any case, dysfunction often remains after the acute injury has healed, leaving the person with a "weak ankle" and prone to further injury.

The expression, "Once a sprain, always a sprain," need not be applicable. Applied kinesiology examination and treatment offer many techniques for returning normal function. Whether the physician treats the acute injury or not it is necessary to understand the mechanics and extent of the original injury when treating residual dysfunction.

Severe ankle sprains are frequently associated with direct trauma, often as a result of athletic endeavors. In athletes, the lateral ankle is the most frequently injured single structure in the body. **(Slimmon & Brukner, 2010)**

The frequency of ankle sprains varies significantly with the sport. Runners have relatively few ankle sprains. They frequently occur when an individual runs at night or otherwise twists an ankle on some substrate factor. **(Slimmon & Brukner, 2010)** Ankle sprains are common in high school basketball. In one study, **(Elkus, 1986)** 70% of eighty-four varsity basketball players from five public high schools had a history of ankle sprain. Of these, 80% had multiple sprains. Sports like football and skiing also have a relatively high incidence of ankle sprains.

There is a constant effort to improve the design of sports equipment to reduce ankle injuries. There has been a dramatic decrease in ankle injuries in alpine skiing as a result of improved slope grooming, a higher level of ski expertise, and better ski equipment. **(Deibert et al., 1998; Leach & Lower, 1985)**

Occasionally various foot articulations can be sprained. These, too, are influenced by footwear and the playing surface. Clanton et al. **(Clanton et al., 1986)** did a study that demonstrates this, revealing that synthetic playing surfaces and lightweight flexible athletic shoes have increased the incidence of metatarsophalangeal sprain. Particularly vulnerable to this trauma are those who have less than 60° dorsiflexion of the metatarsophalangeal articulation, especially of the 1st ray. It is suggested that the joint be protected by more solid shoes or shoes that have less than 6 millimeters descent from heel to forefoot (barefoot running technologics). **(Abshire, 2010)** Thick heels in footwear can result in increased dorsiflexion of the ankle while walking and running, adding more stress to the ankle and throughout the rest of the body. **(Bishop et al., 2006)**

Shoe fit and quality (as discussed in **chapter 4**) is important in reducing the chance of ankle sprain. A shoe that has a small sole as opposed to the individual's foot predisposes one to ankle sprain. **(Paiva de Castro et al., 2010)**

Although playing surfaces, equipment, footwear, and support to the ankle (such as taping) have been studied in depth, little effort has been directed toward intrinsic factors of the injured individual. In sports and everyday living, one must wonder why an ankle is sprained with a routine cut while running or when stepping off a curb. Granted, this type of sprain is usually not as severe as having another player fall on one's leg, causing a deltoid ligament tear. Nevertheless, the almost spontaneous ankle twist in everyday activity is a constant concern to some individuals. One must be able to evaluate, classify, and treat both the severe sprains and the common recurrent "twisted ankle." **(Kohne et al., 2007)**

Types and severity of sprains

In understanding the common types of ankle injuries, it is necessary to observe the following anatomical features. **(Bergmann & Peterson, 2010; Hammer, 1999; Cailliet, 1997)** The lateral malleolus is longer than the medial malleolus. Its distal tip extends further down the lateral articulating surface of the talus, and the bulk of the bone is much less than that of the shorter, bulkier medial malleolus. The tibia and fibula are bound together by the anterior and posterior tibiofibular ligaments, which are thickened expansions of the interosseous membrane that fastens the two bones together throughout their length. The deltoid or medial collateral ligament is heavy and has strong attachments to the internal malleolus, with many bands extending down to unite with ligaments supporting the arch of the foot; thus, in addition to giving medial ankle support it helps support the arch. The lateral collateral ligament has three main branches — the anterior talofibular, posterior talofibular, and calcaneofibular.

In inversion injuries there is push against the strong, stubby internal malleolus that acts as a fulcrum to cause pull to the lateral ligament, which more easily yields to the force. The ankle mortise separates and causes tearing of the lateral collateral ligament, with the severity depending on the amount of force.

In eversion injuries, the talus again pushes against the strong medial malleolus. Compressive force is applied to the more distal aspect of the lateral malleolus, which is thinner, and the bone yields to the force. At the medial side, the talus rides down on the lateral malleolus and causes a pull against the ligaments, which may tear.

There is a considerable difference in the seriousness of sprains to the medial structures as opposed to the lateral ones. Medial sprain is more serious but less frequent. It is caused by forced pronation, while the lateral sprain is caused by forced supination. Obviously a foot and an ankle with excessive pronation are more likely to suffer a medial sprain, and a supinated one a lateral sprain. **(Slimmon & Brukner, 2010; Gray's Anatomy, 2007)** Extended pronation may also cause a person to be more prone to lateral sprain, with the lateral ligaments sustaining the initial impact; **(Adirim & Cheng, 2003; Michaud, 1987)** this is discussed later.

In a medial sprain the structures torn are the deltoid ligament, anterior tibiofibular ligament, and the interosseous membrane. **(Slimmon & Brukner, 2010; Gray's Anatomy, 2007)** The deltoid ligament is so strong that it may fail to tear, but it avulses from the medial malleolus **(Cailliet, 1997)** or significantly tears with fracture of the fibula. **(Gray's Anatomy, 2007)** The medial sprain is frequently accompanied by a fracture of the medial malleolus, distal fibula, the posterior aspect of the tibia, or a combination of these. The fibula can be fractured almost anywhere along its length; consequently, x- rays should be taken of the leg to the knee. **(Yochum & Rowe, 2004)** The medial sprain usually occurs in athletic endeavors in which the foot is fixed to the ground and the leg is forced medially by another player hitting or falling upon it.

Professional soccer (football) players frequently sustain these types of ankle injuries, graphically displayed by watching popular soccer channels around the world. Ozetkin et al. **(Ozetkin et al., 2009)** collected data on 66 players with severe ankle and foot injuries. The most common diagnosis was ankle sprain (30.3%) with anterior talofibular ligament injury. Most (55%) hindfoot injuries were Achilles tendinopathy with or without rupture. Treatment was surgical in 23 patients (35%). The mean time lost from play for players with severe foot and ankle injuries was 61 days (range 21-240 days); after Achilles tendon ruptures, the mean time lost was 180 days. Injury severity was severe (>28 days lost from play) in 64% patients and moderate (8-28 days lost from play) in 36% patients. Serious ankle and foot injuries in this study resulted in players being out of professional competition for about 2 months.

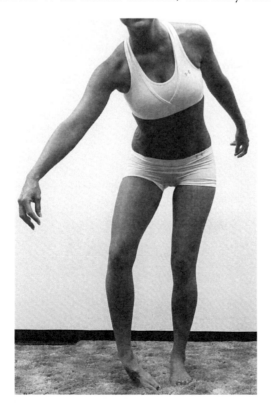

Inversion sprain dynamic

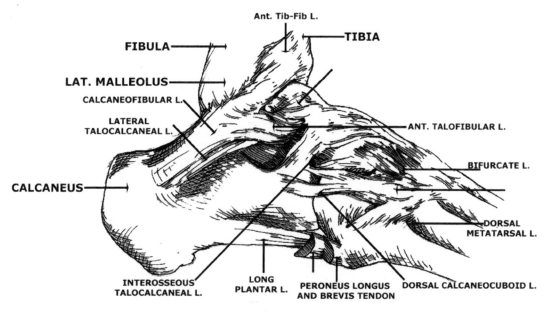

Lateral foot ligaments

The literature has many classifications indicating the extent of an ankle sprain. In general there are three degrees of severity listed as Grades I through III, or mild, moderate, or severe. The grading may rank the injury by the amount of soft tissue damage, which ligaments are torn, or the amount of motion torn ligaments allow on stress tests.

The lateral ankle sprain is much more common, comprising approximately 85% of ankle sprains. **(Slimmon & Brukner, 2010; Cailliet, 1997; Hertling & Kessler, 1996)** It frequently results from forced adduction and inversion and most often plantar flexion, which make up all the motions of supination. There is a predictable sequence of ligament tearing, **(Slimmon & Brukner, 2010; Levangie & Norkin, 2001)** the extent of which provides one type of grading system. The first to tear is the anterolateral capsule and anterior talofibular ligament, which is a Grade I injury. Progressively, the talocalcaneal ligament — a Grade II injury — is included. **(Merck Manual, 2010; Levangie & Norkin, 2001)** The posterior talofibular ligament may tear, but it is rare.

A classification of soft tissue damage evidence and instability can be given based on this literature. This classification of sprains is as follows: (1) local signs of inflammation with no significant swelling or ecchymosis or joint instability; (2) intermediate as moderate swelling, ecchymosis, and some instability on ligamentous testing but a definite end point on stress examination; and (3) severe, with local signs of inflammation and marked instability of the specific ligaments tested.

The sinus tarsi syndrome is characterized by pain and tenderness over the lateral opening of the sinus tarsi. The patient has a feeling of ankle instability. This condition is secondary to an inversion strain intense enough to traumatize the interosseous talocalcaneal ligament. The condition is usually prevented by proper treatment of the sprain, with follow-up to obtain proper bone interaction and foot function. When the condition fails to respond to conservative care, surgical excision of the pathological tissue is performed. **(Pisani et al., 2005)**

Examination

Injury to the ankle is evaluated for severity by areas of tenderness, swelling and ecchymosis, and by motion tests to evaluate ligament integrity. The diagnosis of ankle trauma is much more accurate if it is accomplished within a few hours of the injury. With time swelling becomes more diffuse, and palpation for tenderness is less specific. **(Slimmon & Brukner, 2010; Hertling & Kessler, 1996; Tropp, 1985)**

Examination should be designed to analyze the mechanism of injury and determine the exact structure damaged. If one fails to recognize the extent of injury and treats a severe one as mild, serious consequences may develop; appropriate treatment will not be as effective at a later date. The patient can usually describe how the ankle was twisted, giving an indication of whether there is medial or lateral involvement. Medial sprain will show tenderness under the medial malleolus at the area of the deltoid ligament and over the anterior aspect of the ankle where the anterior tibiofibular ligament and the interosseous membrane are torn. **(Slimmon & Brukner, 2010; Cailliet, 1997)**

It is very important that the nature of the injury not be underestimated. There may be more extensive ligament damage than is observable by range of motion testing, because pain and swelling may limit the examination. **(Slimmon & Brukner, 2010; Hertling & Kessler, 1996)**

First, examine the uninjured ankle to determine range of motion and its characteristics for comparison. After observing the injured side for gross deformity, swelling, and circulatory disturbances, gently move the ankle passively through range of motion stress tests. It is particularly important to observe for lateral motion, which indicates ligament disruption. If there is no lateral motion observed but there is pain on attempting to elicit it, ligament damage is present but not necessarily disruption.

A mild sprain indicates partial tearing of the ligaments,

Applied Kinesiology

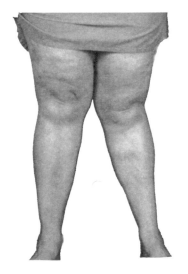
Severely pronated (knock) knees

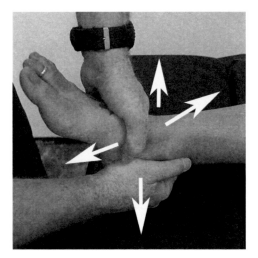
Stress tests

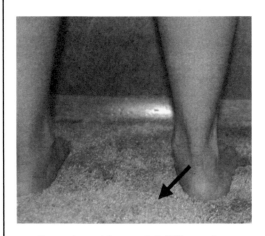
Pronation with curved Achilles tendon

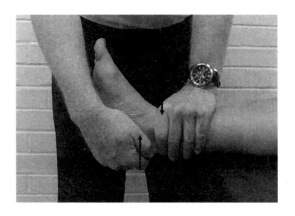
Anterior Drawer sign

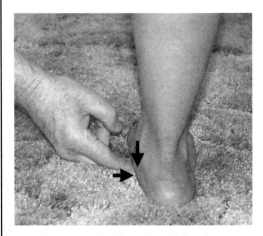
Palpation of collapsed medial arch

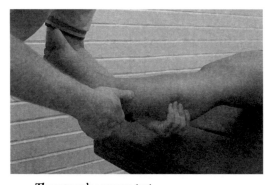
Thompson's squeeze test

Signs and tests of ankle and leg dysfunctions.

Leg and Ankle

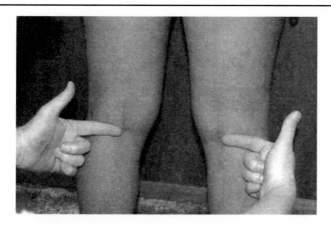

Unlevel knee folds, short tibia and anatomical short leg

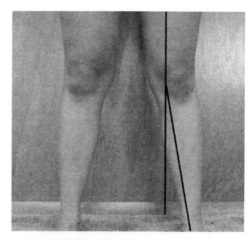

Valgus left knee

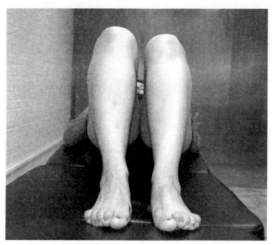

Unlevel knee heights may indicate short tibia and leg

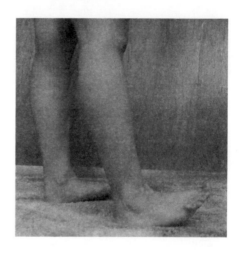

Heel walk for L4-L5 nerve root involvements

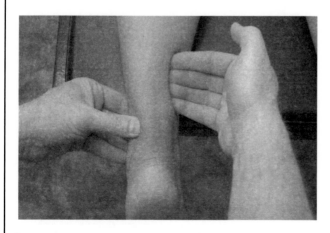

Achilles Tendon palpation

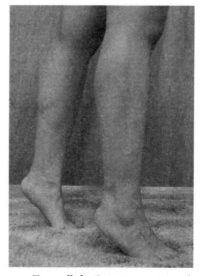

Toe walk for S-1 nerve root involvement

Signs and tests of ankle and leg dysfunctions.

but without weakening the ankle structure. There will be local tenderness, swelling, and mild disability. There will be no pain on normal motion, and only a moderate degree when re-applying the stress that caused the injury.

A moderate sprain is an extension of the mild sprain, still without major loss of joint integrity but with greater ligament damage.

A severe ankle sprain is one in which there has been complete loss of ligament integrity. In most cases, this type of injury requires surgical intervention with appropriate non-weight bearing follow-up. This Grade III ankle sprain is suspected when there is 1) a history of a snap or a pop at time of injury; 2) ecchymosis over the lateral ankle and/or hindfoot; 3) tenderness over the anterior deltoid ligament in addition to more severe tenderness over the lateral ligaments.

The severity of a lateral sprain is further indicated by whether it is a single or a double ligament tear. This is sometimes difficult to differentiate since the anterior talofibular and calcaneofibular ligaments are very close in their origin on the fibula. As the ligaments traverse to the talus and calcaneus respectively, they are more easily palpated for differentiation. (**Slimmon & Brukner, 2010; Logan, 1995**)

The integrity of the anterior talofibular ligament can be assessed by the anterior drawer sign. (**Slimmon & Brukner, 2010; Evans, 2008**) As much relaxation of the ankle as possible should be obtained. This can be assisted by having the knee flexed with the patient seated, taking tension off the ankle from the triceps surae. The tibia and fibula are stabilized, and the foot is gently brought forward in the sagittal plane. If the talus moves forward, it indicates sprain of the anterior talofibular ligament. A posterior drawer maneuver is also described (**Slimmon & Brukner, 2010; Evans, 2008**) that tests for integrity of the posterior talofibular ligament; however, it is rarely injured except in cases of complete ankle dislocation.

When there is an actual tear in a lateral ankle sprain, there is greater inversion as the talus separates from the lateral malleolus. In a simple sprain no gap can be palpated, and the talus remains in its normal position. (**Slimmon & Brukner, 2010; Logan, 1995**)

The twisting injury that sprains an ankle can also sprain the ligaments of the midfoot. (**Adirim & Cheng, 2003; Levangie & Norkin, 2001**) This is usually easily differentiated from an ankle sprain by the location of pain, tenderness, and swelling.

Dislocation of the peroneal tendons is not a frequent occurrence. When it is presented in a general practice, it is often misdiagnosed as an ankle sprain. The dislocation usually develops from forced foot eversion with dorsiflexion and contraction of the peroneus longus and brevis. (**Hertling & Kessler, 1996; Martens et al., 1986; Cox, 1985**) This condition is discussed later.

X-ray examination

X-ray examination is important in severe ankle sprains, and it is done in two phases. (**Yochum & Rowe, 2004; Logan, 1995**) First, routine views are taken to rule out fracture. The projections include AP, lateral, and oblique views. If the trauma has caused foot deformity, one or more x-rays should be taken that display the deformity to provide information that might not otherwise be available about the degree of injury. (**O'Donoghue, 1958**) The ankle may then be manipulated into position for the standard views.

The AP view is done with the plantar surface of the foot perpendicular with the film and the leg internally rotated approximately 10° so the line through the malleoli is parallel with the film. This provides a true AP view. (**Yochum & Rowe, 2004; Logan, 1995**)

The lateral view projects transversely through the malleoli. It is usually taken with the lateral malleolus adjacent to the film. This usually causes poor visualization of the ankle joint space. A variation of technique places the medial malleolus adjacent to the film, with the medial knee on the x-ray table. This slightly everts the forefoot and usually provides a clear view of the joint. (**Yochum & Rowe, 2004; Cox, 1985**) The tube is centered on the lateral malleolus.

A common fracture site is the posterior tibial tip; it is demonstrated with slight external rotation in an off-lateral view. (**Yochum & Rowe, 2004; Logan, 1995**)

The oblique view is taken with the intermalleolar line forming an angle of 35-45° with the surface of the film. This view may reveal fractures not otherwise seen. If clinical examination indicates probable fracture and it is not observed on x-ray, additional oblique views should be taken. (**Logan, 1995**)

Fracture of the talus and 5th metatarsal may be present. A talus fracture should be suspected if symptoms persist after the normal healing time for ligamentous injuries. There are two types of fractures of the 5th metatarsal. Less serious is an avulsion fracture of the base of the metatarsal at the insertion of the peroneus brevis tendon. In the immature skeleton, it must be differentiated from the developing apophysis of the metatarsal. The secondary center generally has a longitudinal axis, whereas the avulsion fracture is transverse. (**Cox, 1985**) A much more severe fracture is one of the diaphysis or proximal shaft of the 5th metatarsal.

In medial sprains, x-ray examination should include the leg to the knee because of possible high fibular fracture. Additional views are necessary for thorough evaluation of the ligaments and the bony structure of the ankle and foot. (**Yochum & Rowe, 2004; Logan, 1995**) Bohler's angle for evaluating calcaneal fracture and other specialized views can aid in evaluating ankle and foot trauma.

The second phase of stress view x-ray examination is done if fractures are ruled out. This can usually be accomplished conservatively if done within the first few hours of injury. If pain is severe or there is significant swelling, it is necessary to anesthetize the area, taking the examination out of the conservative realm. (**Yochum & Rowe, 2004; Logan, 1995**)

The uninjured ankle is routinely examined with stress x-rays for comparison. It is necessary to rule out the effect of normal anatomic variation among individual patients. If the opposite ankle has been involved in an earlier trauma, this information will provide only limited comparison. Either the anterior drawer or the talar tilt measurements are adequate to determine whether one or both of the anterior talofibular or the calcaneofibular ligaments have been damaged. If only one set of measurements is done, the talar tilt series is preferred.

An anterior drawer maneuver is positive for a Grade III injury when a ten-pound force applied for two minutes causes displacement of the talus 2 mm greater than the non-injured ankle. **(Smith & Reischl, 1986)** The optimal position of the ankle and foot for x-ray examination for anterior drawer sign is internal rotation and 10% of plantar flexion. This gives maximum displacement if there is an anterior drawer sign from rupture of the anterior talofibular ligament and others. **(Liu et al., 2001)** If the anterior drawer test is positive, the talar tilt test is not done.

Talar tilt is viewed on the mortise view with the foot in inversion stress. As with the anterior drawer view, 10% of plantar flexion is optimal. This relaxes the muscles and allows maximum displacement if ligaments are torn. **(Larsen, 1986)** Comparison with the uninjured side is necessary since there is a wide range of normal motion. An evaluation of normal ankles revealed talar tilts ranging from 3° to 23°. Smith and Reischl **(Smith & Reischl, 1986)** prefer that the physician perform the talar tilt test. An inversion force is applied slowly to avoid peroneal muscle tension caused by pain. "Instability on the talar tilt maneuver is indicated by a 3 mm or greater separation of the lateral tibiotalar surfaces compared to the noninjured side or a 6° or greater tibiotalar angle compared to the noninjured ankle."

When there is a clinically serious ankle injury but normal ankle stress x-rays and a positive anterior drawer stress x-ray, further evaluation with a subtalar arthrogram should be done to evaluate for a subtalar sprain. **(Liu et al., 2001)** This condition is thought to be more prevalent than generally recognized. "Particular attention should be paid to athletes involved in indoor activities, such as volleyball or basketball. They are prone to this type of injury because of the high adherence between playing surfaces and modern footwear. Unrecognized and untreated subtalar sprains can lead to serious consequences, such as chronically painful ankles or the sinus tarsi syndrome. These problems may be avoided only if the exact extent of the lesion is appreciated." **(Meyer et al., 1988)**

Smith and Reischl **(Smith & Reischl, 1986)** find several disadvantages regarding the general use of arthrography. It requires fluoroscopy, there may be an allergic reaction to the iodine dyes, and ankle arthrography may not be able to identify ligament rupture just several days after injury because the capsule seals itself with a weak layer of fibrin. This prevents the escape of dye, and possibly presents a false-negative conclusion in terms of a ruptured ligament.

Trauma to the distal thickened portion of the tibiofibular interosseous membrane may allow separation of the syndesmosis with resulting ankle instability. Tibiofibular widening at the area of the syndesmosis can be evaluated by two methods of measurement. The "clear space" is the distance between the medial border of the fibula and the lateral border of the posterior tibia as it extends into the incisura fibularis. The tibiofibular overlap is the maximum overlap of the distal fibula and the anterior tibial tubercle. The measurement is made 1 cm above the plafond of the tibia. Normal clear space is less than approximately 6 mm on the AP view of the ankle. Normal tibiofibular overlap is approximately 6 mm, or 42% of fibular width. According to Harper and Keller, **(Harper & Keller, 1989)** the width of the tibiofibular clear space appears to be the most reliable parameter for detecting early syndesmotic widening.

Nerve injury accompanying ankle sprain

Severe ankle sprains may be accompanied by a high percentage of peroneal and tibial nerve injuries. Nitz et al. **(Nitz et al., 1985)** did a clinical and electromyographic study to determine the extent of this type of nerve trauma. They found no nerve damage in sprains that only involved the lateral complex. When the deltoid ligament was involved, the peroneal and tibial nerves had 17% and 10% injury, respectively. When the sprain included the lateral complex, deltoid ligament, and distal anterior tibiofibular ligament, 83% incurred posterior tibial nerve injury. They hypothesized that the mechanism of nerve injury was from stretch at the ankle at time of injury. This may cause entrapment of the common peroneal nerve that can be the cause of persistently weak ankles. **(Staal et al., 1999)** There is a poor prognosis for quick recovery when the peroneal and/or tibial nerves are involved. This should be taken into consideration for return to work or sports activities.

Richie, **(Richie, 2001)** in a comprehensive review of functional ankle instability, suggests that the instability is not the result of hypermobility, but rather the diminishment of proprioception, muscle strength, muscle reaction time, and postural control generally. For the majority of patients in Richie's review, the functional instability of the ankle results from a loss of neuromuscular control. "Proprioceptive deficits lead to a delay in peroneal reaction time…Balance and postural control of the ankle appear to be diminished after a lateral ankle sprain and can be restored through training that is mediated through central nervous mechanisms."

After injuries where muscular strength, timing, velocity and control are decreased, the sensory input, central processing and motor output time may all be increased. **(Hurley, 1999)** This may delay the activation of neuromuscular protective reflexes and is another impediment to the shock-absorbing function of muscles, exposing the joint to damage. **(Colledge et al., 1994)** In addition, improvement of neuromuscular functions as a result of applied kinesiology treatment and rehabilitation regimes for the ankle may decrease the reaction times, improve functional joint stability, strength, and proprioception, **(Hurley & Scott, 1998)** which may be important in restoring the shock-absorption capacity of muscle and protect against further ligament and joint damage.

Treatment

Treatment of mild to moderate ankle sprains ranges from **Rest**, **Ice**, **Compression**, and **Elevation** (RICE) to surgical intervention for severe sprains. The simplest treatment is by RICE therapy.
- **Rest.** Reduce activity and get off your feet.
- **Ice.** Apply ice in a plastic bag or towel over the injured area, following a cycle of 15 minutes on, 40 minutes off.
- **Compression.** Wrap a bandage around the area, but be careful not to pull it overly tight.

- Elevation. Position the foot in an elevated position on a bed, couch or chair, higher than your waist, to reduce swelling and pain.

Initially the patient rests with a compression bandage, usually for 24 hours. Ambulation is then allowed as tolerated with an Unna boot or Elastoplast(R) for 7 to 10 days. **(Safran et al., 1999)** This is usually referred to as mobilization treatment. Care should be taken not to override pain with medication that allows the patient to perform physical activities that would normally be limited by the discomfort. "Injections in these injuries are contraindicated since this can mask significant pathology and may result in a more severe injury." **(Clanton et al., 1986)**

Non-steroidal anti-inflammatory drugs (NSAIDs) do not appear to improve healing rate or effectiveness. A double-blind study of treating ankle sprains with ibuprofen was statistically insignificant over a 28-day period. **(Dupont et al., 1987)** In another double-blind study, Fredberg et al. **(Fredberg et al., 1989)** evaluated ibuprofen in the treatment of acute ankle joint injuries. The subjects had ankle sprain or fracture within 24 hours prior to hospital admission. They were administered a placebo, or 600 mg ibuprofen tablets taken four times daily. There was no reduction in ankle swelling in those taking the medication. The subjects were allowed to take additional analgesic medication if they felt the need. Those taking ibuprofen required as much additional medication as those on the placebo, indicating that ibuprofen has no effect on pain. Their conclusion was that "...the results indicate, no difference between relief of pain and reduction of swelling was demonstrated. We cannot recommend routine treatment with ibuprofen for acute ankle joint injuries." In the treatment of patients with excess inflammation there is an extensive body of literature describing as many as 10,000-20,000 fatalities and 103,000 hospitalizations produced yearly by non-steroidal anti-inflammatory drugs (NSAID) medications, frequently the treatment given for low-back and other joint pains in medical settings. **(Wolfe et al., 1999; Gabriel et al., 1991)** These figures are similar to the number of deaths from HIV and far more than the number of deaths from cervical cancer, myeloma or asthma. If the deaths from gastrointestinal toxicity resulting from the use of NSAIDs were tabulated separately, these would constitute the fifteenth most common cause of death in the United States. **(Wolfe et al., 1999)**

Immobilization treatment consists of casting the ankle, which is usually positioned to bring the torn ligaments together. The cast is left on for 4-6 weeks. When more intensive treatment is desired, as with professional athletes, removable immobilization is applied, such as taping or use of a type of cast brace. This is removed daily for ice therapy initially, followed by contrast whirlpool hydrotherapy. **(Cox, 1985)**

Generally, Grades I and II ankle sprains can effectively be treated by mobilization or immobilization methods. Initial swelling must be controlled by ice, elevation, and pressure bandaging. Healing time is in direct relation to the amount of swelling. The presence of effusion favors the formation of adhesions, which further delay healing. Apart from the toxic nature of NSAIDs, untreated joints have been found to remain in better condition than those treated with NSAIDs. **(Pizzorno et al., 2007; Werbach, 1996)** Protection by wrapping or a cast may be needed for ten days to three weeks, depending on the extent of the damage. **(Hertling & Kessler, 1996)** With strapping and mobilization of lateral ankle sprain where talar tilt is less than 10°, return to full working capacity is expected within one month. In a Grade III sprain, the ankle is often immobilized in a walking cast for six weeks. **(Oztekin et al., 2009; Beskin et al., 1987)** When stress x-rays are negative but clinical examination points to a severe injury, the patient is still managed for a Grade III sprain. **(Oztekin et al., 2009; Hontas et al., 1986)**

Surgical repair of torn ligaments is followed by cast immobilization. There is no common agreement over the indications for surgical repair. The criteria for operative repair are (1) significant talar tilt indicative of a double ligament tear (anterior talofibular and calcaneofibular ligaments), (2) a history of functional instability (multiple sprains) with acute injury, (3) an acute injury with a bony ossicle or avulsion fracture from the lateral malleolus, and (4) associated osteochondral fracture of the talus. Using these indications, 12% of cases resulted in surgery. **(Brand & Collins, 1982)**

The severity of the sprain dictates the intensity of treatment, but there are several considerations that must be made in comparing and determining the type of treatment. These include (1) healing time, (2) stability of ankle after healing, (3) life-style of the patient, (4) cost of treatment, (5) whether surgery can be successfully done even many years after the failure of conservative care, **(Safran et al., 1999)** and (6) avoiding unnecessary surgical complications.

In 1985 Cass et al. **(Cass et al., 1985)** did a long-term follow-up study to determine if ankle stability improves with immediate surgery, as opposed to conservative treatment followed by reconstructive surgery if the conservative treatment fails. They evaluated both the subjective results via patient questionnaire and objective results by clinical examination, stress roentgenology, and biomechanical gait study. A satisfactory result can be expected in 80% of the Grade III patients treated conservatively. Surgical reconstruction can be applied to the remaining 20% at a later time, with the confidence of obtaining patient satisfaction and a reasonable certainty of obtaining a satisfactory objective result as well. Their study does indicate that for those who require high performance and functional perfection of the extremities, such as professional athletes, acute repair gives slightly improved objective functional results. For these individuals, early surgery may be the appropriate choice. Otherwise, they find it is proper to attempt conservative treatment first.

Freeman **(Freeman, 1965)** compared randomly selected patients for effectiveness of lateral ankle sprain treatment done conservatively by strapping and mobilization, immobilization in plaster for six weeks, and suture of the ligament with immobilization for six weeks. In ankles treated by mobilization and immobilization, none displayed more than 8° of relative talar tilt. Duration of disability after strapping and mobilization, immobilization, or surgical treatment was 12, 22, and 26 weeks, respectively. One year after injury, 58% and 53% of patients treated by mobilization and immobilization, respectively, were symptom-free, but only 25% of patients treated by suture and immobilization had become symptom-free. It is concluded, "For these reasons, and because simple sprains are satisfactorily treated by mobilization, it is suggested that mobilization may be the treatment of choice for most, perhaps all, ruptures of the lateral ligament of the ankle." **(Freeman, 1965)**

Finally, Van der Linden et al. **(2004)** reviewed 292 cases of Achilles tendon rupture treated either surgically or non-surgically (with splinting) for ankle stability. For both groups mean follow-up time was 3 to 6 years. There were 14 re-ruptures, ten after surgical repair and four after non-surgical treatment. In the surgical group there were seven major wound problems, 11 minor wound complications and six patients with complaints related to the sural nerve. In the non-surgical group one patient suffered a pulmonary embolism after a re-rupture, 3 months after the initial rupture. There was no difference in mean ankle score and patient-satisfaction score between groups. Only 52% regained their original sports activity level, slightly better in the surgically treated group. With a non-significant difference in re-rupture rate but relatively more complications after surgical repair, non-surgical treatment is the recommended procedure. With a slightly better recovery of sports activity after the surgical repair however, this might be used as an argument for surgical treatment in young athletes.

Non-surgical treatment is usually the choice method of treatment, even though surgical intervention for some patients may give better results. Martin et al. **(Martin et al., 2007)** offer a number of reasons: (1) surgical complications are avoided, (2) the surgical scar can cause future problems of altered sensibility and traumatic neuroma, (3) surgical intervention can be selected, even several years after the injury if initial results are unsatisfactory, (4) residual symptoms after non-surgical treatment are usually mild and not disabling, (5) recovery time is greater after operation than after strapping, and (6) routine surgical treatment of these common injuries would greatly increase the work of surgical units and overall cost of treatment.

Surgical intervention is usually most appropriate for fractures about the ankle, especially for proper reduction of fractures that include the articulating surfaces. **(O'Donoghue, 1958)** Calcaneofibular ligament injury confirmed by stress x-ray is a good surgical candidate. **(Rijke et al., 1988)**

A medial sprain is potentially much more serious than the more common lateral sprain. Primary concern is whether the deltoid ligament sprain is accompanied by interosseous membrane tearing and diastasis of the distal tibia and fibula. If there is diastasis, it is sometimes treated by closed reduction and casting after 48 hours of ice application and compression to decrease the swelling. Surgery may be necessary to remove particles of the medial deltoid ligament in order to reduce the diastasis. **(McBryde, 2007)**

Children present a somewhat different picture in ankle sprains. There is more elasticity in the ligaments of children than in adults; consequently, stretching injuries are more common than severe tears. When there are small flecks of bone from the tip of the lateral malleolus or, rarely, from the tip of the medial malleolus, there is indication that the bone gave way rather than tearing of the ligament. These lesions can be treated as sprains rather than fractures. Immediate application of cold and compression bandaging are important. Unfortunately, these injuries often occur when there is no supervision and immediate measures are not applied.

Healing time in children is shorter than in adults, probably due more to lack of frank tears rather than the age of the child. **(Adirim & Cheng, 2003)** If there is ligamentous tearing, a below-knee plaster cast is the treatment of choice. Generally, in this age group, the cast can be removed after three weeks, with protection for an additional three weeks by an adhesive wrapping or elastic bandages. Surgery is rarely applicable in a child, with the exception of a severely torn deltoid ligament. In this case there is an increased propensity toward surgery. Young patients have a greater percent of residual ankle stability after conservative care than do older patients. They are good candidates for primary surgical repair, especially those who need perfect ankle function, such as athletes. **(Slimmon & Brukner, 2010; Levangie & Norkin, 2001)**

Tarsal-metatarsal sprain in children is generally caused by forced plantar flexion and inversion. The injury is usually not severe, and compression bandaging with local cold applications and crutches is sufficient treatment.

Rehabilitation

Rehabilitation begins with the initial treatment of an ankle sprain and continues until optimal function is obtained. Failure to accomplish the latter may leave the patient with a propensity to additional ankle sprains. Proper treatment helps eliminate the often-quoted phrase, "Once a sprain, always a sprain."

The body's initial reaction to the soft tissue trauma of a sprain is acute inflammation. This is appropriate in the case of an infection, but it is an overreaction to closed soft tissue trauma; **(Marymont et al., 1986)** it promotes edema, which delays healing. **(Fredberg & Stengaard-Pedersen, 2008)** It should be remembered that chronic inflammation may be aggravated by a number of factors including adrenal stress syndrome, nutritional deficiencies and hormonal imbalance, **(Schmitt & McCord, 2010; Pizzorno et al., 2007)** each of which may be diagnosed using applied kinesiology methods. **(Leaf, 2010; Walther, 2000; Goodheart, 1998-1964)**

The first step of local treatment is to reduce the inflammation by the RICE method. At this stage compressive bandaging is important. Care in application avoids circulatory deficiency, but the patient should be made aware of the signs of circulatory embarrassment and told where to cut the tape or remove an elastic support if disturbance develops.

At the initial examination the patient should be tested for lymphatic function, including neurolymphatic reflexes and retrograde lymphatic technique. Since muscles cannot usually be tested in the acute phase, therapy localizing the reflex is an adequate testing method. Retrograde and anterograde lymphatic is tested in the usual manner **(Walther, 2000; Goodheart, 1998-1964)** to ensure normal lymphatic drainage from the lower extremities.

When myofascia is permitted to remain immobile for

extended periods of time, its ground substance solidifies, leading to the loss of the collagen fibers' ability to slide across one another and this permits the development of adhesions. Akeson et al. (**Akeson et al., 1977**) have shown the changes that occur in connective tissue with prolonged immobilization. In Achilles tendinosis, there is a significant decrease in the total collagen content and altered collagen cross-linking that result in an aberrant collagen network similar to fibrosis. (**De Mos et al., 2007**) With easy exercise, torn ligaments heal to be stronger, larger, and have a higher collagen content than when immobilized. (**Yang et al., 2005**)

Most ankle injuries can be treated with the mobilization method. It is best, if applicable, because of the reasons indicated earlier; also, it provides stress on the ligaments to properly align the developing collagen fiber synthesis in the direction of function. (**Solomonow, 2009; Marymont et al., 1986**) The strength becomes greater than the ligament- osseous junction, whether sutured or not. Movement encourages the collagen fibers to align themselves along the lines of structural stress as well as improving the balance of glycosaminoglycans and water, thereby lubricating and rehydrating the damaged connective tissues. (**Lederman, 1997**) Immobilization has the disadvantage of producing muscle atrophy if done at or shorter than the resting length. There is also a decrease in protein synthesis within six hours of immobilization. (**Amiel et al., 1983**)

Weight bearing is not allowed if there is diastasis of the distal tibiofibular syndesmosis, which should be adequately supported for proper healing; if severe, surgical repair is indicated.

There are numerous methods of providing support to the ankle while healing takes place with mobilization. Figure-of-8 taping combined with basket weave is often used and has been shown effective. (**Hubbard & Cordova, 2010**) A classic type of taping is the Gibney boot. (**Hamill et al., 1986**) More extensive taping is often used for athletes as a prophylaxis, and it is effective in providing support during the healing of a sprain. (**Morrissey, 2001**) More rigid support can be supplied by splinting material (Orthoplast, Johnson & Johnson, Inc.) that is softened in hot water and molded to the patient; it then becomes semi-rigid when cooled. This is combined with foam cushioning and tape to provide a support that allows flexion and extension, but controls supination and pronation. (**Pope et al., 1987; Cox, 1985**) A similar support on a more elaborate basis can be provided by modifying a brace designed for fracture treatment. (**Cooke et al., 2009**)

Rehabilitation time after ankle sprain can be reduced by combining cryotherapy, compression massage, and exercise. In appropriately equipped training facilities, this can be accomplished by a pneumatic boot, which cycles pressure around the ankle and foot at 1.5 pounds, alternating 15 seconds on and 15 seconds off. The water bath is maintained around 37° F., and the athlete exercises the ankle with gentle dorsiflexion and plantar flexion. The movement is allowed because of the anesthetic effect of the cold. The cycling pressure assists the venous system in draining congestion from the injured site to help control edema. In addition, elevation enhances the drainage effect. (**Starkey, 1976**)

Prevention and optimal function

The comment, "Once a sprain always a sprain, need not apply" depends on returning normal function — and perhaps improving it — over that present before the injury. Muscle rehabilitation exercises, discussed later, are standard procedures used by many physicians and trainers, but if the muscles are limited by neurologic factors producing arthrogenic inhibition, (**McVey et al., 2005**) the exercises will probably fail to return optimal function. There are many AK examination and treatment techniques that should be routinely applied in the rehabilitation of ankle sprains to return normal function, and possibly enhance it beyond that present prior to the accident. This phase of rehabilitation is particularly important. Ferguson, (**Ferguson, 1973**) who emphasizes the importance of muscle rehabilitation, indicates that an "…athlete who has sustained a knee or ankle ligamentous tear is 15 times more susceptible to reinjury than one with no previous injury." One must question whether the increased susceptibility is mandated by the injury or is due to inadequate treatment and rehabilitation. The importance of specific muscle and joint impairments to optimal gait and ankle function has been carefully described. (**Santilli et al., 2005; Perry et al., 1986**) The use of the manual muscle test for assessing the specific muscle and joint impairments that have been described in many textbooks regarding ankle function and gaiting adds specificity to the evaluation and treatment of these disorders of the ankle.

Applied kinesiology examination can begin as soon as acute injury healing allows muscle testing. All muscles crossing the ankle are tested and corrected, if necessary, by the usual methods. This may require any of the five factors of the IVF. Particular attention is given to any foot subluxations or fixations. In athletic injuries, the neuromuscular spindle cell, Golgi tendon organ, and cutaneous receptors may need treatment. Do not forget to consider remote possibilities, such as spinal subluxations, meridian involvement, or dysfunction of the cranial-sacral primary respiratory system. Extended pronation is a common problem associated with ankle sprains. Michaud (**Michaud, 1997, 1987**) attributes recurrent ankle sprains to failure of the mid-tarsal locking mechanism during gait, i.e., the foot fails to become a rigid lever during the stance phase as is present in extended pronation. It is corrected in the usual manner, as discussed in the previous chapter. Tarsal coalition should be considered in any active adolescent who has recurrent ankle sprains or strains. (**Clanton et al., 1986**)

When an athlete sprains an ankle when making a sharp turn, an activity frequently done, there is the possibility of a reactive muscle. He may have injured a muscle on a previous play that is now inhibiting an important ankle stabilizing muscle. For example, the tibialis anterior may become hypertonic from injury, and by reciprocal inhibition inhibit the antagonistic peroneus longus and brevis that are necessary for lateral ankle support. (**Santilli et al., 2005**)

Another consideration when there is injury for no apparent reason is the ligament stretch reaction. (**Walther,**

2000) This condition is thought to be associated with adrenal gland stress in applied kinesiology. (**Goodheart, 1976**) When the ligaments of a susceptible joint are stretched, associated muscles of that joint temporarily weaken. (**Sprieser, 2003**) Treatment is directed toward the adrenal glands. This is done nutritionally and by treating any of the implicated five factors of the IVF for the particular patient. (**Walther, 2000**)

Selye points out that an athlete is often under considerable stress during competition, even though this may be eustress. (**Selye, 1974**) If there is an underlying adrenal stress disorder, the ligament stretch reaction may be present and cause the muscle function so necessary to perform optimal joint motion to fail. Deutsch and Durlacher suggest that the ligament stretch reaction is a systemic problem, affecting all the ligaments of the body. (**Durlacher, 1977; Deutsch, 1975**) Sprieser and Blaich report on the specificity of the ligament stretch reaction to the knee in acute and chronic conditions. (**Sprieser, 2003; Blaich, 1980**)

It was suggested as far back as the turn of the century that ligamento-muscular reflexes exist from sensory receptors in ligaments to muscles that modify the load imposed on the ligament and joint. Goodheart first discussed the law of the ligaments in 1973. (**Goodheart, 1973**) Goodheart found that pressure applied to the ends of ligaments towards the belly of the ligament tightens it. The opposite force will elongate the ligament.

It has been shown that ligamento-muscular reflexes exist in most extremity joints. (**Solomonow et al., 1998; Freeman & Wyke, 1967**) The ligaments associated with each extremity are richly endowed with afferents that produce reflex activation of the many muscles associated with the extremity's movement. The muscles therefore are a major component in maintaining the stability of the extremity's ligaments, bursae and capsules. (**Solomonow, 2009**)

Goodheart also suggested that when there is a chronically weak muscle there will usually be a ligament involvement (the ligament is stretched) that provides stability in the same direction as the muscle. Conversely, a stretched ligament will cause a weakness in a muscle that provides stability in the same direction. (**Solomonow, 1987**) For extremity dysfunctions, detection of ligament injuries through these ligamento-muscular reflexes can be specifically assessed with the MMT.

Particularly in over-trained athletes who demonstrate higher cortisol levels as well as adrenal gland enlargement due to hypertrophy and hyperplasia, (**Maffetone, 2010; Stallknecht et al., 1990; Barron, 1985**) the ligament stretch reaction of applied kinesiology may be critical in the prevention of one of the causes of recurrent ligament sprains.

In the unusual case in which muscle(s) fails to respond to applied kinesiology techniques, one should consider possible nerve damage as a result of a severe sprain, as previously indicated.

Exercise

In most cases it is advisable to rehabilitate the muscles with exercise. This begins when AK examination indicates the muscles are functioning normally, and as soon as pain and swelling allow brisk walking, usually just several days into the rehabilitation program. Clinical experience has shown that muscles functioning poorly on an AK examination do not respond well, in general, to exercise. This is because the specific arthrogenic impairments resulting from structural subluxations, fixations, remote factors, and other neurological influences on the muscle are still present following the exercise program. Once the AK examination shows normal facilitation of the muscles in the area, peroneal and ankle dorsiflexor muscle strengthening becomes the foundation of the rehabilitation program, because these muscles are responsible for actively resisting an inversion/plantar flexion injury. (**Smith & Reischl, 1986**)

Several have described the use of thick-walled, non-metallic rubber surgical tubing for rehabilitation exercises. (**Kemler et al., 2011**) The U.S. Naval Academy (**Cox, 1985**) used this method wherein thirty-six inches of 3/16" tubing were used to provide isotonic exercise that increases in resistance as the tubing is stretched. The longer the tubing, the less resistance it provides. Tubing exercise has the advantage of being inexpensive, versatile, and easily carried so the patient can exercise frequently throughout the day.

In addition to the home-made method of simply tying a loop on one end of the tubing, a kit called the "Thera-Ciser" is available from Foot Levelers, Inc. (**Foot Levelers, Inc.**) The kit provides a cloth loop, tubing, and a booklet describing the exercises. For ankle sprains, the anterior and lateral compartment muscles are usually exercised.

The anterior compartment muscles (dorsiflexors) are exercised by anchoring one end of the tubing to a piece of furniture; the forefoot is put through a loop at the other end of the tubing. With the knee in extension, the patient dorsiflexes the foot against the resistance of the tubing through the full range of motion. Resistance is increased by starting the procedure with greater stretch on the tubing.

The lateral compartment muscles (peroneal group) are exercised by everting the foot against the resistance of the tubing. One end of the tubing is anchored with the non-involved foot. The forefoot of the exercised foot is put

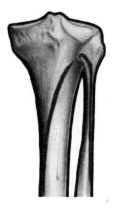

Soleus muscle's origin means that stretching the muscle with the knee bent effects it more than the gastrocnemius

through the loop of the tubing, while the heel is kept firmly on the floor to prevent lower leg rotation. The exercise starts with the foot in inversion and slight plantar flexion. The forefoot is abducted against the resistance into eversion and slight dorsiflexion.

Usually the plantar flexors (gastrocnemius and soleus) need to be stretched rather than exercised. In case strengthening is needed, they are exercised by putting the loop over the distal foot and holding the tube in the hand. The foot is plantar flexed against the tubing resistance. With the knee in full extension, the gastrocnemius and soleus are exercised. With the knee in flexion, the activity is directed more toward the soleus.

With dorsiflexion, the ankle joint mortise has a gradual increase in distance between the malleoli. The average increase in this distance is 1-1/2 mm from full plantar flexion to full dorsiflexion. Even with full plantar flexion, there is no looseness of the talus in the ankle joint mortise. This is attributed to ligament function (**Norkus & Floyd, 2001**) and to tibialis posterior muscle action by Kapandji. (**Kapandji, 2010**) The difference between the transverse dimensions of the anterior and posterior portions of the trochlea of the talus is as much as 4.7 mm. There is less distance change between the malleoli because no point on the tibial articular surface moves along more than one-half the length of the trochlear surface. Close AK examination and correction should be given to the tibialis posterior muscle. When it needs to be exercised, the foot is held in inversion during the plantar flexion exercise described above. (**Ramsak & Gerz, 2002**) More specific tubing exercise is done by starting with the forefoot abducted in full plantar flexion; the heel position is maintained while the forefoot is adducted against resistance.

Proprioceptive training

Prevention of ankle sprains depends on proper organization of the muscles supporting the ankle and foot. Often the cause of the sprain in the first place is poor muscle control when an inversion force is applied to the foot while walking or running. A significant amount of research has shown ankle sprains to be associated with muscle imbalances, most particularly weakness of the dorsiflexors and invertors of the foot. (**Baumhauer et al., 1995**) Chronic ankle sprains have been associated with arthrogenic muscle weakness. (**McVey et al., 2005**). The arthrogenic weakness of these ankle muscles can produce inhibition of the peroneals (**Santilli et al., 2005**) and the hip abductors (**Zampagni et al., 2009; Bullock-Saxton, 1994**)

Disorganization present before or developed from the current injury must be corrected before function is returned to normal. Clinical experience has shown that foot subluxations are a common cause of neurologic disorganization throughout the body, apparently because improper stimulation to the joint proprioceptors of the foot causes dysponesis. AK examination and correction are the prerequisites to proprioceptive training. The training procedures are enhanced when there are no foot/ankle subluxations or fixations and the muscles perform correctly in an AK examination.

The need for proprioceptive training is indicated by decreased ability to stand on one foot with stability. This often decreases following ankle sprain. Freeman et al. (**Freeman et al., 1965**) suggest "…that functional instability of the foot is usually due in the first place to motor incoordination consequent upon articular de-afferentation…." In particular, the mechanoreceptors in the human foot and ankle (among other receptors) control the instantaneous and qualitatively precise contractions of the calf muscles that must occur if the foot is to remain stable on uneven ground. (**Kavounoudias et al., 2001**)

An objective method of evaluating stability, called stabilometry recording, has been described by Lopez-Rodriguez et al. (**Lopez-Rodriguez et al., 2007**) When there is instability on one foot standing, as indicated by stabilometry recording, there is a significantly higher risk of sustaining an ankle injury than when there are normal stabilometric values. Manipulation of the talocrural articulation improved the stabilometric values and distribution of weight on the feet in these patients. Further evidence of this was determined by Tropp et al. (**Tropp et al., 1985**) when they studied one-legged stability standing in a control group and applied that information to 127 soccer players. The soccer players showing an abnormal stabilometric value ran a significantly higher risk ($p < 0.001$) of sustaining an ankle injury during the following season compared to players with normal values. Players with a history of previous ankle joint injury, but a normal stabilometric value, did not run a higher risk compared to players without previous injury. Furthermore, they did not find persistent functional instability in those who had previous ankle joint injury; when instability was present, it increased the risk of future ankle joint injury regardless of whether the person had sustained a previous injury. This indicates that a person with an ankle joint injury is not destined to instability.

Freeman et al. (**Freeman et al., 1965**) evaluated patients who had recent sprains of the foot and ankle by a modified Romberg test, comparing stability when standing on the injured extremity with the uninjured one. They found that 34% of their patients had an objective or subjective proprioceptive deficit. No deficit was detected in 16%. The remainder were unstable, but the cause could not be determined because of pain or stiffness. Thus in at least 34%, and at most 84% of the patients studied, there was instability due to proprioceptive deficit. They attribute this deficit to causing the symptom of the foot "giving way."

If instability continues after muscles have been tested and corrected by applied kinesiology techniques, the patient can be trained on unstable surfaces such as a balance board, or rocker and wobble boards, and ankle disk. **(Waddington & Adams, 2004)** The balance board is about 15" square; at the bottom, a 2" board is centered to balance on. Several boards with higher balancing points can be used for increasing exercise levels. The ankle disk has a spherical undersurface so that it tips in any direction.

The balance board is generally used first. Balancing anterior to posterior exercises the muscles and develops proprioception in flexion and extension. Lateral balancing effectively improves action of inversion and eversion. First both ankles are exercised together. Balancing is done to bring the edge of the board close to the floor, but to control it from touching the floor by muscle action. Movement of the board is then reversed to bring the opposite edge of the board close to the floor before stopping action. As control is optimized, one ankle at a time can be exercised on the board.

With exercise on the ankle disk, there is greater proprioceptive and muscle demand. This activity exercises one ankle at a time. First the ankle is moved in circumduction both clockwise and counterclockwise, keeping the edge of the disk in contact with the floor. As proficiency is obtained, the movement is done to just keep the edge of the disk from touching the floor.

Exercise with the balancing board or an ankle disk is done for about ten minutes per day. After optimal function is obtained the exercise can be reduced to about five minutes per day, three times a week to maintain performance.

The inversion sprain in athletic competition is most common. It may be reduced by a conditioning program to strengthen the peroneal muscles, and by teaching the players to land with a relatively wide-based stance. The latter places the foot slightly lateral to the falling center of gravity, making an inversion stress on the ankle unlikely. **(Smith & Reischl, 1986)**

The work of Vladimir Janda (with its emphasis on treating muscle imbalance, and particularly muscle inhibition, with physiotherapy) has become popular. Morris and Page have presented the Janda approach (particularly using the rocker and wobble board apparatus) comprehensively. **(Page et al., 2010; Morris et al., 2006)**

It should be noted that in Janda's model of the diagnosis and treatment of muscle imbalances, Sherrington's Law of reciprocal innervation operates primarily in one direction: muscle hypertonicity/tightness/spasm generates inhibition in its antagonists and so spasm is treated first. For this reason muscle spasm and tightness are considered the etiological factors of articular dysfunction. In Janda's approach hypertonic muscles are treated with physiotherapeutic means such as massage, stretching, proprioceptive neuromuscular facilitation, electrotherapy and other methods that do not usually include manipulative therapy. **(Page et al., 2010; Chaitow & DeLany, 2002)** In Janda's classic text on MMT **(Janda, 1983)** there is no mention of spinal or other joint manipulation options for the muscle inhibitions found; nor were correlations observed between muscle inhibitions found on examination and cranial, meridian, nutritional, or psychological treatments.

In Janda's model the inhibited (weak) muscles are treated with exercise, rocker boards, wobble boards, balance shoes, and mini-trampolines among others. The principles of this physical therapy approach to muscular imbalances were based on the work of Bobath and Bobath who developed physiotherapy programs for children with cerebral palsy. **(Bobath & Bobath, 1964)**

Muscle imbalance as conceived by Janda was mainly embraced by the physiotherapy community, though in recent years it has lost some of its popularity to the concept of core function and motor control. **(Wallden, 2007)** One common reason for the decline in interest in muscle imbalance in the physiotherapy and manual medicine communities is that, as with nearly all clinical entities, to find a textbook patient showing the muscle imbalance syndromes described by Janda is less common than finding only a partial case. This makes the diagnosis of muscle imbalance syndromes confusing. Further, the approach to diagnosing a muscle imbalance (for practical purposes) is based primarily on subjective assessment, such as the visual observation of standing posture. Based on this visual diagnostic method the prescription of treatment preferred by Janda – corrective stretching, corrective mobilization, corrective exercises and other nutrition and life style advice – may be somewhat non-specific. Additionally, aside from subjective symptomatology, progress is difficult to gauge with this subjective approach.

The visual diagnosis of a specific muscle's problem is difficult. The different elements within the chain of events of any particular movement that a patient undergoes in front of the examiner occur within a fraction of a second, too rapidly to be accessed separately in the absence of laboratory tools. Dananberg makes this clear in the failure of the great toe to move in the 100 msec time-frame it must during the stance phase. **(Dananberg, 2007)** Visual detection of this muscular impairment is essentially impossible. Therefore, what is actually observed by the examiner is the grand total of how rapidly and smoothly a person can change between two activities.

For these reasons it is strongly urged that to evaluate any muscle imbalance, clinical measurement tools, including but not limited to the manual muscle test, dynamometers, and digital cameras should be used.

Tape and support

Tape and support are used at the ankle for two purposes: (1) to provide compression and support for the injured ankle, and (2) to prevent injury of the normal ankle. **(Hubbard & Cordova, 2010; Morrissey, 2001)** Tape should be applied to the ankle over skin that has been prepared with a tape adherent tincture. Tape adhesion has been improved considerably. Some believe the more complicated woven or basket-weave strapping lasts no longer than simpler types.

There is controversy regarding the use of support to prevent ankle injuries. It seems obvious that ankle support muscles, especially the peroneus group and the tibialis posterior, must be functioning normally and there should be no extended pronation. Simultaneous electromyography and stop-action movies show the peroneus brevis and longus and the tibialis posterior to be the most important stabilizing muscles. **(Langer, 2007)** Langer offers a useful test to determine if the tibialis posterior muscle is injured. If patients feel pain or an increase in pain when shifting their weight to the affected foot and rising up onto their toes, then it is likely that this muscle is dysfunctional.

Ferguson (**Ferguson, 1973**) points out, "Only the muscle and its tendon attached to the bones can, by contraction, prevent excessive motion of an individual joint." In his paper, *The Case Against Ankle Taping*, he points out that the subtalar articulation acts as a "safety valve" for the ankle joint and knee. A cleated shoe and taped ankle prevent motion at this safety valve. As the trunk rotates with the foot and ankle fixed firmly to the ground, the strain of rotation becomes more manifested in the knee, which is a frequently injured structure in the lower extremity in sports. They cite studies conducted by the New York State Public High School Athletic Association based on 61,000 reports of high school athletic injuries, as well as studies of high school athletes in Philadelphia indicating that in preventing injury a low shoe, disc heel, and ankle wrap (not tape) or a low shoe, short cleat, and no ankle support produced statistically significant lower rates of serious injuries. An extensive study in college football (**Rovere et al., 1987**) showed that laced ankle stabilizers and low-top shoes prevented injury better than tape support. In general, schools that have adopted a policy of non-taping ankles have found satisfaction in the resulting decrease in knee injuries and ankle sprain.

The use of tape as opposed to elastic support is questionable. Rarick et al. (**Rarick et al., 1962**) found that 40% of the net support of taping is lost after 40 minutes of exercise. It would appear that proper shoes, correct muscle and foot function, and elastic support provide the optimal protection against ankle injury.

There is not universal support for non-taping of ankles to prevent sprains. (**Maffetone, 2003**) A study by Garrick and Requa (**Garrick & Requa, 1973**) of 2,562 intramural basketball players indicated that prophylactic ankle taping and high-top shoes provide the best protection against ankle sprains. This group had 6.5 ankle sprains over 1000 player games (6.5/1000). The rate progressively increased with less support: low-top shoes and taped, 17.6/1000; high-top shoes and untaped, 30.4/1000; low-top shoes and untaped, 33.4/1000. There was no evidence produced in this study that taping increased the propensity toward knee strain.

An electrogoniometric method of evaluating motion between the tibia and the calcaneus-shoe complex revealed significant support to the ankle by a closed basket weave with heel locks and a half figure-eight taping technique over the untaped ankle complex. The evaluation was done before and after figure-eight walking exercise, concluding that ankle taping appears to be an effective technique to help prevent ankle injuries. The tape's effectiveness is more dependent on its tensile strength than its adhesive properties. (**Laughman, et al., 1980**) Stress x-ray reveals Gibney tape support significantly reduces talar tilt before and after 30 minutes of exercise, but elastic bandages do not. (**Vaes et al., 1985**)

Regardless of one's interpretation of the literature on prophylactic taping, it should be obvious that optimal muscle, foot, and leg function and balance are necessary to reduce the frequency and severity of ankle sprains. The use of taping may have an advantage other than direct support to the ankle. From simultaneous electromyography and stop-action movies, Glick et al. (**Glick et al., 1976**) concluded that "...the advantage of taping is probably attributable to its stimulating effect on the peroneus brevis muscle or in short to a dynamic action." In athletes with optimal lower extremity function tape may not provide the same prophylactic value as in those with dysfunction. (**Stoffel et al., 2010**)

Performance is another factor that may be important in deciding whether to support the ankle with tape. With increased ankle support limiting ROM in the ankle and subtalar joints, performance is significantly and progressively decreased. (**Stoffel et al., 2010**)

Peroneal Tendon Dislocation

The peroneal longus and brevis tendons are located in a common synovial sheath that lies in the retromalleolar groove and is held there by the superior peroneal retinaculum. The strong retinaculum rarely ruptures. It either strips away from the periosteum of the lateral malleolus or avulses a thin cortical shell. (**Ferran et al., 2009; Raikin et al., 2008**) The depth of the groove varies; when it is shallow there is predisposition to dislocation of the peroneal tendons. In this case little or no trauma may be necessary for dislocation. In some cases, the tendons can even dislocate with a squat. (**Martens et al., 1986**)

Traumatic dislocation of the peroneal tendons over the lateral malleolus is an uncommon injury. It primarily occurs with various athletic endeavors, particularly skiing, soccer, and ice-skating. It usually occurs with violent dorsiflexion, followed by strong reflex contraction of the peroneus longus and brevis. This often happens from ski tips digging in, with the skier being pitched forward by the sudden deceleration. (**Ferran et al., 2009; Raikin et al., 2009**) When the ankle is in maximum dorsiflexion, the tendons are "bow-stringed" against the soft tissues, maximizing chance of dislocation. A dislocation of the tendon causes a sharp pain and snapping.

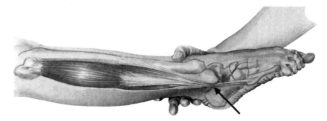

Peroneus longus dislocation

The frequency of peroneal tendon dislocation from skiing has decreased since the introduction of high, rigid ski boots. This boot protects against peroneal tendon injury because edging can be accomplished by the knee levering the rigid boot rather than requiring active foot eversion by the peroneal muscles. (**Oden, 1987**) Even so, there is a higher incidence of peroneal tendon dislocation in a practice that has a high percentage of skiers.

The condition can be acute or chronic. When acute it is often misdiagnosed as a simple ankle sprain, probably because it is so uncommon. The diagnosis becomes more difficult if there is some delay, with swelling over the lateral malleolus. There is tenderness to digital pressure in the retromalleolar sulcus, slightly more proximal than in the usual ankle sprain. (**Martens et al., 1986**) Tensing the peroneus longus and brevis muscles causes pain. After dislocation the tendons may spontaneously relocate. Putting the patient's foot in the position of injury against resistance will elicit retromalleolar pain and often dislocation. (**Evans, 2008**) It is important to examine the other ankle because a congenital luxation of the uninjured ankle is often found, leading to the diagnosis of the injured ankle. The tendon will frequently dislocate when the patient's ankle and foot are placed in plantar flexion and eversion against resistance. (**Oden, 1987**)

Varus stress x-rays of the ankle reveal no lateral instability and are not particularly painful to the patient. (**Yochum & Rowe, 2004; Marti, 1977**) A chip fracture of the lateral malleolus may occasionally be observed on x-ray. (**Yochum & Rowe, 2004; Logan, 1995**) This increases the possibility of surgery since the tendons cannot be held while the fracture unites. (**Safran et al., 1999**)

In the chronic state, there may be snapping or popping of the tendons with activity. When there are chronic ankle and subtalar problems with lateral ankle instability, peroneus longus and brevis tendon lesions should be considered. (**Sammarco & DiRaimondo, 1988**) Sometimes the peroneal retinacula are relatively weak; with time they tear loose, allowing the tendons to dislocate anteriorly over the lateral malleolus, with full dorsiflexion of the ankle. This may initially be painful and eventually develop to the point that the tendon dislocation and relocation are present with circular motion of the foot. The patient may complain that his ankle keeps "popping out." When the tendons are dislocated, pronation is weak or impossible, and the lateral ankle ligaments are unusually vulnerable to injury. (**Ramamurti, 1979**) Chronic cases often have to discontinue sports activities because of the feeling of the ankle giving way.

Conservative treatment of ankle taping, non-weight bearing progressing to weight bearing and an ice/heat regime is applicable. (**Cox, 1985**) An exception to this may be the professional athlete or anyone who puts excessive strain on the tendons, predisposing to future trauma. Most patients have little functional loss if subluxation persists and open reduction can be done at a later date if necessary. (**Safran et al., 1999**)

Surgical treatment for recurrent dislocation of the peroneal tendons consists of re-routing the tendons under the calcaneofibular ligament. This surgical procedure is effective in returning the individual to previous athletic activities. (**Martens et al., 1986**) Another surgical method is to deepen the groove, (**Colville, 1998**) the Kelly procedure, which deepens the retromalleolar sulcus by a sliding fibular veneer bone graft. (**McLennan & Ungersma, 1986**)

The peroneus brevis tendon lies behind the fibular malleolus medially and deeper than the peroneus longus tendon. An unusual occurrence is for the peroneus brevis tendon to move laterally and anteriorly around the peroneus longus tendon, but still remain within the tendon's common sheath. This creates a clicking, and it may prevent participation in sports and other activities. This uncommon injury has been successfully treated by surgically interposing two strips of peroneal retinaculum between the two tendons. This is done under local anesthesia so the patient can demonstrate the dynamic activity of the tendons; it also aids in the assessment of the integrity of the reconstruction. (**Safran et al., 1999**)

Achilles Tendon

The Achilles tendon is the thickest and strongest tendon in the human body. It begins approximately in the middle of the leg, extending to attach to the posterior surface of the calcaneus. Contributing to the Achilles tendon are the triceps surae's muscles: the soleus, gastrocnemius, and the almost vestigial plantaris. The soleus contributes muscle fibers almost to the lower level of the tendon. (**Gray's Anatomy, 2007**) The soleus portion of the Achilles tendon twists to insert on the medial portion of the calcaneus.

Gray's Anatomy (**Gray's Anatomy, 2007**) also notes that the Achilles tendon plays an important part in

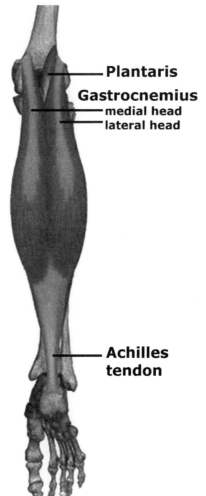

Muscle insertions onto Achilles tendon

reducing the energy cost of locomotion by storing energy elastically and releasing it at a subsequent point in the gait cycle.

The two heads of the gastrocnemius and the soleus are referred to by some as the triple aspect of the triceps surae. The soleus has been shown to have a bipartite behavior, with the lateral portion active in eversion and the medial portion in inversion, (**Gray's Anatomy, 2007; Levangie & Norkin, 2001**) making what could be called the quadriceps surae. If there is foot or gait dysfunction, Achilles tendon stress can develop as the gastrocnemius and soleus muscles attempt to control foot inversion and eversion.

Clement et al. (**Clement et al., 1984**) point out that the Achilles tendon is surrounded by a peritenon, which is comprised of the epitenon and paratenon. The paratenon functions as an elastic sleeve, permitting free movement against the surrounding tissue while maintaining continuity with adjacent structures. The Achilles tendon receives its vascular supply primarily through the paratenon. There is a region of diminished circulation 2-6 cm above the Achilles tendon insertion. They speculate that reduced vascularity at this area may play an etiological role in rupture of the Achilles tendon.

There are many terms used in the literature to describe disorders of the Achilles tendon, e.g. tendinitis, tenosynovitis, tendinosis, tendinopathy, rupture, and partial rupture, among others. In general, there are three conditions associated with the Achilles tendon — tendinitis, bursitis, and rupture. The terms used should describe the tissue disturbed and degree of involvement. Tendinitis does not involve the tendon; rather, it is an inflammatory reaction of the paratenon. Since the tendon does not have a synovial sheath, Clement et al. (**Clement et al., 1984**) recommend that the term "Achilles tenosynovitis" be re-classified to recognize the peritenon. Their classification scheme of Achilles tendon disorders is as follows: (1) peritendinitis, (2) peritendinitis with tendinosis, (3) tendinosis, (4) partial rupture, and (5) total rupture. Since the primary vascularization to the tendon is from the paratenon, most of the inflammatory process takes place there. The term "tendinosis" indicates disruptive lesions of the tendon in the absence of peritenon alterations.

Achilles tendinitis

Inflammation of the Achilles peritenon, which will simply be called tendinitis, is the most common cause of chronic posterior heel pain and it is the most commonly reported form of tendinitis. (**Langer, 2007; Levangie & Norkin, 2001**) The etiology of Achilles tendinitis is varied requiring individualized treatment. The condition usually has a gradual onset with pain and swelling along the Achilles tendon, 2-6 cm proximal to the calcaneus insertion. Examination reveals focal tenderness, induration, and frequently crepitus at this area. (**Cailliet, 1997; Logan, 1995**) Chronicity of the condition can develop into tendenosis and eventually partial or complete rupture. The tendon often becomes strained by extended pronation aggravated by a short triceps surae. (**Langer, 2007**) In a study of 1,192 injured runners, there were 20% with Achilles tendinitis. Of the 20%, 71% had pronated feet and 89% had a valgus heel. (**Lutter, 1983**) According to Johansson and Lysholm and Wiklander, (**Johansson, 1986; Lysholm & Wiklander, 1987**) the annual incidence of Achilles disorders is between 7-9% among top-level runners); however, 28% of 694 runners had other injuries as well to the lower extremity. (**Van Middelkoop et al., 2008**)

Training errors for runners and other athletes in their 30s and 40s are major contributory factors. (**Abshire, 2010; McDougall, 2009; Kubo et al., 2007**) Busseuil et al. (**Busseuil et al., 1998**) reviewed the various factors making runners more susceptible to injuries in a comparative study between a healthy control group (n=216) and runners (n=66) suffering from overuse pathologies of the foot and ankle. Specifically, the researchers looked at runners who needed treatment for iliotibial band syndrome, Achilles tendonitis, stress fracture of the tibia, tibial periostitis and plantar fascitis. The significant correlation between these problems and runners with foot dysfunction demonstrates the importance of thorough investigation of the feet and ankles and the rest of the leg and hip in runners with injuries. Their analysis showed that the injured subjects (among numerous other factors) have a more pronated foot than control group subjects. The many factors that come into play for normal foot function are purposely incorporated into the applied kinesiology examination of the feet.

Clement et al. (**Clement, 1984**) evaluated running with slow-motion high-speed cinematography, which revealed a whipping action or bowstring effect of the Achilles tendon with extended pronation. This occurs when the foot continues to pronate as the tibia begins external rotation. At this point, the foot should be supinating and becoming a rigid lever. (**Clement et al., 1984**) They speculate that the combination of torsional forces transmitted through the tendon due to extended pronation, along with vascular impairment, contributes to degenerative changes in the Achilles tendon. In Achilles tendonitis there is a significant decrease in the total collagen content and altered collagen cross-linking producing a dysfunctional collagen network similar to fibrosis. (**de Mos et al., 2007**) Athletic training may cause the triceps surae to become overstrengthened and short, limiting dorsiflexion and contributing to extended pronation. (**James et al., 1978**)

An inflexible midfoot contributes to Achilles strain. In extended pronation, the integrity of the midfoot typically fails. If it doesn't, there is a longer lever arm to put strain into the Achilles tendon. (**Levangie & Norkin, 2001**) In a similar manner, the cavus foot is intolerant of running and even walking. (**Ogon et al., 1999**)

Training errors are responsible for up to 75% of the cases. (**Chang et al., 2010; Van Middelkoop et al., 2008; Clement et al., 1984**) Included are failure to develop triceps surae flexibility, a sudden increase in training mileage, a severe competitive or training session (such as a marathon), sudden increase in training intensity, repetitive hill running, initiating training after an extended period of inactivity, and running on uneven, slippery terrain. The most common cause of chronic heel pain as the result of skiing is Achilles tendinitis. (**Oden, 1987**)

Diagnosis

Careful differential diagnosis is necessary because symptoms of Achilles tendinitis may simulate some of the symptoms of other running-induced injuries, such as tarsal tunnel syndrome, rupture of portions of the triceps surae, tibial stress fracture, tibialis posterior tendinitis, posterior compartment syndromes, retrocalcaneal bursitis, and plantar fasciitis. (**Langer, 2007; Levangie & Norkin, 2001**) In children, Achilles tendinitis can be mistaken for Sever's disease. (**Logan, 1995; Ellfeldt, 1983**)

Achilles tendinitis pain associated with running typically occurs early in the run and diminishes, but then it worsens again three to four hours after exercise. The patient may complain of pain in the Achilles tendon when getting out of bed in the morning, but the pain subsides during the day. Normal walking typically does not cause pain, while climbing stairs does. (**Langer, 2007; Ogon et al., 1999; Logan, 1995**)

The symptomatic pattern indicates the severity of the condition. In the earlier stage, pain is present only during strenuous activity, which is usually athletic. As the condition advances, the pain increases during sports activities and may force an athlete to stop or cut down on his activity. There may be pain during normal daily activities. With the advanced stage of tendinosis or partial rupture, there is also pain at rest. Severe morning stiffness indicates an advanced condition. (**Langer, 2007; Nelen et al., 1989**)

The location of pain is 2-6 cm proximal to the Achilles tendon insertion into the calcaneus. When maximum tenderness is on the medial border of the tendon, extended pronation as a contributing factor is indicated. Leaf and Lowdon et al. (**Leaf, 2000; Lowdon et al., 1984**) attribute this to bowing of the Achilles tendon from extended pronation, putting tension on the medial aspect of the tendon.

There may be some soft tissue swelling and/or crepitation with active motion. The Thompson squeeze test can be used initially to quickly test for the integrity of the Achilles tendon. On squeezing the belly of the gastrocnemius and soleus muscles, the foot should plantar flex. Diffuse swelling indicates thickening of the paratenon. With nodular swelling, tendinosis or partial tearing is probably present. (**Nelen et al., 1989**)

Treatment

Conservative treatment consists of immediate withdrawal from activities that put strain on the Achilles tendon. Ice massage and contrast baths usually give rapid relief in the absence of stress to the tendon. Heel pads for cushioning (sponge rubber, Sorbothane) are often used, (**Jorgensen, 1985**) but they have been found insignificant in aiding recovery. (**Lowdon et al., 1984**)

Ultrasound is recommended by some, (**Langer, 2007; Ogon et al., 1999; Schepsis & Leach, 1987**) but Clement et al. (**Clement et al., 1984**) state: "Ultrasound has been used to disrupt adhesions between the tendon and peritendinous sheath. However, because ultrasonic waves may disturb leukocytes in traumatized tendon and potentiate an inflammatory reaction, we feel that the beneficial role of ultrasound in the treatment of peritendinitis is questionable."

Clement et al. (**Clement et al., 1984**) recommend that cast immobilization not be used in Achilles tendinitis, even in stubborn cases. Impaired function of the immobilized joints, tendons, ligaments, and muscles may result.

Transverse friction massage, recommended by Cyriax, (**Nagrale et al., 2009; Chaitow & DeLany, 2002; Cyriax & Cyriax, 1983**) is done by grasping and moving the skin over the injured stretched tendon. This is done without lotion and must be directly over the damaged portion of the tendon across the longitudinal fibers. The lesion typically lies on the medial, lateral, or both borders of the tendon. Care is taken to avoid movement between the therapist's fingers and the patient's skin. Cleaning the patient's skin and the therapist's fingers with alcohol helps remove any oil and prevent slipping. Hammer (**Hammer, 2008**) presents a case report of the successful treatment of Achilles tendinosis, plantar fasciosis, and supraspinatus tendinosis using the Graston instrument. The Graston instrument is used to mechanically mobilize scar tissue, increasing its pliability and loosening it from surrounding healthy tissue. It is hypothesized that for degenerated connective tissue, the Graston instrument re-initiates the inflammatory process by introducing a controlled amount of microtrauma to the affected area.

Some use heel lifts to take strain off the tendon. (**Clement et al., 1984**) If this is done, it should only be during the acute stage. As the acute stage is resolved, it is important to correct extended pronation, if present, and lengthen the triceps surae to obtain adequate dorsiflexion. The proper exercise for gastrocnemius-soleus stretch is discussed in the previous chapter. To the wall lean exercise, Clement et al. (**Clement et al., 1984**) add toe raises by having the patient stand on the edge of a stair and hold onto a handrail for balance; the ankle is moved through its full range of motion. As applicable, shoulder weights are added to increase resistance. Increasing the speed of the heel drop adds an eccentric component to the exercise, as recommended by Stanish et al. (**Stanish et al., 1986**) In addition to having adequate length and function of the triceps surae, one must recognize that the leg functions as a unit. If the hamstrings are also tight, there will be additional tension on the triceps surae complex. (**Myers, 2001; Levangie & Norkin, 2001**)

It is important to correct all aspects of extended pronation. It is the most common factor, along with the short triceps surae, that is present with Achilles tendon disorders.

In most cases, significant reduction of training loads is necessary during rehabilitation. In some cases, it is necessary to discontinue running completely. Under these circumstances, other activities such as swimming or bicycling can be done to maintain the training program. (**Maffetone, 2010; Langer, 2007; Ogon et al., 1999; Schepsis & Leach, 1987**) In cycling, the heel should be put on the pedal to eliminate strain to the Achilles tendon. Clement et al. (**Clement et al., 1984**) recommend continued rest for seven to ten days post-symptomatically, before progressive return to pre-injury activity is attempted. Running should not be resumed until structural balance is indicated by an applied kinesiology examination. Failure to get adequate rest and obtain structural balance often results in an exacerbation of Achilles tendinitis.

When running is resumed, it should be reduced and done only on soft, smooth surfaces, avoiding hills. Shoes should have a stable, rounded, flared heel and be properly fitted. A sturdy heel counter cannot be overemphasized as a preventive measure. (**MacLean et al., 2008; Krissoff & Ferris, 1979**) A too-rigid sole does not allow the metatarsophalangeal joints to adequately flex; it lengthens the foot's lever arm and increases strain on the Achilles tendon. (**Maffetone, 2003; Levangie & Norkin, 2001; Clement et al., 1984**)

Souza (**Souza, 1994**) reviews the conservative care of various orthopedic conditions in the lower extremities, including disorders involving the Achilles tendon. Co and Pollard (**Co & Pollard, 1997**) describe the care of a patient with heel pain due to Achilles tendonitis, retrocalcaneal bursitis, and partial rupture of the tendon. Brantingham et al. (**Brangingham et al., 1994**) offer an excellent review of the successful chiropractic care (i.e., with 70% reduction of pain) for a patient with Achilles tendinopathy following 3 years of failed medical care. Carter and Carter (**Carter & Carter, 1997**) are advocates for chiropractic care (using a multidisciplinary approach) for patients with Achilles tendinopathy. Gaymans and Till as well as Cook et al. (**Gaymans & Till, 2003; Cook et al., 2002**) investigate the efficacy of manipulating the foot and ankle joint fixations compared to placebo in the treatment of chronic Achilles tendonitis. Kobsar & Alcantara (**Kobsar & Alcantara, 2009**) describe the post-surgical chiropractic care and rehabilitation of a professional ballet dancer following surgical calcaneal exostectomy and debridement with re-attachment of the Achilles tendon. Ramelli (**Ramelli, 2003**) also describes the successful chiropractic post-surgical care of a 25 year-old male to repair a complete Achilles tendon separation.

Recognizing Achilles tendinitis and not trying to "run through" the problem, along with proper conservative treatment, corrects most of these conditions. It is the competitive distance runner who is likely to have surgery for Achilles tendinitis when conservative methods fail. (**McDougall, 2009; Langer, 2007; Ogon et al., 1999; Schepsis & Leach, 1987**) When in doubt about whether one is dealing with peritendinitis or partial rupture of the Achilles tendon, conservative treatment is recommended. If not successful, surgical exploration often reveals a partial rupture. (**Nelen et al., 1989**)

Achilles tendon rupture

Achilles tendon rupture is not often seen in an applied kinesiology practice. The Achilles tendon is the strongest one in the body and only ruptures with degenerative changes, e.g. gout, (**Beskin et al., 1987**) syphilis, training errors, or medication. (**Hess, 2010; Veves et al., 2002**) Other predisposing factors can be prolonged corticosteroid therapy, overuse of performance enhancing drugs (the bodybuilder steroids), and direct trauma to the area. (**Newnham et al., 1991**) The incidence increases if the practice has a large number of middle-aged white-collar workers who engage in part-time athletic activities. (**Kubo et al., 2007; Langer, 2007; Leppilahti, 1998**) It occurs more often in athletes over 30 years of age, less frequently in younger athletes, and rarely in the elderly (**Kubo et al., 2007; Langer, 2007; Ramelli, 2003; Leppilahti, 1998**) unless they pursue past their prime. (**Hess, 2010; Veves et al., 2002**) The vulnerable person is the weekend, but otherwise sedentary, athlete who participates in sports that put a quick impulse stretch on the Achilles tendon. (**Leppilahti, 1998**)

The etiology of Achilles tendon rupture is controversial. (**Hess, 2010; Ramelli, 2003; Veves et al., 2002**) One concept indicates there is repetitive microtrauma in an aging tendon, possibly relating to deficient blood supply. Another theory is that there is an anomaly in the normal inhibitor mechanism of the musculotendinous unit monitoring force to prevent excessive tension on the tendon. Both mechanisms of Achilles rupture may be applicable, with the former relating to an older person's injury and the latter to younger, more active well-conditioned athletes.

One must consider the neuromuscular spindle cell and Golgi tendon organ interrelationship in the control of functional stability of the muscles of the ankle. Afferent information for functional stability of the ankle comes from receptors in the muscle-tendon system. This is in accord with research on other joints. (**Macefield, 2005**)

Murphy provides an excellent review of current concepts of proprioception, but notes that a great deal remains to be learned about proprioception in the human neuromuscular system. (**Murphy, 2000**) The muscle spindles and Golgi tendon organs are two of the most important mechanoreceptors in muscles. The integration of the information coming from the muscle spindle and the Golgi tendon organs in the Achilles tendon complex must be congruent. If they are not, injuries to this complex are likely. (**Loram, 2009**)

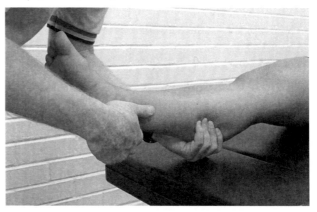

Thompson's squeeze test

Steroids, whether injected for tendinitis or administered parenterally for another condition, increase the likelihood of Achilles tendon as well as plantar aponeurosis rupture. (**Brinks et al., 2010**) In one study (n=48), 63% of the subjects had prior corticosteroid injections for tendinitis ruptures. (**Kellam et al., 1985**) Hydrocortisone injected into the tendon causes separation of collagen fibers within 45 minutes. (**Olchowik et al., 2008**) The medication may relieve the symptoms, allowing the individual to increase activity that might lead to a partial or complete rupture.

Steroids also interfere with the healing process. **(Clement et al., 1984)**

The effect of parenteral steroid treatment is emphasized in a case report of a woman who had nine months of steroid treatment for rheumatic fever and experienced bilateral simultaneous Achilles tendon ruptures while simply walking up stairs. Her ankles gave way in an inversion-type mechanism. She experienced no snapping or popping at the time. **(Price et al., 1986)** Hersh and Heath **(Hersh and Heath, 2002)** point out that spontaneous Achilles tendon rupture associated with long-term oral steroid use is not uncommon, particularly in older patients who use these drugs daily to treat systemic diseases. The long-term use of oral steroids poses problems: immunosuppression, cardiovascular disease, depression, insulin resistance, tissue degeneration, mood swings, slow tissue healing and susceptibility to infection have been noted by clinicians managing patients with chronic pain using steroids. **(Wilson, 2002)**

Achilles rupture due to skiing is declining because of the high-ridged boot and multiple mode release ski bindings. **(Oden, 1987)**

Examination

Examination must be done carefully, because up to 25% of ruptures are initially missed. **(Slimmon & Brucker, 2010; Ferran, 2009)** This may be due to insignificant pain. It may initially be a sharp, sudden pain that diminishes to slight or even disappears. In addition, the precipitating violence may be fairly mild. **(Hess, 2010; Hersh & Heath, 2002)** Achilles rupture may be missed when there is other obvious trauma. A case is reported in which the rupture was missed until it was recognized during ankle fracture surgery. **(Martin & Thompson, 1986)**

Partial rupture causes a stabbing, pricking pain with activity; there are localized regions of tendon thickening and nodules. With partial rupture there is some loss of plantar flexion due only to pain. **(Clement et al., 1984)** Palpation may disclose a small gap in the tendon, with a tender, rolled portion of the tendon on the proximal edge. The damage to the tendon may be both transverse and longitudinal. **(Hess, 2010)**

With total rupture there may be no immediate pain, but there is an immediate functional loss. The patient may experience an audible snap or a sensation of being hit in the back of the leg. There is usually no prior indication that the Achilles tendon is vulnerable to rupture. A definite gap may be palpated in the Achilles tendon within 6–12 hours after rupture. After 24 hours, the gap becomes filled with a hematoma and is difficult to distinguish. **(Rolf & Movin, 1997)**

There is only weak plantar flexion provided by the peroneus longus and brevis and tibialis posterior muscles, **(Clement et al., 1984)** with no ability to plantar flex against resistance. This causes a lack of normal push-off during walking. **(Levangie & Norkin, 2001; Thompson & Doherty, 1962)** When the patient's ankle is passively dorsiflexed by the examiner, there is no firm feeling of the Achilles tendon. **(Ramamurti, 1979)**

Because weak plantar flexion is maintained by the tibialis posterior, peroneus longus and brevis, and the toe flexors, **(Travell & Simons, 1992)** diagnosis may be obscured or delayed. **(Cox, 1985)** The Thompson squeeze test is pathognomonic **(Logan, 1995; Turco & Spinella, 1987; Thompson & Doherty, 1962)** of complete rupture of the Achilles tendon. The test can be done with the patient prone, kneeling on a stool, or on all fours with the feet clear of the table or chair. With the patient relaxed, the physician squeezes the calf muscles. This normally creates passive plantar flexion of the foot. Lack of continuity from the muscles to the calcaneus because of Achilles tendon rupture causes no foot movement from squeezing the gastrocnemius–soleus muscles. Contact for the squeeze should be at the bulk of the muscles. When it is applied proximal to the apex of the soleus curve, there is normally no plantar flexion. **(Clement et al., 1984)**

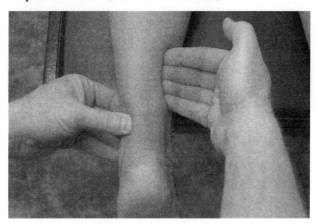

Anterior Achilles tendon can be assessed

Soft tissue changes are observable on x-ray to identify Achilles tendon rupture. **(Slimmon & Brucker, 2010; Yochum & Rowe, 2004; Clement et al., 1984)** A lateral view, with the foot at 90°, should be taken of both the normal and the traumatized foot for comparison. The fat pad triangle between the Achilles tendon and bone is encroached upon by bleeding and edema, causing its sharp border to disappear. Likewise, the fat pad posterior to the tendon is also encroached upon and narrowed. The proximal end of the Achilles tendon may be difficult to identify because of retraction and the surrounding blood and edema. The distal end is usually visible because it is surrounded by the heel fat pad. A bright transillumination light to view the x-ray may help identify the soft tissue structures.

Magnetic resonance imaging (MRI) can help diagnose between partial, severe, and complete rupture of the Achilles tendon. It accurately predicts the condition of the tendon found at surgery. **(Tourne et al., 2010)**

Treatment

There is controversy about whether surgical or non-surgical treatment is optimal. **(Ramelli, 2003)** Those in favor of non-surgical treatment, which includes casting for up to eight weeks, cite the incidence of high surgical complications, increased cost, and good results of non-surgical treatment to support their viewpoint. Those in favor of surgical treatment indicate that the tendon is returned to a more normal length, strength, and range of motion, and the incidence of re-rupture is lower. Complication rate is low, but not as low as with conservative care. Regardless of whether surgical or non-surgical methods treatment is given, Achilles tendons are susceptible to re-rupture. Carden et al's review (**Carden et al., 1987**) of 106 patients with ruptured Achilles tendons showed that in patients treated within 48 hours of injury the result was very similar in conservatively and in operatively treated patients. The incidence of major complications was higher after operation (17%) than in those treated conservatively (4%). However conservatively treated cases have a re-rupture rte of 17.9% compared with a rate of 2.2% in the surgically treated cases.

In a study evaluating results published in the literature in the previous twenty-five years, Wills et al. (**Wills et al., 1986**) found that the rate of re-rupture for surgically treated patients was 1.54%, while for non-surgically treated patients it was 17.7%. Complications as a result of surgical treatment were 20.0%; non-surgical complications were 10%. The surgically treated patients lost thirteen weeks from work, while non-surgically treated patients lost nine weeks. When the re-rupture factor was considered, cost and loss of work advantage for the non-surgically treated patients tended toward balancing.

Molloy and Wood, (**Molloy & Wood, 2009**) in a recent review of the medical literature regarding the treatment of the Achilles tendon, show that tendon re-rupture, sural nerve morbidity, wound healing problems, changes in tendon morphology, venous thromboembolism, elongation of the tendon, complex regional pain syndrome, and compartment syndromes are not uncommon after surgical repair of the ruptured Achilles tendon.

Lea and Smith (**Lea & Smith, 1972**) recommend non-surgical treatment of Achilles tendon rupture. They point out that the tendon has been shown to regenerate itself, and there is a high percentage of complications in surgical repair. They present a study of 66 cases of Achilles tendon rupture treated non-operatively to emphasize the value of non-surgical treatment. The procedure consists of immobilizing the ankle in gravity equinus position in a walking cast for eight weeks. The patient then uses a 2.5 cm heel lift for four weeks, while mobilizing the ankle with gentle resistance exercises that also strengthen the triceps surae. During this period, the patient is advised to avoid sudden jerky movements of the ankle. The results in this series were very good, with re-rupture of the Achilles tendon occurring primarily in cases that were immobilized for a period less than eight weeks. Eleven of the cases were Achilles lacerations, in which the Achilles tendon was completely severed. These had an effective correction similar to the spontaneous ruptures.

Percutaneous repair of Achilles tendon rupture consists of suturing the tendon together without surgically exposing the tendon. It has the cosmetic advantage of not leaving a large scar, but there is the disadvantage of occasional re-rupture over the better results of open repair. (**Nilsson-Helander et al., 2008**)

Surgical repair takes precedence in high performance athletes and those in ballistic sports, e.g. tennis, basketball, racquetball, and volleyball. (**Deangelis et al., 2009; Langer, 2007**) According to Kellam et al., (**Kellam et al., 1985**) non-surgical treatment of Achilles tendon rupture should be considered in a non-athletic patient over the age of fifty because of the high incidence of post-operative wound problems in this age group. In making a decision, an individual should be informed that there is a higher incidence of re-rupture in non-surgically treated Achilles tendon ruptures. In a comparative follow-up study of surgical and non-surgical treatment, 9 of 23 patients treated non-surgically had re-rupture of the Achilles tendon. (**Inglis et al., 1976**) When surgery is to be done, it should not be delayed; follow-up revealed 20% less strength, as tested by the Cybex II dynamometer, when diagnosis and surgical treatment were delayed more than one month.

There are numerous surgical techniques described for repairing a ruptured Achilles tendon. When neglect has caused tendon shrinkage, the surgeon can join the tendon by suturing a supporting substance such as Marlex mesh. (**Ozaki et al., 1989**) The tendon can be supported by suturing in other tendon material, e.g., the peroneus brevis (**Turco & Spinella, 1987**) or plantaris tendon. (**Hohendorff et al., 2009**)

Of the three approaches to Achilles tendon rupture, open surgical repair should be applied to the elite, active athlete to prevent re-rupture; percutaneous repair should be used in the recreational athlete and those concerned with cosmetics; and — finally — inoperative conservative management is suggested for sedentary older patients (over age 50), or chronically ill or debilitated patients. (**Leppilahti, 1998**)

Trauma to the Achilles tendon can be prevented by correcting faulty foot and ankle biomechanics as discussed in this and the previous chapters. It is particularly important that the triceps surae be long enough to provide adequate ankle dorsiflexion. Runners and other athletes should subscribe to a daily program of lower leg strength and flexibility exercise and avoid potentially aggravating training surfaces. The training program should be consistent, without sudden changes in intensity. Athletic and casual shoes should be properly designed and fitted. (**Abshire, 2010; Maffetone, 2010, 2003; Langer, 2007**)

Achilles bursitis

There are two bursae associated with the Achilles tendon. The pre-calcaneal bursa is located between the skin and the Achilles tendon. The retrocalcaneal bursa is between the Achilles tendon and the bone.

The pre-calcaneal bursa is most often involved, and is usually secondary to tight or otherwise ill-fitting shoes. Diagnosis is made by palpating the bursa. The pre-calcaneal bursa is located by lifting the skin posterior to the tendon between the thumb and forefinger. The retrocalcaneal bursa is palpated between the thumb and forefinger anterior to the Achilles tendon. Any palpable thickening or tenderness in either area suggests the presence of bursitis. (**Cailliet, 1997; Logan, 1995**)

Treatment is usually as simple as changing shoe size or style to prevent irritating the bursae. It is rare that a prominence on the calcaneus needs surgical removal. (**Brody, 1980**) This is accomplished by molding the posterior calcaneus. (**Michels et al., 2008**)

Tibialis posterior tendinitis

Tendinitis can develop at the proximal or distal tibialis posterior muscle tendons. When proximal, it is a type of "shin splint" that will be discussed later. Either type is often due to excessive foot pronation that may be observed statically or as extended pronation in mid support. **(Yuill & MacIntyre, 2010; Langer, 2007; Maffetone, 2003)**

Discussion here is limited to distal tibialis posterior tendinitis. The condition can first become evident in poorly conditioned athletes or novice runners running on hard surfaces. It usually occurs during the early portion of a training period. Improper shoes or running on a banked or uneven surface can also be causative. When the condition is not associated with increased activity as in athletics, it is usually insidious in its development and associated with uncorrected extended pronation.

The location of pain in distal tibialis posterior tendinitis is posterior to the medial malleolus, localized along the tibialis posterior tendon, extending to its major insertion at the under surface of the navicular. The pain may also extend proximally along the posterior medial border of the tibia to the origin of the tibialis posterior muscle. Pain is present after long runs, especially on uneven surfaces when there is inadequate foot support. Because of the location of pain along the posterior inferior edge of the medial malleolus, this condition must be differentiated from distal tibial stress fracture. **(Conti, 1994)** Rarely the tibialis posterior tendon can dislocate; it may relate with a history of previous minor ankle trauma. The diagnosis is obvious and the treatment is surgical.

Physical examination may reveal crepitant tenosynovitis in the tarsal tunnel and proximally along the posterior and medial tibial borders. There may be swelling posterior to the medial malleolus. Pain is often present on supination or inversion of the foot against resistance; it is often produced during the manual muscle test. **(Yuill & MacIntyre, 2010; Perry et al., 1986)**

It is important to properly diagnose this condition in its early stages because the tendon may become elongated, leading to the necessity of surgery. When the tibialis posterior fails, extended pronation worsens. This produces a vicious circle, because extended pronation was probably responsible for the tibialis posterior tendinitis in the first place.

Most cases of tibialis posterior tendinitis can be resolved by correcting extended pronation, including any primary factors that are contributing to the condition. If the condition continues without adequate resolution, the tendon can become elongated and cause severe foot distortion. Johnson and Strom **(Johnson & Strom, 1989)** have described three stages of the dysfunction. They consider only the first stage applicable to conservative treatment. In Stage 1 the pain is mild to moderate, and the patient is not incapacitated

A test and a sign are described by Johnson and Strom. **(Johnson & Strom, 1989)** They use the single-heel-rise test to assess tibialis posterior tendon function. Many of the positive characteristics of this test are due to extended pronation. Motion of the foot is assessed as the patient rises on the ball of one foot, with the other foot held off the ground. (While doing this the patient may stabilize against a door or wall for balance.) They state, "The normal sequence for a single-heel rise is as follows. First, the TPT [tibialis posterior tendon] is activated, which inverts and locks the hindfoot, thus providing a rigid structure. Next, the gastrosoleus muscle group pulls up the calcaneus and the heel rise is completed. With elongation of the TPT, however, the initial heel inversion is weak and the patient either rises up incompletely without locking the heel or does not get up on the ball of the foot at all." In Stage 1, the patient is usually able to rise up on the ball, but it will be more painful than the normal side.

In Stage 2, the tendon becomes elongated and foot distortion develops. Pain increases, and the patient finds it difficult to walk and perform normal activities. Swelling and tenderness, located in the same area, increase. The single-heel-rise test becomes more abnormal. The patient stands with his forefoot abducted, giving rise to the "too many toes" sign. This is where the examiner stands behind the comfortably-standing patient and observes for forefoot abduction by counting the number of toes seen on the deformed side. In addition to forefoot abduction, the calcaneus rotates laterally into a valgus position (Helbing's sign).

In Stage 3, the pain may transfer to the lateral aspect of the hindfoot and be located over the sinus tarsi; the hindfoot goes further into a valgus position. The talus impinges on the sinus tarsi, producing the patient's pain symptoms. As this condition progresses into Stages 2 and 3, a surgical tendon transfer of the flexor digitorum longus may be needed to provide stability to the foot and ankle. **(Johnson & Strom, 1989)**

Yuill & Macintyre **(Yuill & MacIntyre, 2010)** present a case report in a 14 year-old soccer player with tibialis posterior tendonopathy who developed right medial foot pain after striking an opponent in the leg while trying to kick the ball 4 months prior to his presentation. Outcome measures of treatment (utilizing Active Release Technique) **(Leahy, 2008)** included subjective pain ratings and manual muscle testing. On initial examination, the tibialis posterior manual muscle test was graded weak. The patient was treated with 4 sessions over 4 weeks with Active Release Technique, particularly over the insertion of the tibialis posterior muscle. Tibialis posterior rehabilitative exercises were also given. After 4 weeks of chiropractic treatment the patient was able to return to playing soccer relatively pain free (1/10 VAS score with jumping and landing), and manual muscle testing of the tibialis posterior showed decreased weakness and improvement.

Shin Splints

The term "shin splints" is misused by most laymen and often by doctors. Common usage indicates any pain in the leg that develops as a result of running, jumping, or other similar activity. There are many factors that can cause this type of pain. An effective therapeutic approach cannot be designed unless the exact cause of the pain is determined, as the etiology of apparently identical symptoms can be considerably different. In addition, there are conditions such as the compartment syndrome, discussed later, that if not treated properly and timely can cause permanent loss of function. Failure to accurately diagnose acute compartment syndrome, and to treat it as a "shin splint" syndrome instead is a common basis for malpractice litigation. (**Bourne & Rorabeck, 1989**) Estimates have shown that shin splints account for 10-15% of all running injuries and up to 60% of all leg pain syndromes. (**Korkola & Amedola, 2001**)

In addition to being non-specific, the term shin splints is misleading. Dictionary definition (**Dorland's Illustrated Medical Dictionary, 2007**) provides the following definitions for shin: 1) The crest or anterior edge of the tibia, and 2) the anterior aspect of the leg below the knee. This would indicate, then, that shin splints is a pain of the anterior leg. In reality, it can be of the anterior or posterior leg. Ideally, the term "shin splints" should be discarded entirely. (**Galbraith & Lavallee, 2009; McBryde, 2007**) Since the term is in common usage and has been defined in Standard Nomenclature of Athletic Injuries, it will be used here with adherence to the definition. Shin splints is restricted to "pain and discomfort in the leg from repetitive running on a hard surface or forcible, excessive use of the foot flexors; diagnosis should be limited to musculotendinous inflammations, excluding fatigue fracture and ischemic disorder."

There are four factors that must be present for the diagnosis of shin splints. (**Galbraith & Lavallee, 2009; Slocum & James, 1968**)

1) Lesion must lie within the plantar flexors or dorsiflexors of the foot. These include but are not exclusive of the tibialis posterior, soleus, flexor hallucis longus and flexor digitorum longus as plantar flexors, and the tibialis anterior, extensor hallucis longus and the extensor digitorum longus as dorsiflexors.

2) Pain develops with repeated rhythmic activity incurred while walking, running, with possible jumping. Untrained, undeveloped, or poorly functioning muscle, or muscle weakened by fatigue is most likely to be involved. Madeley et al. (**Madeley et al., 2007**) show that athletes with shin splints have significant endurance deficits in the ankle joint plantar flexor muscles, including the tibialis posterior. The applied kinesiology Repeat Muscle Activation Patient Induced testing of the tibialis posterior is commonly positive in these patients. (**Gangemi, 2011; Walther, 2000**)

3) Tenderness is most often at attachment of the muscle to the periosteum of the bone or interosseous membrane. The tenderness is most often accompanied by classical signs of mild inflammation.

4) Conditions resulting from direct trauma or disease are excluded, including stress fractures, compartment syndrome, peripheral arterial disease, peripheral nerve entrapments, medial tibial syndrome and soleus syndrome producing periostitis. (**Gerow et al., 1993**)

Shin splints result from abnormal stress upon the soft tissue structure in the anterior or posterior muscle group. The stress results from running or otherwise using the leg in such a manner as to repeatedly activate the muscles. The condition develops with less exogenous stress when there is muscular dysfunction or structural imbalance. The endogenous and exogenous stresses combine to create the condition. The strain or stress can be localized to the muscle itself, as well as the attachments of the muscle to bone, in which case a periostitis develops. (**Gerow et al., 1993; Subotnick, 1991, 1975**) The tight, aching feeling develops, then, from tendinitis, myositis, and/or pereostitis, which should all be evaluated during examination.

During the phases of gait, the anterior muscle group, which includes the tibialis anterior, extensor digitorum longus, and extensor hallucis longus, comes into play at heel strike to decelerate the rest of the foot coming into contact with the ground. Thus, if there is hard heel contact with a solid substrate, the muscles become more active and thus increase the strain. There is continued action of these muscles to prevent the forefoot from slapping the ground.

Normally the anterior muscle group is active during the swing phase, toe-off, and heel contact. When the muscles must continue to be active during the stance phase of gait, they are active during much of the total cycle, thus being overactivated and becoming vulnerable to inflammatory reactions. Edema develops which cannot escape from the basically closed anterior compartment, and aching and pain result.

Athletic training often develops an imbalance of function between the anterior and posterior groups, with the anterior group being weaker. Balance must be regained between the muscles. This often requires stretching of the posterior group to regain adequate dorsiflexion of the foot.

Shin splints of the posterior compartment involve the tibialis posterior, soleus, flexor digitorum longus, and the flexor hallucis longus muscles. This is particularly true for those with mild claw toe deformity. (**Garth & Miller, 1989**) These muscles are active during the stance phase from just after heel contact to just prior to heel-off. If the foot fails to resupinate from its pronated stance at the proper time, the muscles are overstressed, setting up posterior shin splints. The tibialis posterior is especially stressed under this condition because its activity is attempting to stabilize the ankle mortise as well as support the medial longitudinal arch. The tibialis posterior muscle may have tendinitis in the tarsal tunnel, and myositis of the total muscle.

The soleus muscle has become more strongly implicated in shin splints by Michael and Holder (**Michael & Holder, 1985**) particularly when the heel is in a pronated position. From evidence produced by 3-phase radionuclide bone scanning, they describe a characteristic scintographic appearance implicating the soleus muscle attachments as stress areas.

The mechanism of soleus stress appears to relate with pronation where the calcaneus is in a valgus position. The soleus has been shown to be a bipartite muscle that has a medial and a lateral division. (**Moore et al., 2009**) The medial head of the soleus is an inverter of the calcaneus. When an individual with extended pronation runs or walks, the medial head of the soleus appears to attempt to counteract the excess heel eversion, thus putting extra strain on origins of the soleus. This is supported by electromyographic activity of the medial soleus muscle when the heel is passively everted in individuals with shin splints but not with normal subjects.

The role of extended pronation being contributory to shin splints is emphasized in a study by Viitasalo and Kvist, (**Viitasalo & Kvist, 1983**) which revealed an increased Achilles tendon angle (Helbing's sign) in athletes who are prone to shin splints over a normal control group. While running, they had an increased Achilles tendon angle at heel strike and throughout the pronation stage of the stance phase of gait. The shin splint group also had an increase in the passive movement of inversion to eversion.

In studying shin splints in the young athlete, Jackson et al. (**Jackson et al., 1978**) found a common type of shin pain to be localized along the distal posterior medial tibia. It is often a specific response of the bone to stress. In a study of 40 young athletes with leg pain, they found 26 stress reactions of the posterior medial aspect of the distal tibia, with the rest distributed among typical stress fracture, musculotendinous inflammations, and other categories. This makes the stress reaction to the distal tibia the most common source of disability in competitive young runners. The pain generally begins gradually and increases with increased activity.

Diagnosis

In early shin splints, symptoms are minimal and well-defined. By the time the individual seeks a physician's help, the condition is often advanced with signs and symptoms much more vague, making localization and differential diagnosis more difficult. (**Galbraith & Lavallee, 2009; Slocum & James, 1968**) Shin splints must be differentially diagnosed from stress fractures, acute and chronic compartment syndromes, fascial hernias, tenosynovitis, and chronic interosseous membrane strains with or without exostosis formation. In addition, specific disease processes, which have distinctive diagnosis such as cellulitis, chronic osteomyelitis, thrombophlebitis, intermittent claudication, infective and varicose pereostitis, and tumors must be considered.

The pain of a stress fracture is usually located along the medial and lower third portion of the tibia. The pain is usually localized in the bone itself and is very painful to palpation of the bony surface around the fracture site; edema and warmth may also be present. X-rays may not detect the stress fracture for several weeks. Treatment for this kind of stress fracture is rest and reduced weight-bearing stress.

Travell and Simons (**Travell & Simons, 1992**) describe a medial tibial stress syndrome which is related to tension placed upon the periosteum of the tibial cortex which may result in its separation. The distal half of the medial tibia will exhibit localized pain at the sites of the overstressed muscular insertions. Pain often extends to a larger area than that found in stress fractures. (**Edwards & Myerson, 1996**)

Palpation determines the area of maximal tenderness, which is either localized to the bone, the muscles, or the tendon. Often bony tenderness is just distal to the muscle insertion (there are no muscle insertions on the posterior medial tibia in the distal one-third of the leg). Stressing the musculotendinous unit, as in a muscle test, shows inflammation or tear in that unit when there is increased pain. Differentiating between motion or weight bearing aggravating pain helps to distinguish between musculotendinous inflammation or involvement of the bone itself. There may be soft tissue swelling in a bony reaction, which should not be confused with inflammation of an overlying tendon. (**Jackson et al., 2001**)

Shin splints can be classified as to severity of pain. Grade 1, pain after athletic activity; grade 2, pain before and after athletic activity, does not affect performance; grade 3, pain during and after athletic activity, affects performance; grade 4, pain so severe the athlete is unable to compete.

Treatment

Treatment should not begin before ischemic conditions, stress fractures, and other non-related conditions are ruled out. The acute condition can be treated by reduced or changed physical activity, and ice before and after any running. The condition is frequently made worse by attempts to stretch the affected intercompartmental musculature. (**Schafer, 1986**) Additionally, it is often important that the examination should occur after the patient has exercised enough to reproduce the symptoms.

Various therapeutic efforts for shin splints are put into perspective by Andrish et al. (**Andrish et al., 1974**) They studied 2,777 first-year midshipmen at the United States Naval Academy using four prophylactic programs in an attempt to prevent shin splints during training. For control, these groups were compared with the men in the usual physical education program. The various approaches were as follows. 1) They used a heel pad in an attempt to decrease the amount of violent muscle pull on the tibial muscle origins and to decrease the amplitude of the stress applied to the tibia. 2) Heel cord stretching exercises were used because observation was that the majority of shin splint victims have tight heel cords. It is not specifically stated that the heel cord stretching exercises were started at the same time as training. 3) There was use of a heel pad and heel cord stretching exercise. 4) A graduated running program instituted for two weeks prior to pursuing the normal physical education routine. None of the prophylactic approaches were effective in preventing shin splints. In fact, the study groups had more incidences of shin splints than the control group.

Those who developed shin splints in this program were assigned to one of five treatment programs. All treatment programs included no running until pain-free. Additionally, there was 1) ice applied to the affected area three times a day; 2) same ice regime and ten grains of aspirin four times daily for one week; 3) same ice regime and 100 mg phenylbutazone four times a day for one week; 4) same ice regime and heel cord stretching exercises; 5) a short walking cast for one week. The group who received only no running until pain-free statistically demonstrated a significant advantage over the other regimes. Additionally, in order of descending effectiveness were rest, ice and phenylbutazone; rest, ice, and heel cord stretching exercises; rest, ice, and aspirin; and, finally, as the worst, casting.

Andrish et al. (**Andrish et al., 1974**) concluded their report with, "Finally, the fact that twice as many of those with shin splints had no previous training immediately before entering the Naval Academy agrees with the commonly held belief that shin splints tend to be more prevalent in the unconditioned athlete or recruit."

Alternate physical activity, such as biking or swimming, can be used to maintain cardiovascular proficiency. Lutter (**Lutter, 1983**) states, "Approximately five weeks healing time is necessary to return to normal running from an injury of shin splints."

Most often, pronation will be present and should be treated as indicated under that subject. All muscles of the leg, ankle, and foot should be evaluated and treated with applied kinesiology methods if necessary. Particular dysfunctions must be determined in the tibialis posterior muscle, which is associated with many types of leg, knee, foot, and ankle dysfunction. Inhibition of this muscle results in excessive pronation because the tibialis posterior is important in maintaining the medial longitudinal arch when walking and running. (**Hamilton, 1985**) The muscle is also a primary one for athletes who must rise upon their toes. Maffetone (**Maffetone, 1999**) notes that tibialis posterior inhibition may be followed by reciprocal tightness of the gastrocnemius and soleus muscles, producing "posterior shin splints". Pincer palpation, muscle stretch reactions, neurolymphatic reflexes associated with the adrenal gland, and other findings will often be found in this muscle in such cases. (**Leaf, 2010; Walther, 2000**)

Bandy (**Bandy, 1978**) offers the hypothesis that pronation is the primary cause of posterior shin splints. Pronation in this model causes stretching beyond the capabilities of the myofascia of the tibialis posterior muscle and this produces weakness followed by tearing in the muscle during physical activity. Bandy finds no more than 30% of the athletes he has treated with inhibition of the tibialis posterior muscle when tested in the clear; however after the muscle stretch reaction (indicating fascial tension in these cases), the muscle then becomes inhibited on MMT. When he treats the fascia of this muscle, using the origin-insertion technique taught in applied kinesiology, the pain from the shin splints in these athletes is relieved quickly. Heavy pressure to the "knot" or "wrinkle" in the fascia near its attachment to the tibia is effective in these cases. This approach combines the methods of Ida Rolf with Dr. Goodheart effectively. (**Rolf, 1977**)

A momentary overload that causes rupture (tearing) of the tibialis posterior muscle or its tendon has been described. (**Holmes et al., 1990**) This kind of micro-avulsion injury is what the applied kinesiology origin-insertion technique was designed for originally. Microavulsions for the pectoralis major, (**Marmor et al., 1961**) peroneal, (**Davies, 1979**) gastrocnemius, (**Froimson, 1969; McClure, 1984**) plantaris, and extensor digitorum longus (**Perlman & Leveille, 1988**) muscles has also been described. As with a muscle bruise, microavulsion in a muscle is generally treated manipulatively to reduce inflammatory reaction and then promote healing. It is rarely repaired surgically. (**Mense & Simons, 2001**) In the case of a ruptured tendon especially in the ankle region, however, surgical repair may be critically important to avoid muscle imbalance and serious disturbance of ambulation.

Mense and Simons (**Mense & Simons, 2001**) note that "muscle spasm as the result of pain in the same muscle is an exceptional occurrence." Because the number of specific muscles that may be producing pain and joint dysfunction in "shin splints" is large, specific manual muscle tests that isolate the muscles of the anterior and posterior compartments of the calf are critical. Reduced muscle strength has been suggested to be a surrogate marker for the progression of shin splints in some patients. (**Wilder & Sethi, 2004**) Differential diagnosis of the specific muscle(s) producing the problem is essential; and specific MMT of each of these muscles is possible with AK methods.

Poor gait activity, modular interaction, or dural tension may be causing increased stress during the gait cycle. A graduated training program should be developed for the athlete, as shin splints are frequent in the early phases of the training season. The best method of dealing with shin splints is treating dysfunctions of the foot, ankle, knee and leg before the athlete begins training. The technique described by Bandy and used by the authors is effective in treating shin splints that are symptomatic, but treatment is more painful. The use of the Percussor instrument is often effective for treatment of this pathophysiological disturbance as well. (**IMPAC, 2012**) The importance of inflammation in cases of shin splints has been reviewed by Krenner; (**Krenner, 2002**) chiropractic management of this factor in a patient with medial tibial stress syndrome was successful. Comprehensive AK treatment approaches for two cases of shin splints have been reported as well. (**Rogowskey, 1990; Boven, 1988**)

Compartment Syndrome

The leg has five compartments that are non-yielding in character. A compartment syndrome appears when a muscle enlarges, usually with exercise, to the point that the pressure increases within the compartment to interfere with tissue perfusion and neurovascular function to the muscle or muscles in the compartment enclosure. Edema within the compartment, hematoma or infection are additional reasons for this syndrome. (**Gerow et al., 1993**) The pressure can rise to a critical state where there is neurovascular ischemia with paralysis and necrosis following if left untreated in the early stages. It can be a surgical emergency to preserve muscle function. This is done by fasciotomy to relieve pressure within the compartment.

The compartments are closed spaces with the exception of perforation of the fascia by blood vessels and nerves and the exit of tendons distally. There are 46 compartments in

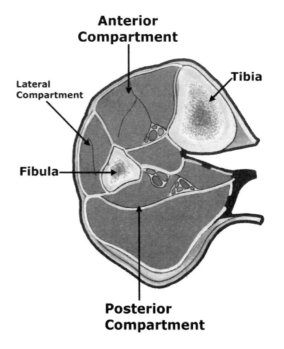

Compartments of leg

the human body with 38 of them in the extremities. While compartment syndrome may occur in a number of areas, over 80% of them occur in the lower extremities. **(Gerow et al., 1993)** Early descriptions **(Carter et al., 1949; Mavor, 1956; Bradley, 1973)** were of the anterior compartment, followed by descriptions of the lateral compartment **(Reszel et al., 1963)** and posterior compartment which was subdivided into deep and superficial compartments. **(Mense & Simons, 2001; Subotnick, 1991)** A fifth compartment has also been described, containing only the tibialis posterior muscle. **(Gerow et al., 1993; Travell & Simons, 1992; Rorabeck, 1986)**

Fascia is firmly attached proximally at the knee. Distally, the three sets of retinacula strengthen it. The interosseous membrane, bone, and septa complete the compartments into closed spaces with the exception of perforation of the fascia by blood vessels and nerves and the exit of the tendons distally. Bradley **(Bradley, 1973)** draws a fascinating analogy that "This anatomical situation would seem to be analogous in many respects to the cranial vault and its contents." Maykel, incidentally, reports on the successful treatment of 12 cases of compartment syndromes of the stomatognathic system using applied kinesiology methods. **(Maykel, 2002)**

The tibialis posterior compartment is described by Davey et al. and confirmed by Travell & Simons **(Travell & Simons, 1992; Davey et al., 1984)** Follow-up studies have confirmed this fifth compartment. **(Pell et al., 2004; Levangie & Norkin, 2001; Edwards & Myerson, 1996)** The initial study was done by injecting radiopaque dye into the tibialis posterior muscle of amputated legs. The dye failed to traverse the osseofascial boundaries of the tibialis posterior muscle, even though the leg was massaged for several minutes prior to roentgenographs being taken. In cadavers they injected lubricating jelly into the tibialis posterior muscle to increase the pressure and then did a fasciotomy, revealing that the pressure was reduced with the procedure. Finally, in human studies, they did before and after pressure measurements in the tibialis posterior muscle and found it to be independent of the deep posterior compartment.

The superficial posterior compartment contains the soleus and gastrocnemius muscle bellies. The deep posterior compartment encloses the posterior tibial, popliteus, flexor hallucis longus, and flexor digitorum longus muscle bellies. The ability of the examiner to specifically isolate these muscles during manual muscle testing examination gives a distinct advantage in the assessment of these compartments and their functional status.

Compartment syndrome etiology can be grouped into three areas. **(Levangie & Norkin, 2001; Platzer, 1992; Bradley, 1973)** Group 1 is idiopathic, associated with strenuous muscle activity and often referred to as the exercise form, or exertional compartment syndrome. **(Edwards & Myerson, 1996)** Group 2 is traumatically induced, including fractures, sprains, and other injuries of the soft tissue. Group 3 is interruption of major vascular supply, including vascular injuries, disease, and thromboembolism. In other than the vascular group of compartment syndromes, the dorsalis pedis pulse is most often present. Group 1 must be differentially diagnosed from shin splints, which is often loosely termed, not being limited to musculotendinous inflammation. Another common cause of a compartment syndrome is external compression of the leg from a cast or anti-shock trousers. **(Lutz & Goodenough, 1989)**

Lueck and Ray **(Lueck & Ray, 1972)** point out that skeletal muscle is able to produce energy during periods of anoxia by the anerobic cycle. This creates an oxygen debt that is replenished later by the local factors that increase local blood flow. However, if this is not replaced, the high concentration of lactic acid and anoxia lead to muscle necrosis. Mense and Simons **(Mense & Simons, 2001)** show that after approximately 4 minutes of hypoxia, muscle spindle discharge become irregular and then stop suddenly. The muscle spindle discharges recovered approximately 3 minutes after restoration of the oxygen supply. But after a second period of hypoxia, the discharge from the muscle spindles failed sooner and the recovery was incomplete. Apparently, the muscle spindle receptor was damaged by the prolonged hypoxia.

When there are multiple muscles in a compartment, they are not all involved to the same degree. In the anterior compartment this appears to be the result of differential strain to the muscles and the variable blood supply. During exercise, greater stress is placed upon the tibialis anterior muscle and the extensor hallucis longus and extensor digitorum longus muscles. The development of muscle necrosis appears to be individual muscle ischemia. Muscles that have the poorest blood supply and are most active during exercise are most involved. **(Mense & Simons, 2001)**

There is no consensus of the exact etiology of compartment syndrome. Pell et al. and Brody **(Pell et al., 2004; Brody, 1980)** associate lateral compartment syndrome with extended pronation in runners with excessively mobile ankles. The pain is in the general lateral ankle area, and the individual often reports his ankle "gives out." It is known that muscles are more active in extended pronation than in the normal foot. **(Snook, 2001)** Correction of extended pronation and its associated muscle dysfunction can be a step toward preventing compartment syndrome. Bradley **(Bradley, 1973)** considers that compartment syndrome is "…but one stage in the continuum of muscular disorders….Until the critical point for the microcirculation is surpassed, lesser and more common forms of this disease are produced."

Travell & Simons **(Travell & Simons, 1992)** suggest there is a strong possibility that in the muscles prone to developing compartment syndrome, myofascial trigger points may make a significant contribution.

The mechanism of injury frequently reported by patients and the literature is a swelling of overworked muscles after unaccustomed heavy exercise or activity. **(Logan, 1995)** This produces a loosening of muscular fascia from the periosteum of the tibia and/or fibula, producing excessive inter- and intra-compartmental pressure. It appears that almost anyone with sufficient exercise can be a candidate for a compartment syndrome. It can appear in an individual's early athletic training, or in the highly trained athlete.

Edwards and Myerson **(Edwards & Myerson, 1996)** describe exertional compartment syndrome in which the tissues confined in the 5 compartments of the leg are adversely influenced by increased pressure which alters circulation and the viability of the tissues. The increased osmotic pressure and muscle swelling raises the intra-compartmental pressures. Swelling and pain may produce sensory deficits or paresthesias and motor weakness or paralysis related to ischemic changes within the compartment. With particular emphasis on the anterior compartment, the first motor signs are loss of strength in the extensor hallucis longus and tibialis anterior muscles. **(Gerow et al., 1993)** Characteristically a chronic anterior compartment syndrome demonstrates weakness of the dorsiflexors of the foot, **(Gibson et al., 1986)** making the use of the manual muscle test a valuable tool in the early stage of this syndrome. Pressure on the nerves within the compartment may result in sensory disturbances as well as motor loss which, in severe cases, may result in foot drop. Onset is gradual and associated directly with the intensity of the exercise that the subject undergoes and is usually relieved by cessation of the activity and rest.

Examination for exertional compartment syndrome must therefore take place after the patient has exercised. Muscle weakness will be evident and tenderness and paresthesias to light touch will be present in the involved muscles in severe cases. Neural and arterial occlusion may produce serious complications; assorted diagnostic tests must be conducted for differential diagnosis, with measurement of intra-compartmental pressure a necessity to confirm the diagnosis of exertional compartment syndrome. **(van den Brand et al., 2005; Edwards & Myerson, 1996)**

Muscle trauma can cause an acute compartment syndrome. Straehley and Jones **(Straehley & Jones, 1986)** report on a fifty-one year old male who felt a "pop" in his calf while playing softball and running bases. Increasing pain required examination within four hours, which revealed elevated compartmental pressure. Fasciotomy was performed. A large tear in the gastrocnemius was identified. Eighteen months post-surgery, there were no significant problems with the leg. Surgery in this case revealed specific muscle trauma as cause of the problem. They go on to point out that "the fact that many of the chronic exertional problems occur in well-trained athletes suggests that muscle hypertrophy may act to decrease the relative space available for the obligatory muscle swelling that occurs following exercise." This has been observed from continuous pressure monitoring of the anterior compartment in race walking, revealing that pressure rises as speed is increased. **(Straehley & Jones, 1986)**

Although compartment syndrome is relatively rare, it is mandatory that it be considered in differential diagnosis of leg pain. The acute syndrome is a surgical emergency. Delay in appropriate treatment allows muscle necrosis and permanent impairment to develop. Early muscle damage occurs within four to six hours, and significant irreversible damage will be present in eighteen hours. **(Pell et al., 2004; Lueck & Ray, 1972)** 90% of patients will make a full recovery within 24 hours if the ischemia is present for less than 4 hours; if ischemia lasts up to 8 hours, only 50% will make a full recovery. **(Gerow et al., 1993)**

Differential diagnosis includes cellulitis, thrombo-phlebitis, tibial stress fractures, osteomyelitis, tenosynovitis, and the group that constitutes shin splints (musculotendinous inflammations). A key diagnostic factor is that in these inflammatory or infectious conditions, a true motor weakness, muscular paralysis, or sensory loss may not always be found. **(Pell, 2004)**

Chronic Form

Compartment syndrome is characterized symptomatically by pain upon exertion. The activity is usually considerable, such as when marching or running. Ordinary walking can sometimes initiate pain, which may be burning, cramp-like, piercing, or contracting. It usually disappears after a few minutes of rest. During the painful period, active dorsiflexion of the ankle is often difficult. The pain resembles the often loosely diagnosed condition of shin splints.

Objective findings may include tenderness and tautness over the compartment. Fascial defects were found in nearly 60% of a group (n=61) studied by Reneman. **(Reneman, 1975)** Muscle abnormality is found in a majority of patients with compartment syndrome where fasciotomy is performed; suggesting that conservative management of chronic compartment syndrome should focus on muscle dysfunctions. The dorsalis pedis pulse is palpable with disturbance of any of the compartments. **(Moore & Friedman, 1989)**

An important diagnostic method is measuring the intracompartmental pressure during exercise. Several methods for doing this have been reported. **(Shadgan et al., 2008)**

Acute Compartment Syndrome

In the acute form there is a progressive course, and structures within the compartment are subject to complete or partial destruction. In the chronic form the intermuscular pressure returns to the initial value fairly quickly after exercise (9-15 minutes). In the acute form the increased pressure remains for a much longer period of time. **(Reneman, 1975)** Pain usually starts immediately after intensive use of the lower limbs. Active or passive movement of the foot increases the pain. Generally the severe pain comes only after the cessation of the physical activities. In the acute form of the anterior compartment syndrome, loss of motion and motor impairment varies from slight limitation of toe and ankle extension to complete foot drop. Eversion of the foot may be limited or impossible, and there may be involvement of the deep peroneal nerve causing weakness of ankle or toe extension. The pain of a compartment syndrome may become severe during the night. This is thought to relate with the diminished blood flow that occurs during rest.

Reszel et al **(Reszel et al., 1963)** point out that "the swelling and tenderness of the leg, fever, and the normal arterial pulses rarely make the first examiner suspect ischemia. The initial diagnosis in practically all cases is cellulitis or thrombophlebitis."

The acute syndrome must be treated as soon as possible by fascial release or there may be permanent damage to the muscles in the compartment. In the acute form, the blood flow is severely disturbed, and the syndrome leads to necrosis of one or several muscles in the majority of cases. Reneman **(Reneman, 1975)** states, "It is not known why patients with the compartmental syndrome of the leg had increased or total intermuscular pressure at rest or higher values than normal upon exercise. It is difficult to understand why in one patient the total intramuscular pressure diminishes soon after exercise (chronic syndrome) whereas in the other it remains above the critical level for a long period of time (acute syndrome)."

Femoral or popliteal artery entrapment may simulate a compartment syndrome. (**Pillai et al., 2008; Bell, 1985**) Pillai et al. and Bell report on cases of popliteal artery entrapment syndrome simulating chronic compartment syndrome. The pain of the two conditions is caused by a muscle ischemia after exercising. At rest there are no positive physical findings. Differential diagnosis in this case was done by measuring the compartment pressure and by an arteriogram. The arteriogram is done after exercise creates pain. In this case report the arteriogram was negative in a neutral position, but showed much narrowing with the foot dorsiflexed. This evaluation can also be done on a non-invasive basis by Doppler ultrasound.

Treatment

Treatment of the acute compartment syndrome is fasciotomy: there is a limited time span in which it can be done and save muscular function. Damage to the muscle begins within four to six hours and is irreversible in eighteen hours. (**Mense & Simons, 2001; Edwards & Myerson, 1996**) Even with fasciotomy, the literature does not report a very good success rate. (**Wall et al., 2007; Bradley, 1973**)

Generally, the condition is initially viewed with an effort toward conservative care. Mubarak et al. (**Mubarak et al., 1989**) warn against unnecessary fasciotomy because it creates muscle weakness. Garfin et al. (**Garfin et al., 1981**) demonstrated this in dogs, showing a 15% muscle strength loss after fasciotomy, but the effectiveness of the surgery was shown by a 50% intracompartmental pressure drop. Deep but careful massage (fascial release, percussion, connective tissue techniques to the area) plus specific stretching for the restricted fascial compartments are good initial therapeutic choices. Manual muscle tests will effectively guide the clinician toward the specific myofascial dysfunctions in this complex scenario.

If applied kinesiology treatment methods are applied to an area which is particularly tense, painful or restricted, the clinician may choose to utilize applied kinesiology approaches which release excessive tone; or one of several versions of strain-counterstrain which releases excessive tone and modifies pain; or myofascial release methods such as the fascial flush, percussion, muscle spindle cell, proprioceptive taping or Golgi tendon organ technique; or if joint restriction is involved and on challenge of the articulations around the compartment proves to be a primary factor in the dysfunctional state of the soft tissues, then an HVLA thrust to a remote area causing the local muscle dysfunction should be employed.

The methods of applied kinesiology assessment and treatment have seamlessly merged with a variety of other methods, techniques and modalities providing the modern clinician with an abundant set of resources with which to handle both acute and chronic somatic dysfunctions.

Fasciotomy in cases of compartment syndrome should only be done after objective tests reveal an abnormal intracompartmental pressure elevation. (**Wall et al., 2007; Mubarak et al., 1989**) Post-operative care involves keeping patients off their feet and their legs elevated following surgery for 2-3 days. Most patients are walking unassisted 2 days and may run or jog in 3 weeks. (**Konstantakos et al., 2007**)

Time is of the essence in this condition. When the patient is seen early, before there is any loss of muscle function, it is reasonable to try to alleviate the pressure by rest and elevation and applied kinesiology treatment. However, with evidence of increasing intracompartmental pressure or with the appearance of a neurologic or muscular deficit, it is recommended that the compartment be explored as early as possible. The muscular deficit present in an acute compartmental syndrome places the applied kinesiologist in an excellent position to recognize the possibility of elevated compartment pressure. Muscles that test weak with manual muscle testing should respond to the usual applied kinesiology techniques. If the muscle or nerve is ischemic, there will be no response, giving a positive indication of compartment syndrome possibility.

Restless Leg Syndrome (RLS)

Restless legs syndrome (RLS) is a common cause of severe insomnia. The treatment of insomnia, when successful, is one of the most satisfying clinical successes in an applied kinesiology practice. "How are you sleeping?" is one of the questions we will often ask new patients. We have found over the years that if patients cannot get a good night's sleep, they are not going to fully recover their health.

Insomnia is a term used to describe the more than 80 million Americans who routinely have trouble falling asleep or staying asleep. For anyone who has ever experienced a few sleepless nights in a row, a feeling of desperation sets in as you struggle to function during the following day. Between 30-40% of the U.S., European, and Japanese populations over the age of 15 reports they've experienced insomnia at least occasionally, (**Ohayon, 2010**), with higher mortality in men related to sleep disturbances. (**Rod et al., 2011**)

It is likely that more than 100,000 motor vehicle crashes are caused annually in the United States by driving while drowsy. Major disasters such as Three Mile Island, Exxon Valdez, Bhopal and Challenger were all officially attributed to sleepiness-related impaired judgment in the workplace. (**National Commission on Sleep Disorders Research, 1992**)

RLS is a neurological sensory/movement disorder affecting 5-15% of the general population. (**Phillips et al., 2000**) A survey conducted in May 2005 of some 1,500 adults reports almost 10 percent of the respondents suffered from restless legs syndrome, and more women than men have it. (**Guillemiault & Framherz, 2005**) Up to 65% of patients with RLS use complementary and alternative medicine. (**Cuellar et al., 2004**) For this reason it is essential that primary contact practitioners be aware of this condition to insure its efficient management. (**Stupar, 2008**)

RLS is characterized primarily by a vague and difficult-to-describe unpleasant sensation in the legs. This discomfort appears primarily during periods of inactivity, particularly during the transition to sleep in the evening.

Patients often have difficulty in describing the unpleasant sensations; they rarely use conventional terms of discomfort such as 'numbness, tingling or pain,' but rather bizarre terms such as 'pulling, searing, drawing, crawling, shimmering or boring,' suggesting that RLS sensations are unlike any experienced by unaffected individuals. These distressing sensations are typically relieved only by movement or stimulation of the legs. It is often difficult for the patient to describe because there is no correlating sensation in normal physiology. It is not the same as the leg going to sleep, such as when sitting for a prolonged period with the knees crossed. It will often be described as a creeping sensation, or that there are small worms in the muscles. The legs may feel tired, heavy, and weak. A common corollary is that the patient feels a need to move his legs. Often he will get up several times during the night to walk around, jump, massage the legs, and otherwise try to obtain relief from the condition. In severe cases, there may be great restriction on the amount of sleep obtained.

"Burning feet" and the "restless legs syndrome" may be caused by entrapment of the tibial nerve at the tarsal tunnel. Symptoms from this condition develop from walking or prolonged standing, and may possibly persist during bedrest. The pain sometimes radiates from the foot to the ankle, or even into the calf. **(Staal et al., 1999)**

The Restless Legs Syndrome Foundation puts the condition this way:

> "Restless Legs Syndrome is an overwhelming desire to move the legs usually caused by uncomfortable or unpleasant sensations in the legs. The sensation can occur during periods of inactivity and become more severe in the evening and night. RLS may often cause difficulty staying or falling asleep, which leads to tiredness or fatigue. Up to 8% of the US population may have this neurologic condition. Many people have a mild form of the disorder, but RLS severely affects the lives of millions of individuals." **(http://www.rls.org)**

Many different techniques have been found by patients for relief: walking about, stomping the feet, rubbing, squeezing or stroking the legs, taking hot showers or baths, or applying ointment, hot packs or wraps to the legs. Although these treatments are effective while they are being performed, the discomfort usually returns as soon as they become inactive or return to bed to try to sleep. The motor restlessness often appears to follow a circadian pattern, peaking between midnight and 4:00 am. **(Trenkwalder et al., 1999)** The prevalence of depression and anxiety found associated with RLS is felt to be secondary to the RLS. The relationship between RLS and insomnia generally and psychiatric conditions are bi-directional: depression may cause insomnia, and insomnia may cause depression. **(Picchietti & Winkelman, 2005)**

The movement can happen just at night, when the jiggling may drive a spouse or significant other up the wall, and/or it can occur during the day, when it will drive everyone else up the wall with the incessant tapping or twitching. In a small percentage of people, the sensations of restlessness and twitching may also be experienced in the arms. The automatic jerks of the legs can disturb their sleep enough to produce a feeling of fatigue the morning after but not enough to fully awaken them. The sleeper's mate, however, is usually fully aroused by their partner's unintended kicks. Both of these people will be in your office asking about this condition together, and the tired expression on their faces and the circles around their eyes can be a clue for you to investigate this problem in a patient.

Many patients with RLS take tranquilizers, muscle relaxers, and over the counter sleep aides to get them to sleep. But most people who use these medications never go into deep restorative sleep, the deep delta-wave sleep that allows the body to get its physiological rest and restorative repair.

One of the most obvious and immediate effects of RLS and its associated insomnia is the increased risk of accidents. As reported in Business Week, "Studies show that someone who has been awake for 24 hours has the same mental acuity as a person with a blood alcohol level of 0.1, which is above the legal limit for driving in most states." **(Business Week, 2004)** But when you consider someone who is a health care worker, pilot, or law enforcement officer, the effects can be catastrophic. Some 39% of health-care workers report that they've had a "near miss or accident" at work due to fatigue in the last year. Further, sleep disorders cost the nation about $45 billion every year in lost productivity, health care and motor vehicle accidents. **(Business Week, 2004)**

Recent studies suggest there may be a susceptibility gene locus in RLS, which would explain why RLS is often familial. **(Desautels et al., 2005)** RLS is commonly seen in pregnancy, hemodialysis or peritoneal dialysis for renal failure and iron-deficiency anemia. Its relationship with iron metabolism abnormalities has led to studies indicating that RLS is associated with abnormal iron metabolism within the central nervous system; **(Allen & Earley, 2001)** the iron deficiency may even require treatment when ferritin levels are normal. **(O'Keeffe, 2005)**

Studies relating this problem with nutritional deficiencies of riboflavin, niacin, and B6 have not been explored in the literature. A search of PubMed showed 10,789 papers on riboflavin; 5,591 papers on niacin; 5,968 papers on niacinamide; and 9,070 papers on vitamin B6. A series of searches was performed, using the terms "restless legs syndrome," "nocturnal myoclonus," "riboflavin," "niacin," "niacinamide," and "vitamin B6". None of these papers explicitly correlated these vitamins with RLS.

One of the conclusions of a case series report by Cuthbert **(Cuthbert, 2007)** was that a nutritional relationship between RLS and these three vitamins may be present.

Recent functional neuroimaging studies have identified thalamic, red nucleus and brainstem involvement in the generation of periodic limb movements in patients with RLS. One PET study found reduced dopamine 2 binding in the caudate and putamen. A subcortical origin of RLS is supported by transcranial magnetic stimulation studies and the successful treatment of the patient cohort in this case series, i.e. RLS has physiological causes . **(Tergau et al., 1999)**

Demonstration of a continuous hypersensitivity to pinprick, but not to light touch, confirms a sensory component to RLS and may explain the efficacy of AK, nutritional, and opiate medications in RLS. **(Stiasny-Kolster et al., 2004)** Patients with RLS will often report that walking, rubbing or massaging the legs, or doing deep knee bends or finger to floor stretches can bring relief, but only briefly. This suggests that there is a structural component to RLS as well.

Medical Treatment of RLS employs a variety of medications. Generally, they choose from dopaminergics, benzodiazepines (central nervous system depressants), opioids, and anticonvulsants. Dopaminergic agents, largely used to treat Parkinson's disease (like Sinemet and Levodopa), have been shown to reduce RLS symptoms and are considered the initial treatment of choice. Good short-term results by treatment with levodopa plus carbidopa have been reported, although patients usually develop augmentation, meaning that symptoms are reduced at night but begin to develop earlier in the day than usual. Dopamine agonists such as pergolide mesylate, pramipexole, and ropinirole hydrochloride may be effective in some patients and are less likely to cause augmentation. The dopaminergic anti-Parkinsonian medications are a particularly common treatment medication now. **(Desautels et al., 2005)**

Benzodiazepines (such as clonazepam and diazepam) may be prescribed for patients who have mild or intermittent symptoms. These drugs help patients obtain a more restful sleep but they do not fully alleviate RLS symptoms and can cause daytime sleepiness. Because these depressants also may induce or aggravate sleep apnea, they should not be used in people with this condition.

For more severe symptoms, opioids such as codeine, propoxyphene, or oxycodone are prescribed for their ability to induce relaxation and diminish pain. **(Limbaugh, 2003)** Side effects include dizziness, nausea, vomiting, and the risk of addiction.

Anticonvulsants such as carbamazepine and gabapentin (Neurontin) are also considered useful for some patients, as they decrease the sensory disturbances (creeping and crawling sensations). Dizziness, fatigue, and sleepiness are among the possible side effects.

The medical and drug company literature shows that no one drug is effective with RLS. What may be helpful to one individual may actually worsen symptoms for another. In addition, medications taken regularly may lose their effect, making it necessary to change medications periodically.

For one of the patients in the study by Cuthbert, **(Cuthbert, 2007)** Sinemet produced urinary incompetence and systemic muscle weakness as well as blurred vision, light-headedness and depression. These kinds of side effects from the medications patients are taking are not uncommon in our experience. 6 of the patients treated successfully with the applied kinesiology protocols in the study by Cuthbert were taking Sinemet and Levodopa (for Parkinson's disease), or muscle relaxers, sedatives, or painkillers for control of this condition. Drugs designed for the treatment of other diseases (Parkinson's disease and seizures) for the treatment of the RLS is not uncommon, and these are examples of a worrisome tactic of Pharmaceutical Companies called "Off-Label Marketing". **(Critser, 2005)**

On May 5, 2005, the FDA approved the first ever drug for treatment of restless legs syndrome: ropinirole (Requip).

The side effects were announced:
"Possible side effects:

SIDE EFFECTS that may occur while taking this medicine include feeling of warmth, dry mouth, sweating, weakness, fatigue, dizziness, drowsiness, lightheadedness, stomach pain, heartburn, gas, nausea, or vomiting. If they continue or are bothersome, check with your doctor. CHECK WITH YOUR DOCTOR AS SOON AS POSSIBLE if you experience swelling of ankles, feet, or hands; unusual fatigue or tiredness; unusual pain; unusual muscle movement; loss of appetite; fast or irregular heartbeat; falling asleep during daily activities; mental or mood changes; impotence; trouble breathing; sore throat; or vision changes. CHECK WITH YOUR DOCTOR IMMEDIATELY IF YOU EXPERIENCE fainting, or chest pain. An allergic reaction to this medicine is unlikely, but seek immediate medical attention if it occurs. Symptoms of an allergic reaction include rash, itching, swelling, severe dizziness, or trouble breathing. If you notice other effects not listed above, contact your doctor, nurse, or pharmacist.

Drug interactions

Drug interactions can result in unwanted side effects or prevent a medicine from doing its job. Use our drug interaction checker to find out if your medicines interact with each other. ADDITIONAL MONITORING OF YOUR DOSE OR CONDITION may be needed if you are taking ciprofloxacin, medicines for anxiety (such as diazepam), medicines for mental or mood problems (such as risperidone), medicines for depression (such as fluoxetine), digoxin, theophylline, levodopa, estrogens, phenothiazines (such as chlorpromazine), butyrophenones (such as haloperidol), thioxanthenes (such as thiothixene), or metoclopramide."

The cost of Requip is $219.96 for 100 tablets (approximately 2 months supply).

Non-pharmacological approaches include decreasing caffeine and alcohol intake, education on sleep hygiene and relaxation methods, moderate exercise and nutritional supplements. **(Stupar, 2008)**

AK treatment for RLS

There is no laboratory test that can identify RLS, and the condition cannot be diagnosed by the physician other than by symptoms reported by the patient. This suggests a functional problem (one of the 5-factors of the IVF) may be the root cause of the disorder.

In each of the cases successfully treated in a case-series report by Cuthbert (n=23), **(Cuthbert, 2007)** pelvic subluxations and weakness of pelvic and leg muscles on MMT were found. Several patients were obese with an anterior sacral base subluxation that increased the lumbo-sacral lordosis. Each of the cases in the series of patients with RLS responded to chiropractic treatment and Cataplex G ™ **(Standard Process Labs, 2012)** vitamin supplementation (a single pill-form combination of riboflavin, niacin, B6, folic acid, PABA, choline, inositol, biotin, and betaine), and this combination of treatment elements was the common factor among all the cases successfully treated (n = 23) for RLS. **(Table 1)**

As part of a thorough whole-body examination, most of the muscles in the body were tested, and the relationship of muscular inhibition on MMT to the patients' primary complaints was explained. Because AK examinations and treatments are interactive, the patient is educated about the relationships between the different areas of the body and how one dysfunction may be creating problems elsewhere. During the course of treatment of patients who presented with more conventional chiropractic problems, the irritation of the RLS was revealed.

The commonality among all the patients with RLS was the finding of facilitation on MMT of inhibited muscle(s) in the legs with oral nutrient testing of Cataplex G. **(Standard Process Labs, 2011)**

Another important part of the successful treatment of RLS in these patients was the reduction of pain in the legs that was present in 14 of the patients. **(LBLP in Table 1)** Treatment of this factor by standard AK methods to the 5-factors of the IVF, including cranial, pelvic category, vertebral, and muscular corrections to the structures found disturbed in the patient proved successful in reducing the low back and leg pain in these cases. In terms of the structural factors, there was no one recurring specific found in a review of these cases with RLS. The heterogeneity of the chiropractic presentation – each patient showing unique and peculiar features – was the rule here.

Eight of the patients who had problems with insomnia, previous to or independent of their RLS, were treated using standard AK evaluation procedures for insomnia problems. **(Walther, 2000)** The patient would be tested in the examination room with the lights off, and the factor that corrected the global inhibitions on MMT with the lights off would be treated.

Additionally, patients who only had trouble falling into sleep often responded to supplementation with Calcium Lactate (SP); **(Goodheart, 1998)** patients who had trouble staying asleep and awoke frequently without cause often responded to Cataplex B (SP); patients with depression, irritable bowel syndrome, and insomnia often improved with 5-HTP (NW) supplementation. **(Nutri-West)** (5-HTP should be taken with 4 ounces of fruit juice, as insulin is required to carry the 5-HTP past the blood brain barrier.) These are consistent findings in an applied kinesiology practice, and were a part of the improvement in the overall picture of some of these patients who had the RLS also.

Twelve of the patients in this case series showed signs and symptoms of adrenal stress disorder (postural hypotension, paradoxical pupillary reaction, Rogoff's sign, weakness of the sartorius, gracilis, or tibialis posterior muscles that showed strengthening upon TL to the relevant reflex, the ligament stress reaction, etc.). In our practice, nutritional support and lifestyle counseling is frequently used for support of the hypoadrenic and/or overstressed patient.

According to several authorities -- Dr. Royal Lee in *Vitamin News* from 1934 and 1952; Dr. George Goodheart in his articles on vitamin B deficiencies; Dr. Wally Schmitt in *Compiled Notes on Clinical Nutritional Products* (1990); and Dr. Philip Maffetone in *Complimentary Sports Medicine* (1999) -- the signs of a vitamin "G" deficiency (or riboflavin, niacin, and B6) include:

1) Cardiovascular -- tachycardia, extra ventricular beats, increased 1st heart sound with a long silence after 1st sound (increased time of diastolic ventricular filling), angina pectoris, and pre-myocardial infarction

2) Psychological -- Excessive worry, apprehension, moodiness, depression, suspicion

3) Digestive -- Insufficient stomach acid production and excess alkalinity, spastic gall bladder

4) Liver -- Cirrhosis of the liver and loss of fat metabolism activity, deficient formation of Yakitron, a physiologic anti-histamine

5) Neurological -- Insufficient acetylcholine activity and cholinesterase activity (for breaking down acetylcholine and for recycling choline), restless, jumpy, or shaky legs, body or limb jerks upon falling asleep, can hear heartbeat on pillow

6) Skin and mucous membranes -- Cheilosis (cracking at corners of mouth), friable skin, especially on face and neck (when shaving), bright red tongue tip, strawberry tongue (purple), loss of upper lip (thin upper lip), irritated mucous membranes of the rectum, vagina, and conjunctiva (frequent tears), excessive oil on face and nose, roughness, cracking and exfoliation of the soles of the feet, and psoriasis

7) Visual -- Burning or itching of eyes, photophobia, blepharospasm, blood shot eyes due to capillary engorgement, seeing only parts of printed words (circumcorneal injection), pallor of the temporal half of optic disc, transient ischemia of retina – like looking through a fish bowl

8) Endocrine – excess estrogen and menstruation, cystic mastitis or gynecomastia, premenstrual tension, and excessive adrenal function.

Dr. Royal Lee, an excellent chronicler of deficiency signs and symptoms in relation to vitamins and minerals, listed the following deficiency signs for the B vitamins that make up the Cataplex G formula, **(Lee, 2007)** which may explain why "Cataplex G" was of value in the cases of RLS reported:

Signs and symptoms for vitamin B2 (riboflavin) deficiency:
a) Myelin degeneration
b) Incoordination
c) Loss of strength in arms or legs
d) Central neuritis, symptoms resembling degeneration of spinal cord

Signs and symptoms for vitamin B3 (niacin or nicotinic acid) deficiency:
a) Myelin degeneration (motor and sensory)

Table 1:
Restless Legs Syndrome Patient List (n=23)

Patient Age and Sex	Problems with Insomnia? (I) Adrenal Stress Disorder? (ASD) Low Back or Leg Pain? (LBLP)	Intensity of RLS: Severe Distracting Moderate Duration of RLS pre-treatment?	Treatment time until resolution of RLS (days)	Did RLS return? (# Months): Maintenance dose of Cataplex G needed? (Yes: No)
80; M	I, LBLP	D: 7 years	12	12: N
50; F	I, ASD	D: (lifetime)	7	17: Y
63; M	LBLP	S: (lifetime)	15	N: N
61; F	I, ASD	M: (3 years)	13	N: Y
77; F	ASD, LBLP	D: (10 years)	4	Every 6 months: Y
69; F	ASD, LBLP	S: (10 years)	8	N; N
68; F	I, LBLP	S: (lifetime)	11	Every 4 months: Y
49; F	ASD	S: (10 years)	11	Every 2 months: Y
64; F	I, ASD	D: 5	6	N: Y
62; F	ASD, LBLP	M: (12 years)	8	N: N
68; F	I	D: (6 years)	7	N: N
47; F	ASD	D: (lifetime)	10	N: N
57; M	LBLP	M: (20 years)	40	N: N
44; F	ASD	S: (10 years)	9	N: N
80; M	LBLP	S: (2 years)	12 (This patient resolved his RLS with Cataplex E2)	N: Y
73; M	LBLP	D: (1 year)	5	Every 4 months: Y
61; M	ASD, LBLP	D: (3 years)	2	Every 4 months: Y
70; F	ASD	S: (2 years); also in arms	4	N: Y
54; M	LBLP	D: (5 years)	24	N: N
66; F	I, ASD, LBLP	S: (10 years)	15	N: N
63; F	LBLP	D: (lifetime)	4	Every 2 months: Y
80; M	I	S: (10 years)	28	N: Y
93; M	LBLP	S: (6 months)	3	N: N

b) Headaches, dizziness, insomnia, depression and impairment of memory
   c) Burning hands and feet
   d) Pain in calves
   e) Numbness and weakness of extremities
   f) Difficulty in walking
   g) Absent knee jerk

Signs and symptoms for vitamin B6 (pyridoxine) deficiency:
   a) Severe sensory neuritis
   i. Numbness and tingling in hands and feet
   ii. Hyperesthesia

In the early investigations of the vitamin B/G complex, the investigators called vitamin B/G an anti-neuritic vitamin. All nerves require B/G vitamins for normal function. Pyridoxine (B6), riboflavin (B3), and folate are especially important. Supplementation should also include B12, biotin and pantothenic acid.

As an example of this fact: inositol has been used in Europe for the past 30 years to lower cholesterol and improve nerve function in diabetics. (**Bloomgarden, 1997; Mayer & Tomlinson, 1983**) Because of its role in cell-membrane function, it has also shown beneficial effects for depression and general neurological function.

Pyridoxal phosphate (vitamin B6) has been implicated as critical in lipid metabolism because its deficiency causes myelin degeneration in man. (**Sauberlich & Canham, 1980**) Impairment of temperature perception is present in a high percentage of RLS patients, and the sensory deficits are at least in part caused by small nerve fiber neuropathy. (**Schattschneider et al., 2004**)

Practically all of the neurotransmitters in the brain are metabolized with the aid of vitamin B6, including dopamine, norepinephrine, serotonin, GABA, histamine, acetylcholine, insulin, growth hormone, follicle-stimulating hormone, luteinizing hormone, aldosterone, glucagon, cortisol, estradiol, testosterone, and epinephrine. Vitamin B6 is also required for the conversion of tryptophan to niacin. Serotonin, a critical factor for normal sleep patterns, is derived from pyridoxal-5 phosphate and 5-hydroxytryptophan, and so another possible improvement in nighttime RLS may have an explanation.

Since dopamine is synthesized with the aid of vitamin B6, and dopaminergic drugs like levodopa and dopamine agonists are the first line of medical treatment choice in idiopathic RLS, the effect of Cataplex G may have another corroboration. However with the use of dopaminergic drugs, augmentation and rebound phenomena are consequences of this type of medical treatment, and must be carefully monitored in long-term treatment.

Niacin has long been used for cardiovascular conditions, especially those involving lipid metabolism. It may also help Raynaud's disease (excessive blood vessel constriction due to cold and symptoms of intermittent claudication) caused by insufficient blood supply to the calf muscles while walking.

According to Carlton Fredericks, Ph.D, "Niacin restores the electrical charge to red blood cells so they don't aggregate and thereby are able to pass through small blood vessels in single file." Deficiency of niacin has been reported to cause weakness, dry skin, lethargy, headache, irritability, loss of memory, depression, insomnia, delirium, and disorientation. Unless the nerve damage or degeneration has gone too far in cases of RLS, the administration of these nutritional factors for healing of the nerve provided quick relief.

Numerous studies have shown that the elderly (18 of the patients in the Cuthbert study were members of the 'senior set') are usually deficient in the nutrients contained in Cataplex G, as well as others. (**Lowenstein, 1982**)

Further study of the relationship between dopamine, central nervous system and vitamin B metabolism will be of great interest and value to neurophysiologists, neurochemists, neuropharmacologists and, of course, to patients with RLS. In the 23 patients with RLS successfully treated in the report, nutritional supplementation was a critical factor.

References

- Abshire D. Natural Running: the simple path to stronger, healthier running. VELO Press: Boulder; 2010.
- Adirim TA, Cheng TL. Overview of injuries in the young athlete. Sports Med. 2003;33(1):75-81.
- Akeson WH, Amiel D, Mechanic GL, Woo SL, Harwood FL, Hamer ML. Collagen cross-linking alterations in joint contractures: changes in the reducible cross-links in periarticular connective tissue collagen after nine weeks of immobilization. Connect Tissue Res. 1977;5(1):15-9.
- Allen R, Earley C. Restless legs syndrome: a review of clinical and pathophysiologic features. J Clin Neurophysiol. 2001;18(2):128-147.
- Amiel D, Akeson WH, Harwood FL, Frank CB. Stress deprivation effect on metabolic turnover of the medial collateral ligament collagen. A comparison between nine- and 12-week immobilization. Clin Orthop Relat Res. 1983;(172):265-70.
- Andrish JT, Bergfeld JA, Walheim J. A prospective study on the management of shin splints. J Bone Joint Surg Am. 1974;56(8):1697-700.
- Bandy J. Posterior shin splints. ICAK USA Collected Papers. ICAK USA: Shawnee Mission, KS; 1978;Summer Meeting:9-13.
- Barron JL, Noakes TD, Levy W, Smith C, Millar RP. Hypothalamic dysfunction in overtrained athletes. J Clin Endocrinol Metab. 1985;60(4):803-6.
- Baumhauer JF, Alosa DM, Renström AF, Trevino S, Beynnon B. A prospective study of ankle injury risk factors. Am J Sports Med. 1995;23(5):564-70.
- Bell S. Intracompartmental pressures on exertion in a patient with a popliteal artery entrapment syndrome. Am J Sports Med. 1985;13(5):365-6.
- Bergmann TF, Peterson DH. Chiropractic Technique: Principles and Procedures, 3rd Ed. Mosby: St. Louis; 2010.
- Beskin JL, Sanders RA, Hunter SC, Hughston JC. Surgical repair of Achilles tendon ruptures. Am J Sports Med. 1987;15(1):1-8.
- Bishop M, Fiolkowski P, Conrad B, Brunt D, Horodyski M. Athletic footwear, leg stiffness, and running kinematics. J Athl Train. 2006;41(4):387-92.
- Blaich RM. Ligament stretch-adrenal stress syndrome related to knee disorders. ICAK USA Collected Papers. ICAK USA: Shawnee Mission, KS; 1980.
- Bloomgarden Z. Antioxidants and Diabetes. Diabetes Care.1997;20:670-673.
- Bobath K, Bobath B: Facilitation of normal postural reactions and movement in treatment of cerebral palsy. Physiotherapy. 1964;50:246-62.
- Bourne RB, Rorabeck CH. Compartment syndromes of the lower leg. Clin Orthop Related Res. 1989;240:97-104.
- Boven LC. A simple assessment for muscle imbalance. Collected Papers International College of Applied Kinesiology: Shawnee Mission, KS;1988:1-3.
- Bradley EL. The anterior tibial compartment syndrome. Surg Gynecol Obstet. 1973;136(2):289-97.
- Brand RL, Collins MD. Operative management of ligamentous injuries to the ankle. Clin Sports Med. 1982;1(1):117-30.
- Brantingham J, Silverman J, Deliman A, Snyder R, Wong J. Chiropractic management of Achilles tendonitis. J Neuromusculoskeletal System. 1994;2(2):52–55.
- Brinks A, Koes BW, Volkers AC, Verhaar JA, Bierma-Zeinstra SM. Adverse effects of extra-articular corticosteroid injections: a systematic review. BMC Musculoskelet Disord. 2010;11:206.
- Brody DM. Running injuries. Clin Symp. 1980;32(4):1-36.
- Bullock-Saxton JE. Local sensation changes and altered hip muscle function following severe ankle sprain. Phys Ther. 1994;74(1):17-28; discussion 28-31.
- Business Week, Jan. 26, 2004. I Can't Sleep: Insomnia and other sleep disorders are wreaking havoc on our health and taxing the economy. Drug companies see an opportunity.
- Busseuil C, Freychat P, Guedj EB, Lacour JR. Rearfoot-forefoot orientation and traumatic risk for runners. Foot Ankle Int. 1998;19(1):32-7.
- Cailliet R. Foot and Ankle Pain, 3rd Ed. F.A. Davis Co: Philadelphia; 1997.
- Carden DG, Noble J, Chalmers J, Lunn P, Ellis J. Rupture of the calcaneal tendon. The early and late management. J Bone Joint Surg Br. 1987;69(3):416-20.
- Carter S, Carter AJ. Chiropractic management of Achilles tendinopathy. Sports Exercise and Injury. 1997;3(3):108–110.
- Carter AB, Richards RL, Zachary RB. The anterior tibial syndrome. Lancet. 1949;2(6586):928-934.
- Cass JR, Morrey BF, Katoh Y, Chao EY. Ankle instability: comparison of primary repair and delayed reconstruction after long-term follow-up study. Clin Orthop Relat Res. 1985;(198):110-7.
- Chaitow L, DeLany JW. Clinical Application of Neuromuscular Techniques: Volume 2, The Lower Body. Elsevier: London; 2002.
- Chang HJ, Burke AE, Glass RM. JAMA patient page. Achilles tendinopathy. JAMA. 2010;303(2):188.
- Clanton TO, Butler JE, Eggert A. Injuries to the metatarsophalangeal joints in athletes. Foot Ankle. 1986;7(3):162-76.
- Clayton ML, Miles JS, Abdulla M. Experimental investigations of ligamentous healing. Clin Orthop Relat Res. 1968;61:146-53.
- Clement DB, Taunton JE, Smart GW. Achilles tendinitis and peritendinitis: etiology and treatment. Am J Sports Med. 1984;12(3):179-184.
- Co V, Pollard H. Management of Achilles tendon disorders. A case review. Australian Chiropractic and Osteopathy. 1997;6(2):58–62.
- Colledge NR, Cantley P, Peaston I, Brash H, Lewis S, Wilson JA. Ageing and balance: the measurement of spontaneous sway by posturography. Gerontology. 1994;40(5):273-8.
- Colville MR. Surgical treatment of the unstable ankle. J Am Acad Orthop Surg. 1998;6(6):368-77.
- Conti SF. Posterior tibial tendon problems in athletes. Orthop Clin North Am. 1994;25(1):109-21.
- Cook JL, Khan KM, Purdam C. Achilles tendinopathy. Manual Therapy. 2002;7(3):121–130.
- Cooke MW et al. Treatment of severe ankle sprain: a pragmatic randomised controlled trial comparing the clinical effectiveness and cost-effectiveness of three types of mechanical ankle support with tubular bandage. The CAST trial. Health Technol Assess. 2009;13(13):iii, ix-x, 1-121.
- Cox JS. Surgical and nonsurgical treatment of acute ankle sprains. Clin Orthop Relat Res. 1985;(198):118-26.

- Critser G. Generation Rx: How prescription drugs are altering American lives, minds, and bodies. Houghton Mifflin Company: Boston; 2005.
- Cuellar N, Galper DI, Taylor AG, D'Huyvetter K, Miederhoff P, Stubbs P. Restless legs syndrome. J Altern Complement Med. 2004;10(3):422-3.
- Cuthbert S. Restless legs syndrome: A case series report. ICAK USA Collected Papers. ICAK USA: Shawnee Mission, KS;2007:45-54.
- Cyriax J, Cyriax P. Illustrated Manual of Orthopaedic Medicine. Butterworths: London;1983.
- Dananberg HJ. Gait style as an etiology to lower back pain. In: Movement, Stability & Lumbopelvic Pain. Ed. Vleeming. Churchill-Livingstone: Edinburgh, 2007.
- Davey JR, Rorabeck CH, Fowler PJ. The tibialis posterior muscle compartment. An unrecognized cause of exertional compartment syndrome. Am J Sports Med. 1984;12(5):391-7.
- Davies JA. Peroneal compartment syndrome secondary to rupture of the peroneus longus. A case report. J Bone Joint Surg Am. 1979;61(5):783-4.
- de Mos M, van El B, DeGroot J, Jahr H, van Schie HT, van Arkel ER, Tol H, Heijboer R, van Osch GJ, Verhaar JA. Achilles tendinosis: changes in biochemical composition and collagen turnover rate. Am J Sports Med. 2007;35(9):1549-56.
- Deangelis JP, Wilson KM, Cox CL, Diamond AB, Thomson AB. Achilles tendon rupture in athletes. J Surg Orthop Adv. 2009;18(3):115-21.
- Deibert MC, Aronsson DD, Johnson RJ, Ettlinger CF, Shealy JE. Skiing injuries in children, adolescents, and adults. J Bone Joint Surg Am. 1998;80(1):25-32.
- Desautels A, et al. Restless legs syndrome: confirmation of linkage to chromosome 12q, genetic heterogeneity, and evidence of complexity. Arch Neurol. 2005;62(4):591-6.
- Deutsch G. An adrenal screen test to determine adrenal insufficiency. ICAK USA Collected Papers. ICAK USA: Shawnee Mission, KS; 1975.
- Dorland's Illustrated Medical Dictionary, 31st Ed. Saunders; 2007.
- Dubbeldam R, Buurke JH, Simons C, Groothuis-Oudshoorn CG, Baan H, Nene AV, Hermens HJ. The effects of walking speed on forefoot, hindfoot and ankle joint motion. Clin Biomech (Bristol, Avon); 2010(8):796-801.
- Dupont M, Béliveau P, Thériault G. The efficacy of antiinflammatory medication in the treatment of the acutely sprained ankle. Am J Sports Med. 1987;15(1):41-5.
- Durlacher JV. Injury susceptibility in hypoadrenia. ICAK USA Collected Papers. ICAK USA: Shawnee Mission, KS; 1977.
- Eccles JC, Eccles RM, Lundberg A. The convergence of monosynaptic excitatory afferents on to many different species of alpha motoneurones. J Physiol. 1957;137(1):22-50.
- Eckhoff DG, Kramer RC, Walkins JJ. Variation in tibial torsion. Clinical Anatomy. 1994;7:76-79.
- Edwards P, Myerson MS. Exertional compartment syndrome of the leg: steps for expedient return to activity. Phys Sportsmed. 1996;24(4):31-46.
- Elkus RA. Tarsal coalition in the young athlete. Am J Sports Med. 1986;14(6):477-80.
- Ellfeldt HJ. Problems of the foot and ankle in professional football and basketball players. In: Symposium on The Foot and Ankle, Ed. Kiene RH, Johnson KA. The C.V. Mosby Co: St. Louis; 1983.
- Evans RE. Illustrated Orthopedic Physical Assessment, 3rd Ed. Mosby; 2008
- Ferguson AB Jr. The case against ankle taping. J Sports Med. 1973;1(2):46-7.
- Ferran NA, Oliva F, Maffulli N. Ankle instability. Sports Med Arthrosc. 2009;17(2):139-45.
- Foot Levelers, Inc. http://www.footlevelers.com/. 518 Pocahontas Ave. N.E., P.O. Box 12611, Roanoke, VA 24027—2611.
- Fredberg U, Stengaard-Pedersen K. Chronic tendinopathy tissue pathology, pain mechanisms, and etiology with a special focus on inflammation. Scand J Med Sci Sports. 2008;18(1):3-15.
- Fredberg U, Hansen PA, Skinhøj A. Ibuprofen in the treatment of acute ankle joint injuries. A double-blind study. Am J Sports Med. 1989;17(4):564-6.
- Freeman MA, Dean MR, Hanham IW. The etiology and prevention of functional instability of the foot. J Bone Joint Surg Br. 1965;47(4):678-85.
- Freeman M, Wyke B. Articular reflexes at the ankle joint: an electromyographic study of normal and abnormal influences of ankle-joint mechanoreceptors upon reflex activity in the leg muscles. J Brit Surg 1967;54(12):990-1001.
- Froimson AI. Tennis leg. JAMA. 1969;209(3):415-6.
- Gabriel SE, Jaakkimainen L, Bombardier C. Risk for serious gastrointestinal complications related to the use of nonsteroidal anti-inflammatory drugs: A meta-analysis. Annals of Internal Medicine. 1991;115(10): 787-796.
- Galbraith RM, Lavallee ME. Medial tibial stress syndrome: conservative treatment options. Curr Rev Musculoskelet Med. 2009;2(3):127-33.
- Gangemi SC. Tibialis posterior torque and stability taping. Collected Papers International College of Applied Kinesiology. 2010-2011.
- Gardner MJ, Demetrakopoulos D, Briggs SM, Helfet DL, Lorich DG. The ability of the Lauge-Hansen classification to predict ligament injury and mechanism in ankle fractures: an MRI study. J Orthop Trauma. 2006;20(4):267-72.
- Garfin SR, Tipton CM, Mubarak SJ, Woo SL, Hargens AR, Akeson WH. Role of fascia in maintenance of muscle tension and pressure. J Appl Physiol. 1981;51(2):317-20.
- Garrick JG, Requa RK. Role of external support in the prevention of ankle sprains. Med Sci Sports. 1973;5(3):200-3.
- Garth WP Jr, Miller ST. Evaluation of claw toe deformity, weakness of the foot intrinsics, and posteromedial shin pain. Am J Sports Med. 1989;17(6):821-7.
- Gaymans J, Till G. The efficacy of manipulation in the management of Achilles tendonitis. J Chiropr Educa. 2003;17(1):9–10.
- Gerow G, Matthews B, Jahn W, Gerow R. Compartment syndrome and shin splints of the lower leg. J Manipulative Physiol Ther. 1993;16(4):245-52.
- Gibson MJ, Barnes MR, Allen MJ, Chan RN. Weakness of foot dorsiflexion and changes in compartment pressures after tibial osteotomy. J Bone Joint Surg Br. 1986;68(3):471-5.
- Ginanneschi F, Del Santo F, Dominici F, Gelli F, Mazzocchio R, Rossi A. Changes in corticomotor excitability of hand muscles in relation to static shoulder positions. Exp Brain Res. 2005;161(3):374-82.
- Glick JM, Gordon RB, Nishimoto D. The prevention and treatment of ankle injuries. Am J Sports Med. 1976;4(4):136-41.
- Goodheart GJ, Jr. Applied Kinesiology 1973 Research Manual, 16th Ed. Privately published: Detroit, MI; 1980.
- Goodheart GJ, Jr. Applied Kinesiology 1973 Research Manual, 12th Ed. Privately published: Detroit, MI; 1976.
- Goodheart GJ, Jr. Applied Kinesiology 1973 Research Manual, 9th Ed. Privately published: Detroit, MI; 1973.
- Goodheart GJ, Jr. The psoas muscle and the foot pronation problem. Chiro Econ. 1967;10(2).
- Goodheart GJ, Jr. Applied Kinesiology Research Manuals. Applied Kinesiology Systems, Inc: Grosse Pt Farms, MI; 1998-1964.

- Gray's Anatomy. Churchill Livingstone: London; 2007.
- Greenman PE. Principles of Manual Medicine. Williams & Wilkins; Baltimore; 2003.
- Guillemiault C, Framherz S. Principles and Practice of Sleep Medicine.Eds. Kryger M, Roth T, Dement W. Elsevier: Philadelphia; 2005:780-790.
- Hamill J, Knutzen KM, Bates BT, Kirkpatrick G. Evaluation of two ankle appliances using ground reaction force data. J Orthop Sports Phys Ther. 1986;7(5):244-9.
- Hamilton WG. Surgical anatomy of the foot and ankle. Clin Symp. 1985;37(3):2-32.
- Hammer WI. The effect of mechanical load on degenerated soft tissue. J Bodyw Mov Ther. 2008;12(3):246-56.
- Hammer WI. Functional Soft Tissue Examination and Treatment by Manual Methods: New Perspectives, 2nd Ed. Aspen Publishers, Inc.: Gaithersburg, MD; 1999.
- Harper MC, Keller TS. A radiographic evaluation of the tibiofibular syndesmosis. Foot Ankle. 1989;10(3):156-60.
- Hersh BL, Heath NS. Achilles tendon rupture as a result of oral steroid therapy. J Am Podiatr Med Assoc. 2002;92(6):355-8.
- Hertling D, Kessler RM. Management of common musculoskeletal disorders, 3rd Ed. Lippincott Williams & Wilkins; 1996.
- Hess GW. Achilles tendon rupture: a review of etiology, population, anatomy, risk factors, and injury prevention. Foot Ankle Spec. 2010;3(1):29-32. Epub 2009 Dec 15.
- Hohendorff B, Siepen W, Staub L. Treatment of acute Achilles tendon rupture: fibrin glue versus fibrin glue augmented with the plantaris longus tendon. J Foot Ankle Surg. 2009;48(4):439-46.
- Holmes GB, Cracchiolo A, Goldner JL, et al. Current practices in the management of posterior tibial tendon rupture. Contemp Orthop. 1990;20:79-108.
- Hontas MJ. Jaddad RJ, Schlesinger LC. Conditions of the talus in the runner. Am J Sports Med. 1986;14(6):486-90.
- Hubbard TJ, Cordova M. Effect of ankle taping on mechanical laxity in chronic ankle instability. Foot Ankle Int. 2010;31(6):499-504.
- Hurley MV, Scott DL. Improvements in quadriceps sensorimotor function and disability of patients with knee osteoarthritis following a clinically practicable exercise regime. Br J Rheumatol. 1998;37(11):1181-7.
- IMPAC. Salem, OR. http://www.impacinc.net/client.
- Inglis AE, Scott WN, Sculco TP, Patterson AH. Ruptures of the tendo achillis. An objective assessment of surgical and non-surgical treatment. J Bone Joint Surg Am. 1976;58(7):990-3.
- Jackson DW, Simon TM, Aberman HM. Symptomatic articular cartilage degeneration: the impact in the new millennium. Clin Orthop Relat Res. 2001;(391 Suppl):S14-25.
- Jackson DW, Jarrett H, Bailey D, Kausek J, Swanson J, Powell JW. Injury prediction in the young athlete: a preliminary report. Am J Sports Med. 1978;6(1):6-14.
- James SL, Bates BT, Osternig LR. Injuries to runners. Am J Sports Med. 1978;6(2):40-50.
- Janda V. Muscle Function Testing. Butterworths: London; 1983.
- Jaskoviak PA. A look at the sprained ankle. Success Express. 1983;6(1).
- Johansson C. Injuries in elite orienteers. Am J Sports Med. 1986;14(5):410-5.
- Johnson KA, Strom DE. Tibialis posterior tendon dysfunction. Clin Orthop Relat Res. 1989;(239):196-206.
- Jones RL. The functional significance of the declination of the axis of the subtalar joint. Anat Rec. 1945;93:151-9.
- Jørgensen U. Achillodynia and loss of heel pad shock absorbency. Am J Sports Med. 1985;13(2):128-32.
- Kapandji IA. The Physiology of the Joints, Vol 2 – Lower Limb, 6th ed. Churchill-Livingstone, New York; 2010.
- Kavounoudias A, Roll R, Roll JP. Foot sole and ankle muscle inputs contribute jointly to human erect posture regulation. J Physiol. 2001;532(Pt 3):869-78.
- Kellam JF, Hunter GA, McElwain JP. Review of the operative treatment of Achilles tendon rupture. Clin Orthop Relat Res. 1985;(201):80-3.
- Kemler E, van de Port I, Backx F, van Dijk CN. A Systematic Review on the Treatment of Acute Ankle Sprain: Brace versus Other Functional Treatment Types. Sports Med. 2011;41(3):185-97.
- Kobsar B, Alcantara J. Post-surgical care of a professional ballet dancer following calcaneal exostectomy and debridement with re-attachment of the left Achilles tendon. J Can Chiropr Assoc. 2009;53(1):17-22.
- Kohne E, Jones A, Korporaal C, Price JL, Brantingham JW, Globe G. A prospective, single-blinded, randomized, controlled clinical trial of the effects of manipulation on proprioception and ankle dorsiflexion in chronic recurrent ankle sprain. . JACA. 2007;44(5):7-17.
- Konstantakos EK, Dalstrom DJ, Nelles ME, Laughlin RT, Prayson MJ. Diagnosis and management of extremity compartment syndromes: an orthopaedic perspective. Am Surg. 2007;73(12):1199-209.
- Korkola M, Amendola A. Exercise-induced leg pain: sifting through a broad differential. Phys Sportsmed. 2001;29(6):35-50.
- Krenner BJ. Case report: comprehensive management of medial tibial stress syndrome. J Chiropr Med. 2002 Summer;1(3):122-4.
- Krissoff WB, Ferris WD. Runners' injuries. Physician Sportsmed. 1979;7(12).
- Kubo K, Morimoto M, Komuro T, Tsunoda N, Kanehisa H, Fukunaga T. Age-related differences in the properties of the plantar flexor muscles and tendons. Med Sci Sports Exerc. 2007;39(3):541-7.
- Langer P. Great Feet for Life: Footcare and Footwear for Healthy Aging. Fairview Press: Minneapolis, MN;2007.
- Larsen E. Experimental instability of the ankle. A radiographic investigation. Clin Orthop Relat Res. 1986;(204):193-200.
- Laughman RK, Carr TA, Chao EY, Youdas JW, Sim FH. Three-dimensional kinematics of the taped ankle before and after exercise. Am J Sports Med. 1980;8(6):425-31.
- Lea RB, Smith L. Non-surgical treatment of tendo achillis rupture. J Bone Joint Surg Am. 1972;54(7):1398-407.
- Leach RE, Lower G. Ankle injuries in skiing. Clin Orthop Relat Res. 1985;198:127-33.
- Leaf D. Static and kinetic visual analysis of the foot and ankle. Collected Papers International College of Applied Kinesiology. 1999-2000;1:109-112.
- Leahy P. Active Release Technique®, soft tissue management system for the lower extremity. Champion Health Activity: Colorado Springs, CO; 2008.
- Lederman E. Fundamental of manual therapy. Physiology, neurology and psychology. Churchill Livingstone: Edinburgh, 1997.
- Lee R. Collected Papers and Reprints. Standard Process Labs: Palmyra, WI; 2007.
- Leppilahti J, Orava S. Total Achilles tendon rupture. A review. Sports Med. 1998;25(2):79-100.
- Levangie PK, Norkin CC. Joint Structure and Function: A Comprehensive Analysis, 3rd Ed. FA Davis Company: Philadelphia; 2001.

- Limbaugh R. Rush Limbaugh announced his addiction to the painkiller OxyContin in 2003 (a form of oxycodone) and proved that, his own anti-drug crusading messages to the contrary, you can indeed "work high."
- Liu W, Siegler S, Techner L. Quantitative measurement of ankle passive flexibility using an arthrometer on sprained ankles. Clin Biomech (Bristol, Avon). 2001;16(3):237-44.
- Logan AL. The Foot and Ankle: Clinical Applications. (A.L. Logan Series in Chiropractic Technique). Gaithersburg, MD: Aspen Publishers, Inc; 1995.
- López-Rodríguez S, Fernández de-Las-Peñas C, Alburquerque-Sendín F, Rodríguez-Blanco C, Palomeque-del-Cerro L. Immediate effects of manipulation of the talocrural joint on stabilometry and baropodometry in patients with ankle sprain. J Manipulative Physiol Ther. 2007;30(3):186-92.
- Loram ID, Maganaris CN, Lakie M. Paradoxical muscle movement during postural control. Med Sci Sports Exerc. 2009;41(1):198-204.
- Lowdon A, Bader DL, Mowat AG. The effect of heel pads on the treatment of Achilles tendinitis: a double blind trial. Am J Sports Med. 1984;12(6):431-5.
- Lowenstein F. Nutritional status of the elderly in the United States of America. Journal of the American College of Nutrition. 1982;1:165-177.
- Lueck RA, Ray RD. Volkmann's ischemia of the lower extremity. Surg Clin North Am. 1972;52(1):145-53.
- Lüscher HR, Clamann HP. Relation between structure and function in information transfer in spinal monosynaptic reflex. Physiol Rev. 1992;72(1):71-99.
- Lutter LD. Forefoot injuries. In: Symposium on The Foot and Ankle. Ed. Kiene RH, Johnson KA. The C.V. Mosby Co: St. Louis; 1983.
- Lutz LJ, Goodenough GK, Detmer DE. Chronic compartment syndrome. Am Fam Physician. 1989;39(2):191-6.
- Lysholm J, Wiklander J. Injuries in runners. Am J Sports Med. 1987;15(2):168-71.
- Macefield VG. Physiological characteristics of low-threshold mechanoreceptors in joints, muscle and skin in human subjects. Clin Exp Pharmacol Physiol. 2005;32(1-2):135-44.
- MacLean CL, Davis IS, Hamill J. Short- and long-term influences of a custom foot orthotic intervention on lower extremity dynamics. Clin J Sport Med. 2008;18(4):338-43.
- Madeley LT, Munteanu SE, Bonanno DR. Endurance of the ankle joint plantar flexor muscles in athletes with medial tibial stress syndrome: a case-control study. J Sci Med Sport. 2007;10(6):356-62.
- Maffetone P. Complementary Sports Medicine: Balancing traditional and nontraditional treatments. Human Kinetics: Champaign, IL; 1999.
- Maffetone P. The Big Book of Endurance Training and Racing. Skyhorse Publishing: New York, NY; 2010.
- Maffetone P. Fix Your Feet. The Lyons Press: Guilford, CT; 2003.
- Marmor L, Bechtol CO, Hall GB. Pectoralis major muscle: function of sternal portion and mechanisms of rupture of normal muscle: case reports. J Bone Joint Surg. 1961;43A:81-87.
- Martens MA, Noyez JF, Mulier JC. Recurrent dislocation of the peroneal tendons. Results of rerouting the tendons under the calcaneofibular ligament. Am J Sports Med. 1986;14(2):148-50.
- Marti R. Dislocation of the peroneal tendons. Am J Sports Med. 1977;5(1):19-22.
- Martin JW, Thompson GH. Achilles tendon rupture. Occurrence with a closed ankle fracture. Clin Orthop Relat Res. 1986 Sep;(210):216-8.
- Martin RL, Stewart GW, Conti SF. Posttraumatic ankle arthritis: an update on conservative and surgical management. J Orthop Sports Phys Ther. 2007;37(5):253-9.
- Marymont JV, Lynch MA, Henning CE. Acute ligamentous diastasis of the ankle without fracture. Evaluation by radionuclide imaging. Am J Sports Med. 1986;14(5):407-9.
- Mavor GE. The anterior tibial syndrome. J Bone Joint Surg Br. 1956;38-B(2):513-7.
- Mayer J, Tomlinson D. Prevention of defects of axonal transport and nerve conduction velocity by oral administration of myo-inositol or an aldose reductase inhibitor in streptozotocin-diabetic rats. Diabetologia. 1983;25(5):433-438.
- Maykel W. Treatment of Bell's palsy by the correction of faults in the stomatognathic system: case history. Collected Papers International College of Applied Kinesiology: Shawnee Mission, KS; 2001-2002;1:145-151.
- McBryde AM, Ed. The Ankle (Selective Exposures in Orthopedic Surgery). Amer Academy of Orthopaedic; 2007.
- McClure JG. Gastrocnemius musculotendinous rupture: a condition confused with thrombophlebitis. South Med J. 1984;77(9):1143-5.
- McDougall C. Born to Run: A Hidden Tribe, Superathletes, and the Greatest Race the World Has Never Seen. Knopf: New York; 2009.
- McLennan JG, Ungersma JA. A new approach to the treatment of ankle fractures. The Inyo nail. Clin Orthop Relat Res. 1986;(213):125-36.
- McVey ED, Palmieri RM, Docherty CL, Zinder SM, Ingersoll CD. Arthrogenic muscle inhibition in the leg muscles of subjects exhibiting functional ankle instability. Foot Ankle Int. 2005;26(12):1055-61.
- Mense S, Simons DG. Muscle Pain: Understanding Its Nature, Diagnosis, and Treatment. Lippincott Williams & Wilkins: Philadelphia; 2001.
- Merck Manuals. www.merck.com; 2010.
- Meyer JM, Hoffmeyer P, Savoy X. High resolution computed tomography in the chronically painful ankle sprain. Foot Ankle. 1988;8(6):291-6.
- Michael RH, Holder LE. The soleus syndrome. A cause of medial tibial stress (shin splints). Am J Sports Med. 1985;13(2):87-94.
- Michaud T. Foot Orthoses and Other Forms of Conservative Foot Care. Williams & Wilkins: Baltimore;1997.
- Michaud T. Pedal biomechanics as related to orthotic therapies: An overview – Part II. Chiro Econ. 1987;29(4).
- Michels F, Guillo S, King A, Jambou S, de Lavigne C. Endoscopic calcaneoplasty combined with Achilles tendon repair. Knee Surg Sports Traumatol Arthrosc. 2008;16(11):1043-6.
- Molloy A, Wood EV. Complications of the treatment of Achilles tendon ruptures. Foot Ankle Clin. 2009;14(4):745-59.
- Moore KL, Dalley RF, Agur AMR. Clinically Oriented Anatomy, Sixth Edition. Lippincott Williams & Wilkins: Philadelphia, PA; 2009.
- Moore RE 3rd, Friedman RJ. Current concepts in pathophysiology and diagnosis of compartment syndromes. J Emerg Med. 1989;7(6):657-62.
- Morris CC. Low Back Syndromes: Integrated Clinical Management. McGraw-Hill: New York; 2006.
- Morrissey D. Unloading and proprioceptive taping. In: Chaitow L. (ed.) Positional release techniques, 2nd ed. Churchill Livingstone, Edinburgh; 2001.
- Mubarak SJ, Pedowitz RA, Hargens AR. Compartment syndromes. Curr Orthop. 1989;3:36-40.
- Murphy D. Conservative management of cervical syndromes. McGraw-Hill: New York; 2000.
- Myers TW. Anatomy Trains: Myofascial Meridians for Manual and Movement Therapists. Churchill-Livingstone; 2001.

- Nagrale AV, Herd CR, Ganvir S, Ramteke G. Cyriax physiotherapy versus phonophoresis with supervised exercise in subjects with lateral epicondylalgia: a randomized clinical trial. J Man Manip Ther. 2009;17(3):171-8.
- National Commission on Sleep Disorders Research. Report of the National Commission on Sleep Disorders Research. Research DHHS Pub. No. 92. Supplier of Documents. U.S. Government Printing Office: Washington, DC; 1992.
- Nelen G, Martens M, Burssens A. Surgical treatment of chronic Achilles tendinitis. Am J Sports Med. 1989;17(6):754-9.
- Neville C, Flemister AS, Houck JR. Deep posterior compartment strength and foot kinematics in subjects with stage II posterior tibial tendon dysfunction. Foot Ankle Int. 2010;31(4):320-8.
- Newnham DM, Douglas JG, Legge JS, Friend JA. Achilles tendon rupture: an underrated complication of corticosteroid treatment. Thorax. 1991;46(11):853-4.
- Nilsson-Helander K, Swärd L, Silbernagel KG, Thomeé R, Eriksson BI, Karlsson J. A new surgical method to treat chronic ruptures and reruptures of the Achilles tendon. Knee Surg Sports Traumatol Arthrosc. 2008;16(6):614-20.
- Nitz AJ, Dobner JJ, Kersey D. Nerve injury and grades II and III ankle sprains. Am J Sports Med. 1985;13(3):177-82.
- Norkus SA, Floyd RT. The anatomy and mechanisms of syndesmotic ankle sprains. J Athl Train. 2001;36(1):68-73.
- Nutri-West. 2132 E. Richards, Douglas, WY 82633.
- Oden RR. Tendon injuries about the ankle resulting from skiing. Clin Orthop Relat Res. 1987;(216):63-9.
- O'Donoghue DH. Treatment of ankle injuries. Northwest Med. 1958;57(10):1277-86.
- Ogon M, Aleksiev AR, Pope MH, Wimmer C, Saltzman CL. Does arch height affect impact loading at the lower back level in running? Foot Ankle Int. 1999;20(4):263-6.
- Ohayon MM. Nocturnal awakenings and difficulty resuming sleep: their burden in the European general population. J Psychosom Res. 2010;69(6):565-71.
- O'Keeffe ST. Iron deficiency with normal ferritin levels in restless legs syndrome. Sleep Med. 2005;6(3):281-2.
- Olchowik G, Siek E, Tomaszewska M, Tomaszewski M. The evaluation of mechanical properties of animal tendons after corticosteroid therapy. Folia Histochem Cytobiol. 2008;46(3):373-7.
- Ozaki J, Fujiki J, Sugimoto K, Tamai S, Masuhara K. Reconstruction of neglected Achilles tendon rupture with Marlex mesh. Clin Orthop Relat Res. 1989;(238):204-8.
- Oztekin HH, Boya H, Ozcan O, Zeren B, Pinar P. Foot and ankle injuries and time lost from play in professional soccer players. Foot (Edinb). 2009;19(1):22-8.
- Page P, Frank CC, Lardner R. Assessment and Treatment of Muscle Imbalance: The Janda Approach. Human Kinetics: Champaign, IL; 2010.
- Paiva de Castro A, Rebelatto JR, Aurichio TR. The relationship between wearing incorrectly sized shoes and foot dimensions, foot pain, and diabetes. J Sport Rehabil. 2010;19(2):214-25.
- Pell RF 4th, Khanuja HS, Cooley GR. Leg pain in the running athlete. J Am Acad Orthop Surg. 2004;12(6):396-404.
- Perlman MD, Leveille D. Extensor digitorum longus stenosing tenosynovitis. A case report. J Am Podiatr Med Assoc. 1988;78(4):198-9.
- Perry J, Ireland ML, Gronley J, Hoffer MM. Predictive value of manual muscle testing and gait analysis in normal ankles by dynamic electromyography. Foot Ankle. 1986;6(5):254-9.
- Phillips, B. et al. Epidemiology of restless legs symptoms in adults. Arch. Intern. Med. 2000;160:2137-2141.
- Picchietti D, Winkelman JW. Restless legs syndrome, periodic limb movements in sleep, and depression. Sleep. 2005;28(7):891-8.
- Pillai J, Levien LJ, Haagensen M, Candy G, Cluver MD, Veller MG. Assessment of the medial head of the gastrocnemius muscle in functional compression of the popliteal artery. J Vasc Surg. 2008;48(5):1189-96.
- Pisani G, Pisani PC, Parino E. Sinus tarsi syndrome and subtalar joint instability. Clin Podiatr Med Surg. 2005;22(1):63-77, vii.
- Pizzorno JE, Murray MT, Joiner-Bey H. The Clinician's Handbook of Natural Medicine, 2nd Ed. Churchill Livingstone: Edinburgh; 2007.
- Platzer W. Color atlas/text of human anatomy. Georg Thieme, Stuttgart; 1992.
- Pope MH, Renstrom P, Donnermeyer D, Morgenstern S. A comparison of ankle taping methods. Med Sci Sports Exerc. 1987;19(2):143-7.
- Price AE, Evanski PM, Waugh TR. Bilateral simultaneous achilles tendon ruptures. A case report and review of the literature. Clin Orthop Relat Res. 1986;(213):249-50.
- Raikin SM, Elias I, Nazarian LN. Intrasheath subluxation of the peroneal tendons. J Bone Joint Surg Am. 2008;90(5):992-9.
- Ramamurti CP. Orthopaedics in Primary Care. Williams & Wilkins: Baltimore; 1979.
- Ramelli FD. Diagnosis, management and post-surgical rehabilitation of Achilles tendon rupture: a case report. J Can Chiropr Assoc. 2003;47(4):261–268.
- Ramsak I, Gerz W. AK Muscle Tests At A Glance. AKSE: Oberhaching; 2002.
- Rarick GL, Bigley G, Karst R, Malina RM. The measurable support of the ankle joint by conventional methods of taping. J Bone Joint Surg Am. 1962;44-A:1183-90.
- Reneman RS. The anterior and the lateral compartment syndrome of the leg due to intensive use of muscles. Clin Orthop Relat Res. 1975;(113):69-80.
- Restless Legs Syndrome Foundation. http://www.rls.org.
- Reszel PA, Janes JM, Spittell JA. Ischemic necrosis of the peroneal musculature, a lateral compartment syndrome: report of case. Proc Staff Meet Mayo Clin. 1963;38:130-6.
- Richie DH Jr. Functional instability of the ankle and the role of neuromuscular control: a comprehensive review. J Foot Ankle Surg. 2001;40(4):240-51.
- Rijke AM, Jones B, Vierhout PA. Injury to the lateral ankle ligaments of athletes. A posttraumatic followup. Am J Sports Med. 1988;16(3):256-9.
- Rod NH, Vahtera J, Westerlund H, Kivimaki M, Zins M, Goldberg M, Lange T. Sleep disturbances and cause-specific mortality: Results from the GAZEL cohort study. Am J Epidemiol. 2011;173(3):300-9.
- Rogowskey TA. Reflex sympathetic dystrophy and lumbar disc involvement. Collected Papers International College of Applied Kinesiology: Shawnee Mission, KS;1990;1:221-227.
- Rolf C, Movin T. Etiology, histopathology, and outcome of surgery in achillodynia. Foot Ankle Int. 1997;18(9):565-9.
- Rolf IP. Rolfing: The Integration of Human Structures. Dennis-Landman Publishers: Santa Monica; 1977.
- Rorabeck CH. Exertional tibialis posterior compartment syndrome. Clin Orthop Relat Res. 1986;(208):61-4.
- Rose GK. Correction of the pronated foot. J Bone Joint Surg. 1962;44B(3).
- Rovere GD, Haupt HA, Yates CS. Prophylactic knee bracing in college football. Am J Sports Med. 1987;15(2):111-6.
- Safran MR, O'Malley D Jr, Fu FH. Peroneal tendon subluxation in athletes: new exam technique, case reports, and review. Med Sci Sports Exerc. 1999;31(7 Suppl):S487-92.

- Safran MR, Zachazewski JE, Benedetti RS, Bartolozzi AR 3rd, Mandelbaum R. Lateral ankle sprains: a comprehensive review part 2: treatment and rehabilitation with an emphasis on the athlete. Med Sci Sports Exerc. 1999;31(7 Suppl):S438-47.
- Sammarco GJ, DiRaimondo CV. Surgical treatment of lateral ankle instability syndrome. Am J Sports Med. 1988;16(5):501-11.
- Santilli V, Frascarelli MA, Paoloni M, Frascarelli F, Camerota F, De Natale L, De Santis F. Peroneus longus muscle activation pattern during gait cycle in athletes affected by functional ankle instability: a surface electromyographic study. Am J Sports Med. 2005;33(8):1183-7.
- Sauberlich H, Canham J. Vitamin B6. In: Modern Nutrition in Health and Disease, 6th Ed. Ch. 61. Eds. Goodhart R, Shils M. Lea & Febinger: Philadelphia; 1980:219-225.
- Schafer RC. Chiropractic management of sports and recreational injuries, 2nd Ed. Williams & Wilkins: Baltimore; 1986.
- Schattschneider J, Bode A, Wasner G, Binder A, Deuschl G, Baron R. Idiopathic restless legs syndrome: abnormalities in central somatosensory processing. J Neurol. 2004;251(8):977-82.
- Schepsis AA, Leach RE. Surgical management of Achilles tendinitis. Am J Sports Med. 1987;15(4):308-15.
- Schmitt WH, McCord KM. Quintessential Applications: A(K) Clinical Protocol (QA). www.quintessentialapplications.com; 2010.
- Selye H. Stress Without Distress. New American Library: New York; 1974.
- Shadgan B, Menon M, O'Brien PJ, Reid WD. Diagnostic techniques in acute compartment syndrome of the leg. J Orthop Trauma. 2008;22(8):581-7.
- Slimmon D, Brukner P. Sports ankle injuries - assessment and management. Aust Fam Physician. 2010;39(1-2):18-22.
- Slocum DB, James SL. Biomechanics of running. JAMA. 1968;205(11):721-8.
- Smith RW, Reischl SF. Treatment of ankle sprains in young athletes. Am J Sports Med. 1986;14(6):465-71.
- Snook AG. The relationship between excessive pronation as measured by navicular drop and isokinetic strength of the ankle musculature. Foot Ankle Int. 2001;22(3):234-40.
- Solomonow M. Ligaments: a source of musculoskeletal disorders. J Bodyw Mov Ther. 2009;13(2):136-54.
- Solomonow M, Zhou B, Harris M, Lu Y, Baratta RV. The ligamento-muscular stabilizing system of the spine. Spine 1998;23(23):2552-62.
- Solomonow M, Baratta RV, Zhou BH, Shoji H, Bose W, Beck C, D'Ambrosia R. The synergistic action of the ACL and thigh muscles in maintaining joint stability. Am J Sports Med 1987;15(3):207-213.
- Souza T. Conservative management of orthopedic conditions of the lower leg, foot and ankle. Topics in Clinical Chiropractic. 1994;1(2):61–75.
- Sprieser P. A new epidemic of knee injuries: anterior cruciate ligament in women athletes. Collected Papers International College of Applied Kinesiology. 2002-2003;1:45-49.
- Staal A, van Gijn J, Spaans F. Mononeuropathies: Examination, Diagnosis and Treatment. WB Saunders: London; 1999.
- Stallknecht B, Kjaer M, Mikines KJ, Maroun L, Ploug T, Ohkuwa T, Vinten J, Galbo H. Diminished epinephrine response to hypoglycemia despite enlarged adrenal medulla in trained rats. Am J Physiol. 1990;259(5 Pt 2):R998-1003.
- Standard Process Labs, Palmyra, WI. www.standardprocess.com.
- Stanish WD, Rubinovich RM, Curwin S. Eccentric exercise in chronic tendinitis. Clin Orthop Relat Res. 1986;(208):65-8.
- Starkey JA. Treatment of ankle sprains by simultaneous use of intermittent compression and ice packs. Am J Sports Med. 1976;4(4):142-4.
- Stiasny-Kolster K, et al. Static mechanical hyperalgesia without dynamic tactile allodynia in patients with restless legs syndrome. Brain. 2004;127(Part 4):773-82.
- Stoffel KK, Nicholls RL, Winata AR, Dempsey AR, Boyle JJ, Lloyd DG. Effect of ankle taping on knee and ankle joint biomechanics in sporting tasks. Med Sci Sports Exerc. 2010;42(11):2089-97.
- Straehley D, Jones WW. Acute compartment syndrome (anterior, lateral, and superficial posterior) following tear of the medial head of the gastrocnemius muscle. A case report. Am J Sports Med. 1986;14(1):96-9.
- Stupar M. Restless legs syndrome in a primary contact setting: a case report. J Can Chiropr Assoc. 2008;52(2):81-87.
- Subotnick S. Sports & Exercise Injuries: Conventional, Homeopathic & Alternative Treatments. North Atlantic Books: Berkeley, CA; 1991.
- Subotnick SI. Compartment syndromes in the lower extremities. J Am Podiatry Assoc. 1975;65(4):342-8.
- Tergau F, Wischer S, Paulus W. Motor system excitability in patients with restless legs syndrome. Neurology,1999;52:1060-1063.
- Thompson TC, Doherty JH. Spontaneous rupture of tendon of Achilles: a new clinical diagnostic test. J Trauma. 1962;2:126-9.
- Tourné Y, Besse JL, Mabit C; Sofcot. Chronic ankle instability. Which tests to assess the lesions? Which therapeutic options? Orthop Traumatol Surg Res. 2010;96(4):433-46.
- Travell JG, Simons DG. Myofascial Pain and Dysfunction: The Trigger Point Manual: The Lower Extremities. Williams & Wilkins: Baltimore; 1992.
- Trenkwalder C, Hening WA, Walters AS, Campbell SS, Rahman K, Chokroverty S. Circadian rhythm of periodic limb movements and sensory symptoms of restless legs syndrome.Mov Disord. 1999;14(1):102-10.
- Tropp H, Askling C, Gillquist J. Prevention of ankle sprains. Am J Sports Med. 1985;13(4):259-62.
- Turco VJ, Spinella AJ. Achilles tendon ruptures--peroneus brevis transfer. Foot Ankle. 1987 Feb;7(4):253-9.
- Turek SL. Orthopaedics – Principles and Their Application, Vol 1, 4th Ed. J.B. Lippincott Co.: Philadelphia; 1984.
- van den Brand JG, Nelson T, Verleisdonk EJ, van der Werken C. The diagnostic value of intracompartmental pressure measurement, magnetic resonance imaging, and near-infrared spectroscopy in chronic exertional compartment syndrome: a prospective study in 50 patients. Am J Sports Med. 2005;33(5):699-704.
- van der Linden-van der Zwaag HM, Nelissen RG, Sintenie JB. Results of surgical versus non-surgical treatment of Achilles tendon rupture. Int Orthop. 2004;28(6):370-3.
- Vaes P, De Boeck H, Handelberg F, Opdecam P. Comparative radiological study of the influence of ankle joint strapping and - - taping on ankle stability. J Orthop Sports Phys Ther. 1985;7(3):110-4.
- Van Middelkoop M, Kolkman J, Van Ochten J, Bierma-Zeinstra SM, Koes BW. Risk factors for lower extremity injuries among male marathon runners. Scand J Med Sci Sports. 2008;18(6):691-7.
- Veves A, Giurini JM, LoGerfo FW. The Diabetic Foot: Medical and Surgical Management. Humana Press: Totowa, NJ; 2002.
- Viitasalo JT, Kvist M. Some biomechanical aspects of the foot and ankle in athletes with and without shin splints. Am J Sports Med. 1983;11(3):125-30.
- Waddington GS, Adams RD. The effect of a 5-week wobble-board exercise intervention on ability to discriminate different degrees of ankle inversion, barefoot and wearing shoes: a study in healthy elderly. J Am Geriatr Soc. 2004;52(4):573-6.

- Wall CJ, Richardson MD, Lowe AJ, Brand C, Lynch J, de Steiger RN. Survey of management of acute, traumatic compartment syndrome of the leg in Australia. ANZ J Surg. 2007;77(9):733-7.
- Wallden M. Rehabilitation and Re-education (movement) approaches. In: Naturopathic Physical Medicine. Ed. Chaitow L. Churchill-Livingstone: Edinburgh; 2007.
- Walther DS. Applied Kinesiology, Volume I – Basic Procedures and Muscle Testing. Systems DC: Pueblo, CO;1981.
- Walther DS. Applied Kinesiology Synopsis, 2nd Ed. International College of Applied Kinesiology, USA: Shawnee Mission, KS; 2000.
- Werbach M. Natural medicine for muscle strain. J Bodyw Mov Ther. 1996;1(1):18-19.
- Wilder RP, Sethi S. Overuse injuries: tendinopathies, stress fractures, compartment syndrome, and shin splints. Clin Sports Med. 2004;23(1):55-81, vi.
- Wills CA, Washburn S, Caiozzo V, Prietto CA. Achilles tendon rupture. A review of the literature comparing surgical versus nonsurgical treatment. Clin Orthop Relat Res. 1986;(207):156-63.
- Wilson JL. Adrenal Fatigue: The 21st Century Stress Syndrome. Smart Publications; 2002.
- Wolfe MM, Lichtenstein DR, Singh G. Gastrointestinal toxicity of nonsteroidal anti-inflammatory drugs. New England Journal of Medicine. 1999 340(24):1888-1899.
- Yang G, Im HJ, Wang JH. Repetitive mechanical stretching modulates IL-1beta induced COX-2, MMP-1 expression, and PGE2 production in human patellar tendon fibroblasts. Gene. 2005;363:166-72.
- Yochum TR, Rowe LJ. Essentials of Skeletal Radiology, 3rd Ed. Lippincott Williams & Wilkins: Baltimore; 2004.
- Yuill EA, Macintyre IG. Posterior tibialis tendonopathy in an adolescent soccer player: a case report. J Can Chiropr Assoc. 2010;54(4):293-300.
- Zampagni ML, Corazza I, Molgora AP, Marcacci M. Can ankle imbalance be a risk factor for tensor fascia lata muscle weakness? J Electromyogr Kinesiol. 2009;19(4):651-9.

> "The opportunity to use the body as an instrument of laboratory analysis is unparalleled in modern therapy; if one approaches the problem correctly, making the proper and adequate diagnosis and treatment, the response is satisfactory to both the doctor and the patient."
> - George J. Goodheart, Jr.

> "The difference between somatovisceral and viscerosomatic reflexes appears to be only quantitative and to be accounted for by the lesser density of nociceptive receptors in the viscera."
> - RC Schafer, 1986

> Head's Law:
> "If a painful stimulus is applied to areas of low sensibility (like an organ) in close central connection with areas of high sensibility (like a muscle), pain may be felt where sensibility is high."

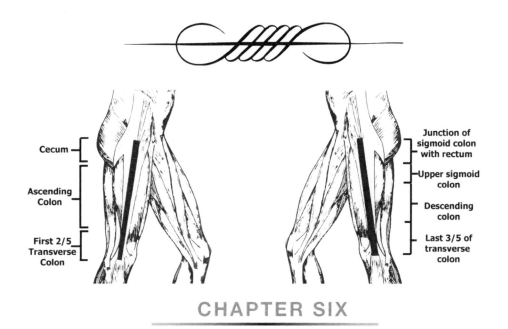

CHAPTER SIX
Applied Kinesiology and Systemic Conditions of the Lower Body

Introduction

Differential diagnosis of systemic disorders of the pelvic region and lower extremities can be extremely complex. It calls on all the skills a clinician has to draw on in order to combine observation and examination together into a coherent whole. Always consider that dysfunction or pathology from any part of the body can refer disorders and pain into the lower body. It should be recognized also that the pathoanatomical model of lower body dysfunction is frequently inadequate. Pain in the lower body often exists in the absence of major findings on diagnostic biomedical tests (CT scan, blood tests, nerve conduction tests, X-ray), and injured tissues can be identified in people who experience no pain. **(Nachemson, 1999)**

Melzack, **(2001)** in his *Neuromatrix Theory of Pain*, echoes Goodheart by showing that it is often necessary to treat the entire person, and not just the painful parts. In evaluating patients with systemic conditions in the lower body, it

must be remembered that biomechanical adaptation may have advanced to a stage where neural, lymphatic, articular, muscular, biochemical and psychological interactions are synchronistic and plentiful. Consideration must additionally be given to the medications a patient is taking, as well as to the symptoms such medications may create, so that the physician can refer the patient back to the drug-prescribing physician should the symptoms be suspected of being linked to the medication(s). A Physician's Desk Reference, which is current and easily searched, is a necessary asset.

The scant or absent hair of the lower leg in patients with arteriosclerosis, especially due to diabetes; **(Veves et al., 2002)** the fine, long leg hair of the anorexic patient; the patient with white flecked nails due to chronic zinc deficiency; **(Shils et al., 1999)** the unilateral leg edema of deep vein thrombosis, thrombophlebitis, pelvic tumor, or cellulitis;

the impaired perception of deep touch, pressure, and vibration, anesthesia, paresthesias, diminished or absent deep tendon reflexes, pathological reflexes, paresis, and ataxia of pernicious anemia; **(Guyton & Hall, 2005)** the dyspareunia due to post-surgical pelvic adhesions, regional ileitis, or pelvic inflammatory disease; **(Chaitow & Jones, 2012)** the left lower quadrant pain of diverticulosis - these and all of the possible differentials make for the high art of clinical practice.

Our purpose here isn't to cover all of that ground, but rather to present tools from applied kinesiology that contribute to distinguishing and resolving systemic disorders that can plague the pelvic region and lower extremities. Often the patient presenting with complex clinical patterns can defy our best efforts at a definitive differential diagnosis even when our tools of history, physical exam, laboratory findings, and imaging have been brought to bear. Visceral hyperalgesia and central sensitization may be predominant features of chronic lower body pains. **(Giamberardino et al., 2010)** Long-term physical, emotional, or biochemical stress may drive long-term potentiation of the CNS so that an individual becomes vulnerable to perceiving sensations that would not normally hurt other individuals. **(Rygh et al., 2002)** In these cases AK (by focusing on the triad of health as it manifests in a particular patient's functional ensemble) can be useful by both helping to integrate and contextualize our existing findings as well as suggesting further orthopedic, neurological, physical exams, laboratory testing, and imaging that may help further establish diagnosis.

The field of clinical nutrition has wide application in treating and preventing systemic disorders, and the evidence supporting this mode of therapy in clinical practice continues to grow. **(Gaby, 2006; Shils et al., 1999)** Applied kinesiology can be useful in adding to the information gathered from history, physical exam, and laboratory findings when applying clinical nutrition for a particular condition. All possibilities for clinical nutrition will not be considered here. The clinical nutrition applications included will be those classically used in applied kinesiology along with the 5-Factor manipulative techniques. **(Goodheart, 1998-1964; Schmitt, 2005, 1981, 1979; Force, 2003; Maffetone, 1999; Boormann, 1979; Deal, 1974)**

Testing Biochemistry with Applied Kinesiology

Ingesting nutrition targeted to a particular organ has been found to more reliably improve muscle strength using AK MMT than placebo, supporting both the organ-muscle relationship model of applied kinesiology and the usefulness of nutrient testing in AK, especially when the information is incorporated with other clinical observations. (Evidence for the AK approach to nutrition is expansively supported in ***Applied Kinesiology Essentials)*** Testing nutrients without ingestion, a method not approved by the International College of Applied Kinesiology, is unreliable. **(Hambrick, 2007; Goodheart, 1998)** AK nutritional testing reflects the nervous system's efferent response to the stimulation of the gustatory and olfactory nerve receptors by various substances. If these receptors are not stimulated (as for instance when a patient holds a pill or has it placed upon the body), the neurological rationale, plausibility and the reliability of sensorimotor pathways for AK testing disappear.

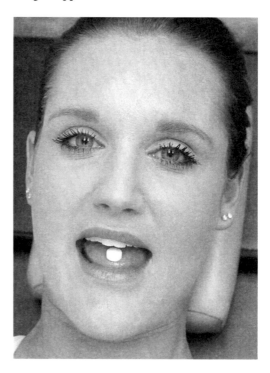

Direct stimulation of the gustatory receptors

AK MMT appears to be a useful method of observing the biochemistry of the organism in that it has shown impressive accuracy for determining food allergies (90.5%) and immunological status when compared to lab tests, though the outcomes of studies have been conflicting. **(Schmitt, 1998; Conable, 2006)** Walther **(2000)** and others have written extensively about this phenomenon; more complete descriptions are beyond the scope of this chapter. However nearly 1,000 case and case series reports over the past 35 years have been given by the members of the ICAK around the world regarding the successful use of nutritional therapies using AK diagnostic means, and are available from several platforms. **(ICAKUSA, 2012; ChiroACCESS, 2012)**

Viscero-somatic Reflexes and Muscle-Organ-Gland Relationships

Viscero-somatic reflex phenomena have been extensively explored in the research **(Pollard, 2004; Masarsky & Masarsky, 2001)** and in the text *Applied Kinesiology Essentials.* It has long been established that visceral pain can be referred not only to skeletal muscles but also to the skin, ligaments, and bone. **(Sinclair et al., 1948)** Travell & Rinzler **(1952)** have shown that pain in the pectoralis muscle can accompany coronary infarction, and this finding has been confirmed. (**Nicholas et al., 1987**) Irritation to specific organs has been observed to result in muscle inhibition, primarily of the muscle(s) associated with the irritated organ and secondarily and to a lesser degree with other muscles of the body. This outcome supports to the muscle-organ relationship model in applied kinesiology. (**Carpenter, 1977**) This finding is further buttressed by research that has shown that stress to an internal organ can result in a viscero-somatic reflex inhibiting both motor and sensory nerves in AK practice. (**Shafton, 2006; Palomar, 2006**) Visceral inflammation has been shown to produce reflex cutaneous leukocyte extravasation. (**Wessellmann, 1997**) Korr presciently observes that viscerosomatic reflex activity may be observed before any symptoms of visceral change are evident and that this phenomenon therefore is of important diagnostic value. (**Korr, 1976**) Finally, Beal (**1985**) elegantly summarizes the research about the "body language" of viscerosomatic disorders by stating that "somatic manifestation is an integral part of visceral disease."

Somato-visceral and viscero-somatic reflexes appear to be interdependent as visceral afferents inhibit the effects of cutaneous afferentation, and cutaneous afferents inhibit the effects of visceral afferentation. (**Pomeranz, 1968**)

Somatoautonomic Nervous System, Somatovisceral Reflexes and Manual Therapies

"The term 'autonomic' is a convenient rather than appropriate title, since the functional autonomy of this part of the nervous system is illusory. Rather its functions are normally closely integrated with changes in somatic activities, although the anatomical basis for such interactions are not always clear...A more realistic notion is that these sets of neurones represent an integrated system for the coordinated neural regulation of visceral and homeostatic function...Rises in blood pressure and pupillodilation may result from the stimulation of somatic receptors in the skin or other tissues."
(Gray's Anatomy, 2004)

The neurosciences are rapidly compiling evidence that manual therapies produce distinct and clinically meaningful effects on visceral functions through somato-visceral reflexes. For a better understanding of this research, we recommend the recent impressive synopses offered by Rome. (**Rome, 2010, Rome, 2009**)

Mechanoreceptors and cutaneoreceptors appear to cause somatovisceral reflex responses mediated by both the parasympathetic and sympathetic nervous systems. (**Sato, 1992**) Soft tissue manipulation to the suboccipital region has been shown to increase upper extremity digital blood flow through down-regulation of sympathetic tone (sympathetic dampening). (**Purdy, 1996**) Stimulation to both cutaneous and subcutaneous afferents has also shown somato-visceral effects. (**Sato, 1995**)

In accord with these research findings, soft tissue reflex techniques such as the neurolymphatic (Chapman's) reflexes and acupuncture meridian stimulation have growing and in some cases impressive support in the literature and appear to be useful, both diagnostically and therapeutically. (**Caso, 2004; Moncayo, 2004**) Soft tissue manipulative techniques for the diaphragm, including neurolymphatic (Chapman) reflexes, have shown increased forced vital capacity. (**Lines, 1990**)

Emerging evidence supports the long-term clinical observations of the chiropractic and osteopathic professions that aberrant spinal mechanics can have an adverse effect on autonomic and visceral function and that spinal manipulation has a modulating effect on autonomic and visceral function. This phenomenon appears to have a sound neurophysiological basis. (**Karason, 2003; Haldeman, 2000; Budgell, 2000; Kimura, 1997; Sato, 1992; Jinkins, 1989**)

We might make reference in particular to the chiropractic profession's heritage of removing interference with normal nerve control of the body. Chiropractic has been successful in developing techniques to evaluate abnormal structure and return it to normal, thus improving both peripheral nerve function and autonomic nerve function. There is, however, a phase of nerve control which the profession has failed to adequately observe and master. This is the neurohumoral control of the autonomic nervous system. The allopathic physician has made considerable investigation into neurohumoral control; in fact, many medications are based on this principle. (**Wilson et al., 1998**) Unfortunately the medications are fraught with side effects, as are most procedures with which we try to control the body rather than returning it to its own natural control. Neural control of visceral and neurohumoral function is a unique coordination of somatic and autonomic motor nervous systems; sensory information and motor control are supplied by both visceral and somatic sensory and motor fiber systems. (**Enck & Vodusek, 2006**)

The most frequent neurohumoral imbalance of the autonomic nerve system is caused by relative hypoadrenia. Norepinephrine from the adrenal medulla is necessary for cholinergic activity in the neurohumoral balance of the autonomic nervous system. The adrenal medulla can be dysfunctional from abnormal nerve supply as a result of a spinal subluxation-fixation, or from prolonged stress as described by Selye. (**1978**) When the general adaptation syndrome (GAS) approaches the third stage of exhaustion, the adrenal is incapable of meeting the demands placed upon it. If stress – physical, mental, thermal, and chemical – is the cause of the relative hypoadrenia, nothing but removal of that stress will bring the patient back to normal. On the other hand, the adrenal can become dysfunctional because of nutritional deficiencies which in themselves are a form of stress. It is, without question, the physician's responsibility to return the nervous system to normal function; however, the correction must be applied to the basic underlying cause of the nervous system's imbalance.

This review of the neurological phenomena underpinning applied kinesiology is not complete, but is presented as a basic context for the applied kinesiology approaches to systemic disorders of the pelvis and lower extremity that follow.—

Myofascial release and somatovisceral effects

The manual therapy called myofascial release technique in applied kinesiology exerts somatovisceral reflex effects as noted by Goodheart, who reported that myofascial release technique to the teres minor resulted in increased axillary temperature. It is hypothesized that this increase in axillary temperature is due to a thyroid mediated increase in metabolic rate resulting from the myofascial release treatment of the teres minor muscle, associated with the thyroid gland. (**Goodheart, 1978**) A number of studies have confirmed this somatovisceral hypothesis. (**Moncayo & Moncayo, 2007; Reuters et al., 2006; Moncayo et al., 2004; Bablis & Pollard, 2004; Duyff et al., 2000**)

Myofascial release technique appears to be a method of systematically modulating visceral function through therapy to myofascia. The usefulness of this application for other organ-muscle reflex patterns besides the teres minor-thyroid pattern has been observed. (**ICAKUSA, 2012**)

Muscle and fascia have been shown to be anatomically and functionally inseparable. Electron microscopy has even demonstrated that fascia has smooth muscle actin cells imbedded within it. (**Schliep et al., 2005; Barnes, 1997**) As muscular dysfunctions are being corrected, it is necessary to understand that fascial structures are also being reorganized and fortified. Myofascia is the sum of connective tissues throughout the body. It provides structural and functional support to the soma and viscera. Fascial adhesions have been a recognized cause of musculoskeletal pain and associated with gross anatomical and histological changes since the early to middle 20th century. (**DeJarnette, 1939; Murray, 1938**)

After joint and muscle spindle input is taken into account, the majority of remaining proprioception comes from the fascial sheaths. Acquired abnormalities in elasticity and afferentation of myofascia due to mechanical strain, trauma, surgery, inflammation, infection, or systemic biochemical stresses can result in aberrant biomechanics and neurological disorganization. (**Lewit, 2004**)

It has been proposed that a limited mechanical model of myofascial adhesion is inadequate to fully explain the inelasticity of myofascial adhesions found upon palpation and the immediate increase in elasticity following manipulation. A neurological model of myofascial manipulation that restores optimal mechanoreception and, thereby, modulates sympathetic tone and myofascial contractility may more fully explain these observations. (**Schleip, 2003**) It has been shown that fascia's gel-like ground substance which invests its collagen and elastic components may be altered to a more liquid state by the introduction of vibration, percussion, heat, movement or manipulation of the tissue, such as that applied in percussion and other soft-tissue techniques used in AK.

Percussion and myofascia

Fulford introduced mechanical percussion into osteopathic soft-tissue therapies in the 1950s. (**Comeaux, 2011**) Percussion alone will frequently remove the Travell and Jones trigger and tender points. (**Cuthbert, 2002**) The "glued" myofascia that Ida Rolf described (**Rolf, 1977**) is also effectively mobilized and corrected by percussion.

Percussor instrument

The immediate effect of percussion is to modify the physical nature of the myofascial matrix. (**Comeaux, 2011**) The greater fluidity in the matrix will allow an improved fiber density, direction, and movement pattern. The fascial sheets will begin to glide across the mobilized and "smoothed out" connective tissue matrix and intramuscular areas. When inner and outer layers of muscle and fascia are in proper balance, the tissue will have proper "tone," the way a piano string that is in perfect tension has perfect pitch. Elasticity of connective tissue between structures

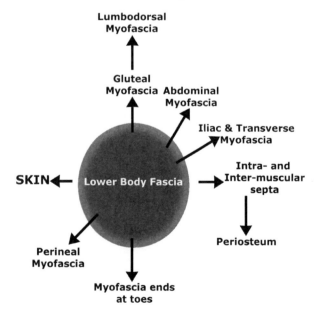

Myofascial continuities in the lower body

is essential for an effective relationship between deep and superficial (intrinsic and extrinsic) muscle layers.

Under conditions of decreased mobility and inadequate tissue fluid dynamics (lymphatic or vascular), collagen undergoes polymerization and other chemical changes, causing shortening or decreased elasticity in myofascial tissues. These changes can sometimes occur over an extended period of time.

The colloids in connective tissue are not rigid; they conform to the shape of their container and respond to pressure even though they are not compressible. Under conditions of accelerated tissue fluid dynamics and increased demand for mobility created by the percussor instrument, the collagen in the myofascia "depolymerizes" and becomes longer and more elastic. A treatment that depolymerizes collagen essentially means that we reduce the molecular size of a colloidal structure. Depolymerization also implies an increase or improvement in the biological activity of the tissue. A fascial matrix with heavy concentrations of fat and gristle and glassy, knotty concentrations in it will have its physical nature modified, smoothed-out, ironed-out, and softened by percussion treatment.

Myofascial tissue that has become short in compensation to a lack of movement elsewhere in the body can be thought of as tissue with a low potential energy and conductivity. Because of the inherent gelosis and tension in these tissues, very little activation energy is required to start the exothermic process of releasing them.

Percussion may also press fluid from the nuclear bag of the muscle spindle cells, reducing the tension in the capsule of the spindles. Positive TL to neuromuscular spindle cells and Golgi tendon organs may also be eliminated with percussor treatment.

Therapeutic manipulation of visceral myofascial adhesions

There are a number of approaches for resolving visceral adhesions and AK MMT is effective in determining the need for a given technique and for the effectiveness of therapy. Visceral manipulation is usually best done in brief therapeutic sessions and re-evaluating and reinforcing treatment on follow-up sessions as needed.

There is a basic rule of thumb in tissue pathology: once a tissue is injured, it often heals by downgrading into a simpler, less specialized form. Early in this century, Dr. Louisa Burns, **(1948)** an extraordinary osteopathic researcher, showed that "somatic dysfunctions" are accompanied by microscopic effusions of blood, with edema and inflammation occurring in the connective tissues of the affected joints and muscles. This "extravasation of blood" resolves over time as these connective tissues thicken. Dr. Burns believed that these structural thickenings of the connective tissues found throughout the body were responsible for the restriction of mobility in joints and for the accompanying pain. The literature of Structural Integration (Rolfing) and massage exhaustively describe these phenomena.

What begins as changed muscle tone as the body adapts as stress progresses to structural changes in the connective tissue elements that surround and supplement the muscle fibers involved in the adaptation. This produces changes in the material properties of the collagenous and elastic connective tissues, including reticular and ground substances that compose the tendons, ligaments, and fascia, as well as the colloidal interstitial fluids. These material tissue changes involve molecular and structural alterations. Embryologically, the entire mesoderm can be caught up in this adaptive response of the body to stress. This is the "body language" that we commonly attempt to read.

In normal connective tissue, the fibers – whether elastin or collagen – are oriented parallel to the forces ordinarily exerted through the tissue, in conformity with Wolff's law. **(Wolff, 1892)** "Every change in the function of a bone is followed by certain definite changes in internal architecture and external conformation in accordance with mathematical laws." When connective tissue is injured or chronically shortened or stretched, there will be an increase in fibroblasts. These new cells produce additional fibers laid out in random matrix fashion, rather than along the original lines of force.

The hard ridges and folds in myofascia that can be felt in patients with myofascial disorders come into being much the way pulling on a corner of a sheet creates deep pleats in the fabric. Similarly, pulls across the back of the upper pelvis and the sacrum give rise to tendon-like structures across the lower back. These ridges, folds, and thickenings of the connective tissue can literally feel like small ropes or cables under the skin. In some patients it feels as though these ropes have knots in them. The structure may even look like a tendon in dissection; it definitely feels like a tendon under the skin. But in anatomy books, no tendons are described in that location.

Common Areas of Myofascial Problems

- Along McGregor's line (from skeletal radiology), which is across the pterygoid and masseter muscle masses and extends back from the jaw to the suboccipital area; the so-called cranial stress bands and switchboards.
- The attachment of the levator scapulae to the upper cervical vertebrae; the upper trapezius and other cervical extensors to the occiput; as well as the ligamentum nuchae.
- The crossover point of the levator scapulae and the supraspinatus at the upper medial point of the scapula (medial thoracic outlet); "the cervicobrachial syndromes."
- The trapezius as it crosses the top of the shoulder near the acromion; the lateral portions of the thoracic outlet.
- The trapezius where it attaches to the scapula below the scapular spine. "It hurts me between the shoulder blades."
- The junction between the lower trapezius and latissimus dorsi behind, and the psoas and diaphragm in front; the dorsolumbar fixation.
- The tight pad created at the lumbosacral junction, and across the top of the pelvis. This

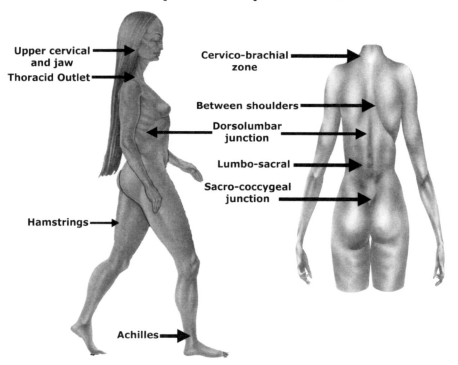

Common Myocascial areas of trouble

myofascial disorder across the sacrum is a major contributor to chronic lower back syndromes. The founder of cranial osteopathy suggests that "sacral sag and fascial drag create old rags."
- The sacro-coccygeal junction.
- The hamstrings.
- The Achilles tendons.

Somatic Fascia

The fascia directly supporting the musculoskeletal systems of the body is the somatic fascia. The somatic fascia includes the subcutaneous structures and the connective tissues surrounding muscles, groups of muscles, blood vessels, and nerves and binds all these structures together into a functional whole.

It has been proposed that the deep fascia of the body depends on the production of hyaluronic acid (HA) within the fascia by fibroblast-like cells (fasciacytes) to maintain free and unrestricted sliding motion between adjacent fibrous fascial layers. Alterations in HA conformation may lead to changes in muscle dysfunction and pain. (**Stecco, 2011**) This is one of the underlying processes creating areas of increased density and decreased mobility found upon palpation that are commonly referred to as fascial (or myofascial) adhesions. Nutritional factors also influence fascial function and structure directly. "Many of the results of deprivation of ascorbic acid [vitamin C] involve a deficiency in connective tissue which is largely responsible for the strength of bones, teeth, and skin of the body and which consists of the fibrous protein collagen." (**Pauling, 1976**) Normally, connective tissue cells are surrounded by a stiff jelly-like or cement-like substance (the ground substance), which in animals deprived of vitamin C becomes watery and thin, powerless to support the cells. It is possible that in cases where there is a failure of the intercellular substances to set to a gel after myofascial treatment, that vitamin C supplementation may provide a solution in many cases. Connective tissue protomorphogens may also be helpful. (**Cuthbert, 2002**) Protection against myofascial tissue damage is also provided by super oxide dismutase (SOD). (**Moncayo & Moncayo, 2007**)

Travell & Simons are even more emphatic that nutritional imbalances must be corrected if myofascial pain is to be adequately resolved: "Nearly half of the patients whom we see with chronic myofascial pain require resolution of vitamin inadequacies for lasting relief... nutritional factors must be considered in most patients if lasting relief of pain is to be achieved." (**1999**)

Over-use syndromes, trauma, surgery, infection, and chronic inflammation may all play a role in histological changes in myofascia that leads to altered myofascial kinetics. (**Stecco, 2008**) These changes in myofascial histology and kinetics may lead to the "propagation of a nociceptive signal even in situations of normal physiological stretch." (**Stecco, 2006**) Many sensory receptors and their axons have a lower tensile strength than the tissues in which they are embedded. Physical trauma to myofascia and their mechanoreceptors and axons may result in localized proprioceptive loss. (**Sharma, 1999**)

Besides the structural role of fascia, it provides

proprioceptive afferentation through Ruffini and Pacini corpuscles, and if normal elasticity of myofascia has been altered the resulting aberrant firing of the afferent receptors may result in abnormal biomechanics and extra-articular pain. (**Stecco, 2009**)

Applied kinesiology includes a number of techniques that address these aspects of myofascial dysfunction. These are fascial release, myofascial trigger point treatment, muscle spindle cell and Golgi tendon organ, percussion and myofascial gelosis techniques. These are outside the scope of this discussion with the exception of the somatovisceral effects of myofascial release technique.

Visceral Myofascia

The visceral myofascia wraps the organs of the body and suspends them. It is less elastic than somatic myofascia. Organ prolapse/ptosis may result from laxity of visceral myofascia, while hypertonicity may limit organ motility.

Hypertonicity, diminished elasticity, or limited shear between planes of visceral myofascia is called myofascial, or visceromyofascial, adhesion. For our purposes here, we will be referring to adhesions of the peritoneum (peritoneal adhesions) which are a relatively common outcome of surgery, injury, and systemic or local inflammatory processes. These often result in intestinal obstruction, infertility, pelvic pain, compartment syndromes, edema of the lower limb, and other functional disorders of the lower body. (**Paoletti, 2006**)

Re-epithelialization of surgically incised peritoneum is rapid (~5-8 days). The principal mediators of the healing process are leukocytes, mesothelial cells, and fibrin. Fibrinolytic enzymes normally modulate proliferation of the fibrin gel matrix associated with the healing peritoneum. (**DiZerega, 1997**) Persistent pelvic tissue inflammation prolongs elevation of epidermal growth factor and transforming growth factor-alpha and promotes the development of pelvic adhesions. (**Chegini, 1994**) A smaller amount of inflammation at the site of peritoneal healing produces less adhesion formation. (**Elkins, 1987; Van Der Wal, 2007**)

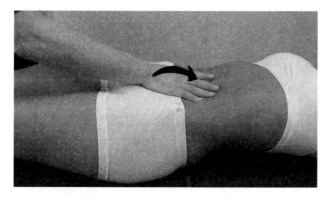

Uterine "Listening" test for organ motility

It has also been proposed that visceral adhesions result, in part, from over-production of reactive oxygen species (ROS) by phagocytes during healing, triggering over-expression of vascular endothelial growth factor. (**Roy, 2004**)

Intra-pelvic adhesions have been associated with a wide range of symptoms that include meteorism, irregular bowel movements, chronic abdominal pain, digestive disorders, infertility, and intestinal obstruction. It has been proposed that 40% of all intestinal obstructions are due to post-surgical adhesions. (**Bruggmann, 2010**)

Visceral Myofascial Therapy by Barral

Barral, a French osteopath, has developed and refined a system of visceral manipulation in the osteopathic tradition. (**Barral, 2005**) Consideration is given for the axes of motility inherent to the individual viscera resulting from embryological development. According to Barral's model, the organs have a motility cycle that is similar to, but independent of, the craniosacral respiratory cycle and, though each organ has unique movement patterns determined by its relationships to the peritoneum, including the mesentery and omenta, the cyclical movement is, ideally, synchronous. Here, motility is defined as the movement inherent to the organ itself, and mobility is the sliding movement between the organs.

This system distinguishes visceral restrictions as being functional and positional. Articular restrictions are functional visceral restrictions that cause loss of organ mobility. Articular restrictions are further classified as adhesions if the restriction results in diminished motility, but leaves mobility intact, and fixations if the restriction results in both diminished motility and mobility. These distinctions become more apparent, it is claimed, with the development of skills necessary to palpate them.

Muscular restrictions (viscerospasms) arise from spasticity or adhesions within the wall of hollow organs such as the small and large intestines, stomach, gall bladder, biliary and pancreatic ducts, and sphincters (e.g., sphincter of Oddi).

Ligamentous laxity, often resulting from lost elasticity due to the chronic restrictive strain of visceral adhesions, can result in ptosis of the organs, a positional restriction.

Barral describes three types of visceral manipulations – direct, indirect, and induction. The direct and indirect techniques are primarily concerned with restoring mobility, though they will indirectly address — at least to some extent – limitations in motility. Induction is primarily concerned with motility and requires highly developed palpatory skills, while indirect technique requires long lever methods utilizing body position to enhance mobility, tissue tension or traction.

Barral offers manipulations for ptosis as well, which will not be described in lieu of the description for the applied kinesiology specific method to follow.

There are two types of direct techniques for articular restrictions – rocking and recoil. In both methods, one should use the distal palmer surface of the hand and fingers and gently traction the tissues perpendicular to the direction of adhesion/restriction for mobility of the soft tissues as determined by palpation and/or a direct AK MMT challenge.

For the rocking method, one should apply short back and forth rocking movements at approximately 10 cycles per minute. The practitioner will feel the underlying tissues become more pliable, elastic, and relaxed under the hands, and when complete, a direct challenge in the direction of previously positive challenge will be negative.

In the recoil method, traction to the tissues is released suddenly and repeated 3-5 times. Again, the tissues should feel softer and more pliable under the hand, and a repeat of the previously positive challenge should now be negative.

For viscerospasm, Barral recommends putting the soft tissues under traction and then mobilizing toward the direction of greatest mobility. Here, too, one can apply an AK MMT challenge in addition to palpation in order to determine when the muscular restriction or viscerospasm has been resolved.

It should be kept in mind that Fryer et al. (2010, 2005) have attempted to detect via palpation tissue irregularities in segmental tissues associated with viscerosomatic dysfunction. Their results failed to show a correlation between palpable changes and irregular motor activity of deep paraspinal muscles. However structural changes in deep and superficial paraspinal muscles and referred hyperalgesia are present in cases of visceral disease and dysfunction. (Giamberardino et al., 2005; Vecchiet et al., 1990) These structural alterations in muscles and functional alterations in strength, measurable with the MMT, may be considered a key factor in the more reliable diagnosis of viscerosomatic dysfunctions.

Visceral Myofascial Therapy by Walker

Walker (2012) describes a method of visceral manipulation utilizing a torquing contact that can be effective for both determining the presence of visceral adhesions and for resolving them.

While holding the hand with the interphalangeal joints in a slightly flexed position ("eagle claw") push lightly into the abdominal wall and turn, or torque, the hand in a counterclockwise direction. The pressure should be firm enough that the force transmits through the abdominal wall and into the underlying viscera, but not be painful. Challenge in this fashion over the suspected area of

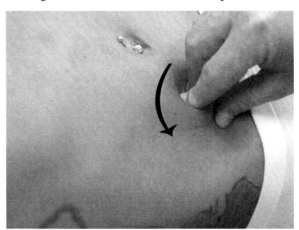

Counter-clockwise sensorimotor challenge followed by MMT

visceral adhesion or screen by challenging the abdominal quadrants.

Challenge with counterclockwise torque over myofascial adhesions will cause a previously strong indicator muscle to weaken, and inspiration or expiration will be found to negate the challenge. Treatment is done with a clockwise torque over the area of positive challenge during the phase of respiration that negates challenge, then easing this pressure with a slight recoil when the patient reaches full inspiration or expiration. Repeat 5-10 times, and re-challenge with the counterclockwise torque to evaluate for effectiveness of correction.

It is useful to include therapy to the segmentally associated area of the spine while treating the area of visceral adhesion to improve results of this therapy. Torque the area of positive visceral challenge on the indicated phase of respiration, and hold the torquing pressure on either full inspiration or full expiration as indicated. Simultaneously, use counter-torque pressure to the skin over the spine opposite the involved viscera. Then continue to hold the torquing pressure on both points while the patient resumes breathing until the clinician and/or the patient feels the visceral tension let go under the anterior (visceral) contact hand.

In chronic cases, it is recommended to include treatment of reflexes associated with the involved viscera/organ as indicated by AK. Finish the manipulation by maintaining pressure over the area of treatment throughout 3-5 complete respiratory cycles. Apply visceral therapy while the patient is in different positions (supine, prone, lateral recumbent, on all fours, and in suspect positions of stress associated with activities of daily living), and simultaneously to surrounding areas of visceral adhesion as indicated by AK MMT challenge.

Walker includes other protocols into his treatment of visceral adhesions, but they are specific to methods taught in association with Neuro-Emotional Technique, (Walker, 2012) which he developed, and are outside the scope of this chapter.

We have found it useful to follow up this protocol with caudal, cephalad, and left and right lateral AK MMT challenges with moderate pressure to the area of visceral adhesion with the distal palmer surface of the hand and fingers. Typically, restriction will be felt that correlates with the direction of positive challenge. Use light thrust in the vector of positive challenge with the same contact as for your challenge and repeat 4-10 times. Re-challenge for correction. Greater elasticity and pliability under the hand during challenge will usually be felt when therapy has been successful.

Visceral Myofascial Therapy by Chaitow (Specific Release Technique)

Chaitow describes a method of visceral therapy emphasizing rapid traction manipulation with a slight rotary component. (Chaitow, 1988) In this form of visceral manipulation, one hand serves as an anchor for the tissues immediately adjacent to the area of adhesion, and the other hand is placed abutting the anchoring hand prior to rapidly pulling away from it.

The active hand makes contact using the ventral pads of the distal aspects of the slightly flexed fingers. Chaitow suggests setting the ventral pad of the middle finger in a slightly forward position relative to the rest of the hand, making it the focus of palpatory diagnosis and manipulation. The lateral aspect of the thumb of the opposite anchoring hand, by pressing into the abdomen posteriorly, serves as a fulcrum point for manipulation by the active hand.

The patient is supine with the knees flexed and the head resting in a slightly flexed position in order to relax the abdominal wall. The doctor stands to the side of the patient opposite the area of adhesion (i.e. if the area of visceral adhesion is in the left lower quadrant of the abdomen, the doctor stands to the right side of the patient). With this example, the doctor will use the right hand as the active hand and the left hand as the anchoring hand. From this position, the doctor can palpate the location and quality of the adhesion by drawing the pads of the active hand across the involved area.

During manipulation, the lateral aspect of the thumb of the anchoring hand is set to the border of the adhesion, and the active hand pulls rapidly away from the anchoring hand. A slight clockwise or counterclockwise torque is added to the linear force.

As an example, if the right hand is the active, manipulating hand, moving the hand laterally and the elbow medially will create a clockwise torque, and moving the hand medially and the elbow laterally will create a counterclockwise torque. Speed will produce the requisite acceleration to develop adequate force necessary to mobilize the adhesion. Apply a very specific, high velocity and low amplitude torsional force to the focal area of the adhesion, analogous to chiropractic adjustment/manipulation to articular fixations/adhesions. Palpation and manipulation is repeated over the area of adhesion until mobility and resiliency to the soft tissues are restored.

Chaitow recommends general neuromuscular treatment to the abdominal region prior to specific release technique and a general soft tissue procedure to lift the viscera into its normal anatomical position. Abdominal and postural exercises are recommended along with diaphragmatic breathing techniques to complement the more specific release technique.

Visceral Repositioning (Ptosis)

Portelli has elegantly combined the concepts of DeJarnette, Rees, Barral, Goodheart and others and described an AK sensorimotor visceral challenge using the MMT for organ position. In this methodology, when an organ (i.e. kidney, large intestine, liver, stomach, pancreas) displaced from its normal position is statically challenged in a vector that increases displacement, an indicator muscle in the organ-muscle relationship of AK will become inhibited. Testing muscles unrelated to the organ will not display a positive challenge, and traction of the organ will not result in a positive challenge unless displacement is present. (**Goodheart, 1992**) Portelli presented this technique for organs of the abdominal and pelvic region originally but later found the protocol applicable to the heart and esophagus as well. (**Portelli, 2001**)

Treatment is implemented with visceral manipulation to replace the organ to its normal anatomical position by tractioning the organ in the direction opposite to the positive challenge. The most common vector for reduction is cephalad since most organs descend due to gravity. (**Smith et al., 2007**) Traction is performed with the distal palmer surface of the hand and fingers increasing to a peak of pressure coincident with the patients' maximum inhalation followed by sudden and sharp coughs. This is repeated several times. Contact pressure is released while the patient exhales, and this maneuver is repeated a few times as needed to negate the positive challenge.

Using the cough to facilitate release of myofascial restriction appears to be very effective, and the increased mobility and elasticity of the tissues can be readily felt. Portelli recommends that the patient reproduce this procedure on oneself before retiring each evening for approximately three months. Testing and correcting for imbalances of the abdominal muscles (rectus abdominus — upper and lower divisions —oblique abdominus, transverse abdominus, and pyramidalis) is recommended.

This technique will commonly be useful for ptosis of the transverse colon and may be associated with recurring ileocecal valve syndrome, weakness of the abdominal wall muscles, premenstrual tension, and constant, nagging pain in the lower abdominal region. (**Portelli, 1987**)

"Soft-Tissue Orthopedics" was developed and researched by Dr. M. L. Rees in conjunction with DeJarnette, (**Rees, 1994**) and was influential in the approaches developed by Portelli. It is based on a unique set of indicators employing techniques involving soft tissue manipulation of the spine, cranium, extremities, organs, glands and other soft tissue, and the balancing of somatopsychic and psychosomatic components. Palpatory indicators were found that correlated spinal, cranial, soft-tissue reflex, acupuncture and other indicator areas that guided the clinician in diagnosing and treating organic disorders. "By non-cutting techniques, to prevent and correct soft tissue deformities, to preserve and improve the function of organs and organ systems and their nerve supply when such function is threatened or impaired by defects, lesions or diseases." (**Rees, 1994**)

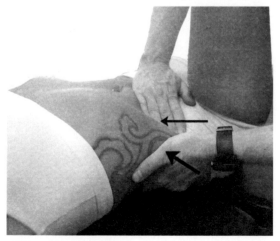

Kidney lift technique
Mobilize the kidney anteriorly, medially, and superiorly

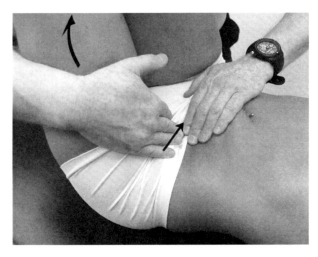

Manipulation of sigmoid colon was shown to produce hypoalgesia in segmentally related spinal structures

An important new investigation by osteopathic researchers in the UK has shown that immediate effects can occur in the paraspinal musculature by improving pressure pain thresholds after manipulation of the sigmoid colon, an approach long used in AK. (**McSweeney et al., 2012**) This study showed that visceral manual therapy produced immediate hypoalgesia that was not systemic but only in somatic structures segmentally related to the organ being mobilized in asymptomatic subjects. The visceral manipulation employed in this study was like the one indicated in AK for the valves of Houston, wherein the supine subject's sigmoid colon was contacted laterally, in the left iliac fossa and drawing it superomedially, and then releasing it, for 1 minute duration.

The Pelvic Region
Chronic Pelvic Pain (CPP)

Though inflammation, infection, and various pathologies must be considered in the differential diagnosis of pelvic pain syndromes, myofascial adhesions must also be considered. The literature now clearly supports the fact that the musculoskeletal system is both a cause and a consequence of CPP. (**Neville et al., 2012; Browning, 2009**) Neville et al. found that women with CPP had "core weakness" in their muscles (particularly the core abdominal and pelvic floor muscles); in these women there was also a higher Total Musculoskeletal Dysfunction (MKSD) Score. Neville et al. continue by reiterating a long-held contention in AK: "If core muscles are weak or the timing of activation and control of the muscles is altered by pain or other disruption, then compensatory mechanisms for muscle imbalances themselves may contribute to the patients' pain complaints." Further, they suggest that "Identifying musculoskeletal factors as a source of CPP would enable earlier and more precise diagnosis, and potentially earlier treatment, of musculoskeletal causes of pelvic pain."

Myofascial adhesions also appear to be a common factor for pelvic pain, with estimates for this etiology being as high as 25%. (**Stones & Mountfield, 2000**) Commonly observed presentations due to myofascial adhesions are deep pelvic pain during certain trunk or pelvic positions, deep respiration, defecation, or sexual intercourse in women (dyspareunia). CPP also appears commonly be a sign of pelvic floor myofascial dysfunction. In general, pelvic floor myofascial trigger points (MTrPs) refer pain to the vagina, penile base, perineum and rectum often with an urgency to urinate. MTrPs in the pelvic floor muscles can mimic coccydynia or levator ani syndromes. (**Simons et al., 1999**)

Extensive coverage of pelvic pain syndromes is presented in *Chronic Pelvic Pain and Dysfunction.* (**Chaitow, 2012**) CPP syndromes include anorectal pain syndrome, bladder pain syndrome, clitoral pain syndrome, endometriosis-associated pain syndrome, interstitial cystitis (IC), pelvic floor muscle pain, pelvic pain syndrome, penile pain syndrome, perineal pain syndrome, post-vasectomy pain syndrome, prostate pain syndrome, pudendal pain syndrome, scrotal pain syndrome, testicular pain syndrome, urethral pain syndrome, vaginal pain syndrome, vestibular pain syndrome, and vulvar pain syndrome. (**Fall, 2010**)

Chaitow cites Anderson (**2006**) concerning the failure of traditional medical therapy to adequately treat CPP "whether involving anti-biotics, anti-androgens, anti-inflammatories, beta-blockers, thermal or surgical therapies, and virtually all phytoceutical approaches."

Masarsky and Browning have developed a model for the pathophysiology of pelvic pain syndromes being potentially related to occult lumber nerve root involvement and resolved through lumbar decompressive therapy (Cox Flexion-Distraction treatment). (**Cuthbert & Rosner, 2012; Browning, 2009, 1988; Masarsky, 2001**) Dr. James Cox, (**personal letter**) founder of the flexion-distraction technique in the chiropractic profession, confirms that "the pudendal plexus is the chiropractor's map in our work to handle genitourinary disease."

Inguinal Hernia

Applied kinesiology treatment for inguinal hernia has been reported to be an effective therapy in both pediatric and elderly patients. Two patients in a case report by Kaufman (**1996**) had considered surgical repair. Treatment consisted of AK therapy to muscles of the pelvic region and groin (origin-insertion, Golgi tendon organ, muscle spindle cell) as well as neurolymphatic reflexes, neurovascular reflexes, ileocecal valve, a category II pelvis, and upper cervical correction. Clinical nutrition directed by AK MMT was included. Patients experienced resolution in 6-8 visits. Though a small case study, the rapid and cost effective resolution of inguinal hernia for patients considering surgical intervention merits consideration of an AK trial in patients with inguinal hernia.

Another case study reported resolution of a right inguinal hernia as well as cryptorchidism after two treatments, one week apart, for a nine month old boy indicating that this approach shows promise. The treatment was chiropractic care to the pelvis and spine directed by AK MMT. (**Maykel, 2003**)

Urinary Tract Dysfunction

Urinary Incontinence

Urinary incontinence (UI) occurs when there is leakage of urine involuntarily, most commonly in older patients. **(Amadhi et al., 2010)** Fantl et al **(1996)** state that incontinence affects 4 out of 10 women and 1 out of 10 men during their lifetime, and about 17% of children below the age of 15. A large post-partum study on the prevalence of UI found that 45% of women experienced UI at 7 years postpartum. Thirty-one percent who were initially continent in the postpartum period became incontinent in the future. **(Wilson et al., 2002)**

Continence depends primarily on the adequate function of two muscular systems – the urethral muscular support system and the sphincteric muscular network of the pelvic floor muscles (PFM). **(Ashton-Miller et al., 2001)** These systems include the levator ani muscle, detrusor muscle, and PFM (coccygeus, obturator externus, obturator internus, gemellus inferior, gemellus superior, and levator ani), as well as the pudendal nerve that emerges from the sacral plexus. The striated muscles of the pelvic floor play an integral role in the closure of the lumen of the urethra and the maintenance of continence. **(DeLancey, 1990)** In women with stress UI, ineffective contraction or control of the pelvic floor muscles permits descent of the bladder neck with inadequate closure of the urethra, resulting in the leakage of urine. **(Steensma et al., 2010)**

The co-morbidities of lumbo-pelvic pain, incontinence and breathing pattern disorders are slowly being elucidated. **(Smith et al., 2008)** Musculoskeletal impairments and specifically muscular imbalances between agonist and antagonist muscles in the pelvis create articular strain and soft tissue stresses which can lead to pain and UI. **(Thompson et al., 2006)**

Current observations suggest that patients with stress incontinence may have imbalances in several lumbo-pelvic muscles which inhibit the PFM and lead to incontinence. **(Smith et al., 2008)** Recent data also indicate that breathing difficulties and incontinence are associated with increased chances for the development of low back pain (LBP), **(Smith et al., 2006)** demonstrating that the interactions between the lumbar and pelvic muscles and joints may be an important consideration in cases of UI.

A recent study assessed strength changes in the PFM using a perineometer (a pressure electromyograph that registers contractions of the PFM) after the application of chiropractic manipulative therapy (CMT). This investigation showed that phasic perineal contraction and basal perineal tonus, force and pressure _increased after CMT_. **(de Almeida et al., 2010)** The duration of these force changes will have to be assessed in subsequent studies of this type.

It was Arnold Kegel who first advocated pelvic

**Urinary incontinence:
Applied Kinesiology case series report**

A retrospective chart review was described on the AK management of 21 patients (age 13 to 90), with a 4 month to 49 year history of daily stress and occasional total urinary incontinence (UI). Each of these patients had associated muscle dysfunction and low-back and/or pelvic pain, with 18 of these patients wearing an incontinence pad throughout the day and night because of unpredictable UI.

Applied kinesiology methods were utilized to diagnose and treat these patients for muscle impairments in the lumbar spine, pelvis and pelvic floor with concomitant resolution of their UI and low back and/or hip pain. Positive MMTs in the pelvis and lumbar spine muscles and particularly the pelvic floor muscles was the most common finding, appearing in every case. Lumbo-sacral nerve root dysfunction was confirmed in 13 of these cases with pain provocation tests (AK sensorimotor challenge); in 8 cases, this sensorimotor challenge was absent. Chiropractic manipulative treatment (CMT) to the soft tissue or articular disturbances related to these inhibitions normalized symptoms in 10 patients, considerably improved 7 cases, and slightly improved 4 cases from between 2 to 6 years.

The following MMT findings were present in this case series of 21 patients with UI:
Inhibition of left, right or bilateral gluteus maximus: 17
Inhibition of left, right or bilateral hamstring: 16
Inhibition of pelvic floor muscles: 14
Inhibition of diaphragm: 11
Inhibition of left, right or bilateral psoas: 11
Inhibition of left, right or bilateral rectus abdominus: 10
Inhibition of left, right or bilateral gluteus medius: 9
Inhibition of left, right or bilateral piriformis: 7
Inhibition of left, right or bilateral rectus femoris: 6

In 13 cases, the AK sensorimotor diagnostic challenge for lumbar disc involvement was positive. When the vertebra above the disc lesion was touched by therapy localization with one hand and the vertebra below with the other hand, therapy localization was positive in these cases. This means that it either strengthened the PFM or other involved muscles or weakened them if they tested initially strong. **(Goodheart, 1975)** This method of localizing disc involvement was added to the routine diagnosis of disc lesions and helped confirm the level of involvement.

An additional applied kinesiology diagnostic approach to disc involvement was the two-handed "AK challenge" of the vertebrae above and below the disc lesion. **(Goodheart, 1975)** The spinous or transverse processes were used as levers, and a separating or compressing force was applied between the two vertebrae. The muscles usually tested with this challenge were the hamstrings if they were initially strong in the particular case under examination. In all 21 patients, this particular sensorimotor challenge was performed; however, the challenge procedure – which guided the use of subsequent Cox flexion distraction decompression treatment – only influenced the function of the involved muscle inhibitions in 13 of the cases. The CMT utilized for the category III and lumbar disc challenges that were present was the flexion-distraction decompression adjustment method. After the Cox flexion distraction decompression treatment of these 13 cases, the positive indicator of a disc lesion using the AK method became negative and improved the UI in all cases.

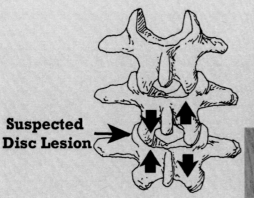

The AK lumbar disc challenge. (Left)
The vertebrae above and below the suspected IVD involvement are challenged by separating or compressing the spinous or transverse processes, then testing a strong indicator muscle for inhibition following the challenge.

Suspected Disc Lesion

Flexion-distraction decompressive treatment corrected the AK sensorimotor challenge to the lumbo-sacral spine. (Right)

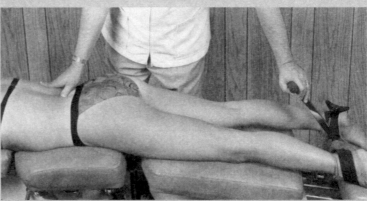

In 8 cases, there was no disc involvement indicated, and yet UI was present — suggesting that lower sacral nerve root dysfunction was not the only etiological cause of UI and PFM inhibitions. In these cases, the treatment for the muscle inhibitions present involved three approaches: (a) CMT for the pelvis, (b) remote treatment to articulations that innervate the tissues involved (the upper cervical spine and the gluteus maximus muscle for example), and (c) percussion to involved myofascial trigger points in the PFM.

Breathing pattern disorders involving the diaphragm muscle were also involved in 10 cases.

In 13 cases, scars from pregnancy, caesarian sections, and one appendectomy were present. Diastasis rectus abdominus was present in 3 of the 13 cases with scars from pregnancy. Gentle pincer palpation of the particular scar, as well as stretching the underlying muscle produced weakening of the same muscle on MMT. AK mechanical treatment to the scar (using a Percussion device) (**IMPAC, 2012**) abolished the pincer palpation finding to the scars and the MMT weakness in the underlying muscle. (**Cuthbert, 2002**)

In 11 cases there was a positive finding of pincer palpation to the PFM. Treatment for the myofascial trigger points (MTrPs) found in the patient's PFM was made with a Percussor instrument. After 2 minutes of percussion upon the MTrPs found in the PFM, the AK "pincer palpation" test became negative, and pressure on the MTrPs that previously produced either referred pain or muscle inhibitions in previously strong indicator muscles no longer occurred. (**Walther, 2000**)

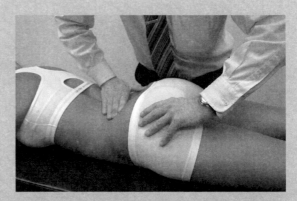

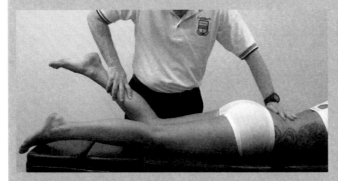

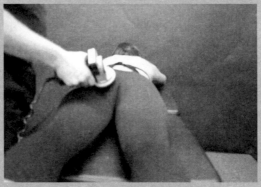

Pincer palpation of the PFM (a) produced weakening of the previously strong hamstrings. (b) Percussion on the PFM (c) corrected this finding and reduced palpation tenderness and referred pain from the MTrPs present in the PFM.

In six cases there had been previous spinal (1), abdominal (1), or pelvic organ (4) surgeries; in 1 of these cases (1), there was normalization of the UI condition; in 3 cases, considerable improvement; and in 2 of these cases, only slight improvement.

The first consultation, examination and treatment for each of these patients lasted for 1 hour. Follow up treatment sessions were required covering 1 to 13 treatment(s) for this cohort to reach maximum improvement for their UI, covering a period spanning 1 day to 6 weeks. The diagnoses of muscular inhibitions related to articular and soft tissue disorders of the pelvis and lumbar spine and their treatment using CMT were effective in attaining the resolution (n=10), considerable improvement (n=7), or slight improvement (n=4) of daily stress and occasional total UI in these cases. These burdens remained corrected as seen by yearly follow up examinations for the past 2 to 6 years.

It should be emphasized that improvement in patients with long-standing UI occurred in 10 cases, even in two patients with UI that had been present for 49 and 40 years, respectively. In these 21 cases, positive MMT of the pelvic and lumbar spinal muscles (particularly the PFM) was the most common manifestation in every case. The AK MMT identified the inhibited PFM or the inhibited muscles related to the PFM anatomically or functionally.

From these considerations and data, there may be reason to suspect that the patterns of muscle weakness associated with UI and disclosed by MMT methods may be of clinical utility in guiding treatment for this disturbing condition.

floor muscle strengthening and retraining for stress incontinence, (**Kegel, 1948**) indicating his recognition of the importance of muscle inhibitions in cases of UI. The usefulness of the AK MMT approach in this model would be the identification of the inhibited muscles involved, and the AK treatment approaches used to immediately address these inhibitions. A number of other reports have been made on the successful use of chiropractic treatment for elderly patients with UI. (**Kreitz & Aker, 1994**) Stude et al reported a case study of a 14-year-old female treated with CMT who recovered completely from traumatically induced UI. (**Stude et al., 1998**) The AK approach to a post-appendectomy induced case of UI has been recently described as well. (**Cuthbert & Rosner, 2011**) Duffy (**2004**) successfully resolved a complex case of UI as well, with treatments including CMT, soft tissue manipulation (uterine lift technique), and AK muscle-reflex therapies, including neurolymphatic reflexes and inactivation of myofascial trigger points for muscles of the pelvis, especially the levator ani. CMT has been shown to be effective in other reports on bladder control problems. (**Reed et al., 1994; Blomerth, 1994; LeBoeuf et al., 1991**) Allen, (**2000**) a Diplomate chiropractic neurologist, has noted a relationship between fixation of the cervico-dorsal spine and inhibition of these bladder related muscles.

Nocturia

Goodheart states that nocturia in children is commonly a result of impaired regulation of CO_2 levels during sleep due to impaired responses of the respiratory centers in the lower brain stem. CMT and AK reflex techniques to the phrenic and intercostal nerves are the recommended therapy. (**Goodheart, 2003**)

Applied kinesiology was successful in treating a four year old male with nocturia. Treatment consisted of chiropractic and craniosacral therapy and supplemental phosphorylated B complex emphasizing thiamin. Correction of foot pronation was part of the therapy, which may potentially influence bladder meridian function. This case study included mention of testing for ADH for possible diabetes insipidus. This reflects good practice in applied kinesiology by including all the clinical tools a clinician has to draw from to arrive at an accurate differential diagnosis. (**Duffy, 2003**)

Additional causative factors for nocturia may be food allergy, (**Mungan, 2005**) sensitivity to artificial flavors, colors, sweeteners, (**Salzman, 1976**) and thiamine deficiency. (**Vanhulle, 1997**) Thiamin deficiency has been reported to be associated with increased nocturnal urination, a short breath holding time, (20 seconds or less) low pulse rate, low body temperature, and failure of the pulse rate to increase as predicted with exercise. (**Goodheart, 1989a**)

Cystitis and interstitial cystitis

Peters calls for evaluation of the pelvic floor when patients diagnosed with interstitial cystitis, urinary frequency and urgency, and pelvic pain have found therapies to the bladder epithelium unsuccessful. (**Peters, 2006**) The evidence for this comes from a study that found an 87% prevalence of pelvic floor dysfunction, as evidenced by levator ani hyperalgesia, in women with IC. (**Peters, 2007**)

Pelvic floor trigger point therapy appears to be effective for patients with chronic interstitial cystitis (IC) and urgency-frequency syndrome. (UFS) A study of 52 subjects (45 women, 7 men) with either interstitial cystitis or urgency-frequency syndrome who were treated at a frequency of 1-2 sessions per week for 8-12 weeks showed moderate to marked improvement or complete resolution of the UFS group (83%) and moderate to marked improvement in the IC group (70%). (**Weiss, 2001**)

Deficiency of vitamin A has been associated with cystitis. Rats fed a vitamin A deficient diet have been found to have a high prevalence of pyelonephritis (68%), cystitis (66%), nephrolithiasis (5%), and squamous metaplasia of the transitional epithelium (100%). (**Munday, 2009**) Another study in rats found that vitamin A deficiency resulted in squamous metaplasia of the urinary bladder and high incidences of cystitis, ureteritis, and pyelonephritis. (**Cohen, 1976**) Consider vitamin A deficiency in patients with chronic cystitis as directed by AK MMT and other clinical findings. Vitamin A deficiency is commonly associated with systemic lymphatic congestion as seen when patients need the AK retrograde lymphatic technique. (**Goodheart, 1979**) Sprieser (**2002**) reports on 50 patients with interstitial cystitis who improved using AK techniques involving treatment to the bladder meridian.

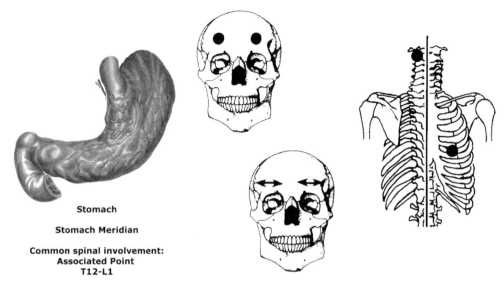

Stomach
Stomach Meridian
Common spinal involvement:
Associated Point
T12-L1

Reflexes for stomach

AK and Digestive Dysfunction

Here we are concerned with the viscera presenting signs and symptoms associated with the pelvic region - small intestine, ileocecal valve, and large intestine. This area has been covered in the AK *Synopsis* by Walther **(2000)** and our intent is to complement that text.

Stomach and small intestine

Hydrochloric Acid

"Too much acid in the stomach" is considered by many to be the characteristic situation when there is a problem with heartburn, or gastroesophageal reflux disease (GERD). **(Bredenoord, 2012)** Actual evaluation of the function of the body reveals that the opposite is true, as has been confirmed by AK analysis and treatment results for decades. Patients with reflux symptoms often do not have excessive esophageal acid exposure, and patients with severe gastroesophageal reflux often do not have reflux symptoms. Reflux characteristics other than acidity, such as the presence of bile, pepsin, liquid, or gas in reflux, and the proximal extent or volume of reflux, may also contribute to symptom perception. Factors contributing to greater esophageal sensitivity may include impaired mucosal barrier function, peripherally mediated esophageal sensitivity (enhanced esophageal receptor signaling), and centrally mediated esophageal sensitivity (physiological stressors, sensitization of spinal sensory neurons).

When there is a disturbance in hydrochloric acid balance, it is usually a *lack* of hydrochloric acid which is present rather than an excessive amount of it. It is well documented that as people age, less hydrochloric acid is produced. Unfortunately, the symptoms of hyperchlorhydria and hypochlorhydria somewhat parallel each other, with both causing a burning-type sensation and discomfort in the epigastrium. A rule of thumb (but by no means totally accurate) is that the person who has burning on an empty stomach has hyperchlorhydria and a person who develops pain after eating has hypochlorhydria. Gas immediately after eating, especially a protein meal, is indicative of deficient hydrochloric acid; gas that develops lower in the abdomen several hours after eating is indicative of a lack of proteolytic-pancreatic enzymes in the intestinal tract. These are strictly rules of thumb whereas a more accurate method is to test, therapy localize and challenge the mechanisms involved.

It is unfortunate that many people who have hypochlorhydria often obtain relief from symptoms by taking antacids. The mechanism for relief is an acid rebound as a result of insulting the body with an alkaline substance. Prolonged treatment of this nature, whether prescribed by a physician or self-administered patent medications, leads to many additional health problems, such as mineral deficiencies and protein imbalance leading to degenerative joint disease. **(Levin, 2004; Jensen & Anderson, 1990)**

Hiatal Hernia

The hiatal hernia can mimic the symptoms of hypo-or hyperchlorhydria. There is usually a substernal burning as the acid regurgitates into the esophagus; however, the burning pain can be in the epigastrium area, radiating to the chest and other places. The evaluation and treatment of hiatal hernia and diaphragm muscle weakness is indicated in all cases of apparent digestive disturbances. Burstein reviewed the case records of 92 patients in his AK-practice and found hiatal hernia dysfunction in 32 out of 92 new patients (34.7%), making it a common finding. **(Burstein, 1990)**

The use of chemical challenging by asking the patient to chew or suck on a tablet is invaluable, not only to determine what nutritional factor is needed, but also to find what is detrimental to body function. The patient in the case report above has no question that the exact answer

Case Report: AK and severe digestive burning

A 53-year old female presented herself for treatment of severe digestive burning. She had previously been a patient for temporomandibular joint disorders and bilateral knee pains. When she originally came to the office for treatment of those conditions, she was extremely skeptical of natural health care because of heavy indoctrination throughout her career as an emergency room nurse that medication was the best solution for any health problem. Our original treatment was successful for her. However, when the digestive disturbance occurred, she did not think of natural health care; she went back to her standard approach of allopathic medicine. Upper GI x-rays were taken with a barium sulfate swallow as well as a gall bladder study. Thorough physical evaluation could find no problem to cause the symptoms she was experiencing. The diagnosis of hyperchlorhydria was made, and the physician told the patient that she had a "nervous stomach," making it necessary for her to take antacid medication with every meal for the rest of her life. The patient did this for about three weeks, and the symptoms completely abated as long as she took the medication. If she skipped the medication with any meal, the symptoms returned – interestingly enough, more severe that before.

Fortunately, this patient retained enough of our previous broad-scope natural health care education to reject taking medication for the rest of her life; she decided to come in for evaluation and treatment. Upon examination, weak bilateral pectoralis major (clavicular division) muscles were found, which were strengthened by half an inspiration held while testing. This indicated a temporal bulge cranial fault (affecting the vagus nerve), which was confirmed by challenge. A parietal descent cranial fault was also found on the opposite side. Furthermore, a diaphragm deficiency was found, correlating with the substernal burning pain experienced by the patient. The key to the diaphragm weakness was found to be a reactive psoas. Because of the difficulty in obtaining the corrections, the patient was asked to walk for a short distance and then re-tested to make certain the bilateral pectoralis major (clavicular division) weakness and the diaphragm weakness did not return. After walking, the condition was present exactly as it had been prior to treatment. Another attempt was made to correct the temporal bulge causing the pectoralis weakness and to correct the reactive psoas-diaphragm complex. Again, correction was very difficult, and the condition returned after the patient walked. When a correction is difficult to obtain and a condition returns rapidly, some problem not yet located is recreating the involvement.

The patient was asked when she had last taken an antacid tablet; she answered approximately three hours earlier. With no further attempt to correct the temporal bulge and reactive psoas-diaphragm complex, she was given betaine hydrochloride to suck on. After she had sucked on the tablet long enough to get a good taste and to swallow some of the material, the pattern was re-tested. The pectoralis major (clavicular division) was now found bilaterally strong, and there was no evidence of temporal bulge or parietal descent cranial faults, and the reactive psoas-diaphragm complex was no longer present. The betaine hydrochloride made all these changes with no other treatment given. Since the patient had some of the antacid tablets she had been taking with her, she was then challenged with this formula. By simply sucking on the antacid tablet for no longer than 15 seconds, the temporal bulge/parietal descent complex returned, as well as the bilateral pectoralis major (clavicular division) weakness and the reactive psoas-diaphragm complex. Again, she sucked on a betaine hydrochloride tablet and the entire abnormal picture was alleviated, and walking no longer caused its return. The patient was given betaine hydrochloride tablets, one with each meal, for two weeks. She now frequently eats chili made with jalapeno peppers (a staple in Pueblo, Colorado) and any other spicy foods she desires. There has been no return of her symptoms of hyper-or-hypochlorhydria.

to her problem was found; she is also grateful that she does not have to take medication designed to override her system for the rest of her life. It is this type of correction and education that has proved itself to be invaluable in the word-of-mouth development of referrals for digestive disturbances in clinical practice.

In hydrochloric acid disturbances, there is typically a bilateral pectoralis major (clavicular division) weakness, as indicated previously. If this bilateral weakness is not present — though the symptomatic picture indicates a hydrochloric acid disturbance — test for a bilateral lower trapezius weakness. This can mask the bilateral pectoralis major (clavicular division) weakness. If there is a bilateral lower trapezius weakness, there will be a dorsal-lumbar fixation. After the fixation is corrected, the weakness of the pectoralis muscle will usually emerge.

There are many functions attributed to hydrochloric acid. One activity that is not widely discussed is the disinfecting action of hydrochloric acid upon bacteria which is swallowed. As the hydrochloric acid concentration in the stomach is lowered, the bactericide effect is diminished. Many germs can pass through the stomach without this protective effect. Hydrochloric acid is important in the first step of protein digestion. It splits the protein molecules and is necessary for the action of the proteolytic enzyme pepsin. Hydrochloric acid is responsible for splitting disaccharides and monosaccharides and for stimulating the flow of bile and pancreatic juice. Bacterial action continues in the stomach as long as the food bolus is alkaline. Basically there is a continuation of salivary digestion in the alkaline medium, where ptyalin (or salivary amylase) continues its work on carbohydrates. Micro-organisms can also continue working in an alkaline stomach to split carbohydrates into gases and organic acids. This contributes to the gassiness and burpiness so often observed in hypochlorhydria patients shortly after eating.

The stomach wall and duodenal area is adequately protected from hydrochloric acid by mucous secretion and an adequate blood supply. If the walls of the stomach and duodenal area are ischemic, there is a lowered resistance to

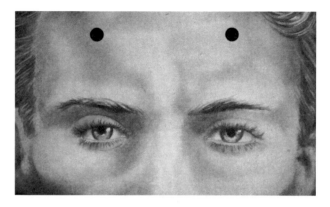
Stomach neurovascular reflex points

breakdown of tissue by hydrochloric acid and the action of pepsin. The neurovascular reflex on the frontal area is often found to be active in these cases. This reflex, of course, correlates with the blood vascular supply to the stomach; it also correlates with general stomach function, including hydrochloric acid production.

Protein Digestion

Many individuals have a protein deficiency, although high levels of protein are eaten in western countries. **(Gaby, 2006)** First, it is necessary to be certain that all the essential amino acids are present in the diet. Goodheart has an illustration of what takes place when one essential amino acid is missing. If you have a wooden barrel made up of staves, it is only functional in holding liquid up to the height of the lowest stave. The staves of the barrel can be likened to the essential amino acids; if one amino acid is deficient, the body will only utilize the amino acids present to the level of the one deficient. With many types of diets, especially vegetarian, it is very difficult to be certain all the essential amino acids are present in sufficient amounts. One method by which this can be evaluated is to have the patient fill out a frequency diet schedule, or keep an actual diary of what he or she eats for one week. The frequency diet schedule asks questions about how often certain types of foods are eaten throughout a month; the diary, of course, evaluates the actual diet of the individual. There are several companies that provide these forms and have computer tie-ins which evaluate the information provided by the patient, giving the amount of amino acids and other nutrients in the particular patient's diet.

Even if there is adequate protein in the diet, inappropriate digestion of protein often occurs, usually because of achlorhydria or hypochlorhydria. Attention should be given to the small intestine and pancreas for the production of the appropriate enzymes, and the liver should be thoroughly evaluated.

Indication of protein deficiency is most readily obtained by observing body language. In protein deficiency, there will be poor growth and quality of fingernails; the hair will be dry and brittle, with cracked ends, and it will tend to fall out; the skin will be dry, hard, and tend to wrinkle, with breaking and splitting. Blood laboratory findings may appear normal to those who are not aware of the body's attempt to maintain normalcy under adverse conditions. We must remember in evaluating blood laboratory that the blood stream is nothing more than a distribution route for nutrients within the body. For example, there is a concentration of 30 times more protein in the joint surfaces than in the blood stream. If the blood stream protein level is down, the body will draw from the joint surfaces to attempt to maintain normal blood protein levels. Rubel discusses the evaluation of the albumin-globulin ratio (AG ratio) as being more significant than evaluating the total protein levels. **(Rubel, 1959)** If protein is pulled from tissues, the AG ratio will change, with globulin rising. If the globulin is above 3 gm%, this indicates that the body is withdrawing protein from tissue to make up the deficient albumin levels.

Protein deficiency is significant in the total digestive process, because enzymes are manufactured from protein. When there is a relative hypoproteinemia, the body goes into a protein-sparing effect in which it reduces various physiologic processes, including enzyme production. Lowered enzyme production causes many phases of the digestive process to be lowered. We see here how one digestive fault – such as hypochlorhydria or achlorhydria – can cause a snowballing effect in the total digestive process. The correction of this problem can be as simple as correcting a temporal bulge cranial fault, or as complicated as searching each and every factor present in a total series of events.

Protein and arthritis

A protein deficiency is involved in many types of arthritis, not just in degenerative arthritis. Before discussing the actual effect on the joint, a brief discussion of protein evaluation and the types of imbalance is paramount.

The many utilizations of protein by every cell, organ, tissue, and function of the human body are well documented. Some specific areas of protein utilization and protein movement are important in the various forms of arthritis. Rubel **(1959)** reviews findings at the University of Illinois indicating that for every gram of protein in the blood stream, a healthy joint surface requires 30 times more. In the presence of a negative nitrogen balance, where there is less protein taken in than the needs of the body and excretion, there is a protein shift from one compartment of the body to another to supply adequate protein.

Frequently the negative nitrogen balance is not caused by a lack of dietary protein intake; it may be caused by the individual's inability to absorb and utilize the protein in the diet. Achlorhydria or hypochlorhydria, especially in aging people, contributes considerably to the inability to absorb protein. The acid content of the stomach has been determined – in the average normal person – to be ideal at the age of 25; at age 40, there is 15% less hydrochloric acid in the stomach; at age 65 there is only 15% of the amount produced at 25 years of age.

Indication of hypochlorhydria or achlorhydria is gaseous distension of the stomach shortly after eating. This pressure is frequently felt high in the abdominal area, perhaps even before the meal is finished, and is relieved by belching. Failure to eliminate the gas often causes symptoms in the diaphragm, resulting in short-windedness and cardiac palpitation. Pruritis ani may be present due to the fact that the fecal material is too alkaline, never having been completely acidified in the stomach. As the food bolus proceeds down into the small intestine, there is action by secretions from the intestinal wall and from the pancreas. The pancreas secretions are stimulated by the food bolus in the duodenum and are influenced by the degree of acidity developed in the stomach. A high acid level will increase the secretion and concentration of the enzymes manufactured and secreted by the pancreas.

Enzymes developed by the pancreas are, of course, protein in nature; when there is a tendency for the body to conserve protein. A deficiency of pancreatic enzymes may be a result of hypoproteinemia.

When considering protein utilization, an important factor is that all the essential amino acids must be present for protein utilization. If there is a deficiency of one amino acid, the other amino acids will be utilized only to the level of the deficient amino acid.

In the presence of a negative nitrogen balance, the body will pull protein from the highly concentrated joint surfaces and synovial membranes. After this withdrawal has taken place to any degree, normal joint movement becomes trauma, with the consequent inflammatory reactions to trauma.

Protein deficiency is one of the mechanisms by which cortisone products and ACTH give relief from arthritis. These medications cause release of albumin from muscle tissue, hence making it available to the blood stream and joint surfaces and decreasing the irritability, inflammation, and arthritic manifestations of the joint. This is accomplished at the expense of muscle tissue; if it is continued for long, it leads to protein deficiency in other tissues, even though the arthritic process has been relieved.

The same mechanism can be applied to bursitis. The negative nitrogen balance causes withdrawal of protein from the synovial caps. Calcium in the body is carried united with albumin; when the albumin has been withdrawn, calcium is left as a precipitate, and consequently calcific bursitis develops.

There is a definite correlation between alkalosis and the protein deficiency mechanism of calcium precipitation and improper disposition. Poor absorption of protein correlates with a systemic alkalosis. The administration of betaine hydrochloride aids in the absorption of protein, and is also significant treatment toward the relative alkalosis.

A very common reason for negative nitrogen balance is excessive gluconeogenesis as the result of a sugar handling stress, such as hyperinsulinism or hypoadrenia. In this condition there is also usually a bowel problem resulting from excessive refined carbohydrates, which creates a malabsorption syndrome affecting protein utilization.

A deficiency in albumin (or other amino acids) will cause an increase of peripheral interstitial tissue fluid, and can frequently be seen in clinical evaluation as a pitting edema. Edema, ascites, thoracic cage fluid, semicircular canal fluid, or any peripheral tissue fluid should be evidence to examine for protein disturbances. **(See Edema section below)**

Laboratory evaluation of protein is helpful; however, more than just the total protein level must be observed. The albumin should be well within the normal range, and the albumin-globulin ratio (A-G ratio) should be normal. When the globulin is high in ratio to the albumin, this indicates that the body is withdrawing albumin from a body compartment, such as the joint surfaces or synovial membrane. If the total protein and the A-G ratio are normal, but the serum globulin is above 3 gm.%, a protein deficiency is indicated. The protein in the blood stream is kept normal at the expense of other tissues, which give up protein. When albumin is withdrawn from tissues to make up blood protein, globulin, which is a much larger molecule, is also pulled, causing a rise in the serum globulin level.

In the presence of protein deficiency, even though the total blood protein may appear normal, there will be many symptoms and conditions – such as anemia, fatigue, sensitivity to cold, myalgia, and possibly atrophy.

As the joint surfaces and synovial membranes are depleted of protein in a protein deficient condition, pathology is more readily developed. The intra-articular fluid balance is disturbed and barometric pressure changes cause the fluid within the joint to painfully expand the joint capsule. As protein is withdrawn, calcium – which has been held in combination with the protein – becomes separated and is precipitated, depositing in bone tissue as hypertrophic arthritis.

Treatment depends upon evaluating the patient's protein intake, including all of the essential amino acids, hydrochloric acid, and production of enzymes from the small intestine and the pancreas. General bowel action must be evaluated and treated if indicated. The usual approaches of applied kinesiology to these organs are indicated, with specific indication to evaluate for the temporal bulge cranial fault and spinal lesions for the improvement of hydrochloric acid production. The liver must be evaluated thoroughly for its role in the production of albumin, and its other roles in protein metabolism.

The patient probably will not gain relief from the joint problem immediately upon obtaining improved absorption of protein. The first protein increase only starts making up the protein deficiency within the tissues, caused by prolonged protein yielding by the tissues. Until sufficient time has been allowed for the tissues to rebuild, there will still be inadequate tissue capability.

When the small intestine is found to be involved, applied kinesiology examination of the pancreas, liver, and gall bladder is indicated. For effective function of the small intestine, it is necessary that an alkaline medium be maintained. This comes primarily from the secretion of sodium bicarbonate and potassium bicarbonate from the pancreas. When there is inappropriate response, either in weakness of response or failure to hold response by the usual nutritional factors for the pancreas, consider adding sodium bicarbonate or potassium bicarbonate one-half hour after meals. The amount should be approximately one-quarter teaspoon after each meal; be certain that the patient does not take it with the meal because it will counteract the effect of hydrochloric acid in the stomach. This alkaline medium in the small intestine is necessary for lipolytic and proteolytic action.

Acid-Alkaline Balance

Acid-alkaline balance within the body is very much stabilized by the buffer system. Actual measuring of the pH of the blood may show very little shift taking place. A shift toward acid or alkaline can take place systemically as the result of varied diets and other metabolic processes, but no major changes are made in the actual pH of the blood, which is held to its close tolerance by the buffer system. A relative alkalinity may cause calcium to be thrown out of solution and deposited into connective tissue, frequently those of the hard-working joints of the body. Deposited calcium and other mineralization are often seen on x-ray in the costal-cartilage of individuals with degenerative joint disease and with a tendency toward calcific bursitis. Previously, measurement of the salivary pH was used as an indicator of the blood pH. It was found that the saliva does not always tend toward an alkaline level when the blood pH varies toward alkaline. There has been, however, a correlation that when the salivary pH is off toward acid or alkaline there is a probability of an acid base balance shift in the blood stream – not necessarily paralleling the shift of the saliva. The normal saliva pH range is about 6.5 – 7.0. The salivary pH can be measured effectively using indicator paper available in rolls from several of the nutritional supply houses and from medical supply stores.

The folk remedy of apple cider vinegar for "arthritis" often improves a relative alkalosis, whether or not a true arthritis condition is present. (**Jarvis, 1985**) It tends to eliminate the joint pains, general achiness upon arising in the morning, etc.

Relative alkalosis commonly found in degenerative joint disease is also correlated with calcific bursitis. Before the diagnosis of calcific bursitis is made, the calcium deposit should be observed on x-ray. To observe this deposit, it is often necessary to do many x-rays of a joint, looking at the articulation from various angles. This is particularly true in the subdeltoid and supraspinatus calcific bursitis.

The shoulder is where more specific bursitis is found than any other area. The strain of muscles working in disharmony can be the possible cause of mechanical stress, complicated by metabolic problems like an acid-alkaline imbalance. The muscles of the shoulder should be evaluated for reactive muscles, weak muscles, hypertonic or shortened muscles, etc. As with many shoulder problems, the entire body's structural balance from the feet to the calvarium must be correlated.

An excellent method to shift an individual from a relative alkalosis to normal is nutritional supplementation in the form of acid calcium. Along with the acid calcium, vitamins A, C, and E are utilized. (**Schmitt, 1990**) When a low potency source of these nutritional products is used, it is given on an hourly basis for the first few days. An indication that the patient is shifting toward a relative acid

SYMPTOMS OF ALKALOSIS	SYMPTOMS OF ACIDOSIS
Heavy, slow pulse	**Suffocation symptoms**
	Frequent sighing
Stiffness of joints	Breathlessness
	Dislike for closed rooms
Symptoms after resting:	High altitude discomfort
Night Cramps	Irregular respiration
Cough	**Hyper-irritability symptoms**
Circulation disturbances	Voice affected during stress
Aching worse upon arising	Tachycardia
in the morning, eases with activity	Photophobia
Skin symptoms	Dysphagia
Dry	Restlessness
Thickened	Insomnia
Burning	Frequent contraction of erector
Itching	papillae
Formication	"Cold sweat" type perspiration
Rapid clotting time	**Dehydration symptoms:**
Feel worse after eating	Dryness of skin
	Dryness of mouth
	Dry, hard stool
	Diminished urination
	Diminished perspiration

Applied Kinesiology and Systemic Conditions of the Lower Body

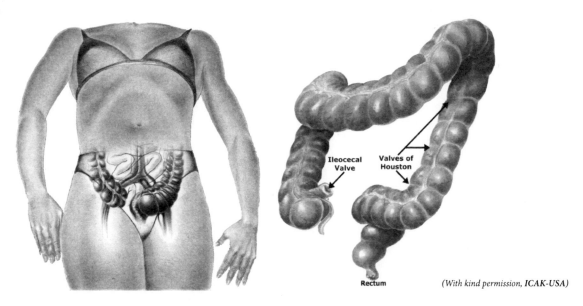

The Large intestine and AK

balance is when they start yawning and shows other indications of shortness of breath. At this time, reduce the acid calcium dosage to one tablet. Continue to reduce it if these breathlessness symptoms do not abate.

If the symptoms are due entirely to calcium precipitation from a relative alkalosis, the patient should get relief within 24 to 48 hours. Within approximately one week, the calcium deposit in the bursa will be improved on x-ray. Calcium deposits observable on x-ray are not solid calcium but are, rather, of a toothpaste consistency *in vivo*.

Applied Kinesiology and the Large Intestine

The large intestine is primarily thought of as a "garbage dump". Treatment is directed to the large intestine only when the garbage is dumped too fast (diarrhea) or too slow (constipation), or when it aches or gives symptoms of discomfort. A close look at large intestine function and dysfunction reveals the etiology for many health problems which are rapidly on the rise today. Every new patient entering natural health care should have his or her large intestine evaluated, at least on the consultation level, by the doctor.

It is interesting to note that the diseases associated with large intestine dysfunction are diseases which have developed in the last century. Their development has paralleled the dietary change, which is the primary cause of large intestine dysfunction. Two primary causes of the large intestine dysfunction are 1) an increased usage of refined carbohydrates, especially white sugar and white flour, and 2) the reduction of dietary roughage. Epidemiological studies of tribal communities in Africa and Oceania, who have retained their original diets, show a low incidence of these diseases. **(Price, 2003)** Both of these groups have diets which are high in fiber and low in refined carbohydrates. When these ethnic groups migrate to Europe or the United States and change their dietary habits, they become just as susceptible to the diseases as the Europeans and the Americans. As we study individual conditions, we find a common etiology. Accordingly we find the treatment to be similar and relatively simple, considering the major implications of the conditions involved.

Large Intestine and Rectal Cancer

The activity in the large intestine varies greatly between those on a high roughage diet and those on a low roughage diet. An individual on a high roughage diet has

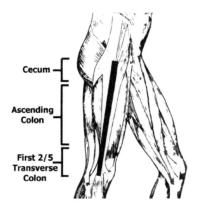
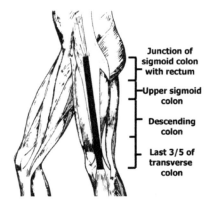

The location of active neurolymphatic reflexes for the tensor fascia lata helps diagnose which area of the large intestine is involved. Bilateral TFL weakness indicates possible hypochromic anemia. Check adductors when TFL is weak.

~ 279 ~

voluminous bowel movements and the material moves rapidly through the large intestine, with 80% of the material removed from the bowel in less than a day and a half. Those on a refined, low roughage, American-type diet have a transit time through the body of about 3 to 4 days, with the material being in the large intestine most of that time.

There are two significant types of bacteria in the large intestine of the individual who consumes the typical low-roughage Standard American Diet, or SAD. These are known as *bacteroides* and *bifidobacteria*. (**Hill et al., 1971**) These cause the bile acid cholic acid to be converted to apcholic acid, and the bile acid deoxycholic acid to be turned into 3-methyl-cholanthrene. Both of these chemicals developed from the bile acids are carcinogens. The basic fact that these carcinogens are developed in the colon, added to the fact that there is stasis of the colon and that the consequent slow movement allows the chemicals to be in contact with the colon walls for long periods of time, leads to the ominous possibility of cancer developing.

The dominant bacteria in individuals with high fiber content diets are *streptococcus* and *lactobacillus*. These bacteria do not break down the bile acids, and these individuals have a much higher amount of bile acids in the feces and serum. (**Antonis & Bershon, 1962**)

As we ponder the tragic increase of large intestine malignancies, it stands to reason that the cause of the condition is contact with carcinogenic substances. It is significant that the incidence of cancer increases along the progression of the large intestine, the highest incidence being at the rectum. The only exception to this is the cecum, which also has a high occurrence of cancer.

The answer to the problem is prevention. Regaining normal bowel chemistry and bacteria, along with the rapid removal of waste products, is the preventative approach to cancers in the large intestine. (**Cheraskin et al., 1968**)

Ileocecal Valve Syndrome

The ileocecal valve is at the very end of the small intestine (ileum) and connects it to the first part of the large intestine. If you imagine a line from the umbilicus to the right anterior superior iliac spine, this valve would be located just below the midpoint of that line. Dysfunction of the ileocecal valve has, in one report, been associated with low back pain (86.6%). (**Pollard, 2006**) A commonly observed sign in AK practice associated with an active ileocecal valve syndrome is acute onset of disabling pain, typically in the low back, but possibly anywhere in the body, with no apparent etiology. (**Shin, 2004**) Commonly, the low back pain will be mistaken for a lumbar disc syndrome. Just as frequently, symptomatology in the digestive system will be the primary symptom.

The valves of Houston, or transverse folds of the rectum (TFR), though different in structure and function from the ileocecal valve, do have a role in regulating transit through the intestinal tract. Most folds extend beyond the middle of the rectal lumen, are thicker at the base, and contain smooth muscle fibers. The TFR divide the rectum into compartments with an alternating side-to-side arrangement that creates a compartmentalization and shelving action, potentially responsible for retardation of stool movement in the rectum. (**Shafik, 2001**) The valves of Houston can produce similar symptoms.

Symptoms that can occur with "open" or "closed" ileocecal valve syndromes include pain in the area of the valve (can be mistaken for right ovary or appendix pain), dizziness, low back pain, shoulder pain (mainly on the right side), nausea, faintness, paleness, sudden thirst, bad breath, black circles under the eyes, and ribbon like stools. Duffy reports on the association of open ileocecal valves in 12 cases of carpal tunnel syndrome. (**Duffy, 1994**) Ileocecal valve syndrome may also aggravate dysmenorrhea and endometriosis and should be evaluated in these conditions. Alis & Alis (**2004**) present a case of symptomatic endometriosis successfully treated with AK methods.

Open ileocecal valves often cause patients to suffer from loose bowels, emotional lability, (**Lebowitz, 1984**) and exhibit vitamin C deficiencies (their vitamin C is exhausted detoxifying the backflow of fecal material). Closed ileocecal valves tend to cause constipation. (**Maykel, 2004**) In many AK case and case series reports, diarrhea caused by the open ileocecal valve syndrome would improve the moment the ileocecal valve was treated.

Schmitt and Morantz (**1981**) observe that the ICV is related to the kidney acupuncture meridian. That is why the use of K-7 and B-58 for ICV problems has been standard treatments within AK for the ICV syndrome.

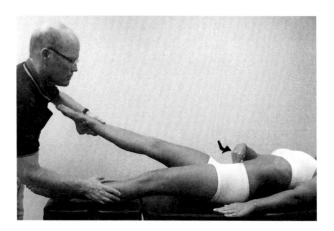 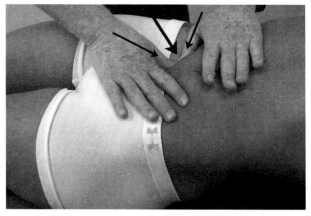

Therapy localization and challenge to the ileocecal valve followed by mobilization of the ICV

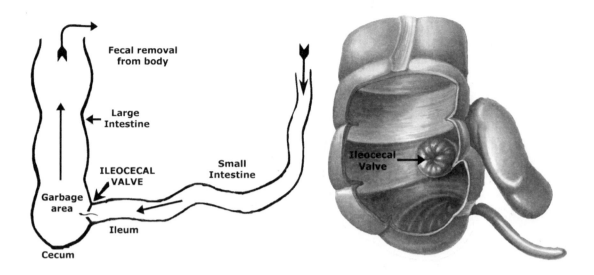

Ileocecal valve

"It therefore often becomes necessary to take the entire acupuncture system into consideration in order to correct the recurrent ICV. For example, occasionally when you back up the acupuncture system using the law of 5 Elements, the problem turns out to be coming from a lung meridian involvement which is related to a T-3 subluxation. This makes no sense whatsoever, neurologically, but on an acupuncture basis, it explains how a T-3 subluxation could cause an ICV problem." **(Schmitt & Morantz, 1981)**

Schmitt **(2002)** also presents the concepts of Michael Gershon, MD (author of *The Second Brain: A Groundbreaking New Understanding of Nervous Disorders of the Stomach and Intestine*) and shows the neurologic rationale behind the AK approach to the enteric nervous system. Fatty acids anywhere in the intestinal lumen stimulate the ENS to decrease peristalsis of the gut at the ileocecal area, what is called in AK the closed ileocecal valve (ICV) syndrome. This reflex exists to keep undigested fat from entering the large intestine, where it may stimulate the growth of unfriendly flora. Carbohydrate also stimulates the enteric nervous system and produces an open ileocecal valve syndrome. The gastrocolic reflex causes increased peristalsis in the large intestine following food intake that stretches the stomach. When the stomach is stimulated by the presence of food, the large intestine is stimulated to empty. These reflexes reflect 3 sensorimotor challenges developed in AK that evaluate whether or not they are functioning properly. It is proposed that these enteric nervous system concepts may be clinically applied by monitoring AK MMT outcomes following various specific sensorimotor challenges.

Normal large intestine function is seldom found in the general population. The most frequent involvement is large intestine stasis, which has many complications that may develop in later life. The significance of these complications ranges from mild general health interference to toxic reabsorption to the death-causing factors such as colon cancer. Every individual coming into the office should be evaluated for large intestine dysfunction, not just those with digestive disturbances. The large intestine, like other aspects of the digestive system, can produce symptoms in remote areas of the body.

There are other organs, glands, and structures which can interfere with normal digestive function. As we examine for any condition, we must remember the interplay which takes place within the body. Mentioned in this chapter are some of the more notable interactions. Quite often, the thyroid is involved with digestive function, especially that of the small intestine and urinary systems. The observant doctor will frequently notice a gurgling and heavy activity of the small intestine immediately after treatment of the thyroid by any one of the 5-factors of the IVF.

Alimentary Canal Flora

Dysbiosis of the intestinal flora has been implicated in the pathogenesis of atherosclerosis and dyslipidemias, hypertension, urolithiasis, pyelonephritis, gallstones, and hepatitis. **(Samsonova, 2010)** Dysbiosis, additionally, has been implicated in autoimmune disorders, allergy, and impaired resistance. **(Prakash, 2011)** Research indicates that dysbiosis may be a factor in the pathogenesis of obesity, inflammatory bowel disease, and gastroesophageal reflux disease (GERD). **(Henao-Mejia, 2012; Kim, 2010; Yang, 2009)** Changes in behavior and brain-derived neurotropic factor have been noted in dysbiotic mice. **(Bercik, 2011)**

Intestinal biota is adversely affected by antibiotics (prescription and as residue in foods), psychological and physical stress, sulfates and sulfites, and diets high in proteins, especially meat, and/or refined or simple carbohydrates. **(Hawrelak, 2004)** Bile appears to modulate the biota of the intestinal tract, as experimental occlusion of the common bile duct results in rapid changes in bowel flora and increased intestinal permeability. **(Fouts, 2012)** Proton pump inhibitors (PPIs) induce dysbiosis indicating that adequate hydrochloric acid production and an acid food bolus entering the small intestine plays a role in normal gastrointestinal biota. (**Wallace, 2011**)

Inflammatory Bowel Disease (Ulcerative Colitis and Crohn's Disease)

Crohn's disease is a chronic, severe lymphatic congestion. There will usually be involvement of the quadriceps or abdominal muscles. Sometimes very prolonged neurolymphatic activity is necessary to bring this condition under control. There is a correlation between narcolepsy and small intestine involvement, especially if it has progressed to the state of Crohn's disease. The severe lymphatic congestion is due to the small intestine's very heavy workload. When there is improper small intestine activity, the area around the villi actually becomes clogged, causing further reduction in intestinal performance. The symptoms of narcolepsy develop as a result of an autointoxication from the activity of the small intestine. This also correlates with malabsorption syndrome and should be evaluated in hypoglycemia cases. The patient will often have an abnormal increase in the sensation of hunger, called bulimia. Sometimes they will literally stand in front of the refrigerator, knowing they want to eat but not knowing what.

Case studies indicate that ulcerative colitis and Crohn's disease are responsive to applied kinesiology therapies. Common findings are disturbances of the lumbar spine, an open ileocecal valve, and a need to eliminate nuts and seeds, grains, and milk from the diet as detected by AK MMT. (**Duffy, 1992**)

Patients with ulcerative colitis are frequently given Azulfidine and other sulfa drugs. A side-effect of the sulfa medication is that it rampantly destroys vitamin C. Patients who have colitis or irritable bowel syndrome or the more severe ulcerated colitis should be evaluated for the tissue levels of vitamin C, since it is one of the necessary natural agents for controlling inflammation. It might be observed that Beauchamp et al. (**2005**) suggest that 3.5 tablespoons of olive oil has the same effect as a 200-mg tablet of ibuprofen. While pharmacological treatment of inflammation remains an option for patients, there are now well-established strategies for modulating inflammation via dietary and nutritional manipulation. (**Gaby, 2006**)

Adrenal involvement in large intestine dysfunctions

Accompanying an increase in refined carbohydrates and a loss of fiber in the Standard American Diet is the problem of a functional hypoadrenia. As a result of functional hypoadrenia, the individual may have a lack of mineralocorticoids and glucocorticoids, the pro-inflammatory and anti-inflammatory hormones produced by the adrenal glands. Colitis often accompanies functional hypoadrenia and can ultimately end in diverticulitis and ulcerative colitis. As well as making dietary changes for normal colon action, it is important to support the adrenal if it is involved with inflammatory changes in the large intestine.

Variations are found in anti-oxidant enzymes in IBD subjects. Low plasma selenium and glutathione peroxidase has been found along with low intestinal mucosal concentrations of zinc and metallothionein, a zinc dependent enzyme. (**Sturniolo, 1998**) Metallothioneins, antioxidants and regulators of the transcription factor nuclear factor (NF)-kappaB, are found to be abnormally expressed in IBD subjects, indicating an inborn error of zinc homeostasis. (**Waeytens, 2009**) Intestinal mucosal copper/zinc isoform of superoxide dismutase (Cu/Zn-SOD) has also been found to be low in IBD subjects. (**Kruidenier, 2003**) N-acetylcysteine and l-carnitine have been found to restore intestinal mucosal glutathione peroxidase levels. (**Cetinkaya, 2006; Nosáľová, 2000**)

The pro-inflammatory metabolites of arachadonic acid, prostaglandins E2 and thromboxane B2, are elevated in the colonic mucosa in ulcerative colitis. (**Pavlenko, 2003**)

Melatonin appears to play a role in moderating the mucosal inflammation of IBD, indicating that abnormal circadian rhythm may be a factor in the pathogenesis. (**Necefli, 2006**)

Methylsulfonylmethane (MSM), an organosulfur compound that appears to enhance glutathione peroxidase levels, has been shown to decrease intestinal mucosa damage in IBD. (**Amirshahrokhi, 2011**) Studies have shown that vitamin E is anti-inflammatory and glutathione peroxidase sparing of IBD intestinal mucosa. (**Tahan, 2011; Beno, 1997**)

Dietary phenethylisothiocyanate, a natural constituent of foods from the brassica family (broccoli, spinach, cabbage, cauliflower, Brussels sprouts, kale, collard greens, bok choi, kohlrabi), improved body weight and stool consistency and decreased intestinal bleeding and mucosal inflammation in IBD mice. (**Dey, 2010**) It is interesting to note that cabbage juice is a traditional remedy for ulcerations of the stomach and intestines.

Always rule out underlying intestinal infection, especially *Entamoeba histolytica*, in patients presenting with IBD. (**Horiki, 1996; Chan, 1995; Patel, 1989**)

Ulcers of the intestinal tract is one of the three conditions of the Triad of Selye, three conditions always found in lab animals if stressed sufficiently to induce adrenal fatigue. The Triad of Selye is (1) hypertrophy of the adrenal glands, (2) atrophy of the lymphatic system, and (3) ulcers in the intestinal tract. (**Selye, 1978**)

Calcium decreases the inflammatory effect of bile acids on intestinal mucosa. (**Wargovich, 1983**) Vitamin D receptor polymorphisms (**Naderi, 2008**) and vitamin D and K deficiencies (**Nakajima, 2011; Kuwabara, 2009**) are common in IBD populations. Some studies have proposed that low vitamin D plays a role in the pathogenesis of IBD. (**Souza, 2012; Ananthakrishnan, 2011**)

Celiac disease is a very specific type of inflammatory bowel disease that is caused by gluten allergy and avoidance of all gluten is the primary therapy. Besides the gluten containing grains rye, wheat, and barley, the patient must be aware of gluten used as a food additive.

Interestingly, restless leg syndrome (RLS) occurs frequently in association with Crohn's disease and may be an extraintestinal manifestation of the disease. **(Gemignani, 2010)** Other diagnostic considerations for the syndrome are uremic or diabetic neuropathy, Parkinson disease, multiple sclerosis, caffeine, tricyclic antidepressants, barbiturates, benzodiazepines, and anemia.

Irritable Bowel Syndrome (Mucous Colitis)

Mucous colitis (IBS) is a functional inflammatory condition of the large intestine. Symptoms can be useful to discriminate between IBD and IBS, with straining upon defecation, diarrhea, and abdominal bloating.

Probiotic support appears to suppress inflammation and colonic barrier disruption induced by stress; the most beneficial strains appear to be the *Lactobacilli* group and *Saccharomyces boulardii* and the *benefits extend to IBD*. **(Agostini, 2012; Goldin, 2008)** Bile acid malabsorption may be a trigger for diarrhea predominant IBS and *Lactobacilli* and *Bifidobacteria* subspecies are able to deconjugate and absorb bile acids. **(Camilleri, 2006; Smith, 2000)**

Food allergy, as measured via IgE and IgG antibodies, is more common in IBS populations when compared to controls, particularly to milk, wheat, and soy. **(Carroccio, 2011)** A food allergy model for IBS is supported by the observation that both numbers of mast cells and their mediators are increased in the intestinal mucosa for patients with IBS. **(Walker, 2008; Kalliomäki, 2005)** Lactose intolerance should be considered in patients with IBS as well. **(Vernia, 2001)**

IBS populations have higher arachidonic acid (AA) levels than control populations, along with the pro-inflammatory metabolites of AA, prostaglandin E(2) (PGE(2)) and leukotriene B(4) (LTB(4)), suggesting that prostaglandin imbalances are a factor in the pathogenesis of IBS and dietary therapy should include decreasing AA rich foods in the diet and increasing omega 3 fatty acids to improve prostaglandin balance. **(Clarke, 2010)** Intake of hydrogenated oils (HOs) in the diet should be limited in IBS patients since HOs impede conversion of the anti-inflammatory prostaglandin groups PG1 and PG3 and aggravate PG2 dominance.

Activation of protease-activated receptors (PARs) in the intestinal tract modulates motility, inflammation, visceral nociception, and epithelial functions (immune, permeability and secretory) suggesting a regulatory role for pancreatic proteolytic enzymes. **(Vergnolle, 2004)**

Peppermint oil is a promising therapeutic agent for IBS, possibly due to anti-spasmodic and choleretic effects. **(Grigoleit, 2005)** The choleretic benefit suggests looking to biliary insufficiency and low ejection fraction as a possible underlying and aggravating conditions in patients with IBS.

A study was also conducted to evaluate the comparative effect of yoga and conventional treatment in diarrhea-predominant irritable bowel syndrome (IBS) in a randomized control design. The patients were 22 males, aged 20-50 years, with confirmed diagnosis of diarrhea-predominant IBS. The conventional group (n = 12, 1 dropout) was given symptomatic treatment with loperamide 2-6 mg/day for 2 months, and the yoga intervention group (n = 9) consisted of a set of 12 asanas along with Surya Nadi pranayama (right-nostril breathing) two times a day for 2 months. All participants were tested at three regular intervals, at the start of study — 0 month, 1 month, and 2 months of receiving the intervention — and were investigated for bowel symptoms, autonomic symptoms,

Warrior Two pose and Utkatasana (Chair Pose) are purported to help with gastric and digestion problems

autonomic reactivity (battery of five standard tests), surface electrogastrography, anxiety profile by Spielberger's Self Evaluation Questionnaire, which evaluated trait and state anxiety. Two months of both conventional and yogic intervention showed a significant decrease of bowel symptoms and state anxiety. This was accompanied by an increase in electrophysiologically recorded gastric activity in the conventional intervention group and enhanced parasympathetic reactivity, as measured by heart rate parameters, in the yogic intervention group. The study indicates a beneficial effect of yogic intervention over conventional treatment in diarrhea-predominant IBS. **(Taneja et al., 2004)**

Many yoga asanas focus on moving the stomach and intestines in all directions to aid the digestive process. **(Shankardevananda, 2003; Iyengar, 1995)** This stimulates the flow of blood to all parts of the intestines and generally helps foods pass through the body.

Constipation

Increased sympathetic activity decreases bowel tone by decreasing peristalsis and tightening the sphincters. Increased parasympathetic activity increases bowel tone by increasing bowel wall peristalsis and relaxing the sphincters. Always consider autonomic tone when treating the patient suffering constipation. **(Lee, 1998)**

When the diet is highly refined, there is a slower passage of the large intestine contents and increased contractions of the large intestine to move the hard, dry material. The effort of peristalsis along the narrow colon stretches and eventually causes the pouches of diverticulosis. It is a short step to diverticulitis as a result of a piece of hardened fecal matter or fecalith blocking the opening to diverticula. Inflammation, infections, and abscesses can occur.

Underlying causative factors to consider in constipation are dehydration (may be aggravated by electrolyte deficiency), serotonin deficiency or receptor resistance, biliary and pancreatic insufficiency, low fiber diet, dysbiosis, food allergies, and hypothyroidism. **(Syrigou, 2011; Jamshed, 2011; An, 2010; Gale, 2009)**

Dehydration is an almost universal nutritional deficiency in our culture. Satisfactory respiration, digestion, assimilation, metabolism, elimination, waste removal, and temperature regulation are bodily functions that can only be accomplished in the presence of water. Water is essential in dissolving and transporting nutrients such as oxygen and mineral salts, vitamins and nucleoproteins via the blood, lymph, and other bodily fluids: these are the constituents of connective tissue. Effective nutritional testing requires that the mouth and tongue be moist, so that the nutrition can mix with the lingual receptors of the tongue. Good lymphatic function also depends on an adequate daily intake of water. For tissue healing to take place, water must be plentiful. A healthy person should consume a minimum of 1/3 ounce of water for every pound of body weight, and double this amount in times of stress or illness. This usually means drinking 6-8 glasses of water every day. This should be purified water, not coffee, tea, fruit juice, sodas, milk and other liquids. The AK test for dehydration can help the patient and the clinician recognize when there is a problem.

Applied kinesiology treatment can correct chronic constipation. A case report of treating lifelong constipation in a thirteen year old boy included chiropractic care to the lumbar spine, ileocecal valve, avoidance of common

Dehydration may be evaluated using the AK hair-pull test.

Ask the patient to tug the hair on their scalp and test a previously strong indicator muscle.

If it weakens, ask the patient to drink some water. Retesting the muscle in a dehydrated patient will usually show strengthening.

allergens (milk, corn, soy, wheat), and inclusion of dietary water-soluble fiber. (**Maykel, 2003**) Treatment of neurolymphatic (Chapman's) reflexes has been reported to be successful therapy for chronic constipation and associated viscerosomatic reflex low back pain in an AK setting. (**Caso, 2001**)

Chiropractic manipulation of the spine, cranium and neck has been shown to beneficially change bowel behavior, even in cases of chronic constipation, sometimes after just one treatment. (**Browning, 2009; Masarsky & Masarsky, 2001; Wagner, 1995**)

Diarrhea

Underlying etiological factors of diarrhea may be hypochlorhydria, pancreatic and biliary insufficiency, open ileocecal valve, and intestinal flora. (**Mossner, 2010; Sarles, 1989**) Consider possible gastrointestinal infection - bacterial, viral, mycotic, or parasitic. (**Rajamanicka, 2009**)

Saccharomyces boulardii has been shown to be an effective probiotic for treating infectious diarrhea. (**Dinleyici, 2012**) Consider activated charcoal as an adsorbent for diarrhea due to infectious, toxic, or inflammatory agents. (**Sergio, 2008**) Research has shown that vitamin A and zinc deficiencies predispose to chronic diarrhea. (**Bhan, 1996; Sinha, 1995**)

Infantile Colic

Research indicates that coliform concentration is higher in infants with colic when compared to controls and *Lactobacillus* cultures are antagonistic to coliform populations. (**Savino, 2011**) Feeding infants symbiotic formulas of both prebiotic and probiotic cultures has been found to be safe and to decrease incidence of respiratory infections during the first two years of life. (**Kukkonen, 2008**)

Breast-fed infants have significantly higher levels of coliform-antagonistic bifidobacteria and salivary IgA when compared to bottle-fed populations. (**Wold, 2000; Yoshioka, 1983**) Insufficient or imbalanced intestinal microbial flora has also been associated with atopic eczema in infants. (**Abrahamson, 2011**)

A study of 316 infants with the cooperation of 38% of the chiropractors in Denmark was conducted by Klougart et al. (**1989**) Based on the diaries of parents, 94% of the patients resolved their colic in 2 weeks or less (an average of 3 visits), with 23% of the patients resolving their colic after a single treatment. The upper cervical area was the most frequently adjusted. The authors noted that spontaneous resolution of infantile colic usually takes 12 to 16 weeks.

Obesity

Although obesity is commonplace today, (**Centers for Disease Control and Prevention, 2012**) it has not always been that way. In fact, even today in some tribal areas of Africa, India, and Oceania, and other areas where an unrefined diet is used, obesity is almost unheard of. According to the CDC, more than one-third of U.S. adults (35.7%) are obese. Approximately 17% (or 12.5 million) children and adolescents aged 2—19 years are obese. In 2010 no state in the United States had a prevalence of obesity less than 20%. Thirty-six states had a prevalence of 25% or more; 12 of these states (Alabama, Arkansas, Kentucky, Louisiana, Michigan, Mississippi, Missouri, Oklahoma, South Carolina, Tennessee, Texas, and West Virginia) had a prevalence of 30% or more.

There are two primary causes of obesity today. First are the so-called "empty calories" which add to the body's calorie count, but provide no material for healthy tissue-building. These items are primarily white sugar, white flour, and alcohol. The second item is a lack of fiber in the diet. A diet comprised primarily of fiber dense foods, such as fruits, vegetables, beans, nuts and seeds, and unprocessed grains requires more chewing, slower intake, and promotes a sense of satiety at lower caloric intake. (**Isaksson, 2012; Beck, 2009**)

A major key to many of the health problems resulting from large intestine dysfunction lies in increasing the fiber content of the diet. Unfortunately with today's refined, ultra-processed foods, it becomes almost impossible to obtain fiber from foods purchased at the supermarket.

An easy answer is to include the so-called "waste products" from wheat in our diet. Bran can be purchased very economically from any health-food store and added easily to the diet. Make it a drink by mixing it with water or juice, or mix it with such foods as meat loaf, cereals, etc.

The amount of bran necessary for normal colon function varies with individuals. Start with one teaspoon three times per day, and increase until the desired results are obtained. Some individuals will get good results with three teaspoons per day; others will require several tablespoons per day. The basic criterion is that the individual have a well-formed, soft stool, preferably on a daily basis. The stool should be completely odor-free. The odor on the stool is the classic criterion for adequate bowel action. This gives indication of the bacteria content as well as whether putrefaction processes are taking place.

Flatulence may occur when first incorporating bran into the diet. The amount of bran ingested can be decreased until the flatulence diminishes and the digestive tract adapts to it, at which point the dosage can then be increased. If the problem persists consider the possibility of gluten allergy or hypochlorhydria.

Post-Antibiotic Effects and Candidiasis

Antibiotics are non-discriminating in their action and commonly result in intestinal dysbiosis. When antibiotics are given for an infection, they also affect the flora of the intestines. William G. Crook, MD, and author of *The Yeast Connection*, (**1986**) states that "Broad-spectrum antibiotics resemble machine gun-shooting terrorists in a crowded airport. While they're killing enemies, they also kill friendly and innocent bystanders. In a similar manner, antibiotics knock out friendly bacteria on the interior membranes of a person's body while they're eradicating enemies. When this happens, yeasts flourish and put out a toxin that affects various organs and systems in the body, including the

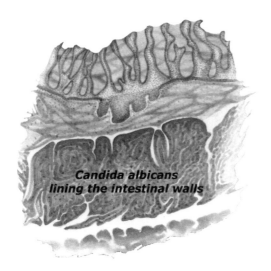
Candida albicans lining the intestinal walls

immune system." A major side effect of chronic candidiasis is persistent inflammation of the intestinal wall and increased intestinal permeability. This permits absorption of multiple antigenic and toxic substances into the bloodstream, resulting in allergies. **(Schmidt et al., 1994)**

Sometimes it is necessary, after antibiotic therapy has been used, to re-implant bacteria or provide a better culture medium for their growth. Support with probiotic and prebiotic formulas may be needed to restore intestinal biosis. Using viable colonizing strains of lactobacillus acidophilus (small intestine) and bifidobacteria (large intestine) will check and inhibit growth of innate intestinal opportunistic yeasts. **(Chaitow & Trenev, 1989)** The dosage is usually ½ teaspoons in a glass of warm water three times per day for children and two to three capsules (one or two teaspoons powder) three times daily for adults. Temporary use of anti-fungal compounds may also be required to control yeast overgrowth in the intestinal tract. Prescription anti-fungals are rarely necessary; the anti-fungal action of caprylic acid, garlic, cinnamon, aloe vera juice, Spore-X (Nutri-West®), pau de arco, and other anti-fungal herbs is usually effective, if the intestinal flora is restored adequately.

Anyone with a great deal of yeast in the system may experience numerous "die-off" symptoms during the remission of the candida infestation. The yeast organisms carry poison within them. As they die, their membranes eventually rupture releasing the poison or toxins into the system. The actual "die-off" can cause some "I-just-don't-feel-good" types of symptoms. This signifies a healing crisis, and is evidence that recovery from candida infestation is progressing. Although patients sometimes get worse before getting better, once the poisons are flushed through, they usually continue to get well.

Limiting soluble carbohydrate intake during the phases controlling yeast overgrowth, restoring intestinal flora, and healing the inflamed and hyper-permeable intestinal wall will improve outcomes. AKMMT, elimination diet, or blood or skin allergy testing can also be used to identify allergenic foods that need to be eliminated from the diet to down-regulate intestinal wall inflammation and promote restoration of optimal permeability. It has been this authors' experience that the broad spectrum of mild and moderate food reactions found upon AK MMT and laboratory food allergy testing will resolve once digestion, intestinal flora, and gut wall permeability are restored.

While mushrooms, yeasts, and fermented foods (i.e. vinegar, soy sauce, miso) do not promote growth of intestinal yeasts, it is common for patients with intestinal yeast overgrowth to become immunologically sensitive to these foods and avoidance until the issue is resolved will promote recovery."

Incorporating healthy lifestyle changes make a big difference in fighting candida. First, the yeast must be killed, then the intestinal garden replanted, the immune system and digestive system strengthened, and diet and lifestyle modified. The most important factor involves withholding the type of food on which the yeast thrives. Using the AK MMT, the elimination diet, or blood or skin allergy tests that identify allergenic foods that need to be eliminated from the diet can down-regulate intestinal wall inflammation and promote restoration of optimal permeability. It is wise to leave out sugar, refined carbohydrates such as white bread and pasta, alcohol, and even honey. This strict diet regime is required only for a period of time – not forever. If successful in controlling candida symptoms within a few months, the patient can slowly reintroduce these allergenic foods into the diet.

Vinegar, soy sauce, and miso are several examples of fermented items that should be avoided by most candida patients. Restrict from the diet all yeast-containing foods for a minimum of three months. While candida is different from baker's yeast or brewer's yeast, patients are advised to adhere to the philosophy that if yeast is growing out of control in the body, any yeast may further upset the digestive system.

Expect to experience more allergic reactions to yogurt, cheese, or kefir containing some type of yeast. Yogurt has been over emphasized in many books. Actually, yogurt doesn't contain the curative properties touted mainly because its amounts of friendly bacteria, *acidophilus*, are miniscule compared to what is truly needed. Additionally, most yogurt products are loaded with synthetic sugars that are one of the basic foods of the candida organism in the first place.

It's smart to avoid any fungus-containing food such as peanuts and mushrooms. It's much easier to overcome candidiasis if the patient simply leaves out fermented and fungal foods – the yeasts, sugars, and alcohol – altogether. Antifungal foods such as garlic and olive oil (oleic acid) are good choices.

Fecal matter matters

We can now see how small and large intestinal dysfunction is endemic to modern life simply by the proliferation of over-the-counter preparations for large intestine or bowel control. There are heavy metal preparations to bring more fluid into the bowel, stimulants, bulks, items for starting the bowel, means for stopping the bowel, gums to chew, and enema bags. There are even high-colonic irrigating machines for home use.

Physicians do not always help these cases, because they reassure patients that it is normal for some people to have a bowel movement every 3rd or 4th day or even once a week; as long as the stool moved is soft, there is no need for concern. For too many years the accepted treatment for conditions such as colitis, diverticulitis, and diverticulosis has been a bland diet – which is the wrong thing for the patient to do, even though the diet is a so-called soothing diet to the bowel.

Because there has been so much misinformation among the lay public, and from doctors as well, a major project of patient education lies before us. We must overcome these improper treatments and lack of concern, and in fact reverse certain types of treatments which, in the past, were thought to be correct, but are in reality making the patient worse.

Quite often obtaining normal large intestine function is extremely important in obtaining correction of the acute problem which brings the patient to the physician. Obviously in these cases we are concerned with getting the condition under control as rapidly as possible. The patient who is perhaps of greatest concern is the individual who has significant large intestine stasis and/or abnormal flora, but has no symptomatic complaint upon initial consultation. This is the person who has problems in the developmental stage. These potential problems are of major significance and early correction may possibly be life-saving. These are the conditions which must be ferreted out and correction obtained while correction is still possible.

Female Reproductive Systems

Infertility and Amenorrhea

In a study by Wurn, 28 infertile women diagnosed with complete fallopian tube occlusion underwent a twenty session trial of manual therapies to address pain and restricted soft tissue mobility due to adhesions in the pelvic region. It resulted in unilateral or bilateral patency (as measured by hystero-salpingography) in 61% of the subjects. Subsequent to treatment, 53% of the subjects in the fallopian tube patient group reported natural intrauterine pregnancy. (**Wurn, 2008**) These findings conformed closely to a previous study by the same researcher. (**Wurn, 2004**)

A number of case reports suggest that AK treatment is effective for resolving infertility. The non-invasive nature, safety, and cost-effectiveness of this conservative approach to infertility when compared to standard fertility medicine should be borne in mind. The AK therapies include chiropractic, craniosacral, meridian (acupuncture), and clinical nutrition. Based upon these reports, consideration should be given to pituitary function (pituitary drive technique), thyroid function (including Grave's disease), adrenal function, iron deficiency, and essential fatty acid deficiency, including wheat germ oil. (**Kaufman, 1996; Duffy, 1993; Heidrich, 1991**) AK combined with clinical nutrition has been successful in resolving amenorrhea. (**Kharrazian, 2007**)

Infertility has been linked to aberrant modulation of the hypothalamic-pituitary-gonadal axis. (**Scott, 1989**) AK methods for normalizing adrenal function, identifying food allergies, and decreasing mechanical stress to the uterus (uterine lift technique) show promise for resolving dysmenorrhea and infertility in some patients. (**Hickey, 2007**)

There appear to be four primary types of amenorrhea - hyperprolactinaemic, hypogonadotrophic, hypergonadotrophic, and normogonadotrophic. Hyperprolactinaemic amenorrhoea is often due to pituitary adenoma, hypergonadotrophic amenorrhoea is due to ovarian failure, and normogonadotrophic amenorrhoea is due to aberrant GnRH secretory patterns and is often secondary to PCOS. Hypogonadotrophic amenorrhoea is usually associated with stress and nutritional deficiency, often as a result of restricted dietary patterns and insufficient body fat. (**Crosignani & Vegetti, 1996**) Stress of various types, such as dieting, heavy training, or intense emotional events, appears to be a pathogenic factor for both normogonadotrophic and hypogonadotrophic patterns of amenorrhea. These forms of amenorrhea are mediated via the hypothalamus and may manifest with or without weight loss. (**Genazzani, 2010**)

Under-eating and overexercising are common stresses leading to amenorrhea. One study showed that twenty percent of ballet dancers studied were amenorrheic. (**Stokic, 2005**) It has been common to associate low body fat with amenorrhea, though some research challenges this common view and has indicated that low overall body weight has a closer correlation. (**Estok, 1991; Sanborn, 1987**) Always consider eating disorders in the low weight amenorrheic patient. (**Gentile, 2011**) Goodheart suggests that to restart the menstrual cycle, a course of increased calcium levels 1,500 mg. per day (the same as postmenopausal females), with 300 mg of zinc and 150 mg vitamin B6 daily may be useful. (**Goodheart, 1992**)

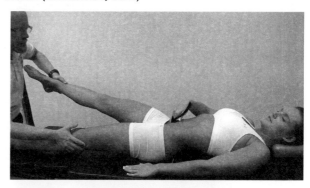

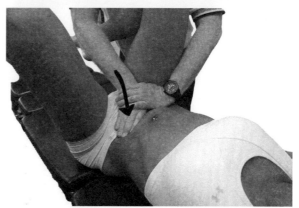

Uterine Lift technique: challenge and visceral manipulation

Dysmenorrhea

The pain of primary dysmenorrhea can be very intense. Pain threshold measurements to electrical stimulation of the skin, subcutaneous tissue and muscle have been seen to be lower than normal in women with dysmenorrhea, particularly in the rectus abdominus muscle. (**Giamberardino et al., 1997**) As reviewed in the text *Applied Kinesiology Essentials*, muscle pain usually creates muscle weakness on the MMT, indicating the relationship between positive MMT findings and women with this condition. These conditions of the female reproductive system are characterized by a generalized hypersensitivity to painful stimuli, suggesting that these patients experience a wide spectrum of functional disorders. It is reasonable to suggest that even if only the muscular impairments (in addition to the other AK approaches) are corrected, significant improvements can be expected in these patients. It is also probable that addressing the physical impairments found in these patients will impact psychosocial disturbances as well, further advancing the goals of treating the patient's total problem.

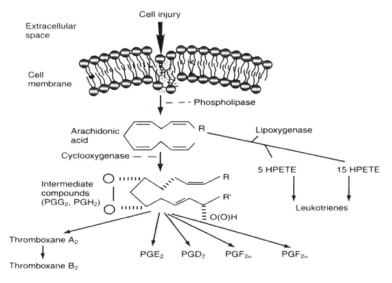

Pain, prostaglandins and dysmenorrhea

Women suffering dysmenorrhea have been shown to have higher 17 beta-estradiol (E2) level, when compared to asymptomatic controls, in the latter half of the menstrual cycle and low progesterone levels, thereby creating a high estrogen/progesterone ratio. (**Zahradnik, 1984; Ylikorkala, 1979**)

Prostaglandin metabolism is an important consideration for management of dysmenorrhea. Dysmenorrhea and dysfunctional uterine bleeding are also associated with up-regulation of arachadonic acid-derived PG2 series prostaglandins, (**Coll Capdevila, 1997**) and management with NSAIDs to down-regulate production of PG2 has proven effective (**Dawood, 2007**)

Nitric oxide (NO) appears to play a role in modulating myometria and may play a role in managing dysmenorrhea. (**Wetzka, 2001**)

The prostaglandin fractions PGE2 and PGF2α have been implicated as the agents producing uterine smooth muscle contraction and vasospasm of the uterine arterioles, leading to ischemia and the cramping sensation of dysmenorrhea. (**Dawood, 1990**) Medications most commonly used to treat dysmenorrhea are the nonsteroidal anti-inflammatory drugs (NSAIDs) ibuprofen, naproxen, and mefenamic acid. These act by inhibiting cyclooxygenase in the inflammatory pathway reflecting arachidonic acid metabolism to the prostaglandins, as shown in the figure. (**Dawood, 1988**) However, the side effects of NSAIDs (gastrointestinal disturbances, nausea, vomiting, constipation, headache, vertigo, fatigue, and allergic reactions) have been documented (**Mehlisch, 1988; Shapiro, 1988; Calesnick, 1987**) and need to be taken into consideration.

A series of studies over the years have suggested that spinal manipulation is capable of relieving the symptoms of dysmenorrhea. (**Snyder, 1996; Boesler et al., 1993; Smith and Rogers, 1992; Kokjohn et al., 1992; Liebl and Butler, 1990; Arnold-Frochot, 1981; Thomason et al., 1979; Hitchcock, 1976**) The most compelling of these has provided a sound biochemical rationale for the analgesic effect for the intervention, demonstrating that spinal manipulation in the lumbar region also reduces the concentration of the metabolite of the PGF2α fraction (15-keto-13,14-dihydroprostaglandin F2α) in plasma fractions taken from patients 1 hour after treatment. (**Kokjohn, 1992**) A larger randomized controlled trial (**Hondras, 1999**) delivered inconclusive results; however, that investigation was compromised by the facts that (1) effleurage was applied to both control and experimental groups of patients and (2) pain upon entry and washouts of exercise and the taking of NSAID medications were eliminated, all reducing or eliminating the possible effects of spinal manipulation.

Higher levels of the prostaglandin, 6-Keto-PGF1 alpha, correlate with increased severity of dysmenorrhea in women with endometriosis. (**Koike, 1992**) Fish oil and olive oil have been shown to down-regulate production of 6-Keto-PGF1 alpha. (**Correa, 2009; Faust, 1989**)

Research indicates that functional bowel disorders may underlie dysmenorrhea. (**Crowell, 1994**)

According to Schmitt, (**1990, 1981**) graphing basal temperature for a complete menstrual cycle can give insight into the relative balance and ratios between progesterone and estrogen. Schmitt also states that scanty menses can indicate high thyroid, low estrogen, and/or high progesterone levels, whereas heavy menses can indicate a low thyroid, high estrogen, and/or low progesterone. Additionally, short menstrual cycles (<26 days) typically have a high estrogen/progesterone ratio, whereas long menstrual cycles (>29 days) typically indicate low estrogen/progesterone ratio. Goodheart suggests that a menstrual cycle of more than 28 days indicates a hypothyroid condition, and one of less than 28 days indicates a hyperthyroid one. These observations have been congruent with our clinical experience and can be a useful complement to physical exam and laboratory findings.

Vitamins B6, B12, and/or folic acid deficiency should be considered in women using hormone replacement or contraceptives. (**Veninga, 1984; Anderson, 1976**)

Carbamide has been researched as a possible useful agent to promote uterine smooth muscle relaxation with a therapeutic effect for dysmenorrhea. (**Novakovik, 2007; Cheuk, 1993**)

Massage therapy to the pelvic region has been shown to be viable therapy for relieving dysmenorrhea in women with chronic endometriosis diagnosed with laparoscopy. (**Valiani, 2010**)

Premenstrual Syndromes

Premenstrual syndrome (PMS) is related to dysmenorrhea, with as many as 150 symptoms. (**Sveinsdottir & Reame, 1991**)

Estrogen imbalances

A high estrogen-progesterone ratio is associated with premenstrual syndromes and is characterized by mood disorders. The PMS group shows water and salt retention, abdominal bloating, mastalgia and weight gain that is associated with high aldosterone levels. The group is also characterized by blood sugar handling stress (premenstrual craving for sweets — especially chocolate) increased appetite, palpitation, fatigue, headache, and even syncope. A rarer type of PMS is associated with polycystic ovary syndrome (PCOS), low estrogen, high progesterone, and high adrenal androgens. (**Abraham, 1983**)

PMS is usually associated with persistence of high estrogen after the normal mid-cycle ovulatory spike. This is usually due to insufficient clearance of circulating estrogen by the Phase II detoxification pathway in the liver. An increasing body of scientific data supports the hypothesis that conjugation (sulfation) and deconjugation (desulfation) of estrogens is important in the regulation of biologically active steroid hormones in target tissues as well. Polymorphisms appear to dramatically influence the rate of these processes. Methylation is the most quantitatively active pathway in the sulfation and desulfation of estrogens. Glutathione (GSH) is a major enzyme in this system and has major antioxidant/anti-inflammatory effects systemically, as well. (**Raftogianis, 2000**) Xenoestrogens can also alter endogenous estrogen signaling and disrupt normal signaling pathways. (**Watson, 2011**) Xenobiotic compounds may accentuate estrogen activity and be an underlying and aggravating factor in PMS.

Xenoestrogens tend to accumulate and persist in adipose tissue for decades and may cause long-lasting, adverse endocrine effects. Xenoestrogen compounds include dichlorodiphnyltrichloroethane (DDT) and its metabolites, bisphenols, alkylphenols, dichlorophenols, methoxychlor, chlordecone, polychlorinated benzol derivatives (PCBs), and dioxins. Besides being receptor competitors and having higher biological activity than endogenous estrogens, some xenoestrogens may interfere with the production and metabolism of ovarian estrogens. (**Lorand, 2010**)

Phytoestrogens and bioflavonoids can be used to down-regulate estrogen activity. Phytoestrogens have a lower level of estrogen activity at receptor sites and can be used to attenuate the estrogenic effects of xenobiotics. (**Zava, 1997**) Procyanidin compounds from red wine and grape seeds have shown suppression of estrogen synthesis through aromatase inhibition. (**Eng, 2003**)

Altered serotonin activity may also be a factor in the causation of PMS. (**Halbreich, 1993**)

The depression commonly associated with PMS may be due to the increased demand on vitamin B6 reserves as a cofactor in conjugation during the luteal phase of the menstrual cycle and the need for B6 as a cofactor in the production of 5-hydroxtryptamine (serotonin). As high as 31% resolution of premenstrual depression with vitamin B6 supplementation has been reported in a group of women taking oral contraceptives. Possible symptoms of vitamin B6 deficiency include hyperirritability, loss of appetite, loss of weight, general weakness, lassitude, confusion, and a hypochromic, microcytic anemia with a high serum iron level. (**Prothro, 1981**) The usefulness of B6 and, to some degree magnesium, has been confirmed by other studies. (**De Souza, 2000; Doll, 1989; Bermond, 1982**) The patient may be advised to eat whole grains, nuts, seeds, beans, lentils, liver, organ meats, wheat germ, and Brewer's yeast. In a literature review, Douglas (**2002**) also supports calcium supplementation in managing PMS.

Prostaglandin imbalance seems to play a role in many cases of premenstrual syndrome and dysmenorrhea with a deficiency of dihomo-γ-linolenic acid (DGLA) being the most common pattern. Support DGLA levels by providing gamma-linolenic acid (GLA) with evening primrose, black currant, borage oils, and/or spirulina. The cofactors magnesium, zinc, vitamin C, B3, and B6 may be needed for conversion of GLA to DGLA.

A crossover randomized controlled trial, (**Walsh et al., 1999**) a case series, (**Wittler, 1992**) and numerous case reports have shown that spinal manipulative therapy may alleviate this condition. (**Masarsky & Masarsky, 2001**)

Breast tenderness is a frequent symptom of PMS. To determine whether AK technique was of benefit to women with breast pain, an open pilot study was conducted at the Hedley Atkins Breast Unit, Guy's Hospital, London, UK. (**Gregory et al., 2001**) Eighty-eight newly presenting women with self-rated moderate or severe mastalgia were recruited for the study. The AK treatment involved rubbing the neurolymphatic reflexes of the TFL muscle while monitoring painful areas of the breasts. The women were predominantly pre-menopausal, and patients with both cyclical and non-cyclical pain were included in the study. Patients' self-rated pain scores, both before and immediately after applied kinesiology, were compared, together with a further score 2 months later. Immediately after treatment there was considerable reduction in breast pain in 60% of patients with complete resolution in 18%. At the visit after 2 months, there was a reduction in severity, duration and frequency of pain of 50% or more in about 60% of cases (P<0.0001). This preliminary study suggests that applied kinesiology may be an effective treatment for mastalgia, without side-effects and merits testing against standard drug therapies.

Menstrual headaches were successfully treated with AK protocols that included craniosacral and chiropractic manipulative therapies, adrenal support, clinical nutrition and the avoidance of aspartame along with good food combining principles. (**Calhoon, 2004**)

Endometriosis

As mentioned previously, endometriosis has been associated with dysmenorrhea and high levels of 6-Keto-PGF1 alpha (a metabolite of PG2), and fish oils have been shown to lower the levels of this prostaglandin. Other approaches like this would be to decrease arachadonic acid intake (less animal fats), avoid trans-fats, and increase the intake of omega-3 and omega-6 fatty acids to improve the ratios between prostaglandin families (Pg1, PG2, and PG3) and to lower levels of pro-inflammatory PG2 and higher levels of the anti-inflammatory PG1 and PG3 families.

AK care has been successful in the treatment of endometriosis in a 25 year old patient with chronic, severe abdominal pain associated with her menstrual cycle and dyspareunia despite laparoscopic surgery for endometriosis six months prior to treatment. Care included chiropractic, dietary modification (primarily avoidance of refined food, use of digestive enzymes, and progesterone cream to address estrogen dominance.) (**Alis & Alis, 2004**)

Polycystic Ovary Disease

Estrogen activity may by high due to insulin resistance and resulting polycystic ovary disease (PCOS). The solution here is improved diet with carbohydrate restriction and intense interval exercise to improve insulin sensitivity. Chromium, B fraction (primarily thiamin), and optimizing mitochondrial function (mostly citric acid/Kreb's cycle and electron transport systems) is often helpful for improving blood sugar regulation. An estimated 90% of the American population does not consume the minimum recommended intake of 50 micrograms of chromium daily. (**Shils et al., 1999**) Researchers at the United States Department of Agriculture have estimated that up to 25% of heart disease in the United States could be prevented merely by consuming adequate quantities of chromium. Human subjects who take 200 micrograms daily of chromium supplements lose more fat and gain more lean tissue. (**Gaby, 2006**) Another study found that chromium provided significant drops in fasting blood glucose in diabetics. (**Evans et al., 1993**) Several studies have shown that chromium supplements lower fasting blood glucose levels in normal healthy individuals. (**Anderson et al., 1983**)

Vaginitis/Vaginosis

Women who have higher than normal estrogen levels, diabetes, and vaginal dryness possess predisposing factors for vaginal candidiasis and vaginosis. (**Dennerstein, 1998**) Soy isoflavones have been reported to be effective management for menopausal vaginal dryness. (**Li, 2010**)

Supplementation of vitamin A and beta-carotene has been reported to decrease the risk for bacterial vaginosis. (**Christian, 2011**) Vaginal application of vitamin C has also shown promise for management of vaginosis. (**Petersen, 2011**)

Oral and vaginal *lactobacilli* probiotic therapy has been shown useful for management and decreasing risk for vaginosis. (**Delia, 2006; Reid, 2001**) The approach appears to be more effective than metronidazole vaginal gel. (**Anukam, 2006**)

Broda Barnes (**1976**) has linked hypothyroidism (and hypoiodinism) with uterine fibroids and vaginitis. In women with thick mucus secretions in the vaginal area, iodine should be considered. (**Goodheart, 1998; Schmitt, 1990**) Goodheart also suggests vitamin B and wheat germ oil and have been successful for many women with vaginitis.

Benign Uterine Fibroid/Leiomyoma

Risk factors for the development of uterine fibroids include obesity, (**Shikora, 1991**) low vitamin A status, (**Martin, 2011**) and zinc deficiency. (**Sahin, 2009**) Cadmium toxicity appears to be a risk factor for both uterine fibroids and endometriosis, (**Jackson, 2008**) and lycopene supplementation, like iodine, may decrease the risk for developing fibroids. (**Sahin, 2004**)

Serum estrogen and progesterone levels do not differ between normal myometria and uterine leiomyometria groups, but uterine leiomyoma tissues are characterized by significantly increased estrogen and progesterone levels and a high estrogen-progesterone ratio. (**Potgieterm, 1995**)

Male Reproductive Systems

Prostate Disease

The prostate gland is part of a man's urinary and sex organs. It is about the size of a walnut, doughnut-shaped, and it surrounds the urethra which exits the bladder. The urethra has two functions in men. The first is to carry urine from the bladder, the second to carry semen during sexual climax.

Over 50% of men will develop an enlarged prostate (or benign prostatic hypertrophy) in their lifetime. A man over 50 having problems urinating is usually suffering from an enlarged prostate. As men get older, there is a tendency for the prostate to grow. As it grows, it squeezes the urethra. Since urine travels from the bladder through the urethra, the pressure from the enlarged prostate may affect bladder control. The symptoms of BPH are:

- A frequent and urgent need to urinate that occurs first at night.
- Trouble starting a urine stream. Straining required to get the urine flowing.
- A weak stream of urine. Takes longer to urinate than when younger.
- Only a small amount of urine flows.
- Feeling that more urine remains, even when finished.
- Leaking or dribbling.

Aside from the symptoms associated with BPH, the condition is a risk factor for developing prostate cancer and merits the clinicians' focused management. The healthcare burden of prostate disorders is evidenced by the 2 million or so physician office visits per year in the US. (**Chaitow & Jones, 2012**) It is often recommended that men over the age of 40 have yearly prostate exams. The exam involves a doctor inserting a gloved finger into the rectum and feeling the lower part of the prostate for any abnormality. However, in the case of BPH, often the prostate has not enlarged to a point that can be recognized by physical exam. Ultrasound measurements are another common diagnostic method, and then a blood test may be used to differentiate BPH from a more serious prostate cancer. Surgery for BPH may have only temporary, but sometimes permanent effects on sexual function. **(Garnick, 2012)** Most men recover sexual function within a year after surgery. The exact length of time depends on how long the symptoms had been present before the surgery was done and on the type of surgery. Side effects include erectile dysfunction and loss of bladder control as well as semen that no longer goes out of the penis during orgasm. Instead it goes backwards into the bladder.

Risk for developing benign prostatic hypertrophy (BPH) has been associated with high testosterone levels, but research indicates that testosterone (TT) has an anti-inflammatory effect on prostate tissues and development of BPH has closer association with dihydrotestosterone (DHT) levels. **(Vignozzi, 2012)** Specifically, there appears to be a direct association between larger prostate volume and higher DHT and DHT/TT levels. **(Liao, 2012)**

The enzyme 5-alpha reductase converts testosterone to dihydrotestosterone and inhibition of 5-alpha reductase has been shown to "block the undesirable effects of T on the prostate, without blocking the desirable anabolic effects of T on muscle, bone, and fat." **(Borst, 2005)** Zinc has been shown to be an important 5-alpha reductase inhibitor that decreases prostate weight in rats without affecting testicular function. **(Fahim, 1993)**

Saw palmetto (Serenoa repens, sometimes referred to as *sabal* in Europe) grows naturally in the southeastern United States, including Georgia, Mississippi, and particularly Florida. Saw palmetto has been shown to inhibit 5 alpha-reductase in some studies, though there is some controversy as to its therapeutic value clinically. **(Habib, 2005)** One study in rats found the combination of serenoa repens, lycopene, and selenium is more effective for preventing hormone dependent prostatic growth than serenoa repens alone. **(Altavilla, 2011)** Other studies have discovered a correlation between BPH and selenium deficiency. **(López Fontana, 2010; Muecke, 2009; Thomas, 1999)** One study concluded that both vitamin E and selenium are potentially protective for BPH. **(Klein, 2003)**

Though testosterones are the focus in the pathogenesis of BPH, estradiol stimulates proliferation of prostate stromal cells suggesting that excessive aromatization of testosterone may be a factor. **(Zhang, 2008)** Prostate size has been found to be associated with estradiol/bioavailable testosterone ratio. **(Roberts, 2004)** Exposure of rats to the xenobiotic estrogenic endocrine disruptor, bisphenol A (BPA), has been shown to induce BPH, suggesting a possible generalized role for xenobiotic compounds in BPH pathogenesis. **(Wu, 2011)**

Dietary flavonoids may have beneficial effects that decrease BPH risk through their modulating effect on the phase I detoxifying enzymes (cytochrome P450 pathway) and activation of UDP-glucuronyl transferase, glutathione S-transferase, and quinone reductase of the phase II pathway. Dietary flavonoids, also, inhibit aromatase activity, thereby decreasing the pathogenic effects of xenobiotic compounds. **(Moon, 2006; Hiipakka, 2002)**

Flavonoids are widely distributed in plants with common dietary sources being citrus, berries, rutin from buckwheat, onion, legumes, tea, red wine, and cocoa. Supplemental sources include quercitin, the isoflavones (genistein, daidzein, glycitein), proanthocyanidins, and anthocyanidins.

Lycopene has support as prevention and therapy for BPH. This carotenoid antioxidant is more biologically active in singlet oxygen quenching than vitamin E (~100x) and glutathione (~125x). Though it is a red carotenoid found in a number of foods, the most significant dietary source is tomato, with the concentration and bioavailability being the most prominent in tomato paste. Research indicates that lycopene is therapeutic for BPH. **(Wertz, 2009; Schwarz, 2008)**

Fats seem to play a role in BPH pathogenesis. Vitamin D appears to be anti-proliferative and promote cellular maturation of prostate cells. **(Feldman, 1995)** The role of vitamin D may be through the down-regulation of COX-2 expression and PG2 production and by the arrest of NF-kappaB. **(Penna, 2009)** Research finds that the essential fatty acid profile in BPH is deficient primarily in the omega3 fatty acids and, secondarily, in the omega6 fatty acids. **(Yang, 1999)**

A literature review showed an inverse relationship between exercise levels in general and the incidence of BPH. **(Sea, 2009)**

One of the most impressive aspects in studies on natural approaches to prostate health has been the improvement in quality-of-life scores. **(Murray & Pizzorno, 1997)** Many men who suffer from an enlarged prostate also suffer from sleep deprivation. By improving the bothersome symptoms such as nocturia and the sleep deprivation it produces, a man's mental outlook is dramatically improved. Usually symptoms resulting from mild-to-moderate prostate enlargement respond more readily to these treatments than symptoms due to severe enlargement. No significant side-effects have been reported in the medical literature from these natural treatment methods.

Aching Pain, Lactic Acidosis, and Mitochondrial Dysfunction

Mitochondrial diseases are a group of potentially life threatening disorders, primarily effecting the nervous system, that represent severe impairment of mitochondrial function due to genetic or mutagenic mitochondrial DNA aberrations. Here we are discussing mitochondrial function relative to sub-optimal tricarboxylic/citric acid (TCA) cycle and electron transport system (ETS) functions.

It is common in some quarters of orthopedics and biomedicine to consider dysfunction of body systems to be an either/or phenomenon where a given system either functions adequately or has pathology. Fundamental to AK is the concept that many illnesses seen in clinical practice are due to dysfunction and dys-ease of systems that results in functional, rather than pathological illnesses. In this model, health is a spectrum and the early stages of many disease processes experience a stage of dysfunction before developing into tissue breakdown and pathology. An example here would be the slow and incremental development of prostate cancer, which may go through a phase of benign prostatic hypertrophy for a long period before the inflamed and dysfunctional prostate tissues cross the threshold into cancerous pathology. In this case too there is a range of mitochondrial dysfunction from the fatigue and tissue soreness from suboptimal cellular energy production to life threatening pathology. Here we will cover functional issues concerning mitochondrial energy production.

Functional systemic lactic acidosis is a state of suboptimal function of the TCA cycle resulting in increased reliance upon anaerobic glycolysis and increased production of lactic acid. Additionally, reduced function of the TCA cycle will result in diminished CO_2 production and put stress on the systemic buffering role of the bicarbonate system. The combination of increased lactic acid production and bicarbonate deficiency disturbs the systemic pH.

Patients with suboptimal mitochondrial function will most commonly complain of fatigue and generalized aching and stiffness in their muscles and sometimes in their joints. An inability to concentrate is common, and patients will often say they feel "foggy headed." This condition is often misdiagnosed as fibromyositis or fibromyalgia and, sometimes, as chronic fatigue syndrome. The serum CO_2 levels will be either in the low range of normal (common) or just below normal range (less common). It has been proposed that though the norm for serum CO_2 is 23-32 mmoI/L, an optimal range is 26-31 mmoI/L, and below this range the patient has a relative systemic lactic acidosis. Low CO_2 can be due to a number of causes, including ketoacidosis, renal failure, and intoxication of organic acids or sulfates, but the most common in a functional medicine practice is lactic acidosis. (**Eidenier, 2007**)

If the lactic acidosis becomes more severe, symptoms may include anxiety, shortness of breath, arrhythmia, tachypnea, tachycardia, nausea, and generalized muscle weakness. (**See Acid-Alkaline section above**) The anxiety associated with mitochondrial dysfunction may be partly due to impaired GABA, since alpha-ketoglutaric acid, produced by TCA cycle function, is a necessary precursor for glutamic acid and GABA production.

Schmitt observes that when patients have functional lactic acidosis muscle weaknesses found with AK MMT will be negated temporarily by having the patient rebreathe their own air in-and-out of a paper bag for a few cycles. This test is suggested to be reliable indicator of low CO_2 reserve and suboptimal TCA cycle function and sometimes an indicator for the need of Vitamin B6. (**Schmitt, 1990**) This approach has been examined by Ozello. (**Ozello, 2006**)

When the rebreathing negates muscle inhibition as determined by AK MMT, one or more of the nutrients essential to the TCA cycle may be sufficient for restoration of adequate TCA cycle function. Thiamine (B1) deficiency is the most common need, but also riboflavin (B2), niacin (B3), pantothenic acid (B5), manganese, magnesium, biotin, iron, sulfur, or phosphorus.

Lipoic acid is also critical for the TCA cycle function and mercury is known to deplete sulfur-containing antioxidants such as L-acetyl-L-cysteine, L-glutathione, and alpha-lipoic acid. (**Houston, 2011**) If lipoic acid tests deficient according to AK MMT, differentially diagnose the patient for possible mercury toxicity. Less commonly, arsenic toxicity will be a factor in the lipoic acid depletion.

Coenzyme Q_{10} (ubiquinone) has been identified as a critical component of the electron transport chain, the next step in the respiratory cycle following the TCA cycle. (**Brandt & Trumpower, 1994; Rich & Wikstrom, 1986**) Its role in preserving mitochondrial integrity cannot be overemphasized.

Once a patient is supported with the nutrients indicated from AK MMT, the symptoms associated with lactic acidosis will diminish within a few days, though a loading dose for one to two weeks may be needed initially. If thiamin is indicated, based upon AK MMT, the phosphorylated form — thiamin pyrophosphate — may be required for optimal clinical response. According to Schmitt, if the patient has a good clinical response initially and then regresses, support of the electron transport system is indicated. (**Schmitt, 1990**) The nutrients essential for electron transport are CoQ_{10}, iron, copper, and phosphorus. CoQ_{10} deficiency as determined by AK MMT is the most common finding. Examine for a CoQ_{10} deficiency in patients taking statin drugs as they inhibit production of mevalonic acid, the precursor in the synthesis of CoQ_{10}. (**Deichmann, 2010**)

When TCA cycle dysfunction is present, anaerobic glycolysis will be up-regulated, putting stress on blood sugar regulating systems, since anaerobic glycolysis produces only two molecules of ATP per each molecule of glucose compared to 38 molecules of ATP per each molecule of glucose. This may result in chronically high cortisol levels in an effort of the body to up-regulate gluconeogenesis and place an excessive demand on glucagon production, gluconeogenesis, and chromium reserves.

The catabolic shift required by the up-regulation of gluconeogenesis is likely to be an underlying factor for low DHEA reserves, impaired protein and collagen metabolism, impaired immune-competency, osteoporosis, slow wound healing, and muscle wasting seen in this group of patients.

Urea-Guanidine Cycle and Deep Aching Pain

Excessive guanidine production has been proposed as a possible causative agent for pain. Guanidine, a metabolite of the urea-guanidine cycle, is a nitric oxide synthase (NOS) inhibitor. (**Ruetten, 1996**) Low tissue NO has been associated with vaso- and broncho-constriction, increased arterial wall inflammation, increased oxidative stress, up-regulation of renin release, impaired macrophage functions, impaired gastric motility, and diminished intestinal mucin secretion. Excessive guanidine is also known to cause muscle spasticity, (**Martensson, 1946**) and Duffy has associated high guanidine tissue levels with the deep, aching pain seen in patients with fibromyositis/fibromyalgia. (**Duffy, 2008**)

With impaired urea cycle function, there will be an adaptive up-regulation of the guanidine cycle to preserve production of the essential metabolite creatine with an associated increase in the production of guanidine. (**Natelson, 1979**) Guanidine inhibits the TCA cycle just discussed, lowering the production of two TCA metabolites, CO_2 and aspartate, which are cofactors for the urea cycle, further down-regulating the TCA cycle and up-regulating anaerobic glycolysis and lactic acid levels. (**Rao, 1992**)

Schmitt has developed a protocol for determining urea cycle functions through AK MMT. An ammonia sniff challenge is used to determine deficient, sufficient, or excessive ammonia tissue levels. With optimal balance of the urea cycle, sniffing ammonia will neither weaken nor strengthen an indicator muscle. **(Schmitt, 1990)**

When an ammonia sniff challenge strengthens a weak indictor muscle, there may be protein deficiency, but the most common reason is deficient pyridoxal-5′-phosphate (P5P), the activated form of vitamin B6. This may be caused by deficiency of the cofactors necessary for conversion of vitamin B6 to the coenzyme P5P. These cofactors are magnesium, zinc, riboflavin, and/or phosphorus. Since transamination reactions depend on P5P, a common finding is low or low-normal alanine transaminase (ALT) and/or aspartate transaminase (AST). **(Lacour, 1982)** A number of polymorphisms can inhibit both production of P5P and activity of P5P on dependent enzyme systems. **(Clayton, 2006)**

When the ammonia sniff challenge is positive, tissue ammonia burden is most likely too high due to impaired synthesis of urea or creatine from ammonia. If the ammonia sniff test is positive with AK MMT, test for the following nutrients: B6 (possibly, P5P), iron, magnesium, manganese, molybdenum, biotin, thiamin, riboflavin, niacin, pantothenic acid, lipoic acid, and a source of the enzyme arginase. Many of these nutrients are cofactors in the TCA cycle and indicate the interdependence between the TCA and urea cycles. In our opinion, Schmitt has developed an elegant and clinically effective method for testing and resolving many functional metabolic faults and his work is both recommended and beyond the scope of this discussion. **(Schmitt, 2005)**

Inactivity Aggravated Pain and Calcium Metabolism

High calcium-phosphorus ratio is associated with a particular pattern of deep aching pain and stiffness that is accentuated by inactivity and attenuated by movement. The relief from a few minutes of movement will often be complete and the soreness and stiffness will return when immobile for more than ~30 minutes.

Typically, the patient will have a serum calcium-phosphorus ratio greater than 2.4 (ideal value is approximately 2.4) and will find almost miraculous relief once the calcium-phosphorus ratio is improved. Stress to calcium-phosphorus metabolism will often be due to over-utilizing calcium supplementation or from a need for more dietary phosphorus. A trial of supplementation with liquid ortho-phosphoric acid will typically bring about a dramatic change in one to two days and confirm the diagnostic suspicion. Sources of phosphorus are milk, egg yolk, meat, grains, nuts, legumes, and lecithin.

Burning pain, NSAIDs, and the prostaglandin system

Chronic burning pain is usually due to imbalanced prostaglandin (PG) metabolism with relative up-regulation of the pro-inflammatory prostaglandin2 (PG2) family. When non-steroidal anti-inflammatory drugs (NSAIDs) provide temporary relief of burning pain, underlying prostaglandin imbalance is strongly indicated. The underlying PG imbalance can be due to excess prostaglandin2 (PG2) levels; deficient anti-inflammatory prostaglandin1 (PG1) levels; or prostaglandin3 (PG3) levels, or both. The patient usually needs less animal fats in the diet (animal fats are the most common source of arachidonic acid, the precursor of PG2) and more vegetable, nuts, seeds, grains, or fish oils.

Clinical disorders due to insufficient PG3 levels appear to be common and account for a wide range of illnesses. **(Simopoulos, 2006; Rudin, 1982)** Prostaglandins may sensitize nerve endings to the pain-producing effects of other compounds, such as bradykinins. Omega3 fatty acids are precursors for PG3 and dietary sources are nuts, seeds, beans, grains, flax, fish, meat, eggs, and milk. Animal fats from grass-fed sources are naturally higher in omega3 fatty acids than grain fed sources. **(Daley, 2010)**

Occasionally, deficiency of PG1 is primary. This is usually seen in patients with disturbances of female reproductive system. Most vegetable oils are naturally high in omega6 fatty acids with black currant, borage, and evening primrose oils being particularly concentrated sources and are unaffected by delta-6 desaturase inhibition. **(Horrobin, 1983)**

Trans-fats (hydrogenated oils) inhibit delta-6 desaturase (D6D) and limit production of the anti-inflammatory PG1 and PG3 families. This process underlies the prevalence of high pro-inflammatory PG2 levels relative to PG1-PG3 levels and the prevalence of chronic inflammatory diseases and pain in cultures consuming refined foods, especially when animal fats are over-indulged. D6D activity is attenuated by hydrogenated fats, alcohol, arachidonic acid, glucose, fructose, smoking, heavy metal burdens, catecholamines, glucocorticoids, and thyroxine and is accentuated by B6, magnesium, zinc, protein, insulin, and ATP.

Eliminate Trans fats from the diet

When trans fat inhibition of D6D is present, patients can be helped with PG1-PG3 precursors that are D6D dependent, using evening primrose, borage, and black currant seed oils for the PG1 family and fish oils for the PG3 family.

In December 2006, New York became the first U.S. city to ban artificial trans fats at restaurants — from the corner pizzeria to high-end bakeries. (**New York Times, 2006**) The city's prohibition on trans fats was a victory for Mayor Michael Bloomberg, an outspoken health advocate, and his activist health commissioner, Dr. Thomas R. Frieden.

When PG2 levels are too high in relation to the PG1-PG3 families, inflammation may be up-regulated and the patient may experience temporary relief when using NSAIDs. (**See discussion in Dysmenorrhea section above**) There will also be facilitation of inhibited muscles upon insalivation (AK sensorimotor challenge) of a powdered mixture of aspirin, acetaminophen, and ibuprofen. Conversely a facilitated indicator muscle may weaken when a patient is challenged with a concentrated source of arachidonic acid (lard is very effective for this purpose). Essential fatty acids and cofactors can be screened for negating the positive arachidonic acid challenge and the patient can be supported appropriately.

When PG imbalances are present, having the patient smell chlorine will cause a strong indicator muscle to weaken upon AK MMT, and testing with insalivation of an indicated essential fatty acid or cofactor nutrient at the same time will negate the positive chlorine challenge. This test was originally described by Schmitt. (**Schmitt, 2005**)

Itching, edematous pain and bradykinin

This type of pain is often associated with edema (bradykinin is a potent vasodilator), itching, and sensitivity to pressure. (**Ständer, 2006; Graven-Nielsen, 2001**) Interestingly, bradykinin excess caused by ACE inhibitors may be associated with a persistent slight, dry cough. (**Dicpinigaitis, 2006**) Kinins appear to be an important mediator of the pain, swelling, and cellular damage associated with inflammatory joint diseases. (**Bhoola, 1992**)

Bradykinin release is triggered by injury, but abnormal modulation may play a role in chronic pain. (**Raja, 1988**) Natural bradykinin inhibitors are bromelain, aloe, and polyphenols (bioflavonoids), as found in citrus, red wine, coffee, tea, and chocolate. (**Bautista-Pérez, 2004; Bouskela, 1997; Lotz-Winter, 1990**)

Systemic inflammatory pain

Oxidation

Reactive oxygen species (ROS) are essential metabolites of oxidative phosphorylation in the mitochondria, and their role has been documented in producing DNA damage, lipid peroxidation, oxidation of amino acids, inactivation of enzymes, degenerative diseases and aging. Antioxidant enzymes, primarily superoxide dismutase, glutathione peroxidase, methionine reductase, and catalase, normally control ROS levels and mitigate their damage.

Poorly modulated ROS metabolism results in up-regulation of cellular and systemic inflammation. High or high normal C-reactive protein (CRP), fibrinogen, and erythrocyte sedimentation rate (ESR) can be used as indirect monitors of ROS-triggered inflammation when informed by the overall clinical picture. Urinary malondialdehyde (MDA) may be a clinically useful monitor of ROS stress.

AK MMT can be used to challenge for the presence of ROS stress by having the patient smell chlorine. Chlorine is a source of hypochlorite free radicals and will cause a strong indicator muscle to weaken when ROS stress is present. Any of the antioxidant promoting nutrients can be tested for negation of this positive chlorine olfactory sensorimotor challenge; support accordingly.

Methylation

Hyperhomocysteinemia due to impaired methylation metabolism has been associated with atherosclerotic cardiovascular diseases, stroke, peripheral arterial occlusive disease, venous thrombosis, autoimmune and neurological disorders, including autism. (**Houston, 2012; James, 2004; Richardson, 2003; Herrmann, 2001**)

AK MMT has shown an association with high homocysteine levels, a common cause of systemic inflammation, and bilateral psoas muscle weakness, a common cause of low back and pelvic pain. (**Rogowsky, 2005**) High homocysteine levels are usually moderated through sufficient methylcobalamine (B-12), 5-methyltetrahydrofolate (MTHF), folic acid, pyridoxyl-5-phosphate (P5P), serine, betaine, and/or arginine. (**Gaby, 2006; Figure below**)

When smelling chlorine causes an indicator muscle to weaken with AK MMT and the positive challenge is negated by taurine, methylation metabolism is impaired. Test methylation cofactors (methionoine, magnesium, methylcobalamine (B-12), folic acid, 5-methyltetrahydrofolate (MTHF), a methyl donor (betaine, choline), and molybdenum, and support as indicated using the nutrient(s) that negated the positive chlorine smell challenge.

Leukotrienes

When the production of PG2 is inhibited by NSAIDs, the production of leukotrienes from arachidonic acid is up-regulated. Leukotrienes are powerful pro-inflammatory compounds that have been associated with the pathophysiology of inflammatory and allergic conditions affecting skin, joints, and respiratory and gastrointestinal systems. (**Haeggström, 2002**) Leukotriene up-regulation from NSAID use accounts for the side effects from this family of drugs. (**Stevenson, 2003; Rainsford, 1993**)

When leukotrienes are improperly modulated in a patient, there will be inhibition of previously strong muscles with insalivation of a powdered mixture of aspirin, acetaminophen, and ibuprofen. Check the patient using AK

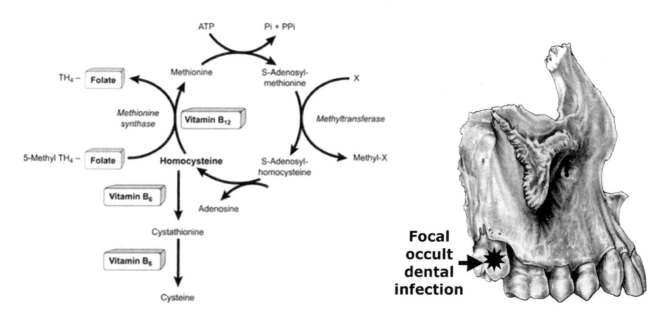

Homocysteine Pathway

MMT for essential fatty acid metabolism precursors and cofactors to improve the ratio between pro-inflammatory PG2 and the anti-inflammatory PG1-PG3s, and guide the patient on how to decrease their arachadonic acid intake through less animal fats in their diet and to avoid NSAIDs. Also consider testing for leukotriene-inhibiting nutrients. These are vitamin E (>200 IU/day), eicosapentaenoic acid (EPA), quercitin, glutathione, selenium, and aloe vera.

Glycation

Impaired glucose metabolism (either diabetes or insulin resistance) results in increased glycation of proteins or fats with sugars to produce advanced glycation endproducts (AGEs) which contribute to ROS stress and inflammation. This is a major mediator of the high incidence of neuropathy, cataracts, and cardiovascular disease seen in diabetics. (**Veves et al., 2002; Kanauchi, 2001**) One researcher has proposed that glycation is pathogenic for a wide array of chronic, degenerative illnesses and may be associated with senescence itself. (**Cárdenas-León, 2009**)

Glucose is a key component of this process, but the efficiency of fructose glycation is approximately ten times greater than that of glucose, leading some researchers to conclude that fructose-dependent diets lead to higher AGEs production and accentuated development of AGEs-mediated diseases. (**Gul, 2009; Gul, 2009a; Mikulíková, 2008; Tokita, 2005; Schalkwijk, 2004**)

Fundamental to control of glycation is improving blood sugar metabolism and optimizing insulin sensitivity through diet and exercise. The dietary issues do not need to be belabored here, but research does indicate that short, intense, interval exercise is more effective for improving insulin sensitivity than longer, non-interval exercise sessions. (**Metcalfe, 2011; Babraj, 2009**)

As for clinical nutrition, lipoic acid has been shown to decrease glycation, and taurine has been shown to alleviate glycation-mediated changes to collagen in fructose-fed rats. (**Thirunavukkarasu, 2005; Nandhini, 2005**) Rutin, a dietary flavonol, has been shown to inhibit glycation and would, at least, suggest the same role for other favonols. (**Muthenna, 2011**) A wide range of polyphenols widely found in fruits and vegetables show inhibition of glycation. (**Saraswat, 2009; Biesalski, 2007**) Citric acid from citrus fruits and ginger has also shown promise. (**Saraswat 2010; Nagai, 2010**)

Focal occult infection

The theory that focal occult infections can mediate systemic inflammatory and degenerative effects was first proposed in the 1920s. (**Lee, 1925; 1923**) Though discredited for many decades, the focal infection theory of systemic disease is becoming once more better established. (**Goymerac, 2004**) The most recognizable example here is periodontal disease being a risk factor for systemic inflammation and related degenerative diseases, such as cardiovascular disease. (**Glickman, 2009**)

One study of periodontal disease indicated a causative role for systemic inflammatory markers through the lowering of CRP, interleukin-6 (IL-6), and LDL cholesterol levels from baseline after two months of periodontal therapy. (**Somma, 2010**) It is possible to trace bacteria recovered from peripheral blood to occult focal infection in tooth apices after a root canal, and it has been suggested that the resultant bacteraemia and circulating endotoxins may have systemic effects. (**Murray, 2000**)

The Medical Journal of Applied Kinesiology (**2012**) has been publishing reports from dentists and other clinicians in the German-speaking world on this subject and has produced an impressive compendium of the AK diagnostic findings in cases of focal occult infections and their treatment.

Sinusitis is another common focal occult infection, with the presenting complaint often being fatigue, in many cases accompanied by the complaint of "foggy-headedness" or some similar symptom. (**Chester, 1996**) A study from the Mayo Clinic of 210 subjects with chronic sinusitis concluded that 96% were also positive for fungal infection. (**Ponikau, 1999**)

Protozoan infection is known to be a causitive agent for irritable bowel syndrome, and *Blastocystis hominis* infection has been reported to produce reactive arthritis, indicating systemic inflammatory and immunoactive effects. (**Stark, 2007; Lakhanpal, 1991**) Emulsified oregano oil is an effective therapy for *Blastocystis hominis* and other protozoan infections. (**Force et al., 2000**) Intestinal bacteria and viruses and food allergens can dysmodulate immune and inflammatory responses, primarily through promotion of IFN-gamma, TNF-alpha and other pro-inflammatory cytokines. (**Peña, 1998**)

Joint pain

Osteoarthritis

Risk factors for osteoarthritis (OA) include arthrogenic muscle inhibition, muscle weakness, joint injury, age, obesity, genetics, and aberrant biomechanics. (**Hurley, 2002**) Patients will usually complain of stiffness, limited range of motion, and feel worse after physical activity. Interestingly, imaging studies for joint pathology often do not correspond to symptoms indicating the critical role of biomechanics. This concurs with the conclusion of Rosomoff, who stated, "It is our opinion that in many cases attributed to disc pathology, arthritis or spinal canal stenosis the symptoms may, in fact, be caused by soft tissue dysfunction. What appears as pathology on an X-ray may be the end result of abnormal biomechanics resulting from this dysfunction." (**Rosomoff, 1989**)

Applied kinesiology plays an important role in the amelioration of osteoarthritis by optimizing muscle strength, coordination and timing to optimize postural and movement patterns, including the gait-cycle. By optimizing muscular and articular biomechanics, joint injury with use is minimized and the mechanical pathogenesis of osteoarthritis minimized.

The degree of synovitis found with osteoarthritis correlates with synovial concentration of both leukotrienes and kinins, indicating a critical role for prostaglandin and bradykinin metabolic faults in OA. (**Nishimura, 2002; Bhoola, 1992**)

Glycation promotion, ROS stress, and low-level chronic inflammation have also been implicated in the pathogenesis of OA. (**Shane-Anderson, 2010; Ziskoven, 2010; Regan, 2008; Hadjigogos, 2003; Verzijl, 2002**) Joint immobilization promotes ROS through diminished superoxide dismutase (SOD), catalase (CAT), and glutathione peroxidase (GSH-Px) in the joint capsule. (**Erdem, 2009**) Yoga and Tai Chi/Qi Gong may be useful for management of OA through decreasing arthritic symptoms and improving joint function and balance sense. (**Ebnezar, 2012; Lee, 2009; Kolasinski, 2005; Song, 2003**)

Rheumatoid Arthritis

Patients with RA show increased ROS stress. (**Seven, 2008; Cai, 2005**) Free-radical oxidation products in serum were significantly elevated in patients with RA compared to normal controls. (**Lunec, 1981**) SOD values were significantly lower in RA patients compared to controls. (**Disilvestro, 1992**) Dysmodulation of DNA methylation has been proposed to have epigenetic effects that may be a factor in the pathogenesis of autoimmune diseases, including RA. (**Strickland, 2008**)

Zinc, copper, and selenium metabolism may be vectors in the pathogenesis of RA as well. In RA patients the erythrocyte sedimentation rate (ESR), acute-phase proteins, interleukin-1 beta (IL1 beta), and tumor necrosis factor alpha (TNF alpha) imbalances correlated negatively with serum zinc and positively with serum copper. (**Zoli, 1998**)

Mean serum copper concentration in RA subjects has been positively correlated with erythrocyte sedimentation rate (ESR) and Ritchie articular index. (**Strecker, 2005**) Low zinc and low Zn/Cu ratio are found in plasma of RA patients. (**Ala, 2009**) Patients with RA may also tend toward dietary patterns that are deficient in pyridoxine, zinc and magnesium. (**Kremer, 1996**)

Bradykinin up-regulation may be a factor in RA as synovial fluids show excessive release of bradykinin. (**Sharma, 1994**)

As with any autoimmune disorder, the triggering mechanism may be an occult locus of infection as previously discussed. For this reason, closely examine the sinuses, teeth and gums, intestinal and urinary tract. Intestinal inflammation has been associated with arthritis. (**Mielants, 1987**) Occult focal apical infections of teeth after root canal have been implicated in the pathogenesis of RA, (**Murray, 2000**) and periodontal treatment was reported to result in RA remission. (**Iida, 1985**)

Total fasting of 7-10 days induces marked reduction of RA pain, inflammation, and edema within a few days and the remission subsides over time after discontinuing the fasting. (**Palmblad, 1991; Sundqvist, 1982**) Intestinal permeability improves with fasting, suggesting a role for food allergy and/or digestive dysfunction in RA. (**Sköldstam, 1991; Panush, 1991**) Direct involvement of food allergy in the pathogenesis of RA was challenged in a study suggesting that the contribution of food allergy to RA pathogenesis may be mediated through impaired intestinal permeability rather than through direct influence on humoral immune response. (**Kjeldsen-Kragh, 1995**)

Fasting followed by a lactovegetarian diet appears to be good management for RA and one study showed significant decreases in platelets, leukocytes, total IgG, IgM rheumatoid factor, and complement components C3 and C4 after one month. (**Müller, 2001; Kjeldsen-Kragh, 1995**) The benefit, at least in part, is likely due to the decreased arachadonic acid intake associated with vegetarian diets and the shift to a lower PG2/PG1-PG3 ratio compounded by an increased consumption of dietary omega3 fatty acids. (**Simopoulos, 2002**)

Electric blankets may be a causative or aggravating factor in arthritis and the adverse effect of electromagnetic radiation can be observed with AK MMT. (**Maykel, 2007**)

Circulatory system

Differential diagnosis is essential for circulatory problems of the lower extremity. Bilateral edema of the lower extremities should always lead to minimally considering heart, kidney, or liver pathologies or dysfunction, while unilateral edema should lead to consideration of deep vein thrombosis, thrombophlebitis, pelvic tumor, and cellulitis. Differential diagnosis of these possible pathologies is outside the scope of this chapter.

Varicose veins

Hemorrhoids and varicose veins are one of the most common afflictions of the Western world, affecting people of all ages and both genders. Worse yet, the problem typically worsens over time. Known causes include constipation, pregnancy and poor toilet habits such as straining while reading a favorite book. Symptoms include itching, bleeding and pain. Genetics play a large role in the predisposition to varicose veins. Don't overlook, however, the role that collagen formation plays in optimal venous structural strength and elasticity and the influence portal congestion on venous pressure in the pelvis and lower extremities have in predisposing to varicose veins and hemorrhoids. Also, note the influence of catecholamines and thiamine for increasing vascular tone.

Bioflavonoids have been shown to decrease capillary fragility and have positive effects on capillary permeability and blood flow. They show promise for the treatment of bruising, varicose veins, edema, and hemorrhoids. (**Martin, 1955**) Additionally, bioflavonoids show antioxidant, antihistamine, and anti-inflammatory effects. Numerous studies have shown bioflavonoids improve outcomes in treatment of hemorrhoids. (**Di Pierro, 2011; Misra, 2000; Ho, 2000**)

The authors have also found that the dietary supplement stone root, also called collinsonia root (Standard Process®), can be helpful. This product has been successfully used for decades for patients with hemorrhoids and/or varicose veins. Two capsules with a glass of water morning and evening on an empty stomach can help even the most stubborn cases. Patients usually take these for a month or two, and then discontinue when symptoms are gone. Patients keep some of this on hand for the first sign of any recurrence, which may indicate the need to assess the diet once more.

Edema

Generalized edema must be considered on the basis of the cellular metabolism of the cell wall which, in effect, is the osmotic transfer control that governs the movement and permeability of all body fluids. If not due to heart failure, consider liver, kidney, and adrenal dysfunction or pathology in the differential diagnosis. In women experiencing edema that occurs during the middle and late stages of the menstrual cycle, excessive estrogen may be suspected. This may be due to overproduction or inadequate deconjugation by the phase II detoxification pathway of the liver.

Edema due to nutrient deficiencies

It is worth noting that three classic nutrient deficiencies have been associated with edema in the lower extremities: kwashiorkor (protein), beriberi (thiamin), and pellagra (niacin). Though the symptoms by standard criteria are limited to severe deficiencies, it is important to consider clinically that nutrient deficiencies and their related symptoms exist in a spectrum of severity, and that sub-clinical deficiencies exist, particularly in the elderly. (**Gaby, 2006; Shils et al., 1999**)

Biochemical individuality and subclinical nutrient deficiency

Cheraskin, Ringsdorf, and Williams have each proposed models of biochemical individuality wherein each individual has an utterly unique biochemistry based on their genetic constitution, epigenetic history, and environment — often resulting in inborn errors of metabolism. These inborn errors can result in resistance to the influence of nutrients on enzyme systems and predispose patients to a need for higher levels of nutritional supplementation than the average population. (**Bucher, 2011; Cheraskin, 1977; Cheraskin, 1976; Williams, 1956**) AK adds to the physician's nutritional knowledge an ability to determine, to a certain extent, the effects of various nutritional products on the biochemical individuality of the patient being evaluated.

Zinc, magnesium, and vitamin D are good examples of this as they are commonly deficient upon laboratory exam and can underlie a wide array of symptoms in patients that would appear to be sufficiently nourished based upon an analysis of their diet. (**Hambidge, 2007; Holick, 2006; Whang, 1987**)

Sub-clinical B12 deficiency is now recognized as a valid clinical entity with subtle manifestations of the classic deficiency pattern and being responsive to B12 supplementation, especially in the elderly. (**Herrmann, 2003**) Based upon this principle, consider sub-clinical protein, thiamin, or niacin deficiency in patients with otherwise idiopathic edema of the lower extremity. (**Padhila, 2011; Igata, 2010; Singleton, 2001**)

Morabia has argued that laboratory testing is inadequate to diagnose these syndromes and that clinicians must still rely on observation of patients' patterns of signs and symptoms for adequate differential diagnosis. (**Morabia, 2011**)

McCarty has proposed a model for subtle nutrient deficiency leading to aberrant physiology and functional illness when he states "many nutritional agents involved in bioenergetics (regulating mitochondrial and antioxidant functions) are often functionally sub-saturated." (**McCarty, 1981**)

Rudin (**1982**) has proposed a model for pellagraform and beriberiform diseases that present as syndromes mimicking true pellagra and beriberi, but lack the developed pathologies associated with these classic nutrient deficiency diseases. Rudin calls these disorders substrate pellagra and substrate beriberi. Rudin proposed that the interdependency of B vitamins and their cofactor substrate essential fatty acids (primarily omega3 fatty acids) for production of structural and regulatory proteins and lipids could be compromised by malnutritional synergy — the combined subclinical deficiencies of proteins, fats, vitamins, and minerals.

Nutrient Deficiencies Associated With Edema

Kwashiorkor	BeriBeri	Pellagra
Pedal edema, ascites, enlarged and fatty liver, thinning hair, loss of teeth, skin depigmentation, dermatitis, irritability, anorexia.	Lower extremity edema, weight loss, emotional disturbances, impaired sensory perception (Wernicke's encephalopathy), parasthesias, arrhythmia, mental confusion/speech difficulties, vasodilation, awakening at night, short of breath, tachycardia, shortness of breath with activity.	Dermatitis (red skin lesions desquamation, erythema, scaling, keratosis of sun-exposed areas), photophobia, aggression, anxiety, mental confusion, insomnia, dementia, sensitivity to odors, neuralgia, neuritis, ataxia, weakness, alopecia, edema, glossitis (smooth, beefy red), diarrhea, dilated cardiomyopathy.
Protein deficiency	**Thiamin deficiency**	**Niacin deficiency**
Adequate hydrochloric acid and pancreatic enzymes are required for absorption; cooking all proteins may result in deficiency of heat-labile amino acids, small intestine absorption must be sufficient, and the urea cycle must be functional to have adequate protein metabolism.	Provide B1 (thiamine); a phosphorylated form may be needed as many patients ineffectively phosphorylate synthetic forms (i.e. thiamin hydrochloride). Refining removes naturally occurring thiamin from foods and sulfites degrade it. Refined carbohydrates increase tissue demand for thiamin.	Provide B3 (niacin); a phosphorylated form may be needed. Avoid corn, corn-derived sugars, and excessive sun exposure. Niacin is synthesized in the liver from tryptophan; tryptophan is heat-labile and a lack of uncooked protein sources may predispose to tryptophan deficiency.

Table 1

Table 1 may help the clinician understand the etiology for some presentations of edema in the lower extremities. It is important to note here that we not claiming that true kwashiorkor, beriberi, and pellagra are common diseases; they are, indeed, very rare, especially in industrialized societies. What we are attempting to make the clear is that functional (rather than pathological) nutrient deficiencies are more prevalent than supposed.

According to Rudin,

"...since substrate essential fatty acids are processed by many B vitamin catalysts, an EFA deficiency will mimic a panhypovitaminosis B, i.e., a mixture of substrate beriberi and substrate pellagra resembling vitamin beriberi and pellagra but exhibiting as even more diverse endemic disease. This would constitute a second stage of the modern malnutrition and explain why some workers now hold the dominant diseases of modernized societies to be new, nutritionally based, pellagraform yet lipid-related and to range, once again, from heart disease to psychosis."

Impaired Microcirculation Sympathetic Vasoconstriction

In patients with chronic cold hands and feet, especially when concurrent with clamminess, look to sympathetic dominance of the autonomic nervous system since sympathetic tone is peripherally vasoconstrictive and diaphoretic. Other considerations for cold extremities are anemia and conditions that lower metabolic rate and body temperature, in general. These include hypothyroidism, hypoglycemia, mitochondrial dysfunction (citric acid cycle and electron transport), and essential fatty acid and/or protein deficiency.

Raynaud's Syndrome

In patients with Raynaud's syndrome secondary to sclerodermal disease, look to high blood viscosity due to increased fibrinogen levels. **(Sergio, 1983)**

Other causes of Raynaud's are increased sympathetic tone, hyperviscosity, and hyperactivation of platelets and erythrocytes. Increased prostaglandins PGE1 improved peripheral blood flow in Raynauds patients. **(Belluci, 1988)** Goodheart noted that patients with essential fatty deficiency commonly complain of being cold despite a normal body temperature. **(Goodheart, 1985)**

Blood Hyperviscosity Syndromes

Increased viscosity increases peripheral resistance of the microcirculation. Blood viscosity is known to increase with increased IgG, IgA, IgM, fibrinogen, and triclyceride levels. These can be found in Waldenström's disease and hypertriglyceridemias. **(Crepaldi, 1983)** There is a strong correlation between metabolic syndrome and increased viscosity and the correlative physiological changes likely to be causative factors for arterial hypertension. **(Zhang, 2006; Hrnciarová, 1995)**

Increased blood viscosity decreases oxygen perfusion, results in decreased total oxygen delivery, and increases risk of veno-occlusive syndromes. **(DeFilippis, 2007; Maeda, 2006)** Interestingly, Qi Gong exercise has shown positive effects on blood viscosity. **(Lee, 2009)** The cardiovascular protective effects of omega3 fatty acids may be in part due to lowering blood viscosity. **(Reiner, 2007)**

Rouleaux Formation

Rouleaux formations are aggregates (clumping) of RBCs. Conditions associated with rouleaux formation are diabetes mellitus, infection, autoimmune and inflammatory disorders, anemia, and cancer. Rouleaux formation results in higher ESR results. Rouleaux formation increases with rising pH and decreases relative to pH lowering influence of CO_2. **(Cicha, 2003)** High plasma triglycerides and platelets increase rouleaux formation. **(Cicha, 2001; Baumler, 1987)** RBC cell membrane charge appears to modulate rouleaux formation. **(Antonova, 2006; De Lorenzo, 2004)**

Blood Viscosity And Rouleaux Interactions

Hyperviscosity syndromes are associated with rouleaux formation of the RBCs. **(Ballas, 1975)**

High blood viscosity, a determinant of total arterial resistance (TAR), and rouleaux formation has been proposed as a causation of arterial hypertension due to increases peripheral resistance of the microcirculation. **(Cicco, 1999; Zannad, 1985)**

Applied Kinesiology's Future in Stress-Related Illness

Most illnesses in industrialized societies are due to functional rather than pathological processes; most pathological illnesses are preceded by a chronic period of functional illness. Health is not an accident; it is the outcome of the interaction of an individual's genetic constitution and environment. Many people "get by" throughout their lives without optimal organic or biomechanical function and yet remain asymptomatic. This may depend on the goddess Fortune as well as the world-view and impulses of the person in addition to their inherited characteristics, nutritional status, psychosocial factors, life history and more – in other words, the entire context within which the applied kinesiology triad of health is experienced and embraced. If one of the objectives of work in this field is to prevent illness and ameliorate the burdens of the living patient and to help them realize their full potential, then what has been discussed in this chapter will become a part of the health care approach physicians and knowledgeable patients around the world embrace.

The Canadian endocrinologist, Hans Selye, (1976) developed the model for functional illness when he published his extensive research concerning the processes underlying this type of disease. He clearly shows that disease results when the homeostasis of a living organism is overcome due the cumulative effect of stressors to which it must adapt, be they physical, chemical, thermal, or mental.

From the beginnings of applied kinesiology, practitioners have observed an association between muscle-joint function and visceral-autonomic dysfunction. It is exciting to see accumulating research and developing models from a wide range of academicians and clinicians converging toward concurrence with the field of applied kinesiology. This development will, ideally, lead to more coordination with physicians from other fields and backgrounds to work synergistically with clinicians utilizing applied kinesiology methods in the treatment of patients with functional illnesses.

Evidence-based medicine, basic science and clinical outcomes data now exists to support the assessment and treatment (frequently co-treatment with other specialist physicians) for patients with disorders of the nervous, autonomic, neurohumoral, immune, respiratory, circulatory, and lymphatic systems using applied kinesiology methods. The objective of this work is to prevent illness, ameliorate suffering, and to help patients reach their full potential. Hopefully, the methods here will be embraced and used toward those ends.

References

- Aaseth J, Munthe E, Førre O, Steinnes E. Trace elements in serum and urine of patients with rheumatoid arthritis. Scandinavian journal of rheumatology. 1978;7(4):237-40.
- Abraham GE. Nutritional factors in the etiology of the premenstrual tension syndromes, The Journal of reproductive medicine. 1983;28(7):446-64.
- Abrahamsson TR, et al. Low diversity of the gut microbiota in infants with atopic eczema. J Allergy Clin Immunol. 2012;129(2):434-440.
- Agostini S. A marketed fermented dairy product containing Bifidobacterium lactis CNCM I-2494 suppresses gut hypersensitivity and colonic barrier disruption induced by acute stress in rats. Neurogastroenterol Motil. 2012;24(4):376-e172.
- Ala S, Shokrzadeh M, Pur Shoja AM, Saeedi Saravi SS. Zinc and copper plasma concentrations in rheumatoid arthritis patients from a selected population in Iran. Pak J Biol Sci. 2009;12(14):1041-4.
- Alis G, Alis S. Endometriosis: a case study. Collected Papers International College of Applied Kinesiology: Shawnee Mission; KS. 2004:1-2.
- Allen M. A Functional Approach To The Treatment Of Urinary Incontinence. Collected Papers International College of Applied Kinesiology. 2001:27-35
- Allen M. Viscerosomatic Reflexes and the Brain That Influences Them. Collected Papers International College of Applied Kinesiology: Shawnee Mission, KS. 2008:23-26.
- Altavilla D, et al. The combination of Serenoa repens, selenium and lycopene is more effective than serenoa repens alone to prevent hormone dependent prostatic growth. J Urol. 2011;186(4):1524-9.
- Ahmadi B, et al. The hidden epidemic of urinary incontinence in women: a population-based study with emphasis on preventive strategies. Int Urogynecol J Pelvic Floor Dysfunct. 2010; 124(2):86-89.
- Amirshahrokhi K, Bohlooli S, Chinifroush MM. The effect of methylsulfonylmethane on the experimental colitis in the rat. Toxicol Appl Pharmacol. 2011;253(3):197-202.
- An HM, et al. Efficacy of Lactic Acid Bacteria (LAB) supplement in management of constipation among nursing home residents. Nutr J. 2010;9:5.
- Ananthakrishnan AN, et al. Higher predicted vitamin D status is associated with reduced risk of Crohn's disease. Gastroenterology. 2012;142(3):482-9.
- Anderson R. Traditional therapy for chronic pelvic pain does not work: What do we do now? Nat. Clin. Pract. Urol. 2006;3(3):145-156.
- Anderson RA, et al. Metabolism. 1983;32:894.
- Anderson KE, et al. Effects of oral contraceptives on vitamin metabolism. Adv Clin Chem. 1976;18:247-87.
- Antonis A, Bershon I. The influence of diet on serum lipids in South African white and Bantu prisoners. Am J Clin Nutr. 1962;10:484-99.
- Antonova N, Riha P. Studies of electrorheological properties of blood, Clinical hemorheology and microcirculation. 2006;35(1-2):19-29.
- Anukam KC, et al. Clinical study comparing probiotic Lactobacillus GR-1 and RC-14 with metronidazole vaginal gel to treat symptomatic bacterial vaginosis. Microbes Infect. 2006;8(12-13):2772-6.
- Arnold-Frochot S. Investigation of the effect of chiropractic adjustments on a specific gynecological symptom: Dysmenorrhea. Journal of the Australian Chiropractic Association. 1981;10(1):6-10, 14-16.
- Ashton-Miller JA, Howard D, DeLancey JOL. The functional anatomy of the female pelvic floor and stress continence control system. Scand J Urol Nephrol Suppl. 2001;(207):1-7; discussion 106-25.
- Bablis P, Pollard H. Hypothyroidism: A New Model for Conservative Management in Two Cases. Chiro J Aust. 2004;34:11-18.
- Babraj JA, et al. Extremely short duration high intensity interval training substantially improves insulin action in young healthy males. BMC Endocr Disord. 2009;9:3.
- Ballas SK. The erythrocyte sedimentation rate, rouleaux formation and hyperviscosity syndrome. Theory and fact. Am J Clin Pathol. 1975;63(1):45-8.
- Barnes M. The basic science of myofascial release. J Bodyw Mov Ther. 1997;1(4):231-238.
- Barral JP. Visceral Manipulation. Eastland Press: Seattle; 2005.
- Bäumler H, Schürer B, Distler J. Rouleau formation of erythrocytes is influenced by thrombocytes. Folia Haematol Int Mag Klin Morphol Blutforsch. 1987;114(4):478-9.
- Bautista-Pérez R, Segura-Cobos D, Vázquez-Cruz B. In vitro antibradykinin activity of Aloe barbadensis gel. J Ethnopharmacol. 2004;93(1):89-92.
- Beal MC. Viscerosomatic reflexes: a review. J Am Osteopath Assoc. 1985;85(12):786-801.
- Beck EJ, et al. Oat beta-glucan increases postprandial cholecystokinin levels, decreases insulin response and extends subjective satiety in overweight subjects. Molecular nutrition & food research. 2009;53(10):1343-51.
- Bellucci S, Kedra W, Ajzenberg N, Tobelem G, Caen J. [New treatments for Raynaud's syndrome]. Nouv Rev Fr Hematol. 1988;30(1-2):103-7.
- Beno I, Staruchová M, Volkovová K. [Ulcerative colitis: activity of antioxidant enzymes of the colonic mucosa]. Presse Med. 1997;26(31):1474-7.
- Bercik P, et al. The intestinal microbiota affect central levels of brain-derived neurotropic factor and behavior in mice, Gastroenterology. 2011;141(2):599-609.
- Bermond P. Therapy of side effects of oral contraceptive agents with vitamin B6. Acta vitaminologica et enzymologica. 1982;4(1-2):45-54.
- Bhan M, et al. Epidemiology & management of persistent diarrhoea in children of developing countries. The Indian Journal of Medical Research. 1996;104:103-14.
- Bhoola KD, Elson CJ, Dieppe PA Kinins — key mediators in inflammatory arthritis? British journal of rheumatology. 1992;31(8):509-18.

- Biesalski HK. Polyphenols and inflammation: basic interactions, Current opinion in clinical nutrition and metabolic care. 2007;10(6):724-8.
- Blomerth PR. Functional nocturnal enuresis. J Manipulative Physiol Ther. 1994;17(5):335–338.
- Boesler D, et al. Efficacy of high-velocity low-amplitude manipulative technique in subjects with low-back pain during menstrual cramping. J Am Osteopath Assoc. 1993;93(2):203-8, 213-4.
- Borrmann WR. Comprehensive Answers to Nutrition. New Horizons Publishing: Chicago; 1979.
- Borst SE, Lee JH, Conover CF. Inhibition of 5alpha-reductase blocks prostate effects of testosterone without blocking anabolic effects, American journal of physiology, endocrinology and metabolism. 2005;288(1):E222-7.
- Bouskela E, Donyo KA. Effects of oral administration of purified micronized flavonoid fraction on increased microvascular permeability induced by various agents and on ischemia/reperfusion in the hamster cheek pouch, Angiology. 1997;48(5):391-9.
- Bredenoord AJ. Mechanisms of reflux perception in gastroesophageal reflux disease: a review. Am J Gastroenterol. 2012;107(1):8-15.
- Barnes BO, Galton L. Hypothyroidism: the unsuspected illness. Thomas Y. Crowell Company: New York; 1976.
- Brandt U, Trumpower B. The protonmotive Q cycle in mitochondria and bacteria. Critical Review of Biochemistry and Molecular Biology. 1994;29(3):165-197.
- Browning JE. Pelvic Pain and Organic Dysfunction: A new solution to chronic pelvic pain and the disturbances of bladder, bowel, gynecologic and sexual function that accompany it. Outskirts Press, Inc.: Denver, CO; 2009.
- Browning JE. Chiropractic distractive decompression in the treatment of pelvic pain and organic dysfunction in patients with evidence of lower sacral nerve root compression. J Manipulative Physiol Ther. 1988;11(5):426-32.
- Brüggmann D, et al. Intra-abdominal adhesions: definition, origin, significance in surgical practice, and treatment options. Deutsches Ärzteblatt international. 2010;107(44):769-75.
- Bucher J, et al. A systems biology approach to dynamic modeling and inter-subject variability of statin pharmacokinetics in human hepatocytes. BMC systems biology, 2011;5:66.
- Budgell BS Reflex effects of subluxation: the autonomic nervous system. J Manipul Physiol Ther. 2000;23(2):104-106.
- Burns L. Pathogenesis of Visceral Disease Following Vertebral Lesions. Journal Printing Co.: Kirksville, MO; 1948.
- Burstein E. The frequency of common AK findings. Collected Papers International College of Applied Kinesiology: Shawnee Mission, KS; 1990:13-15.
- Cai WC, et al. [Determination of oxidation-reduction level in patients with rheumatoid arthritis], Di 1 jun yi da xue xue bao. (Chinese) 2005;25(6):749-50.
- Calesnick B, Dinan AM. Prostaglandins and NSAIDS in primary dysmenorrhea. American Family Physician. 1987;35:223-5.
- Calhoon J. Applied Kinesiology Management Of Menstrual Headaches; A Case History. Collected Papers International College of Applied Kinesiology: Shawnee Mission, KS. 2005:3-4.
- Camilleri M Probiotics and irritable bowel syndrome: rationale, putative mechanisms, and evidence of clinical efficacy, Journal of clinical gastroenterology. 2006;40(3):264-9.
- Cárdenas-León M, et al, [Glycation and protein crosslinking in the diabetes and ageing pathogenesis], Revista de investigación clínica; organo del Hospital de Enfermedades de la Nutrición, 2009;61(6):505-20.
- Carroccio A, et al, Fecal assays detect hypersensitivity to cow's milk protein and gluten in adults with irritable bowel syndrome, Clinical gastroenterology and hepatology: the official clinical practice journal of the American Gastroenterological Association. 2011;9(11):965-971.e3.
- Carpenter S.A., et al 1977. Evaluation of Muscle-Organ Association, Part I and II. Clin Chiro. II(6):22-33 and III(1):42-60.
- Caso M. 2004. Evaluation of Chapman's neurolymphatic reflexes via applied kinesiology: a case report of low back pain and congenital intestinal abnormality. J Manipulative Physiol Ther. 27(1):66.
- Caso M. Case History: Chapman's Neurolymphatic Reflexes, Congenital Intestinal Abnormality, And Bowel Evacuation Time. Collected Papers International College of Applied Kinesiology:Shawnee Mission, KS; 2002:7-9.
- Centers for Disease Control and Prevention, 2012. U.S. Obesity Trends. http://www.cdc.gov/obesity/data/trends.html.
- Cetinkaya A, et al. Effects of L-carnitine on oxidant/antioxidant status in acetic acid-induced colitis. Digestive diseases and sciences. 2006;51(3):488-94.
- Chaitow L, Jones RL. Chronic pelvic pain and dysfunction: practical physical medicine. Elsevier: Edinburgh; 2012.
- Chaitow L, Trenev N. Probiotics. Harper Collins: New York; 1989.
- Chaitow L, 1988, Soft~Tissue Manipulation, Healing Arts Press:142-147.
- Chan KL, et al. The association of the amoebic colitis and chronic ulcerative colitis, Singapore medical journal. 1995;36(3);303-5.
- Chegini N, et al. Identification of epidermal growth factor, transforming growth factor-alpha, and epidermal growth factor receptor in surgically induced pelvic adhesions in the rat and intraperitoneal adhesions in the human. Am J Obstet Gynecol. 1994;171(2):321-7.
- Cheraskin E, Ringsdorf WM, Jr. Predictive Medicine. Keats Publishing, Inc.: New Canaan, CT; 1977.
- Cheraskin E, et al. The "ideal" daily niacin intake. International journal for vitamin and nutrition research. 1976;46(1):58-60.
- Cheraskin E, Ringsdorf WM, Jr., Clark JW. Diet and disease. Keats Publishing, Inc.: New Canaan, CT; 1968.
- Chester AC. Chronic sinusitis. American family physician. 1996;53(3):877-87.
- Cheuk J, et al. Inhibition of contractions of the isolated human myometrium by potassium channel openers. American Journal of Obstetrics and Gynecology. 1993;168(3 Pt 1):953-60.
- Cicco G, Pirrelli A. Red blood cell (RBC) deformability, RBC aggregability and tissue oxygenation in hypertension, Clinical hemorheology and microcirculation. 1999;21(3-4);169-77.

- Cicha I, et al. Changes of RBC aggregation in oxygenation-deoxygenation: pH dependency and cell morphology. American journal of physiology. Heart and circulatory physiology. 2003;284(6):H2335-42.
- Cicha I, Suzuki Y, Tateishi N, Maeda N. Enhancement of red blood cell aggregation by plasma triglycerides. Clin Hemorheol Microcirc. 2001;24(4):247-55.
- Clarke G, et al. Marked elevations in pro-inflammatory polyunsaturated fatty acid metabolites in females with irritable bowel syndrome. Journal of lipid research. 2010;51(5):1186-92.
- Cohen S., Wittenberg J. Effect of avitaminosis A and hypervitaminosis A on urinary bladder carcinogenicity of N-(4-(5-nitro-2-furyl)-2-thiazolyl)formamide. Cancer Research. 1976;36(7 PT 1):2334-9.
- Coll Capdevila C. Dysfunctional uterine bleeding and dysmenorrhea. The European journal of contraception & reproductive health care: the official journal of the European Society of Contraception. 1997;2(4):229-37.
- Comeaux Z. Dynamic fascial release and the role of mechanical/vibrational assist devices in manual therapies. J Bodyw Mov Ther. 2011:15(1):35-41.
- Conable K, et al. Comparison of applied kinesiology neuromuscular screening and laboratory indicators of adverse reactions to foods. Collected Papers International College of Applied Kinesiology: Shawnee Mission, KS; 2007:35-44.
- Correa JA, et al. Virgin olive oil polyphenol hydroxytyrosol acetate inhibits in vitro platelet aggregation in human whole blood: comparison with hydroxytyrosol and acetylsalicylic acid. The British journal of nutrition. 2009;101(8):1157-64.
- Crepaldi G, et al, [Blood hyperviscosity syndromes]. La Ricerca in clinica e in laboratorio. 1983;13 Suppl 3:89-104.
- Christian P, et al. Maternal vitamin A and β-carotene supplementation and risk of bacterial vaginosis: a randomized controlled trial in rural Bangladesh. The American journal of clinical nutrition. 2011;94(6):1643-9.
- Clayton PT. B6-responsive disorders: a model of vitamin dependency. Journal of inherited metabolic disease. 2006;29(2-3):317-26.
- Crook WG. The yeast connection. Vintage Books: New York; 1986:133.
- Crosignani PG, Vegetti W. A practical guide to the diagnosis and management of amenorrhoea. Drugs. 1996;52(5):671-81.
- Crowell MD, et al. Functional bowel disorders in women with dysmenorrhea, The American journal of gastroenterology. 1994;89(11):1973-7.
- Cuthbert SC, Rosner AL. Conservative chiropractic management of urinary incontinence using applied kinesiology: a retrospective case-series report. J Chiropr Med. 2012;11(1):49-57
- Cuthbert SC, Rosner AL. Conservative management of post-surgical urinary incontinence in an adolescent using applied kinesiology: a case report. Altern Med Rev. 2011 Jun;16(2):164-71.
- Cuthbert S. Restless Leg Syndrome: A Case Series Report. Collected Papers International College of Applied Kinesiology: Shawnee Mission, KS;2006:45-54.
- Cuthbert S. The piriformis muscle and the genito-urinary system: the anatomy of the muscle-organ-gland correlation. Collected Papers International College of Applied Kinesiology: Shawnee Mission, KS; 2004:125-140.
- Cuthbert SC. Applied Kinesiology and the Myofascia. Int J AK and Kinesio Med. 2002;14.
- Daley CA, et al. A review of fatty acid profiles and antioxidant content in grass-fed and grain-fed beef, Nutrition journal. 2010;9:10.
- Dauphine D. Case report of applied kinesiology treatment and allergic cutaneous vasculitis. Collected Papers International College of Applied Kinesiology: Shawnee Mission, KS. 1994:217-221.
- Dawood MY Khan-Dawood FS. Clinical efficacy and differential inhibition of menstrual fluid prostaglandin F2alpha in a randomized, double-blind, crossover treatment with placebo, acetaminophen, and ibuprofen in primary dysmenorrhea. American journal of obstetrics and gynecology. 2007;196(1):35.e1-5.
- Dawood MY. Dysmenorrhea. Clinical and Obstetrical Gynecology 1990;33:168-178.
- Dawood MY. Nonsteroidal anti-inflammatorydrugs and changing attitudes toward dysmenorrhea. American Journal of Medicine 1988;84:23-9.
- Deal SC. New Life Through Nutrition. New Life Publishing Co.: Tucson, AZ; 1974.
- de Almeida BS, Sabatino JH, Giraldo PC. Effects of high-velocity, low-amplitude spinal manipulation on strength and the basal tonus of female pelvic floor muscles. J Manipulative Physiol Ther. 2010;33(2):109-116.
- DeFilippis, AP, et al. Blood is thicker than water: the management of hyperviscosity in adults with cyanotic heart disease. Cardiology in review. 2007;15(1):31-4.
- DeJarnette MB. Technic and practice of bloodless surgery. Major Bertrand DeJarnette: Nebraska City, NB; 1939.
- Deichmann R, Lavie C, Andrews S. Coenzyme Q10 and statin-induced mitochondrial dysfunction. The Ochsner journal. 2010;10(1):16-21.
- Delia A, et al. [Effectiveness of oral administration of Lactobacillus paracasei subsp. paracasei F19 in association with vaginal suppositories of Lactobacillus acidofilus in the treatment of vaginosis and in the prevention of recurrent vaginitis]. Minerva ginecologica. 2006;58(3):227-31.
- DeLancey JO. Anatomy and physiology of urinary continence. Clin Obstet Gynecol. 1990;33(2):298-307.
- Dennerstein G. Pathogenesis and treatment of genital candidiasis. Aust Fam Physician. 1998;27(5):363-9.
- De Lorenzo A, et al. Resting metabolic rate incremented by pulsating electrostatic field (PESF) therapy. Diabetes, nutrition & metabolism. 2004;17(5):309-12.
- De Souza MC, et al. A synergistic effect of a daily supplement for 1 month of 200 mg magnesium plus 50 mg vitamin B6 for the relief of anxiety-related premenstrual symptoms: a randomized, double-blind, crossover study, Journal of women's health & gender-based medicine. 2000;9(2):131-9.
- Dey M, et al. Dietary phenethylisothiocyanate attenuates bowel inflammation in mice. BMC chemical biology. 2010;10:4.
- Dicpinigaitis PV. Angiotensin-converting enzyme inhibitor-induced cough: ACCP evidence-based clinical practice guidelines. Chest. 2006;129(1 Suppl):169S-173S.
- Dinleyici EC, Eren M, Ozen M, Yargic ZA, Vandenplas Y. Effectiveness and safety of Saccharomyces boulardii for acute infectious diarrhea. Expert Opin Biol Ther. 2012;12(4):395-410.

- Di Pierro F, et al. Clinical effectiveness of a highly standardized and bioavailable mixture of flavonoids and triterpenes in the management of acute hemorrhoidal crisis. Acta Bio-Medica : Atenei Parmensis. 2011;82(1):35-40.
- DiSilvestro RA, Marten J, Skehan M. Effects of copper supplementation on ceruloplasmin and copper-zinc superoxide dismutase in free-living rheumatoid arthritis patients. Journal of the American College of Nutrition. 1992;11(2):177-80.
- diZerega GS. Biochemical events in peritoneal tissue repair. Eur J Surg Suppl. 1997;577:10-6.
- Doll H, Brown S, Thurston A, Vessey M. Pyridoxine (vitamin B6) and the premenstrual syndrome: a randomized crossover trial. The Journal of the Royal College of General Practitioners. 1989;39(326):364-8.
- Dobrik I. Disorders of the iliopsoas muscle and its role in gynecological diseases. J Manual Medicine. 1989;4:130-133.
- Douglas S. Premenstrual syndrome. Evidence-based treatment in family practice, Canadian family physician Médecin de famille canadien, 2002;48:1789-97.
- Duffy C. Applied kinesiology management of urinary incontinence in a pediatric patient: a case history. Collected Papers International College of Applied Kinesiology: Shawnee Mission, KS. 2005:21-22.
- Duffy C. Applied kinesiology management of nocturnal enuresis: a case study. Collected Papers International College of Applied Kinesiology: Shawnee Mission, KS. 2004:27-29.
- Duffy D. Chiropractic cost effectiveness in the treatment of bladder symptoms. Collected Papers International College of Applied Kinesiology: Shawnee Mission, KS. 1994:41-44.
- Duffy D. Chiropractic cost effectiveness in the treatment of infertility: a case study. Collected Papers International College of Applied Kinesiology: Shawnee Mission, KS. 1994:45-48.
- Duffy D. Chiropractic cost effectiveness in carpal tunnel syndrome. Collected Papers International College of Applied Kinesiology: Shawnee Mission, KS.1994: 24-33.
- Duffy D. Crohn's disease and ulcerative colitis respond to chiropractic care. Collected Papers International College of Applied Kinesiology: Shawnee Mission, KS. 1992:21-21-26.
- Duffy D. Clinical tips on structure and nutrition for the alternative physician. Balancing Body Chemistry. 2008:197.
- Duyff RF, Van den Bosch J, Laman DM, van Loon BJ, Linssen WH. Neuromuscular findings in thyroid dysfunction: a prospective clinical and electrodiagnostic study. J Neurol Neurosurg Psychiatry. 2000;68(6):750-5.
- Ebnezar J, et al. Effect of integrated yoga therapy on pain, morning stiffness and anxiety in osteoarthritis of the knee joint: A randomized control study. International journal of yoga. 2012;5(1):28-36.
- Eidenier H. More Than Just a Bunch of Numbers - Making Sense of Blood Chemistry Results. Balancing Body Chemistry With Nutrition Seminars: Union Lake, MI; 2007.
- Elkins TE, et al. A histologic evaluation of peritoneal injury and repair: implications for adhesion formation. Obstetrics and gynecology. 1987;70(2):225-8.
- Ellinger S, et al. [Tomatoes and lycopene in prevention and therapy — is there an evidence for prostate diseases?], Aktuelle Urologie. 2009;40(1):37-43.
- Enck P, Vodusek DB. Electromyography of pelvic floor muscles. J Electromyogr Kinesiol. 2006;16(6):568-77.
- Eng ET, et al. Suppression of estrogen biosynthesis by procyanidin dimers in red wine and grape seeds. Cancer research. 2003;63(23):8516-22.
- Erdem M, et al. [Joint immobilization increases reactive oxygen species: an experimental study]. Acta orthopaedica et traumatologica turcica. 2009;43(5):436-43.
- Estok PJ. Rudy EB, Just JA. Body-fat measurements and athletic menstrual irregularity, Health care for women international. 1991;12(2):237-48.
- Evans GW, et al. J. Inorganic Biochem. 1993;49:177.
- Fahim MS, et al. Zinc arginine, a 5 alpha-reductase inhibitor, reduces rat ventral prostate weight and DNA without affecting testicular function. Andrologia. 1993;25(6):369-75.
- Fall M, et al. EAU guidelines on chronic pelvic pain. Eur Urol. 2010;57(1):35-48.
- Fantl JA, Newman DK, Colling J et al. Managing acute and chronic urinary incontinence, clinical practice guideline, no. 2. US Department of Health and Human Services. Rockville MD; 1996.
- Faust T, et al. Effects of fish oil on gastric mucosal 6-keto-PGF1 alpha synthesis and ethanol-induced injury. The American journal of physiology. 1989;257(1 Pt 1):G9-13.
- Feldman D, Skowronski RJ, Peehl DM. Vitamin D and prostate cancer. Advances in experimental medicine and biology. 1995;375:53-63.
- Force M. Choosing Health: Dr. Force's Functional Selfcare Workbook. Health Knowledge; 2003.
- Force M, Sparks WS, Ronzio RA. Inhibition of enteric parasites by emulsified oil of oregano in vivo. Phytotherapy research: PTR. 2000;14(3):213-4.
- Fouts DE, et al. Bacterial translocation and changes in the intestinal microbiome in mouse models of liver disease. J Hepatol. 2012.
- Gaby AR. The natural pharmacy, 3rd Ed. Three Rivers Press: New York; 2006.
- Gale JD. The use of novel promotility and prosecretory agents for the treatment of chronic idiopathic constipation and irritable bowel syndrome with constipation. Advances in therapy. 2009;26(5):519-30.
- Garnick MB. The great prostate cancer debate. Sci Am. 2012;306(2):38-43.
- Gemignani F. Can restless legs syndrome be generated by interacting central and peripheral abnormal inputs? Sleep Med. 2010;11(6):503-4.
- Genazzani AD, et al. Hypothalamic amenorrhea: from diagnosis to therapeutical approach. Annales d'endocrinologie. 2010;71(3):163-9.
- Gentile MG, et al. Resumption of menses after 32 years in anorexia nervosa, Eating and weight disorders : EWD. 2011;16(3):e223-5.
- Giamberardino MA, et al. Viscero-visceral hyperalgesia: characterization in different clinical models. Pain. 2010;151(2):307-22.
- Giamberardino MA, et al. Relationship between pain symptoms and referred sensory and trophic changes in patients with gallbladder pathology. Pain. 2005;114(1-2):239-49.

- Giamberardino MA, Berkley KJ, Iezzi S, de Bigontina P, Vecchiet L. Pain threshold variations in somatic wall tissues as a function of menstrual cycle, segmental site and tissue depth in non-dysmenorrheic women, dysmenorrheic women and men. Pain. 1997;71(2):187-97.
- Glickman LT, et al. Evaluation of the risk of endocarditis and other cardiovascular events on the basis of the severity of periodontal disease in dogs. Journal of the American Veterinary Medical Association. 2009;234(4):486-94.
- Goebel A, et al. Altered intestinal permeability in patients with primary fibromyalgia and in patients with complex regional pain syndrome. Rheumatology (Oxford, England). 2008;47(8);1223-7.
- Goldin BR, Gorbach SL. Clinical indications for probiotics: an overview. Clinical infectious diseases: an official publication of the Infectious Diseases Society of America. 2008;46 Suppl 2: S96-100; discussion S144-51.
- Goodheart GJ, Jr. AK classic case management: enuresis. Int J AK and Kinesio Med. 2003;16: 22-23.
- Goodheart GJ, Jr. Being a family doctor. ICAKUSA: Shawnee Mission, KS; 1992.
- Goodheart GJ, Jr. Applied Kinesiology 1989 Workshop Procedural Manual. Self-published: Detroit; 1989: 70-86.
- Goodheart GJ, Jr. Applied Kinesiology 1989 Workshop Procedural Manual. Self-published: Detroit. 1989:59.
- Goodheart GJ, Jr. Applied Kinesiology 1985 Workshop Procedural Manual. Self-published: Detroit; 1985:16-51.
- Goodheart GJ, Jr. Applied Kinesiology 1978 Workshop Procedure Manual, 14th Ed. Self-published: Detroit; 1978.
- Goodheart GJ, Jr. Applied Kinesiology 1979 Workshop Procedure Manual, 15th ed. Self-published: Detroit; 1979.
- Goodheart GJ, Jr. Applied Kinesiology Workshop Procedure Manual, 11th Ed. Self-published: Detroit, MI; 1975.
- Goymerac B, Woollard G. Focal infection: a new perspective on an old theory. General dentistry. 2004;52(4):357-61; quiz 362, 365-6.
- Graven-Nielsen T, Mense S. The peripheral apparatus of muscle pain: evidence from animal and human studies. The Clinical journal of pain. 2001;17(1):2-10.
- Gray's Anatomy: The Anatomical Basis of Clinical Practice. Churchill Livingstone: Edinburgh; 2004.
- Gregory WM, Mills SP, Hamed HH, Fentiman IS. Applied kinesiology for treatment of women with mastalgia. Breast. 2001;10(1):15-9.
- Grigoleit HG, Grigoleit P. Pharmacology and preclinical pharmacokinetics of peppermint oil. Phytomedicine: international journal of phytotherapy and phytopharmacology. 2005;12(8):612-6.
- Gul A, Rahman MA, Hasnain SN. Influence of fructose concentration on myocardial infarction in senile diabetic and non-diabetic patients. Experimental and clinical endocrinology & diabetes : official journal, German Society of Endocrinology [and] German Diabetes Association. 2009;117(10):605-9.
- Gul A, Rahman MA, Hasnain SN. Role of fructose concentration on cataractogenesis in senile diabetic and non-diabetic patients. Albrecht von Graefes Archiv für klinische und experimentelle Ophthalmologie. 2009;247(6):809-14.
- Guyton AC, Hall JE. Textbook of Medical Physiology, 11th ed. W.B. Saunders Co: Philadelphia; 2005.
- Habib FK, et al. Serenoa repens (Permixon) inhibits the 5alpha-reductase activity of human prostate cancer cell lines without interfering with PSA expression, International journal of cancer. Journal international du cancer. 2005;114(2):190-4.
- Haeggström JZ, Wetterholm A. Enzymes and receptors in the leukotriene cascade. Cellular and molecular life sciences: CMLS. 2002;59(5):742-53.
- Hadjigogos K. The role of free radicals in the pathogenesis of rheumatoid arthritis. Panminerva medica. 2003;45(1):7-13.
- Halbreich U, Tworek H. Altered serotonergic activity in women with dysphoric premenstrual syndromes. International journal of psychiatry in medicine. 1993;23(1):1-27.
- Haldeman S. Neurologic effects of the adjustment. J Manipulative Physiol Ther.2000;23(2):112-114.
- Hambidge KM, Krebs NF. Zinc deficiency: a special challenge. The Journal of nutrition. 2007;137(4):1101-5
- Hambrick T. Inter-examiner Reliability of Manual Muscle Testing for Heavy Metal Toxicity: A Blinded Evaluation of Non-Standard Methods of applied kinesiology testing compared with laboratory findings. Collected Papers International College of Applied Kinesiology: Shawnee Mission, KS. 2008:97-111.
- Hawrelak JA, Myers SP. The causes of intestinal dysbiosis: a review. Alternative medicine review. 2004;9(2):180-97.
- Heidrich J. Female Infertility: Case Histories Of Applied Kinesiology Response. Collected Papers International College of Applied Kinesiology: Shawnee Mission, KS. 1992:47-49.
- Henao-Mejia J, et al. Inflammasome-mediated dysbiosis regulates progression of NAFLD and obesity. Nature. 2012;482(7384):179-85.
- Herrmann W. The importance of hyperhomocysteinemia as a risk factor for diseases: an overview. Clinical chemistry and laboratory medicine: CCLM / FESCC. 2001;39(8):666-74.
- Herrmann W, et al. Functional vitamin B12 deficiency and determination of holotranscobalamin in populations at risk. Clinical chemistry and laboratory medicine : CCLM / FESCC. 2003;41(11):1478-88.
- Hickey B. Infertility.Collected Papers International College of Applied Kinesiology: Shawnee Mission, KS. 2008:115-116.
- Hiipakka RA, et al. Structure-activity relationships for inhibition of human 5alpha-reductases by polyphenols. Biochemical pharmacology. 2002;63(6):1165-76.
- Hill MJ, Drasar BS, Hawksworth G, Aries V, Crowther JS, Williams RE. Bacteria and aetiology of cancer of large bowel. Lancet. 1971;1(7690):95-100.
- Hitchcock ME. The manipulative approach to the management of primary dysmenorrheal. Journal of the American Chiropractic Association. 1976;75:97-100.
- Ho Y, et al. Micronized purified flavonidic fraction compared favorably with rubber band ligation and fiber alone in the management of bleeding hemorrhoids: Randomized controlled trial. Diseases of the Colon and Rectum. 2000;43(1):66-9.
- Holick MF. High prevalence of vitamin D inadequacy and implications for health, Mayo Clinic proceedings. Mayo Clinic. 2006;81(3):353-73.

- Hondras MA, Long CR, Brennan PC. Spinal manipulative therapy versus a low force mimic maneuver for women with primary dysmenorrhea: a randomized, observer-blinded, clinical trial. Pain. 1999;81(1-2):105-14.
- Honkanen V, et al. Serum zinc, copper and selenium in rheumatoid arthritis, Journal of trace elements and electrolytes in health and disease. 1991;5(4):261-3.
- Horiki N, et al. [Case report of colitis associated with Blastocystis hominis infection].The Japanese journal of gastro-enterology. 1996;93(9):655-60.
- Horrobin DF. The role of essential fatty acids and prostaglandins in the premenstrual syndrome. The Journal of reproductive medicine. 1983;28(7):465-8.
- Houston MC. What your doctor may not tell you about heart disease. Grand Central Life & Style: New York; 2012.
- Houston MC. Role of mercury toxicity in hypertension, cardiovascular disease, and stroke, Journal of clinical hypertension (Greenwich, Conn.). 2011;13(8):621-7. http://
- Hurley M. Muscle, exercise and arthritis. Ann Rheum Dis. 2002;61(8):673–675.
- Hrnciarová M, et al. [Insulin resistance and arterial hypertension], Vnitrní lékarství. 1995;41(2):111-6.
- ICAKUSA, 2012. www.icakusa.com/research/.
- Igata A. Clinical studies on rising and re-rising neurological diseases in Japan — a personal contribution. Proceedings of the Japan Academy. Series B, Physical and biological sciences. 2010;86(4):366-77.
- Iida M, Yamaguchi Y. [Remission of rheumatoid arthritis following periodontal treatment. A case report]. Nihon Shishubyo Gakkai kaishi. 1985;27(1):234-8.
- IMPAC, 2012. http://www.impacinc.net/client/index.php. IMPAC: Salem, OR.
- Isaksson H, et al. Whole grain rye breakfast - sustained satiety during three weeks of regular consumption. Physiology & behavior. 2012;105(3):877-84.
- Iyengar BKS. Light on Yoga, Revised Edition. Schocken: 1995.
- Jackson LW, Zullo MD, Goldberg JM. The association between heavy metals, endometriosis and uterine myomas among premenopausal women: National Health and Nutrition Examination Survey 1999-2002. Human reproduction (Oxford, England). 2008;23(3):679-87.
- James SJ, et al. Metabolic biomarkers of increased oxidative stress and impaired methylation capacity in children with autism. The American journal of clinical nutrition. 2004;80(6):1611-7
- Jamshed N. et al. Diagnostic approach to chronic constipation in adults. American family physician. 2011;84(3):299-306.
- Jarvis DC. Folk Medicine: A New England Almanac of Natural Health Care From A Noted Vermont Country Doctor. Fawcett; 1985.
- Jensen B, Anderson M. Empty harvest: understanding the link between our food, our immunity, and our planet. Avery Publishing Group, Inc.: Garden City Park, NY; 1990.
- Jinkins JR. The anatomical basis of vertebral pain and autonomic syndrome associated with lumbar disc extrusion. Am J Roentg. 1989;152(6):1277-1289.
- Kalliomäki MA. Food allergy and irritable bowel syndrome. Current opinion in gastroenterology. 2005;21(6):708-11.
- Kanauchi M, Tsujimoto N, Hashimoto T. Advanced glycation end products in nondiabetic patients with coronary artery disease. Diabetes care. 2001;24(9):1620-3.
- Karason AB. Drysdale IP. Somatovisceral response following osteopathic HVLAT: A pilot study on the effect of unilateral lumbosacral high-velocity low-amplitude thrust technique on the cutaneous blood flow in the lower limb. J Manipulative Physiol Ther. 2003;26(4):220-225.
- Kaufman A, et al. Reflex changes in heart rate after mechanical and thermal stimulation of the skin at various segmental levels in cats. Neuroscience. 1977;2(1):103-9.
- Kaufman S. Case History: Correction Of Inguinal Hernia By Applied Kinesiology Management. Collected Papers International College of Applied Kinesiology: Shawnee Mission, KS; 1997:57-58.
- Kaufman S. Infertility: Successful Management By Applied Kinesiology After Failure Of Medical Treatment. Collected Papers International College of Applied Kinesiology: Shawnee Mission, KS. 1997:59-60.
- Kegel AH. Progressive resistance exercise in the restoration of the perineal muscles. Am J Obstet Gynecol. 1948;56:238-248.
- Kenton K, Brubaker L. Relationship between levator ani contraction and motor unit activation in the urethral sphincter. Am J Obstet Gynecol. 2002;187(2):403-6.
- Kharrazian D. Amenorrhea Related To Autoimmune Demyelinating Disease - A Case Study. Datis. Collected Papers International College of Applied Kinesiology: Shawnee Mission, KS. 2007:125-129.
- Kim JM. [Inflammatory bowel diseases and enteric microbiota]. The Korean journal of gastroenterology . 2010;55(1):4-18.
- Kimura A, Sato A. Somatic regulation of autonomic functions in anesthetized animals – neural mechanisms of physical therapy including acupuncture. Jpn J Vet Res. 1997;45(3):137-145.
- Kjeldsen-Kragh J, et al. Changes in laboratory variables in rheumatoid arthritis patients during a trial of fasting and one-year vegetarian diet. Scandinavian journal of rheumatology. 1995;24(2):85-93.
- Kjeldsen-Kragh J, et al. Antibodies against dietary antigens in rheumatoid arthritis patients treated with fasting and a one-year vegetarian diet. Clinical and experimental rheumatology. 1995;13(2):167-72.
- Klein EA, et al. SELECT: the selenium and vitamin E cancer prevention trial. Urologic oncology. 2003;21(1);59-65.
- Klougart N, Nilsson N, Jacobsen J. Infantile colic treated by chiropractors: a prospective study of 316 cases. J Manipulative Physiol Ther. 1989;12(4):281-8.
- Koike H, et al. Correlation between dysmenorrheic severity and prostaglandin production in women with endometriosis. Prostaglandins, leukotrienes, and essential fatty acids. 1992;46(2):133-7.
- Kokjohn K, Schmid DM, Triano JJ, Brennan PC. The effect of spinal manipulation on pain and prostaglandin levels in women with primary dysmenorrhea. J Manipulative Physiol Ther. 1992;15(5):279-85.
- Kolasinski SL, et al. Iyengar yoga for treating symptoms of osteoarthritis of the knees: a pilot study. Journal of alternative and complementary medicine . 2005;11(4):689-93.

- Korr IM. Spinal cord as organizer of disease process. Academy of Applied Osteopathy Yearbook: Newark, OH; 1976.
- Kreitz BG, Aker PD. Nocturnal enuresis: treatment implications for the chiropractor. J Manipulative Physiol Ther. 1994;17(7):465-473.
- Kremer JM, Bigaouette J. Nutrient intake of patients with rheumatoid arthritis is deficient in pyridoxine, zinc, copper, and magnesium. The Journal of rheumatology. 1996;23(6):990-4.
- Kruidenier L, et al. Differential mucosal expression of three superoxide dismutase isoforms in inflammatory bowel disease. The Journal of pathology. 2003;201(1):7-16.
- Kukkonen K, et al. Long-term safety and impact on infection rates of postnatal probiotic and prebiotic (synbiotic) treatment: randomized, double-blind, placebo-controlled trial. Pediatrics. 2008;122(1):8-12.
- Kuwabara A, et al. High prevalence of vitamin K and D deficiency and decreased BMD in inflammatory bowel disease. Osteoporosis international : a journal established as result of cooperation between the European Foundation for Osteoporosis and the National Osteoporosis Foundation of the USA. 2009;20(6):935-42.
- Lacour B, et al. [In vitro supplementation of pyridoxal phosphate for the optimisation of the determination of the catalytic activity of alanine aminotransferase and aspartate aminotransferase in kidney transplant patients (author's transl)]. Clinica chimica acta: international journal of clinical chemistry. 1982;120(1):1-12.
- Lakhanpal S, et al. Reactive arthritis from Blastocystis hominis. Arthritis Rheum. 1991;34(2):251-3.
- Leboeuf C, Brown P, Herman A, Leembruggen K, Walton D, Crisp TC. Chiropractic care of children with nocturnal enuresis: a prospective outcome study. J Manipulative Physiol Ther 1991;14(2):110–115.
- Lebowitz M. Body mechanics. MMI Press: Harrisville, NH; 1984.
- Lee CT, et al. Abnormal vagal cholinergic function and psychological behaviors in irritable bowel syndrome patients: a hospital-based Oriental study. Digestive diseases and sciences. 1998;43(8):1794-9.
- Lee HJ, et al. Tai Chi Qigong for the quality of life of patients with knee osteoarthritis: a pilot, randomized, waiting list controlled trial. Clinical rehabilitation. 2009;23(6):504-11.
- Lee MS, et al. Qigong for type 2 diabetes care: a systematic review. Complementary therapies in medicine. 2009;17(4):236-42.
- Lee R. Dental Infections and related Degenerative Diseases. J Am Med Assoc. 1925;84(4):254-261.
- Lee R. Dental Infections, Oral and Systemic. Penton publishing company: Cleveland, OH; 1923.
- Levin P. Perfect Bones: A Six-Point Plan for Healthy Bones. Ten Speed Press; 2004.
- Lewit K, Olsanska S. Clinical importance of active scars: abnormal scars as a cause of myomyofascial pain. J manipul physiol therap. 2004;27(6):399-402.
- Li Y, et al. [Effect of soy isoflavones on peri-menopausal symptom and estrogen], Wei sheng yan jiu = Journal of hygiene research,. 2010;39(1):56-9
- Liao CH, et al. Significant association between serum dihydrotestosterone level and prostate volume among Taiwanese men aged 40-79 years. The aging male : the official journal of the International Society for the Study of the Aging Male. 2012;15(1):28-33.
- Liebl NA, Butler LM. A chiropractic approach in the treatment of dysmenorrhea. J Manipulative Physiol Therap. 1990; 13(3): 101-106.
- Lines D, et al. Effects of Soft Tissue Technique and Chapman's Neurolymphatic Reflex Stimulation on Respiratory Function. J Aust Chiro Assoc. 1990;20(1):17-22.
- López Fontana CM, et al. [Relation between selenium plasma levels and different prostatic pathologies]. Actas urologicas españolas. 2010;34(7):625-9.
- Lóránd T, Vigh E. & Garai J. Hormonal action of plant derived and anthropogenic non-steroidal estrogenic compounds: phytoestrogens and xenoestrogens. Current medicinal chemistry. 2010;17(30):3542-74.
- Lotz-Winter H. On the pharmacology of bromelain: an update with special regard to animal studies on dose-dependent effects. Planta medica. 1990;56(3):249-53.
- Lunec J, et al. Free-radical oxidation (peroxidation) products in serum and synovial fluid in rheumatoid arthritis. The Journal of rheumatology. 1981;8(2):233-45.
- Maeda N, et al. Triglyceride in plasma: prospective effects on microcirculatory functions. Clinical hemorheology and microcirculation. 2006;34(1-2):341-6.
- Maffetone P. Complementary Sports Medicine: Balancing traditional and nontraditional treatments. Human Kinetics: Champaign, IL; 1999.
- Martensson J. Effect of guadidine and synthalin on the citric acid metabolism. Acta Medica Scandinavica. 1946;55(Fasc. I).
- Martin W. Treatment of capillary fragility with soluble citrus bioflavonoid complex. Int Rec Med Gen Pract Clin. 1955;168(2):66-9.
- Martin CL, et al. Serum micronutrient concentrations and risk of uterine fibroids. Journal of women's health (2002). 2011;20(6):915-22
- Masarsky CS, Masarsky MT. Somatovisceral Aspects of Chiropractic: An Evidence-Based Approach. Churchill Livingstone: Philadelphia; 2001.
- Maykel W. Case Study: Cryptorchidism Correction With Conservative Chiropractic Applied Kinesiology. Collected Papers International College of Applied Kinesiology: Shawnee Mission, KS. 2004:75-76.
- Maykel W. Case Study: Chronic Severe Constipation Caused By Asymptomatic L3-4 Intervertebral Disc Syndrome And Closed Ileocecal Valve. Collected Papers International College of Applied Kinesiology: Shawnee Mission, KS. 2004:69-70.
- Maykel W. Case Cohort Study: Two Cases of Crippling Arthritis Cured With The Removal of An Electric Blanket. Collected Papers International College of Applied Kinesiology: Shawnee Mission, KS. 2008:143-144.
- McCarty MF. Toward a "bio-energy supplement" — a prototype for functional orthomolecular supplementation. Medical hypotheses. 1981;7(4):515-38.
- McSweeney TP, Thomson OP, Johnston R. The immediate effects of sigmoid colon manipulation on pressure pain thresholds in the lumbar spine. J Bodyw Mov Ther. 2012. (In Press)
- Medical Journal of Applied Kinesiology, 2012. http://www.daegak.de/literatur-materialien.html.
- Mehlisch DR. Ketoprofen, ibuprofen, and placebo in the treatment of primary dysmenorrheal: A double-blind crossover comparison. Journal of Clinical Pharmacology. 1988;28:S29-S33.

- Melzack R. Pain and the neuromatrix in the brain. J Dent Educ. 2001;65(12):1378.
- Metcalfe RS, et al. Towards the minimal amount of exercise for improving metabolic health: beneficial effects of reduced-exertion high-intensity interval training. European journal of applied physiology; 2011.
- Mielants H, et al. Late onset pauciarticular juvenile chronic arthritis: relation to gut inflammation. The Journal of rheumatology. 1987;14(3):459-65.
- Misra MC, Parshad R. Randomized clinical trial of micronized flavonoids in the early control of bleeding from acute internal haemorrhoids. The British Journal of Surgery. 2000;87(7), 868-72.
- Moncayo R, Moncayo H. A musculoskeletal model of low grade connective tissue inflammation in patients with thyroid associated ophthalmopathy (TAO): the WOMED concept of lateral tension and its general implications in disease. BMC Musculoskelet Disord. 2007;8:17.
- Moncayo R, Moncayo H, Ulmer H, Kainz H. New diagnostic and therapeutic approach to thyroid-associated orbitopathy based on applied kinesiology and homeopathic therapy. J Altern Complement Med. 2004;10(4):643-50.
- Moon YJ, Wang X, Morris ME. Dietary flavonoids: effects on xenobiotic and carcinogen metabolism. Toxicology in vitro : an international journal published in association with BIBRA. 2006;20(2):187-210.
- Morabia A. Until the lab takes it away from epidemiology. Preventive medicine. 2011;3(4-5):217-20.
- Mössner J, Keim V. Pancreatic enzyme therapy. Deutsches Ärzteblatt international. 2010;108(34-35):578-82.
- Muecke R, et al. Whole blood selenium levels (WBSL) in patients with prostate cancer (PC), benign prostatic hyperplasia (BPH) and healthy male inhabitants (HMI) and prostatic tissue selenium levels (PTSL) in patients with PC and BPH. Acta oncologica (Stockholm, Sweden). 2009;48(3):452-6.
- Mikulíková K, Eckhardt A, Kunes J, Zicha J, Miksík I. Advanced glycation end-product pentosidine accumulates in various tissues of rats with high fructose intake. Physiological research / Academia Scientiarum Bohemoslovaca. 2008;57(1):89-94.
- Müller H, de Toledo FW, Resch KL. Fasting followed by vegetarian diet in patients with rheumatoid arthritis: a systematic review. Scandinavian journal of rheumatology. 2001;30(1):1-10.
- Munday J, McKinnon H. Cystitis, pyelonephritis, and urolithiasis in rats accidentally fed a diet deficient in vitamin A. Journal of the American Association for Laboratory Animal Science: JAALAS. 2009;48(6), 790-4.
- Mungan NA, et al. Nocturnal enuresis and allergy. Scandinavian journal of urology and nephrology. 2005 ;39(3):237-41.
- Murray M, Pizzorno J. Encyclopedia of Natural Medicine, 2nd Ed. Three Rivers Press;1997.
- Murray C. Myofascial Adhesions In Pain Low In The Back And Arthritis. JAMA. 1938;111(20):1813-1818.
- Murray CA, Saunders,WP. Root canal treatment and general health: a review of the literature. International endodontic journal. 2000;33(1):1-18.
- Muthenna P, et al. Inhibition of advanced glycation end-product formation on eye lens protein by rutin. The British journal of nutrition. 2011:1-9.
- Nachemson A. Back pain: delimiting the problem in the next millennium. Int J Law Psychiatry. 1999;22(5-6):473-90.
- Naderi N, et al. Association of vitamin D receptor gene polymorphisms in Iranian patients with inflammatory bowel disease. Journal of gastroenterology and hepatology. 2008;23(12):1816-22.
- Nagai R, et al. Citric acid inhibits development of cataracts, proteinuria and ketosis in streptozotocin (type 1) diabetic rats. Biochemical and biophysical research communications. 2010;393(1):118-22.
- Nakajima S, et al. Association of vitamin K deficiency with bone metabolism and clinical disease activity in inflammatory bowel disease. Nutrition. 2011;27(10):1023-8.
- Nandhini AT, Thirunavukkarasu V, Anuradha CV. Taurine prevents collagen abnormalities in high fructose-fed rats. The Indian journal of medical research. 2005;122(2):171-7.
- Natelson S. Clinical Chemistry. 1979;25(7):1343-1344.
- Necefli A, et al. The effect of melatonin on TNBS-induced colitis. Digestive diseases and sciences. 2006;51(9):1538-45.
- Neville CE, et al. A preliminary report of musculoskeletal dysfunction in female chronic pelvic pain: a blinded study of examination findings. J Bodyw Mov Ther. 2012;16:50-56.
- New York Times, 2006. http://www.nytimes.com/2006/12/06/nyregion/06fat.html.
- Nicholas AS, et al. A somatic component to myocardial infarction. J Am Osteopath Assoc. 1987;87(2):123-9.
- Nishimura M, et al. Relationships between pain-related mediators and both synovitis and joint pain in patients with internal derangements and osteoarthritis of the temporomandibular joint. Oral surgery, oral medicine, oral pathology, oral radiology, and endodontics. 2002;94(3):328-32
- Nosáľová V, et al. Effect of N-acetylcysteine on colitis induced by acetic acid in rats. General pharmacology. 2000;35(2):77-81.
- Novakovic R, et al. The effect of potassium channel opener pinacidil on the non-pregnant rat uterus. Basic & Clinical Pharmacology & Toxicology. 2007;101(3):181-6.
- Ozello R. Aerobic Muscle Weakness - A Case History. Collected Papers International College of Applied Kinesiology: Shawnee Mission, KS. 2007:29-30.
- Padilha EM, et al. Epidemiological profile of reported beriberi cases in Maranhão State, Brazil, 2006-2008, Cadernos de saúde pública / Ministério da Saúde, Fundação Oswaldo Cruz, Escola Nacional de Saúde Pública. 2011 ;27(3):449-59.
- Palmblad J, Hafström I, Ringertz B. Antirheumatic effects of fasting. Rheumatic diseases clinics of North America. 1991 ;17(2):351-62.
- Palomar J. Visceral Parietal Pain (VPP). Collected Papers International College of Applied Kinesiology: Shawnee Mission, KS. 2007:73-77.
- Panush RS. Does food cause or cure arthritis? Rheumatic diseases clinics of North America. 1991;17(2):259-72.
- Paoletti S. The Fasciae: Dysfunction and Treatment. Eastland Press: Seattle; 2006.

- Patel AS, DeRidder PH. Amebic colitis masquerading as acute inflammatory bowel disease: the role of serology in its diagnosis. Journal of clinical gastroenterology. 1989;11(4):407-10.

- Pauling L. The commond cold and flu. WH Freeman: New York; 1976.

- Pavlenko VV, Iagoda AV. [Synthesis of eicosanoids in the colonic mucosa in patients with ulcerative colitis]. Klinicheskaia meditsina. 2003;81(10):39-43.

- Peña AS, Crusius JB. Food allergy, coeliac disease and chronic inflammatory bowel disease in man, Vet Q, 20 Suppl. 1998;3:S49-52.

- Penna G, et al The vitamin D receptor agonist elocalcitol inhibits IL-8-dependent benign prostatic hyperplasia stromal cell proliferation and inflammatory response by targeting the RhoA/Rho kinase and NF-kappaB pathways. The Prostate. 2009;69(5):480-93.

- Peters KM, Carrico DJ. Frequency, urgency, and pelvic pain: treating the pelvic floor versus the epithelium. Current urology reports. 2006;7(6):450-5.

- Peters KM, et al Prevalence of pelvic floor dysfunction in patients with interstitial cystitis. Urology. 2007;70(1):16-8.

- Petersen EE, et al. Efficacy of vitamin C vaginal tablets in the treatment of bacterial vaginosis: a randomised, double blind, placebo controlled clinical trial. Arzneimittel-Forschung. 2011;61(4):260-5.

- Pollard HP, et al. Can the Ileocecal Valve Point Predict Low Back Pain Using Manual Muscle Testing? Chiropr J Aust. 2006;36:58-62.

- Pollard H. The Somatovisceral Reflex: How Important for the "Type O" Condition? Chiropr J Aust. 2004;34: 93-102.

- Pomeranz B, et al. Cord cells responding to fine myelinated afferents from viscera, muscle and skin. J Physiol.1968;199(3):511–532.

- Ponikau JU, et al. The diagnosis and incidence of allergic fungal sinusitis, Mayo Clinic proceedings. Mayo Clinic. 1999;74(9):877-84.

- Portelli V. Viscero-somatic Reflexes: Their Clinical Manifestations Using Muscle Testing To Evaluate The Heart And Esophagus. Collected Papers International College of Applied Kinesiology: Shawnee Mission, KS. 2001;1:155-175.

- Portelli V. Ptosis of the Transverse Colon. Collected Papers International College of Applied Kinesiology, Winter: Shawnee Mission, KS. 1987:409.

- Potgieter HC, Magagane F, Bester MJ. Oestrogen and progesterone receptor status and PgR/ER ratios in normal and myomatous human myometrium, East African medical journal. 1995;72(8):510-4.

- Prakash S, et al. Gut microbiota: next frontier in understanding human health and development of biotherapeutics. Biologics: targets & therapy. 2011;5:71-86.

- Price WA. Nutrition and Physical Degeneration, 6th Edition. Keats Publishing, Inc.: New Canaan, CT; 2003.

- Prothro J. Any depression from OC-altered vitamin B6 levels? [Answer to question of Jan Marquand], Contracept Technol Update. 1981;2(9):121-3.

- Purdy WR., et al. Suboccipital dermatomyotomic stimulation and digital blood flow. Am Osteopath Assoc. 1996;96(5):285-9.

- Raftogianis R, et al. Estrogens as Endogenous Carcinogens in the Breast and Prostate, Chapter 6: Estrogen Metabolism by Conjugation. J Natl Cancer Inst Monogr. 2000;(27):113-124.

- Rainsford KD. Leukotrienes in the pathogenesis of NSAID-induced gastric and intestinal mucosal damage, Agents and actions. 1993;39 Spec No:C24-6.

- Raja SN, Meyer RA, Campbell JN. Peripheral mechanisms of somatic pain, Anesthesiology. 1988;68(4):571-90.

- Rajamanickam A, et al. Chronic diarrhea and abdominal pain: pin the pinworm. Journal of hospital medicine: an official publication of the Society of Hospital Medicine. 2009;4(2):137-9.

- Rao SV, Reddy GV, Indira K. Effect of guanidine on carbohydrate metabolism of rat. Biochemistry international.1992; 26(2):377-80.

- Reed WR, Beavers S, Reddy SK, Kern G. Chiropractic management of primary nocturnal enuresis. J Manipulative Physiol Ther. 1994;17(9):596–600.

- Rees ML. The art and practice of chiropractic, 3rd Revision. International Systemic Health Organization, Inc.: Sedan, KS. 1994.

- Regan EA, Bowler R.P, Crapo J.D. Joint fluid antioxidants are decreased in osteoarthritic joints compared to joints with macroscopically intact cartilage and subacute injury. Osteoarthritis and cartilage / OARS, Osteoarthritis Research Society. 2008;16(4):515-21.

- Reid G, et al. Oral probiotics can resolve urogenital infections. FEMS immunology and medical microbiology. 2004;30(1):49-52.

- Reiner E, et al. [The role of omega-3 fatty acids from fish in prevention of cardiovascular disease]. Liječnički vjesnik. 2007;129(10-11):350-5.

- Reuters VS, et al. Clinical and muscular evaluation in patients with subclinical hypothyroidism. Arq Bras Endocrinol Metab. 2006;50(3):523-31.

- Rich FR, Wikstrom M. Evidence for a mobile semiquinone in the redox cycle of the mammalian cytochrome bc1 complex. FEBS Letters. 1986;194(1):176-182.

- Richardson B. DNA methylation and autoimmune disease. Clinical immunology (Orlando, Fla.). 2003;109(1):72-9.

- Roberts RO, et al. Serum sex hormones and measures of benign prostatic hyperplasia. The Prostate. 2004;61(2):124-31.

- Rogowskey T. The Connection Between Homocysteine, The Psoas Minor Muscle, And Low Back Pain. Collected Papers International College of Applied Kinesiology: Shawnee Mission, KS. 2006:151-156.

- Rolf I. Rolfing. Dennis-Landman Publishers: Santa Monica; 1977.

- Rome P. Neurovertebral Influence Upon The Autonomic Nervous System: Some Of The Somato-Autonomic Evidence To Date. Chiropr J Aust. 2009;39(1):2-17.

- Rome P. Neurovertebral Influence on Visceral and ANS Function: Some of the Evidence To Date - Part II: Somatovisceral. Chiropr J Aust. 2010(40:1):9-33.

- Rosomoff HL, et al. Physical Findings in Patients with Chronic Intractable Benign Pain of the neck and/or back. 1989;37:279.

- Roy S, et al. Reactive oxygen species and EGR-1 gene expression in surgical postoperative peritoneal adhesions. World journal of surgery. 2004;28(3):316-20.

- Rubel LL. The GP and the endocrine glands. Self-Published; 1959.
- Rudin DO. The dominant diseases of modernized societies as omega-3 essential fatty acid deficiency syndrome: substrate beriberi, Medical hypotheses. 1982;8(1):17-47.
- Ruetten H, Southan GJ, Abate A, Thiemermann C. Attenuation of endotoxin-induced multiple organ dysfunction by 1-amino-2-hydroxy-guanidine, a potent inhibitor of inducible nitric oxide synthase. British journal of pharmacology. 1996;118(2):261-70.
- Rygh LJ, Tjølsen A, Hole K, Svendsen F. Cellular memory in spinal nociceptive circuitry. Scand J Psychol. 2002;43(2):153-9.
- Sahin N, et al. Zinc picolinate in the prevention of leiomyoma in Japanese quail. J Med Food. 2009;12(6):1368-74.
- Sahin K, et al. Lycopene supplementation prevents the development of spontaneous smooth muscle tumors of the oviduct in Japanese quail. Nutrition and cancer. 2004;50(2):181-9.
- Salzman LK. Allergy testing, psychological assessment and dietary treatment of the hyperactive child syndrome. The Medical journal of Australia. 1976;2(7):248-51.
- Samsonova NG, et al. [Intestinal dysbiosis and atherogenic dyslipidemia]. Eksp Klin Gastroenterol. 2010;(3):88-94.
- Sanborn CF, et al. Athletic amenorrhea: lack of association with body fat. Medicine and science in sports and exercise. 1987;19(3):207-12.
- Saraswat M, et al. Antiglycating potential of Zingiber officinalis and delay of diabetic cataract in rats. Molecular vision. 2010;16:1525-37.
- Saraswat M, et al. Prevention of non-enzymic glycation of proteins by dietary agents: prospects for alleviating diabetic complications. The British journal of nutrition. 2009;101(11):1714-2.
- Sarles J, et al. [Chronic diarrhea with hypergastrinemia and achlorhydria without gastritis in a 5-year-old child]. Annales de pédiatrie. 1989;36(4):259-61.
- Sato A. The reflex effects of spinal somatic nerve stimulation on viscera function. J Manipulative Physiol Ther. 1992;15(1):57-61.
- Savino F, et al. Antagonistic effect of Lactobacillus strains against gas-producing coliforms isolated from colicky infants. BMC microbiology. 2011;11:157.
- Schalkwijk CG, Stehouwer CD, van Hinsbergh VW. Fructose-mediated non-enzymatic glycation: sweet coupling or bad modification. Diabetes/metabolism research and reviews. 2004;20(5):369-82.
- Schleip R, Klingler W, Lehmann-Horn F. Active fascial contractility: Fascia may be able to contract in a smooth muscle-like manner and thereby influence musculoskeletal dynamics. Med Hypotheses. 2005;65(2):273-7.
- Schleip R. Myofascial plasticity – a new neurobiological explanation: Part 1. J Bodyw Mov Ther. 2003;7(1):11-19
- Schleip R. Myofascial plasticity – a new neurobiological explanation: Part 2. J Bodyw Mov Ther. 2003;7(2):104-116.
- Schmidt MA, Smith LH, Sehnert KW. Beyond antibiotics: 50 (or so) ways to boost immunity and avoid antibiotics. North Atlantic Books: Berkeley, CA; 1994.
- Schmitt WH, Jr. Techniques based on concepts of the enteric nervous system. Collected Papers International College of Applied Kinesiology: Shawnee Mission, KS. 2002:177-180.
- Schmitt W, Leisman G. Correlation of Applied Kinesiology Muscle Testing Findings with Serum Immunoglobulin Levels for Food Allergies. International Journal of Neuroscience. 1998;96:237-244
- Schmitt WH, Jr. Compiled Notes On Clinical Nutrition Products. 1990:120-181.
- Schmitt WH, Jr., McCord T. Quintessential Applications: AK Clinical Protocol. HealthWorks!: St. Petersburg, FL; 2005.
- Schmitt WH, Jr. Common Glandular Dysfunctions in the General Practice – An Applied Kinesiological Approach. AK Study Program: Chapel Hill, NC;1981.
- Schmitt WH, Jr., Morantz J. Advanced AK (seminar notes). Self-published; 1981:8-10.
- Schmitt WH, Jr. Compiled Notes on Clinical Nutrition Products. AK Study Program: Chapel Hill, NC; 1979.
- Schwarz S, et al. Lycopene inhibits disease progression in patients with benign prostate hyperplasia, The Journal of nutrition. 2008;138(1):49-53.
- Scoop A. An Experimental Evaluation of Kinesiology in Allergy and Deficiency Disease Diagnosis. Journal of Orthomolecular Psychiatry. 1979;7(2):137-8.
- Scott MG, et al. Hormonal evaluation of female infertility and reproductive disorders. Clinical chemistry. 1989; 35(4):620-9
- Sea J, Poon KS, McVary KT. Review of exercise and the risk of benign prostatic hyperplasia. The Physician and sports medicine. 2009;37(4):75-83.
- Selye H. The Stress of Life, 2nd Ed. McGraw-Hill: New York;1976
- Sergio G, et al. Hemorrheology and neurovascular syndromes of the extremities, La Ricerca in clinica e in laboratorio. 1983;13 Suppl 3:481-5.
- Sergio GC, et al. Activated charcoal to prevent irinotecan-induced diarrhea in children. Pediatric blood & cancer. 2008;51(1):49-52.
- Seven A, et al. Lipid, protein, DNA oxidation and antioxidant status in rheumatoid arthritis. Clinical biochemistry. 2008;41(7-8):538-43.
- Shafik A, et al. Transverse folds of rectum: anatomic study and clinical implications. Clinical anatomy. 2001;14(3):196-203.
- Shafton AD, et al The visceromotor responses to colorectal distension and skin pinch are inhibited by simultaneous jejunal distension. Pain. 2006;123(1-2):127-36.
- Shane Anderson A, Loeser RF. Why is osteoarthritis an age-related disease? Best practice & research. Clinical rheumatology. 2010;24(1):15-26.
- Shankardevananda S. The Practice of Yoga for the Digestive System. Yoga Publications Trust; 2003.
- Shapiro SS. Treatment of dysmenorrheal and premenstrual syndrome with non-steroidal anti-inflammatory drugs. Drugs. 1988;36:475-90.
- Sharma JN, Buchanan WW. Pathogenic responses of bradykinin system in chronic inflammatory rheumatoid disease. Experimental and toxicologic pathology: official journal of the Gesellschaft für Toxikologische Pathologie. 1994;46(6):421-33.
- Sharma L. Proprioceptive impairment in knee osteoarthritis. Rheum Dis Clin N Am. 1999;25(2):299-314.

- Shikora SA, et al. Relationship between obesity and uterine leiomyomata, Nutrition. 1991;7(4):251-5.
- Shils ME, et al. Modern nutrition in health and disease, 9th Ed. Lippincott, Williams & Wilkins: Baltimore; 1999.
- Shin B. Case history: ileocecal valve and hidden cervical disc. Collected Papers International College of Applied Kinesiology: Shawnee Mission, KS. 2003:41-42.
- Simons D, Travell J, Simons L. Myofascial pain and dysfunction: The trigger point manual, Vol. 1: Upper half of the body, 2nd Ed. Williams & Wilkins: Baltimore; 1999.
- Simopoulos, AP. Omega-3 fatty acids in inflammation and autoimmune diseases, Journal of the American College of Nutrition. 2002;21(6):495-505.
- Simopoulos AP. Evolutionary aspects of diet, the omega-6/omega-3 ratio and genetic variation: nutritional implications for chronic diseases. Biomedicine & pharmacotherapy. 2006;60(9):502-7.
- Sinclair DC, Weddell G, Feindel WH. Referred pain and associated phenomena. Brain. 1948; 71(2):184-211.
- Singleton CK, Martin PR.. Molecular mechanisms of thiamine utilization, Current molecular medicine. 2001;1(2):197-207.
- Sinha N. Maternal and child health in India: A critical review. Health for the Millions. 1995;21:37-46.
- Sköldstam L, Magnusson KE. Fasting, intestinal permeability, and rheumatoid arthritis, Rheumatic diseases clinics of North America. 1991;17(2):363-71.
- Smith MD, Russell A, Hodges PW. Is there a relationship between parity, pregnancy, back pain and incontinence? Int Urogynecol J Pelvic Floor Dysfunct. 2008;19(2):205-11.
- Smith MD, Coppieters MW, Hodges PW. Postural response of the pelvic floor and abdominal muscles in women with and without incontinence. Neurourol Urodyn. 2007;26(3):377-85.
- Smith MD, Russell A, Hodges PW. Disorders of breathing and continence have a stronger association with back pain than obesity and physical activity. Aust J Physiother. 2006;52:11-16.
- Smith MJ, et al. Bile acid malabsorption in persistent diarrhoea, Journal of the Royal College of Physicians of London. 2000;34(5):448-51.
- Smith VC, Rogers SR. Premenstrual and postmenstrual syndrome, its characteristics and chiropractic care. Am D Chiro. 1992;14(3):4-6.
- Snyder BJ, Sanders GE. Evaluation of the Toftness system of chiropractic adjusting for subjects with chronic back pain, chronic tenshion headache, or primary dysmenorrhea. Chiropractic Technique. 1996;8:3-9.
- Somma F, et al. Oral inflammatory process and general health. Part 1: The focal infection and the oral inflammatory lesion. European review for medical and pharmacological sciences. 2010;14(12):1085-95.
- Song R, Lee EO, Lam P, Bae SC. Effects of tai chi exercise on pain, balance, muscle strength, and perceived difficulties in physical functioning in older women with osteoarthritis: a randomized clinical trial. The Journal of rheumatology. 2003;30(9):2039-44.
- Souza HN, et al. [Low levels of 25-hydroxyvitamin D (25OHD) in patients with inflammatory bowel disease and its correlation with bone mineral density]. Arquivos brasileiros de endocrinologia e metabologia. 2008;52(4):84-91.
- Sprieser PT. The treatment of urinary tract disorders and interstitial cystitis. Collected Papers International College of Applied Kinesiology: Shawnee Mission, KS. 2002: 49-50.
- Standard Process Labs, 2012. www.standardprocess.com/
- Ständer S, Schmelz M. Chronic itch and pain — similarities and differences. European journal of pain. 2006;10(5):473-8.
- Stark D, et al. Irritable bowel syndrome: a review on the role of intestinal protozoa and the importance of their detection and diagnosis. Int J Parasitol. 2007;37(1):11-20.
- Stecco C, et al. Hyaluronan within myofascia in the etiology of myomyofascial pain. Surgical and radiologic anatomy. SRA. 2011;33(10):891-6.
- Stecco C, et al. The ankle retinacula: morphological evidence of the proprioceptive role of the myofascial system. Cells, tissues, organs. 2010;192(3):200-10.
- Stecco A, et al. Anatomical study of myomyofascial continuity in the anterior region of the upper limb. J Bodyw Mov Ther. 2009;13(1):53-62.
- Stecco C, et al. Histological study of the deep myofasciae of the limbs. J Bodyw Mov Ther. 2008;12(3):225-30.
- Stecco C, et al. Histological characteristics of the deep myofascia of the upper limb. Italian journal of anatomy and embryology. 2006;111(2):105-10.
- Steensma AB, Konstantinovic ML, Burger CW, de Ridder D, Timmerman D, Deprest J. Prevalence of major levator abnormalities in symptomatic patients with an underactive pelvic floor contraction. Int Urogynecol J Pelvic Floor Dysfunct. 2010;21(7):861-7.
- Stevenson DD, Zuraw BL. Pathogenesis of aspirin-exacerbated respiratory disease. Clinical reviews in allergy & immunology. 2003;24(2):169-88.
- Stokić E, et al. Body mass index, body fat mass and the occurrence of amenorrhea in ballet dancers, Gynecological endocrinology: the official journal of the International Society of Gynecological Endocrinology. 2005; 20(4):195-9.
- Stones RW, Mountfield J. Interventions for treating chronic pelvic pain in women. Cochrane Database Syst. Rev. 2000;(4),CD000387.
- Strecker D. [Copper content in rheumatoid arthritis patients]. Annales Academiae Medicae Stetinensis, 51 Suppl. 2005;1:129-33.
- Strickland FM, Richardson BC. Epigenetics in human autoimmunity. Epigenetics in autoimmunity - DNA methylation in systemic lupus erythematosus and beyond. Autoimmunity. 2008;41(4):278-86.
- Stude DE, Bergmann TF, Finer BA. A conservative approach for a patient with traumatically induced urinary incontinence. J Manipulative Physiol Ther. 1998;21(5):363-367.
- Sturniolo GC, et al. Altered plasma and mucosal concentrations of trace elements and antioxidants in active ulcerative colitis, Scandinavian journal of gastroenterology. 1998;33(6):644-9.
- Sundqvist T, et al. Influence of fasting on intestinal permeability and disease activity in patients with rheumatoid arthritis. Scandinavian journal of rheumatology. 1982;11(1):33-8.

- Sveinsdottir H, Reame N. Symptom patterns in women with premenstrual syndrome complaints: a prospective assessment using a marker for ovulation and screening criteria for adequate ovarian function. J Adv Nurs. 1991;16(6):689-700.
- Syrigou EI, et al. Food allergy-related paediatric constipation: the usefulness of atopy patch test. European journal of pediatrics. 2011;170(9):1173-8.
- Tahan G, et al. Vitamin E has a dual effect of anti-inflammatory and antioxidant activities in acetic acid-induced ulcerative colitis in rats, Canadian journal of surgery. Journal canadien de chirurgie. 2011;54(5): 333-8.
- Taneja I, et al. Yogic versus conventional treatment in diarrhea-predominant irritable bowel syndrome: a randomized control study. Appl Psychophysiol Biofeedback. 2004;29(1):19-33.
- Tarp U. et al. Glutathione peroxidase activity in patients with rheumatoid arthritis and in normal subjects: effects of long-term selenium supplementation. Arthritis and rheumatism. 1987;30(10):1162-6.
- Thirunavukkarasu V, Anitha Nandhini AT, Anuradha CV. Lipoic acid improves glucose utilisation and prevents protein glycation and AGE formation. Die Pharmazie. 2005;60(10):772-5.
- Thomas JA. Diet, micronutrients, and the prostate gland. Nutrition reviews. 1999;57(4):95-103.
- Thomason PR, Fisher DL, Carpenter PA, Fike GL. Effectiveness of spinal manipulative therapy in treatment of primary dysmenorrheal. J Manipulative Physiol Thera. 1979;2(3):140-145.
- Thompson JA, O'Sullivan PB, Briffa NK, Neumann P. Altered muscle activation patterns in symptomatic women during pelvic floor muscle contraction and Valsalva manouevre. Neurourol Urodyn. 2006;25(3):268-76.
- Tlaskalová-Hogenová H, et al. Commensal bacteria (normal microflora), mucosal immunity and chronic inflammatory and autoimmune diseases. Immunology letters. 2004;93(2-3):97-108.
- Tokita Y, et al. Fructose ingestion enhances atherosclerosis and deposition of advanced glycated end-products in cholesterol-fed rabbits. Journal of atherosclerosis and thrombosis. 2005;12(5):260-7.
- Travell J, Rinzler SH. The myomyofascial genesis of pain. Postgrad Med. 1952;11(5):425-34.
- Valiani M, et al. The effects of massage therapy on dysmenorrhea caused by endometriosis. Iranian journal of nursing and midwifery research. 2010;15(4):167-71.
- van der Wal J.B, Jeekel. J Biology of the peritoneum in normal homeostasis and after surgical trauma, Colorectal disease. The official journal of the Association of Coloproctology of Great Britain and Ireland, 9 Suppl. 2007; 2:9-13.
- Vanhulle C, et al. Antineoplastic chemotherapy and Wernicke's encephalopathy. Archives de pédiatrie : organe officiel de la Sociéte française de pédiatrie. 1997;4(3):243-6.
- Vecchiet L, et al. Referred muscular hyperalgesia from viscera: clinical approach. Advances in pain and research therapy. 1990;13:175-182.
- Veninga KS. Effects of oral contraceptives on vitamins B6, B12, C, and folacin, J Nurse Midwifery. 1984;29(6):386-90.
- Vergnolle N. Modulation of visceral pain and inflammation by protease-activated receptors. British journal of pharmacology. 2004;141(8):1264-74.
- Vernia P, et al. Lactose malabsorption, irritable bowel syndrome and self-reported milk intolerance. Digestive and liver disease: official journal of the Italian Society of Gastroenterology and the Italian Association for the Study of the Liver. 2001;33(3):234-9.
- Verzijl N, et al. Rosslinking by advanced glycation end products increases the stiffness of the collagen network in human articular cartilage: a possible mechanism through which age is a risk factor for osteoarthritis, Arthritis and rheumatism. 2002;46(1):114-23.
- Veves A, Giurini JM, LoGerfo FW. The Diabetic Foot: Medical and Surgical Management. Humana Press: Totowa, NJ; 2002.
- Vignozzi L, et al. Testosterone protects from metabolic syndrome-associated prostate inflammation: an experimental study in rabbit. The Journal of endocrinology. 2012;(1):71-84.
- Waeytens A, et al. Evidence for a potential role of metallothioneins in inflammatory bowel diseases. Mediators of inflammation. 2009;729172.
- Wagner T. Irritable bowel syndrome and spinal manipulation. Chiropractic Technique. 1995;7:139-140.
- Walker MM, Talley NJ. Functional gastrointestinal disorders and the potential role of eosinophils, Gastroenterology clinics of North America. 2008;37(2):383-95.
- Walker S. Neuro Emotional Technique. http://www.netmindbody.com/.
- Wallace JL, et al. Proton pump inhibitors exacerbate NSAID-induced small intestinal injury by inducing dysbiosis. Gastroenterology. 2011;141(4);1314-22, 1322.e1-5.
- Walsh BJ, Pollus BI. The frequency of positive common spinal clinical examination findings in a sample of premenstrual syndrome sufferers. J Manipulative Physiol Thera. 1999;22(4):216-230.
- Wargovich MJ, et al. Calcium ameliorates the toxic effect of deoxycholic acid on colonic epithelium. Carcinogenesis. 1983;4(9);1205-7.
- Watson CS, Jeng YJ, Guptarak J. Endocrine disruption via estrogen receptors that participate in nongenomic signaling pathways. The Journal of steroid biochemistry and molecular biology. 2011;127(1-2);44-50.
- Weiss JM. Pelvic floor myomyofascial trigger points: manual therapy for interstitial cystitis and the urgency-frequency syndrome. The Journal of urology. 2001;166(6):2226-31.
- Wertz K. Lycopene effects contributing to prostate health, Nutrition and cancer. 2009;61(6):775-83.
- Wesselmann U, Lai J. Mechanisms of referred visceral pain: uterine inflammation in the adult virgin rat results in neurogenic plasma extravasation in the skin. Pain. 1997;73(3):309-17.
- Wetzka B, et al. Effects of nitric oxide donors on the contractility and prostaglandin synthesis of myometrial strips from pregnant and non-pregnant women. Gynecological endocrinology: the official journal of the International Society of Gynecological Endocrinology. 2001;15(1):34-42.
- Whang R. Magnesium deficiency: pathogenesis, prevalence, and clinical implications. The American journal of medicine. 1987;82(3A):24-9
- Williams R. Biochemical Individuality. McGraw-Hill: New York; 1956.

- Wilson PD, Herbison P, Glazener C, McGee M, MacArthur C. Obstetric practice and urinary incontinence 5-7 years after delivery. ICS Proceedings of the Neurourology and Urodynamics. 2002;21(4):284–300.
- Wilson JD, Foster DW, Kronenberg HM, Larsen PR. Williams textbook of endocrinology, 9th Ed. WB Saunders Company: Philadelphia; 1998.
- Wittler MA. Chiropractic approach to premenstrual syndrome (PMS). Journal of Chiropractic Research and Clinical Investigation. 1992;8:26-9.
- Wold AE, Adlerberth I. Breast feeding and the intestinal microflora of the infant — implications for protection against infectious diseases. Advances in experimental medicine and biology. 2000;478:77-93.
- Wu JH, et al. Oral exposure to low-dose bisphenol A aggravates testosterone-induced benign hyperplasia prostate in rats. Toxicology and industrial health. 2011;27(9):810-9.
- Wurn BF, et al. Treating fallopian tube occlusion with a manual pelvic physical therapy. Alternative therapies in health and medicine. 2008;14(1):18-23.
- Wurn BF, et al. Treating female infertility and improving IVF pregnancy rates with a manual physical therapy technique. Med Gen Med. 2004;6(2):51.
- Wurn LJ, et al. Increasing orgasm and decreasing dyspareunia by a manual physical therapy technique. Med Gen Med. 2004;6(4):47.
- Yang L, et al. Inflammation and intestinal metaplasia of the distal esophagus are associated with alterations in the microbiome. Gastroenterology. 2009;137(2):588-97.
- Yang YJ, et al. Comparison of fatty acid profiles in the serum of patients with prostate cancer and benign prostatic hyperplasia. Clinical biochemistry. 1999;32(6):405-9.
- Ylikorkala O, et al. Serum gonadotrophins, prolactin and ovarian steroids in primary dysmenorrhea. British journal of obstetrics and gynaecology. 1979;86(8):648-53.
- Yoshioka H, et al. Development and differences of intestinal flora in the neonatal period in breast-fed and bottle-fed infants. Pediatrics. 1983;72(3):317-21.
- Zannad F, et al. [Blood hyperviscosity syndrome in essential arterial hypertension. Characterization and clinical effects]. Archives des maladies du coeur et des vaisseaux. 1985;78(11):1706-9.
- Zahradnik HP, Breckwoldt M. Contribution to the pathogenesis of dysmenorrhea. Archives of gynecology. 1984;236(2):99-108.
- Zava DT, Blen M, Duwe G. Estrogenic activity of natural and synthetic estrogens in human breast cancer cells in culture, Environmental health perspectives. 1997;105 Suppl 3:637-45.
- Zhang Z, et al. The proliferative effect of estradiol on human prostate stromal cells is mediated through activation of ERK. The Prostate. 2008;68(5):508-16.
- Zhang L, et al. Blood rheological properties are strongly related to the metabolic syndrome in middle-aged Chinese. International journal of cardiology. 2006;112(2):229-33.
- Ziskoven C, et al. Oxidative stress in secondary osteoarthritis: from cartilage destruction to clinical presentation? Orthopedic reviews. 2010; 2(2):E23.
- Zoli A, et al. Serum zinc and copper in active rheumatoid arthritis: correlation with interleukin 1 beta and tumour necrosis factor alpha. Clinical rheumatology. 1998;17(5):378-82.